ADVANCES IN
MOLECULAR AND
CELL BIOLOGY

Volume 28 • 1999

THE ADHESIVE INTERACTION OF CELLS

ADVANCES IN MOLECULAR AND CELL BIOLOGY

THE ADHESIVE INTERACTION OF CELLS

Series Editor: E. EDWARD BITTAR
Department of Physiology
University of Wisconsin—Madison
Madison, Wisconsin

Guest Editors: DAVID R. GARROD
ALISON J. NORTH
MARTYN A.J. CHIDGEY
School of Biological Sciences
University of Manchester
Manchester, England

VOLUME 28 • 1999

JAI PRESS INC.

Stamford, Connecticut

100 Prospect Street
Stamford, Connecticut 06901

ISBN: 0-7623-0495-2

Transferred to digital printing 2006

CONTENTS

LIST OF CONTRIBUTORS

Avri Ben-Ze'ev	Department of Molecular Cell Biology Weizmann Institute of Science Rehovot, Israel
Martyn A.J. Chidgey	Division of Medical Sciences University of Birmingham Queen Elizabeth Hospital Birmingham, England
Sandra Citi	Dipartimento di Biologia University of Padova Padova, Italy
Michelangelo Cordenonsi	Dipartimento di Biologia University of Padova Padova, Italy
Elisabeth A. Cox	Department of Cell and Structural Biology University of Illinois at Urbana-Champaign Urbana, Illinois
Jay Cross	Samuel Lunenfeld Research Center Mount Sinai Hospital Toronto, Canada
Caroline H. Damsky	Department of Stomatology University of California San Francisco, California
Xiaoping Du	Department of Pharmacology University of Illinois Chicago, Illinois

Gwynneth M. Edwards	School of Biological Sciences University of Manchester Manchester, England
Susan J. Fisher	Department of Stomatology University of California San Francisco, California
David R. Garrod	School of Biological Sciences University of Manchester Manchester, England
Olga Genbacev	Department of Stomatology University of California San Francisco, California
Mark H. Ginsberg	Department of Vascular Biology Scripps Research Institute La Jolla, California
Linda J. Green	School of Biological Sciences University of Manchester Manchester, England
Ian R. Hart	Richard Dimbleby Department of Cancer Research/ICRF St. Thomas' Hospital London, England
Dorian O. Haskard	National Heart and Lung Institute Imperial College School of Medicine at Hammersmith Hospital London, England
John J. Hemperly	Becton Dickinson Research Center Research Triangle Park, North Carolina
Alan F. Horwitz	Department of Cell and Structural Biology University of Illinois at Urbana-Champaign Urbana, Illinois
Martin J. Humphries	School of Biological Sciences University of Manchester Manchester, England

Anna Huttenlocher	Department of Cell and Structural Biology University of Illinois at Urbana-Champaign Urbana, Illinois
Rolf Kemler	Max-Planck Institute für Immunobiologie Frieburg, Germany
Douglas A. Lauffenburger	Department of Chemical Engineering Massachusetts Institute of Technology Cambridge, Massachusetts
Xuan Li	Department of Immunology Duke University Medical Center Durham, North Carolina
Justin C. Mason	National Heart and Lung Institute Imperial College School of Medicine at Hammersmith Hospital London, England
John F. Marshall	Richard Dimbleby Department of Cancer Research/ICRF St. Thomas' Hospital London, England
Julie McHale	National Heart and Lung Institute Imperial College School of Medicine at Hammersmith Hospital London, England
Alison J. North	School of Biological Sciences University of Manchester Manchester, England
Sean P. Palecek	Department of Chemical Engineering Massachusetts Institute of Technology Cambridge, Massachusetts
Sarah K. Runswick	School of Biological Sciences University of Manchester Manchester, England
David L. Simmons	Department of Neurosciences SmithKline Beecham New Harlow, Essex, England

Jörg Stappert	Max-Planck Institute für Immunobiologie Frieburg, Germany
Douglas A. Steeber	Department of Immunology Duke University Medical Center Durham, North Carolina
Charles H. Streuli	School of Biological Sciences University of Manchester Manchester, England
Thomas F. Tedder	Department of Immunology Duke University Medical Center Durham, North Carolina
Chris Tselepis	School of Biological Sciences University of Manchester Manchester, England
Sarah R. Wallis	Unipath Bedford, England
Yan Zhou	Department of Stomatology University of California San Francisco, California

PREFACE

Our aim in editing "The Adhesive Interaction of Cells" has been to assemble a series of reviews by leading international experts embracing many of the most important recent developments in this rapidly expanding field. The purpose of all biological research is to understand the form and function of living organisms and, by comprehending the normal, to find explanations and remedies for the abnormal and for disease conditions. The molecules involved in cell adhesion are of fundamental importance to the structure and function of all multicellular organisms. In this book, the contributors focus on the systems of vertebrates, especially mammals, since these are most relevant to human disease. It would have been equally possible to concentrate on developmental processes and adhesion in lower organisms.

A major function of adhesion molecules is to bind cells to each other or to the extracellular matrix, but they are much more than "glue". Adhesions in animal tissues must be dynamic—forming, persisting, or declining in regulated fashion—to facilitate the mobility and turnover of tissue cells. Moreover, the majority of adhesion molecules are transmembrane molecules and thus provide links between the cells and their surroundings. This gives rise to another major function of adhesion molecules, the capacity to transduce signals across the hydrophobic barrier imposed by the plasma membrane. Such signal transduction is crucially important to many aspects of cellular function including the regulation of cell motility, gene expression, and differentiation.

The work in this book progresses through four sections. Part 1 discusses the four major families of adhesion molecules themselves, the integrins (Green and

Humphries), the cadherins (Stappert and Kemler), the selectins (Tedder et al.) and the immunoglobulin superfamily (Simmons); part 2 considers junctional complexes involved in cell interactions: focal adhesions and adherens junctions (Ben Ze'ev), desmosomes (Garrod et al.), and tight junctions (Citi and Cordenonsi). The signaling role of adhesion molecules is the focus of part 3, through integrins and the extracellular matrix (Edwards and Streuli), through platelet adhesion (Du and Ginsberg), and in the nervous system (Hemperley). In part 4, the aim is to show how adhesive phenomena contribute to important aspects of cell behavior and human health. Leukocyte trafficking (Haskard et al.), cancer metastasis (Marshall and Hart), cell migration (Paleck et al.), and implantation and placentation (Damsky et al.) are the topics considered in depth.

The different sections are, of course, not mutually exclusive: it is both undesirable and impossible to separate structure from function when considering cell adhesion. Each chapter has its unique features, but some overlap is both inevitable and valuable since it provides different perspectives on closely related topics. We hope that the whole contributes a valuable and stimulating consideration of this important topic.

Production of a volume of this nature must involve some delay between the original conception and publication. We have been able to do some updating at the proof state and believe that consideration of major concepts has not been adversely affected.

David R. Garrod
Martyn A.J. Chidgey
Alison J. North
Guest Editors

PART I

ADHESION MOLECULES AND THEIR LIGANDS

THE MOLECULAR ANATOMY OF INTEGRINS

Linda J. Green and Martin J. Humphries

Advances in Molecular and Cell Biology
Volume 28, pages 3-26.

ISBN: 0-7623-0495-2

I. INTRODUCTION

The integrins are a major family of cell adhesion receptors involved in adhesive interactions either between cells or between cells and the extracellular matrix. These adhesive events are vital for cell migration, environmental sensing and tissue organization, and play key roles in many processes, e.g. immune system function, platelet aggregation, tissue regeneration during wound healing, migration of epithelial cells during epigenesis, peripheral nerve regeneration, and the spatiotemporal organization of cells during embryonic development. In addition to its role in normal cellular function, adhesion is also implicated in a wide variety of human diseases such as thrombosis, inflammation, tumor metastasis, osteoporosis, ulceration, rheumatoid arthritis, and microbial and parasitic infection. Resolving the mechanisms underlying adhesion are likely to help in developing strategies for intervention in these diseases. This chapter focuses primarily on recent advances in the elucidation of ligand and integrin active sites and also integrin activation. Readers are referred to the following reviews for an overview of the field and citations of earlier work (Humphries, 1990; Hynes, 1992; Juliano and Haskill, 1993; Sonnenberg, 1993; Loftus et al., 1994).

II. INTEGRIN–LIGAND INTERACTIONS

All integrins are α/β heterodimers, composed of non-covalently associated subunits, and in mammals 16 α and 8β subunits have been shown to combine into 22 different dimers. The known αβ combinations, given in figure 1, show that most α subunits associate with only one or two β subunits, although αV is an exception in combining with β1, β3, β5, β6 and β8. Historically, integrins have been grouped on the basis of a common β subunit (β1, β2 and β3) integrins and although the more recently identified receptors fall outside this classification, the ligands bound by the three families do share common features as shown in Figure 1. β1 and β3 integrins are expressed by most cell types and predominantly mediate cell-matrix adhesion, while the β2 integrins, restricted to leukocytes, are involved in homotypic cell interactions and leukocyte-endothelial cell adhesion. Furthermore, the β1 integrins bind mainly to connective tissue macromolecules such as fibronectin, collagens and laminins, while the β3 family are involved in adhesion to "vascular" ligands, for example fibronectin, fibrinogen, vitronectin, thrombospondin, and von Willebrand factor.

The ligand binding specificities of the integrins show two interesting features. First, some integrins are able to bind different ligands whose structures appear dissimilar, for example α2β1 binds to collagens and laminins and αVβ3 binds to bone sialoprotein, fibrinogen, fibronectin, laminin, thrombospondin, vitronectin, and von Willebrand factor. Second, some ligands bind to several integrins, for example fibronectin binds to α4β1, α5β1, αVβ1, αllbβ3, αVβ6, and αVβ8. The molecular basis for these interactions is unclear, although in the case of an integrin which

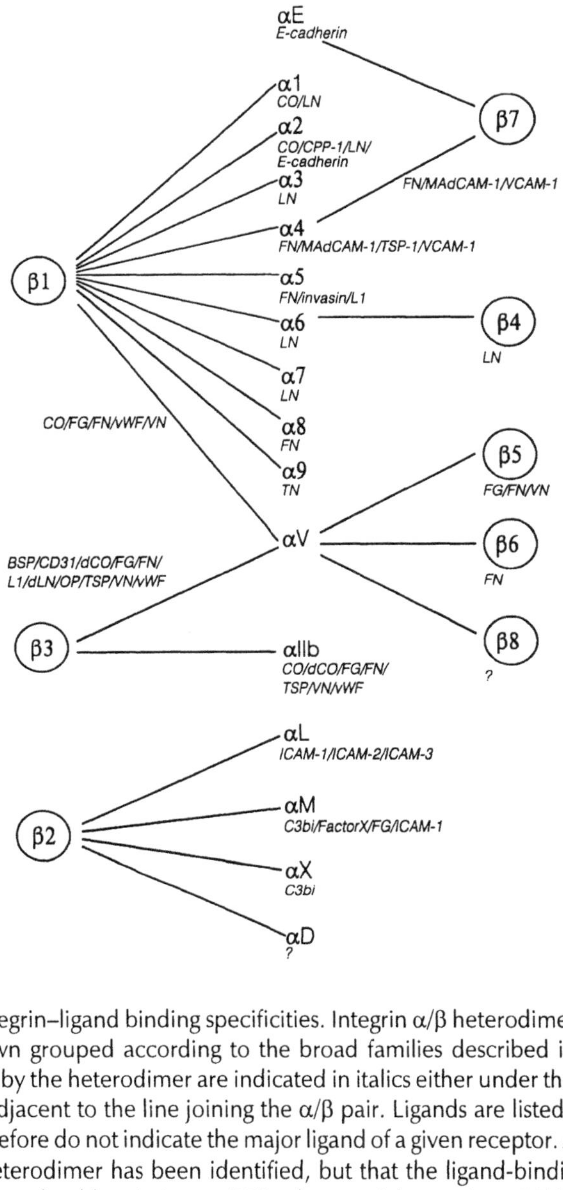

Figure 1. Integrin–ligand binding specificities. Integrin α/β heterodimers, identified to date, are shown grouped according to the broad families described in the text. The ligands bound by the heterodimer are indicated in italics either under the α subunit, the β subunit or adjacent to the line joining the α/β pair. Ligands are listed in alphabetical order and therefore do not indicate the major ligand of a given receptor. A ? sign denotes that an α/β heterodimer has been identified, but that the ligand-binding specificity is unknown. Ligand abbreviations: BSP, bone sialoprotein; C3bi. complement component; CD31, cluster of differentiation antigen 31; CO, collagens; dCO, denatured collagens; CPP-1, collagen C-propeptide (type 1); FG, fibrinogen; FN, fibronectin; ICAM, intercellular adhesion molecule; L1 is a recently discovered cell adhesion molecule; LN, laminins; dLN, denatured laminin; MAdCAM-1, mucosal addressin cell adhesion molecule-1; OP, osteopontin; TN, tenascin; TSP thrombospondin; VCAM-1, vascular cell adhesion molecule-1; VN, vitronectin; vWF, Jon Willebrand factor.

binds multiple ligands, it is conceivable that the ligands may share a common receptor-binding sequence. Different integrins are not restricted to one cell type, suggesting that the multiplicity of ligand recognition is not a reflection of simple cell type-specific differences. The binding of multiple receptors by single ECM macromolecules, such as fibronectin, implies that different integrins mediate different functions or different aspects of the same function by transducing different signals. Differences in the ability of integrins to promote migration have been reported for the fibronectin receptors $\alpha4\beta1$ and $\alpha5\beta1$ (Dufour et al., 1988; Akiyama et al., 1989). This may be a reflection of differences in the ligand binding affinity of each receptor (a low affinity interaction allowing migration and a stronger interaction causing immobilization).

III. ACTIVE SITES IN INTEGRIN LIGANDS

The identification of binding sites in integrin ligands has involved the generation of fragments by proteolytic and chemical cleavage and the testing of the fragments in cell adhesion assays. The use of progressively smaller fragments and synthetic peptides has led to the identification of short peptides as being the minimal active sequences involved in the interaction.

The prototype short peptide motif, RGD, was first identified in the central cell binding domain of fibronectin (Pierschbacher and Ruoslahti, 1984; Yamada and Kennedy, 1984, 1985) and has since been found in many adhesive extracellular macromolecules including fibrinogen. bone sialoprotein. collagens, von Willebrand factor, thrombospondin and nidogen (for a review see Humphries, 1990). RGD is therefore a shared motif and provides an explanation for how one integrin can recognize many different extracellular macromolecules. The R and D residues also occur in the recognition sites on collagen IV that have been identified for $\alpha1\beta1$ and $\alpha2\beta1$ integrins (Eble et al., 1993; Kern et al., 1993; Kühn and Eble, 1994). The essential amino acids of the $\alpha1\beta1$ recognition site were found to lie on different polypeptide chains and the conformation of the triple helix holds them in the correct position. The RGD sequence of fibronectin is located within the 10th type III repeat and recently a pentapeptide, PHSRN, has been identified in the 9th type III repeat that synergistically enhances the cell-adhesive activity of the fibronectin RGD sequence (Aota et al., 1994).

The integrin $\alpha4\beta1$ has been shown to recognize sites in the alternatively spliced IIICS region of fibronectin and the adjacent heparin-binding domain. The minimal active sequence LDV has been identified as a high affinity site (Komoriya et al., 1991), while the peptides REDV and IDAPS represent lower affinity sites (Mould et al., 1991; Mould and Humphries, 1991). The LDV sequence represents a second common integrin-binding motif, since homologous sequences have also been identified as binding sites in vascular cell adhesion molecule-1 (VCAM-1), intercellular cell adhesion molecules-1, -2, and -3 (ICAM-1, -2 and -3) and mucosal addressin cell adhesion molecule (MAdCAM-1)(for a review see Newham and

Humphries, 1996). However, unlike the invariant RGD sequence, this second motif varies within a definable consensus sequence, L/I-D/E-S/T/V-P/S.

Studies on the 3-D structure of the fibronectin type III repeat using NMR and X-ray crystallography (Main et al., 1992; Leahy et al., 1992) have shown it to resemble an immunoglobulin (Ig) domain. The RGD of the 10th type III repeat occupies a position in a flexible loop region of the molecule thus exposing it to adhesion receptors. The recently solved crystal structure of the two N-terminal Ig domains of VCAM-1 (Jones et al., 1995) has shown that the VCAM-1 IDSP sequence lies in a connecting loop between two β strands (C and D) of the most membrane distal Ig fold and is also in an exposed position. Homology modeling studies on the other integrin-binding Ig-domain-containing cell adhesion molecules (Ig-CAMs) suggest that these molecules might fold in a similar way and that the LDVP motif would lie in an exposed C-D loop, although it is likely that local conformations are different between the Ig-CAMS. The similarity between the location of the RGD and LDVP motifs and their presentation to the integrin suggests that Integrin–Ligand binding sites may share a broadly related topology.

Further evidence for a common structure of the integrin-binding region of different ligands has come from studies on the synergy sites required for optimal binding by the different integrins. Homology modeling studies on the synergy site, PHSRN, found in the ninth type III repeat of fibronectin suggest that it too is located in a loop between two β stands (Aota et al., 1994). A recent study has identified lower affinity synergy sites on the α4 ligands VCAM-1 and MAdCAM-1. In each case they lie on the adjacent Ig domain and are also predicted to lie in exposed loops (Newham et al., 1997). This requirement by the integrin for two binding sites of different affinities may at least partly explain how one ligand can bind many integrins. Although the integrin αllbβ3 requires a synergy site in addition to the RGD site for optimal binding to fibronectin, the peptide recognition process appears to differ significantly (Bowditch et al., 1994; Aota et al., 1994).

IV. INTEGRIN STRUCTURE

Both the α and β subunits are type I transmembrane glycoproteins. Most integrins have a large extracellular domain (approximately 1200 amino acids in α subunits and 800 amino acids in β subunits) and a much shorter cytoplasmic domain (50 residues or less). The exception is β4 which has a cytoplasmic domain of over 1000 residues.

Sequence analysis of the α subunits has revealed that the N-terminal moieties contain a seven-fold repeat of an homologous segment of approximately 60 ammo acids (Tuckwell et al., 1994). The three or four of these that are the most C-terminal contain divalent cation binding sites, which bear sequence homology to the well characterized cation-binding EF-hand motif found in calmodulin and parvalbumin (Kretsinger and Nockolds, 1973; Strynadka and James, 1989). A recent study that predicts how these seven repeated domains are folded is described below.

The α subunits that associate with β2 (αL, αM, αX, and αD) and also α1, α2 and αE have a 200 amino acid segment inserted between the second and third terminal repeats. This inserted or I-domain is related in sequence to the A-domain found in von Willebrand factor, cartilage matrix protein, type VI collagen and complement factor B and will be termed the αA-domain in this review. Other α subunits (α3, α5, α6, α7, α8, αllb and αV), are post-translationally cleaved at a conserved site to give a 25-30 kDa transmembrane chain disulphide-bonded to a larger extracellular chain as shown in Figure 2 (Hynes, 1992). Interestingly, possession of an αA-domain or a cleavage site appears to be mutually exclusive and corresponds with the grouping of integrin α subunits on the basis of sequence identity (Hynes, 1992). The αE subunit is unusual in that it has a proteolytic cleavage site located in a unique extra domain of 55 residues located just N-terminal to the αA-domain.

The α4 subunit and the more recently discovered α9 do not possess an A-domain nor are they proteolytically cleaved in the same way as the other α subunits. However, α4 does possess a proteolytic cleavage site in the extracellular domain (see Figure 2). Cleavage occurs between K557, R558, and S559 resulting in two non-covalently associated fragments of 80 and 70 kDa. Cleavage of the a4 subunit is partial and variable, in that either form or a mixture of both can be expressed by different cell types. The significance of the cleavage is unclear, since mutation of the cleavage site has no detectable effect on ligand binding (Teixido et al., 1992).

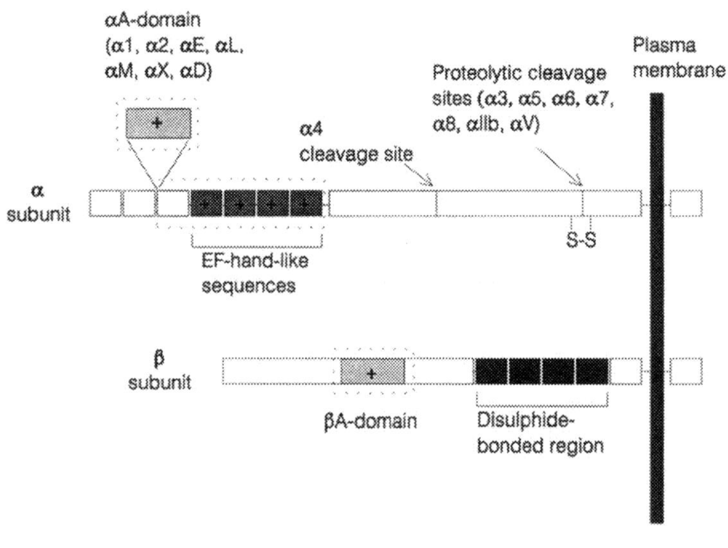

Figure 2. The domain structure of a generic integrin. The structure of the α and β subunits are shown with the extracellular domains to the left of the plasma membrane. The structural features of the different α subunits described in the text are indicated. Sites that have been shown to bind divalent cations are denoted by +. Regions implicated in ligand binding are enclosed by dashed rectangles.

The cytoplasmic tails of the α subunits are approximately 30 amino acids long and in certain subunits, α3, α6 and α7 (Tamura et al., 1991; Cooper et al., 1991; Kramer et al., 1991) have been found to undergo alternative splicing. The cytoplasmic domain isoforms are expressed in a cell type-specific manner and may be relevant in the regulation of integrin function.

Certain integrin β subunits can also undergo alternative splicing of the cytoplasmic domain (β1, Altruda et al., 1990; β3, van Kuppervelt et al., 1989; and β4, Tamura et al., 1990). The β4 cytoplasmic domain, which contains four fibronectin-like type III repeats, also undergoes proteolytic processing (Giancotti et al., 1992). The C-terminal moiety of the extracellular domain of the β subunits contains four cysteine-rich domains which are believed to be internally disulphide-bonded (Calvete et al., 1991) and a "conserved" domain (see Figure 2). This latter domain is found near the amino terminus of the β subunit and by comparison of consensus hydropathy plots or secondary structure prediction appears to be related to the von Willebrand factor A-domain (termed the βA-domain in this review; Tuckwell and Humphries, 1997).

Electron microscopy of integrins shows a globular head, apparently comprising parts of both subunits, and two stalks extending into the membrane (Nermut et al., 1988). A model for the structure of the heterodimer has been proposed in which the head is composed of the seven N-terminal domains of the α subunit, and the N-terminal region of the β subunit, while the stalks are the remainder of the extracellular α subunit, and the cysteine-rich repeats of the β subunit.

V. INTEGRIN ACTIVE SITES

Until recently. little has been known about the nature of integrin binding sites. Poorly defined regions of both subunits have been implicated by a variety of techniques such as chemical cross-linking of peptides, epitope mapping of inhibitory antibodies and mutational analysis. However, structural analysis of domains by both X-ray crystallographic and structure prediction methods, together with information from more detailed mutational analysis is beginning to reveal how the ligand-binding pocket on integrins is formed.

A. The Integrin αA-domain

The knowledge that the von Willebrand factor A-domains mediate protein–protein interactions with collagen led to the investigation of the integrin α subunit A-domains as potential sites of ligand binding. Initial studies using inhibitory monoclonal antibodies (mAbs) and site-directed mutagenesis implicated the αM A-domain in ligand binding (Diamond et al., 1993; Michishita et al., 1993). More direct evidence has come from the expression of recombinant A-domains from the αL, αM and α2 subunits (Randi and Hogg, 1994; Ueda et al., 1994; Zhou et al., 1994; Tuckwell et al., 1995). These were all shown to bind ligand in a cation-dependent manner and the ligands bound were the same as those bound by the parent integrin.

The solution of the crystal structure of the recombinant A-domain of the αM subunit (Lee et al., 1995a) shows that this structure is folded independently of the remainder of the α subunit and suggests a possible mechanism of ligand binding. The αM A-domain is an alternating α/β type structure composed of six β strands surrounded by seven α helices which is typical of a Rossman fold structure found in a number of enzymes.

At one end of the molecule, a divalent cation is coordinated to a DxSxS motif and two other non-contiguous oxygenated residues either directly or via a water molecule. These five amino acids are absolutely conserved in all integrin αA-domains. Dimerisation between A-domains was also observed in the Mg^{2+}-bound form, with a glutamate side chain from one αA-domain completing the octahedral coordination sphere of the metal in the crystal pair. The integrin-binding motifs in ligands often contain acidic residues, usually aspartate or glutamate, consequently the mode of cation coordination employed by the A-domain was postulated to mimic a ligand-occupied conformer of the αA-domain. This was termed a metal-ion-dependent adhesion site (MIDAS) motif. Evidence to support the presence of a cation bridge between the integrin αA-domain binding site and the ligand binding sites is still only circumstantial (i.e., the requirement of cations for binding and use of acidic peptide motifs by integrin ligands). The ligands bound by the αA-domains of the β2 integrins contain an LDV-type motif (IETP in ICAM-1 and LETS in ICAM-2 and -3), while collagen IV, which binds the αA-domains of α1 and α2, employs a three dimensional, multichain sequence involving R and D residues to bind to α1β1 integrin, although these motifs have not yet been shown to be required for binding to the αA-domain. However, a number of studies using inhibitory mAbs and mutational analysis have shown that the cation-binding face of the αA-domain is involved in ligand binding (for review see Humphries. 1996).

B. The Putative βA-domain

Chemical cross-linking of RGD peptides showed binding to residues 61-203 of the β subunit of αvβ3 (Smith and Cheresh, 1988) and to residues 107-171 of the β subunit of αllbβ3 (D'Souza et al., 1988). Analysis of the mutant αllbβ3 receptor from patients with Glanzmann's thrombasthenia showed that a point mutation at position 119 (D to Y) of the β3 subunit abrogated the binding of αllbβ3 to fibrinogen (Loftus et al., 1990), and the introduction of this mutation into the β3 subunits of αllbβ3 and αvβ3 integrins abolished ligand binding in vitro (Loftus et al., 1990). Alignment of this region of the β3 and β1 subunits with sequences from the other known β subunits showed that this aspartate was absolutely conserved (Bajt and Loftus, 1994) and that other adjacent oxygenated residues in this region were also conserved and formed a DxSxS motif. Mutation of the equivalent residue in the β1 subunit (D130 A) abolished the binding of α5β1 to fibronectin, underlining the importance of this region (Takada et al., 1993). Alanine substitution of the conserved S121 and S123 in the β3 subunit produced an αllbβ3 receptor that failed to bind to fibrinogen (Bajt and Loftus, 1994) and similar studies on the β2 subunit of αLβ2 and αMβ2 showed a requirement for the equivalent aspartate (D134) and first serine (S136) in ligand binding (S138A failed to be expressed; Bajt et al., 1995).

As previously described, the recently elucidated crystal structure of the αM A-domain shows that a DxSxS motif plays a role in cation coordination. This DxSxS motif has been shown by alignment to be a common feature of A domains in general and the A domains of the α subunit in particular, and mutations of the conserved D140, S142 and S144 in the α subunit abrogated the binding of αMβ2 to iC3b (Michischita et al., 1993). Lee et al. (1995a) suggested the presence of an A-domain-like structure within the β subunit based on sequence homologies and hydropathy plot similarities. Several recent studies have analyzed the role played by residues in the β subunit that are equivalent to those implicated in the α subunit MIDAS. The metal binding site co-ordinating residues in the αM A-domain include T209 and D242 in addition to the DxSxS motif. Puzon-McLaughlin and Takada (1996) used alanine scanning mutagenesis and subsequent expression to analyze candidate residues in the β1 subunit and identified S132, N224, D226, E229, D233, D267, and D295 in addition to D130. Goodman and Bajt (1996) found that D232 and E235 in the β2 subunit were required for ligand binding, while Tozer et al., (1996) showed that residues D217 and E220 in the β3 subunit were essential in addition to the DxSxS motif. The differences in amino acid requirement between the individual β subunits and the αM A domain suggests that the structure of this region of the β subunit is very similar but not identical to the αM MIDAS domain. Further evidence for the presence of a modified A domain in the β subunit has come from recent structure prediction data (Tuckwell and Humphries, 1997). Confirmation of the importance of this region in the β3 subunit has come from a recent study whereby functional variants of αllbβ3 caused by chemical mutagenesis were assessed (Baker et al., 1997). In addition to an effect of mutations at D119, D117, and E220 on ligand binding, P219 was also found to be sensitive. This residue is highly conserved in β subunits and may be required for the correct conformation of the ligand binding site.

C. The Integrin α Subunit

Ligand-receptor crosslinking experiments have also implicated regions of the α subunit in ligand-binding. Smith and Cheresh (1990) identified two sites within the region 139-348 on the α subunit as being RGD binding sites in the vitronectin receptor αVβ3. A peptide from the C-terminus of the fibrinogen γ chain has been shown to crosslink to residues 294-314 on the allb subunit of αllbβ3 (D'Souza et al., 1990). In each case binding was close to the EF-hand-like, cation-binding domains of the N-terminal repeats. Mutational analysis of the α subunits has proved extremely difficult, since most mutations of critical residues in the EF hand-containing cation-binding domains leads to loss of expression of the integrin. However, Masumoto and Hemler (1993) showed that conservative substitution of key residues led to a significant decrease in ligand binding. Using in vitro translation products, Stanley and colleagues (1994) demonstrated that domains V and VI of the αL subunit of αLβ2 contain ICAM-1 binding sites and that these map close to the EF hand-like domains.

More recent studies have implicated regions outside these cation-binding domains as being important for ligand binding. Kamata and co-workers (1995) localized the putative binding sites within the α4 subunit by mapping inhibitory anti-α4 mAbs using interspecies chimeras. A region N-terminal to the α4 cation binding sites (107-268) was localized as the putative binding site for the ligands VCAM-1, the CS1 peptide of fibronectin and MAdCAM-1. Moreover, alanine-scanning mutagenesis of this region indicated that the critical residues are clustered in a predicted β-turn structure (181-190) of the third N-terminal repeat of the α4 subunit (Irie et al., 1995). Mutations in the predicted β-turn structure of the third N-terminal repeat of the αllb subunit also blocked binding of αllbβ3 to soluble fibrinogen (Kamata et al., 1996). Using αllb/αV chimeras, Loftus and co-workers (1996) have identified the first 334 residues of the αllb subunit as regulating the ligand recognition specificity of β3 integrins. Chimeras that omitted the N-terminal 140 residues or the first two cation-binding domains failed to change ligand specificity.

The nature of the α subunit ligand-binding site and its interaction with that on the β subunit to form the ligand-binding pocket cannot be fully understood until the structure of both the separate subunits and their association to form the heterodimer is known. A recent structure prediction study (Springer, 1997) has suggested that the seven N-terminal repeats of the α subunits are folded into a β-propeller domain. The domains contain seven four-stranded β-sheets arranged in a torus around a pseudosymmetry axis, with the sequences implicated in ligand recognition being found within loops on the upper surface of the propeller. The juxtaposition of this β-propeller domain to the putative β subunit A-domain is as yet unknown. However, a recent study by Mould and colleagues (1997) has attempted to map the fibronectin-binding interface on α5β1 using inhibitory mAbs. The results indicated that the synergy region of fibronectin is recognized primarily by the N-terminal repeats of the α5 subunit, and that the RGD site is bound by the β subunit. Since the synergy site and the RGD site lie on the same face of the fibronectin molecule, the binding sites (i.e., the top faces) of both the α subunit β-propeller and the β subunit A-domain must be co-planar when binding ligand.

This cooperation between the α and β subunits in ligand recognition, whereby the β subunit binds to a common peptide motif and the α subunit provides specificity by recognizing a synergy site, may be a general mechanism for integrin–ligand interactions. Support for this hypothesis has come from a recent investigation into α4β1 binding which showed that the CS1 peptide, which contains the LDV motif, attenuated the mAb 13 binding site on the $\beta1$ subunit but did not perturb any $\alpha4$ epitopes (Newham et al., 1998). This suggests that the RGD and LDV motifs are functionally and, to some extent, structurally analogous.

VI. MODULATION OF INTEGRINS VIA THE EXTRACELLULAR DOMAINS

Integrins can exist in both active and inactive states on the cell surface. There is evidence that in the inactive state the ligand binding sites are masked by interactions between the subunits and that activation of the integrin causes a conformational change and exposure of the binding site (reviewed by Mould, 1996). The regulation of integrin activation is complex and although the conformational changes that facilitate ligand binding take place outside the cell, regulatory signals from within the cell play a key role in activation (this will be discussed fully in a later section). Divalent cations are essential for integrin function and recent studies have given an insight into their role in integrin activation.

A. The Role of Divalent Cations

There is evidence of direct binding of cations to integrins. Smith and Cheresh (1991) demonstrated the covalent coupling of Co (111) to $\alpha V\beta3$, while Gulino and co-workers (1992) showed Ca^{2+} binding to a recombinant fragment of αllb that spanned the EF-hand-like domains. In addition, modeling studies comparing calmodulin EF-hands with integrin EF-hand-like sequences predict that integrins can coordinate divalent cations (Tuckwell et al., 1992). Studies using a synthetic peptide from the putative A-domain region of $\beta3$ (residues 118-131) demonstrated binding to both the calcium analogue terbium and RGD peptides, with ligand binding causing displacement of cation from this peptide (D'Souza et al., 1994). The authors proposed a mechanism of integrin–ligand binding termed the "cation displacement hypothesis" in which the ligand is initially bridged to the integrin through the cation to form a ternary complex and cation is subsequently displaced from the ligand-binding site. The use of a cation as a bridge between the integrin and ligand was also suggested by Lee and colleagues (1995a) following the elucidation of the crystal structure of the αM A-domain.

Integrin α subunits contain three or four divalent cation binding domains within the seven N-terminal repeats and there is evidence that multiple sites can regulate ligand-binding function. Masumoto and Hemler (1993) showed that conservative mutations of the cation-coordinating residues in each of the three EF-hand-like domains of $\alpha4$ resulted in greatly reduced ligand binding. They also demonstrated an order of preference for different divalent cations during ligand binding, with $Mn^{2+} > Mg^{2+} > Ca^{2+}$ for CS1 and $Mn^{2+} > Mg^{2+} = Ca^{2+}$ for VCAM-1. Smith and co-workers

(1994) investigated the role of cations in the binding of αIIbβ3 and αVβ3 to fibrinogen. Ca^{2+} supported the binding of αIIbβ3 but not αVβ3, while Mn^{2+} supported the binding of both integrins to fibrinogen. For αVβ3, low Ca^{2+} concentrations increased the affinity of Mn^{2+} for integrin, but high concentrations completely inhibited Mn^{2+}-induced binding of fibrinogen (Smith et al., 1994). This mixed type inhibition suggests a mechanism in which two separate cation-binding sites regulate ligand binding to β3-integrins, a ligand-competent site that binds Mn^{2+} and Ca^{2+} and an effector site that binds Ca^{2+}. Studies on the cation regulation of α5β1 interactions with fibronectin indicated the presence of at least three binding sites each with distinct cation preferences (Mould et al., 1995a). Two of these sites were predicted to be ligand-competent, a high-affinity Mn^{2+} site and a low-affinity Mg^{2+} site. The third site is similar to the effector site described for αVβ3 (Smith et al., 1994). Ca^{2+}-binding to this site slightly increases the affinity of Mn^{2+}, and greatly increases the affinity of Mg^{2+} for their respective sites (Mould et al., 1995a).

Some stimulatory mAbs have epitopes on integrins that are sensitive to divalent cations. The binding of anti-β1 mAbs 12G10 (Mould et al., 1995b) and 9EG7 (Bazzoni et al., 1995) and the anti-β3 mAb AP5 (Honda et al., 1995) show a cation dependency similar to that of the ligands. The binding of these mAbs is promoted by Mn^{2+} (and to a lesser extent Mg^{2+}) and inhibited by Ca^{2+}. These data suggest that the divalent cations cause conformational changes in the integrin that expose previously masked epitopes. Cation-induced conformational changes were also observed when crystal structure data obtained for the αM A-domain with bound Mn^{2+} (Lee et al., 1995b) was compared with the Mg^{2+}-bound form (Lee et al., 1995a). Based on the observation of two distinct conformational states of the αM A-domain, Lee and co-workers (1995b) proposed a two-state model for the regulation of integrin function where there is a conformational equilibrium between the inactive and ligand-bound forms of the α A-domain.

Based on evidence that integrins undergo further conformational changes in response to ligand binding (see below), Mould (1996) has recently developed this model to include a third conformational state. The model is based on the assumption that integrins are allosteric proteins that can exist in distinct conformational states and that there is an equilibrium between the inactive, the active, and ligand-bound states. In this model, divalent cations that support ligand binding (Mn^{2+} and Mg^{2+}) shift the conformational equilibrium from the inactive to the active state and expose the ligand-binding site, whereas inhibitory cations (Ca^{2+}) drive the equilibrium in the opposite direction (see Figure 3).

As yet, the physiological relevance of cations for integrin activation has not been demonstrated and indeed it is difficult to envisage the subtle changes in cation concentration that would be required taking place in the extracellular environment. The relatively high extracellular Ca^{2+} concentration should cause most integrins to be inactive according to the model. It is more likely that in vivo integrin activation is regulated either by inside-out signals or by unknown extracellular factors or a combination of both.

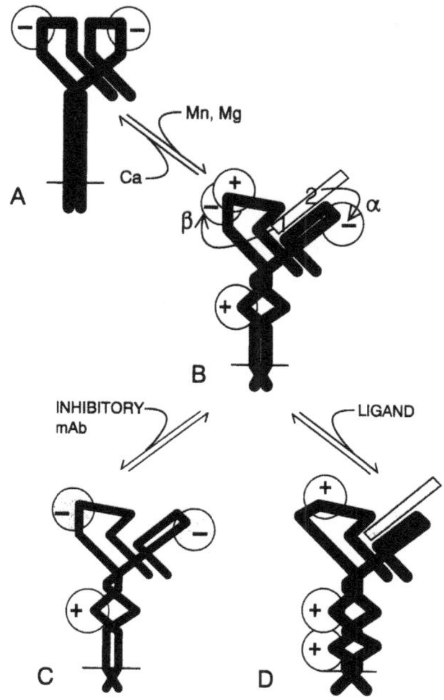

Figure 3. A model of the conformational changes that take place in integrin extracellular domains during divalent cation occupancy and ligand binding. Both α and β subunits undergo conformational changes during activation and subsequent ligand binding. The globular head of the β subunit is drawn on the left and that of the α subunit on the right in each case. The epitopes for both inhibitory/anti-LABS (–) and stimulatory/anti-LIBS (+) mAbs are depicted by circles and the ligand-binding pocket by a rectangle. The potential for both mAb and ligand binding is portrayed by open circles or rectangles respectively, while the occupied integrin is shown by filled circles or rectangles. (**A**) The Ca^{2+}-occupied inactive integrin expresses epitopes for inhibitory, but not stimulatory, mAbs on both the α and β subunits and is unable to bind ligand. (**B**) The binding of Mn^{2+} or Mg^{2+} causes a conformational change resulting in an active integrin that has the potential to bind ligand. Some ligands have been shown to have a major active site (1) and a synergy site (2) and, in the case of α5β1 site 1, (RGD) binds to the β subunit and site 2 (PHSRN) to the α subunit as detected by anti-LABS mAbs. The conformational changes expose epitopes for stimulatory mAbs on the β subunit, and there is overlap between the epitopes for stimulator and inhibitory mAbs in the βA-domain region. Activation may also lead to a reorientation of the cytoplasmic domains and a concomitant change in basal signaling. (**C**) The binding of inhibitory/anti-LABS mAbs to the active integrin induces a shape change that either precludes ligand binding by removing the ligand-binding pocket or prevents a full ligand response by blocking secondary conformational changes induced by ligand binding. (**D**) The binding of ligand to the active integrin destroys the epitopes for the inhibitory/anti-LABS mAbs and induces further stimulatory/anti-LIBS epitopes. Conformational changes that occur in the globular heads as a result of ligand binding are propagated via the stalks to the cytoplasmic domains to begin the signaling cascade.

B. Mechanism of Action of anti-LIBS and anti-LABS mAbs

The conformational changes of the integrin that occur as a consequence of ligand binding lead to the exposure of epitopes known as ligand-induced binding sites or LIBS (reviewed by Williams et al., 1994; Faull and Ginsberg, 1995). A subset of the antibodies that have been shown to stimulate integrin function react preferentially with the ligand-occupied form of the integrin and are termed anti-LIBS mAbs. The stimulation of integrin function by these antibodies can be explained by the allosteric model of integrin activation proposed by Mould (1996). The anti-LIBS mAb would bind to and stabilize the ligand-occupied conformation thus promoting ligand binding (see Figure 3). Clearly, the mapping of these ligand-induced epitopes will lead to a greater understanding of the mechanisms of integrin activation. Anti-LIBS mAbs that recognize the β1 subunit have been mapped to three distinct regions. 9EG7 maps to an area within the disulphide-bonded region (residues 495-602) of the extracellular stalk (Bazzoni et al., 1995), the 12G10 epitope lies in the putative β A-domain (Mould et al., 1995c), and the 15/7 epitope lies in between (Puzon-McLaughlin et al., 1996). A recent report has identified a novel activating anti-β1 antibody (Faull et al., 1996). The epitope recognized by QE.2E5 is highly conserved and lies in residues 426-587 of the cysteine-rich repeats. β3 integrin LIBS have been shown to map to the extreme N-terminus (residues 1-6) as detected by AP5 (Honda et al., 1995) and to residues 602-690 of the disulphide-bonded region as detected by LIBS2 (Du et al., 1993). The mapping of anti-LIBS mAbs to the stalk region as well as to known binding regions suggests that a number of conformational changes follow ligand engagement and some of these may play a role in signal transduction.

Many activating (8A2, TS2/16 and A1A5) and inhibitory (4B4, 4B5, 13, AIIB2 and P4C10) mAbs have been mapped to a region (residues 207-218) of the β1 A-domain (Takada and Puzon, 1993). Using competitive ELISA experiments, the anti-LIBS mAb, 12G10, was found to bind at or very close to this region of the β1 subunit (Mould et al., 1995c). Unlike 12G10, the TS2/16 and 8A2 epitopes are unaffected by ligand occupancy suggesting that 12G10 may recognize a conformation that is naturally induced in the β1 subunit after cation or ligand binding and that TS2/16 and 8A2 activate the integrin by changing its shape in a non-physiological manner.

A recent kinetic study has investigated the effect of ligand recognition by α5β1 on the binding of the function blocking anti-β1 antibody, mAb 13 (Mould et al., 1996). Ligand decreased the binding of mAb 13 to α5β1 and the concentration of ligand required for half-maximal inhibition of antibody binding was independent of antibody concentration suggesting that ligand acts as an allosteric inhibitor of antibody binding. Hence, although mAb 13 has been mapped to a discrete part of the putative ligand-binding region of the β1 subunit (Takada and Puzon, 1993), it does not compete directly with ligand for binding to α5β1 but recognizes an epitope that is attenuated by ligand occupancy. mAb 13 binds preferentially to the unoccupied conformation of $\alpha5\beta1$ and may inhibit ligand binding by either stabilizing this unoccupied state or preventing a conformational change that needs to take place for full ligand binding (see Figure 3). Since other inhibitory antibodies have also been mapped to the ligand-binding domain of the $\beta1$ subunit (Takada and Puzon, 1993) they may also recognize sites that become attenuated during ligand occupancy. Mould et al. (1996) have termed these epitopes ligand-attenuated binding sites (LABS).

VII. MODULATION OF INTEGRINS VIA THE CYTOPLASMIC TAILS

The activity of integrins can also be controlled by inside-out signaling via the cytoplasmic domains (for reviews see Ginsberg et al., 1992; Hynes, 1992; Sastry and Horwitz, 1993). Individual α and β cytoplasmic domains are highly conserved across species and both groups of subunits have conserved sequence motifs suggesting that both the α and β cytoplasmic tails play unique and important roles in the regulation of adhesion.

A. β Cytoplasmic Domains

A comparison of the sequences of the cytoplasmic tails of the $\beta1$, $\beta2$, and $\beta3$ subunits shows three clusters of amino acids that are highly conserved (see Figure 4). A membrane-proximal stretch of 11 amino acids consists mainly of charged residues, three of these are identical and eight out of the 11 are highly conserved. The other two conserved clusters are both NPxY motifs, although the most C-terminal of these has variations in each of the three subunits. These three conserved regions are also found in the cytoplasmic domains of other β subunits (Sastry and Horwitz, 1993) with the second NPxY motif being the most poorly conserved. These regions have been implicated in the binding or association of many cytoskeletal and regulatory proteins (for review see Dedḥar and Hannigan, 1996). Outside-in integrin signaling will be discussed fully in another chapter of this book.

Early studies involving the expression of integrins with mutations or deletions in the β cytoplasmic tail showed at least part of it to be essential for integrin function. A point mutation in the cytoplasmic domain of the $\beta3$ subunit (S752P) caused defective activation of the platelet integrin $\alpha IIb\beta3$ in a variant of Glanzmann's thrombasthenia (Chen et al., 1992). Mutants containing deletions in the chicken $\beta1$ cytoplasmic domain at the C-terminus neither promoted adhesion nor localized in focal contacts (Hayashi et al., 1990) and truncations in the cytoplasmic tail of the β subunit of $\alpha L\beta2$ eliminated binding to ICAM-1 and sensitivity to phorbol esters (Hibbs et al., 1991).

More recent studies, particularly on the $\beta1$ and $\beta3$ subunits, have helped to define the sequences within the cytoplasmic tails that are required for integrin activation.

β1A KLLMIIHD**RREFAKFEKEK**MNAKWDTGE**NPIY**KSAVTTVV**NPKY**EGK

β1B ....SYKTSKKQSGL

β1C

..........................SLSVAQPGVQMCDISSLQPLLTSRFQQFSCLSIPSTW
DYRVKILFFIRVP

β1D ..........................Q.......PINNFK..N.GR.AGL

β2 .A.IHLS.L..YRR.....LKSQWNND-..LF...T...M...FAES

β3 ...IT....K......E.RAR.....AN..L..E.TS.FT.IT.RGT

β3' .........K......E.RAR.....VRDGAGRFLKSLV

Figure 4. Sequences of the cytoplasmic domains of the β1, β2, and β3 subunits. Identity with the β1A sequence is indicated by dots. A dash is inserted in the β2 sequence to optimize alignment. The sequences for the cytoplasmic variants of the β1 and β3 subunits described in the text are shown. The three conserved clusters of residues described in the text are underlined and shown in bold in the β1A sequence. (Adapted from Sastry and Horwitz, 1993; Dedhar and Hannigan, 1996).

O'Toole et al. (1994) constructed chimeric integrins in which the extracellular and transmembrane domains of αllb and β3 were joined to the cytoplasmic domains of α5 and β1 respectively, resulting in a constitutively active receptor when expressed in CHO cells. This system was used to assess the effects of mutations and truncations of both the a subunit (see below) and the β subunit. The β subunits to be tested were co-transfected with high affinity α subunits and their ability to alter the affinity of the integrin was measured by the binding of PAC1, an activation specific mAb. In addition to the previously identified point mutation, S752P, the β3 truncation mutant ending at D723 also abolished high affinity binding. This truncation removes both of the NPxY motifs and most of membrane-proximal conserved motif from the cytoplasmic tail emphasizing their importance in integrin activation.

Subsequent studies (O'Toole et al., 1995; Hughes et al., 1995) that used this method of testing the effect of mutant β subunits on constitutively active α subunits identified further sequences in both the β1 and β3 subunits that are involved in integrin activation. Point mutations at residues N785 and Y788 of the first NPxY motif of the β1 subunit and F727, F730, E733, and Y747 in the β3 subunit were found to abolish PAC1 binding (O'Toole et al., 1995). A construct encoding an alternatively spliced form of the β3 cytoplasmic domain, β3B (van Kupperfelt et al., 1989), also resulted in a low-affinity receptor. This variant lacked all but two residues of the membrane proximal conserved sequence as well as the two NPxY motifs. A similar variant has been identified for the β1 subunit which has been shown not to localize at focal contacts (Balzac et al., 1993). Hughes et al. (1995) found that deletion of the four C-terminal residues of the β3 cytoplasmic tail was enough to abolish PAC1 binding and that this low-affinity state was maintained in all truncations up to R724, which corresponds to the β1 and β3 spliced variants, suggesting that an almost complete β3 tail is required for inside-out signaling. However, deletion of the entire β3 cytoplasmic tail, Δ717, resulted in the expression of a constitutively active receptor that was independent of both the α subunit cytoplasmic domain and of the cellular signaling machinery. The overexpression of an isolated β3 cytoplasmic domain, previously shown to be a potent inhibitor of integrin activation (Chen et al., 1994), had no effect on the activation of the Δ717 construct (Hughes et al., 1995).This suggests that residues 717-724 of the β3 cytoplasmic tail maintain αllbβ3 in a default low-affinity state.

Puzon-McLaughlin and co-workers (1996) used the conformation-dependent and activation-specific anti-β1 mAb 15/7 to show that truncation of more than 16 residues of the β1 cytoplasmic domain resulted in a constitutively low-affinity state in agreement with the consensus that nearly all of the β cytoplasmic tail is required for integrin activation. A recent report by Baker et al. (1997) has confirmed the requirement for a near full length β1 cytoplasmic domain. In this study the constitutively active αllbβ3 chimeras previously described (O'Toole et al., 1994) were subjected to chemical mutagenesis before expression and activation defective mutants were characterized. Several of the mutants contained deletions in the β1 cytoplasmic tail and another, P761S, disrupted the first NPxY motif, indicating that the N, P and Y residues of this motif are all required for activation in β1.

B. α Cytoplásmic Domains

Like the β subunits, the cytoplasmic domains of α subunits are highly conserved across species but unlike the β subunits show little homology to each other, apart from a membrane proximal sequence, GFFKR, that is found in all α subunits (for review see Sastry and Horwitz, 1993).

Parallel studies on α4β1 (Kassner and Hemler, 1993) and α2β1 (Kawaguchi and Hemler, 1993) showed that truncations just after the conserved GFFKR sequence caused a loss of adhesive activity and sensitivity to phorbol esters when the constructs were expressed in cells in which the wild type integrin was active. However, exchange of the α subunit cytoplasmic domain with that of another integrin (α2 or α5 for α4 and α4 or α5 for α2) had no effect on the constitutive activity of the integrin or its response to phorbol ester. The loss of adhesive activity of the truncated mutants could be overcome by the addition of a stimulatory anti-β1 mAb, TS2/16 or by Mn^{2+}, proving that the integrins were not irreversibly inactive and that the α cytoplasmic domain-dependent mechanism controlling mutation in the activation could be bypassed by external agonists.

Interestingly, an α2 construct that was deleted just prior to the conserved GFFKR sequence was not expressed at the cell surface. Conversely, O'Toole and colleagues (1991) expressed an αllb cytoplasmic mutant that was truncated just before the GFFKR sequence and found it to be constitutively active for PAC1 and fibrinogen binding. An αIIb mutant with a truncation just after this conserved sequence did not bind PAC1 suggesting that the GFFKR sequence is maintaining a default low affinity state (O'Toole et al., 1994). In contrast to the constitutively active αIIb/α5, β3/β1 chimeras discussed previously, the activation of the GFFKR deletion mutants was energy independent and the β cytoplasmic domain had no effect on the activation state.

Both the α and β-subunit cytoplasmic domains have highly conserved, membrane proximal sequences the deletion of which results in the expression of a constitutively active receptor. This led Hughes and co-workers (1996) to propose the presence of a salt bridge between adjacent residues in these sequences which would stabilize a default inactive conformation. Disruption of this bridge by the binding of other molecules may result in activation of the integrin. Mutational analysis has demonstrated an absolute requirement for the aspartate of the β subunit sequence and the arginine and two phenylalanines of the α subunit sequence (Hughes et al., 1996).

C. Molecular Interactions with the Cytoplasmic Tails

The studies outlined above show that, at least for the β subunit, nearly all of the cytoplasmic tail is required for physiological activation of the integrin and that activation involves the binding of cellular components to one or both of the cytoplasmic tails resulting in a conformational change that can be propagated to the extracellular domains. Integrin cytoplasmic tails have been shown to bind to or associate with many cytoskeletal and also regulatory and signal-transducing proteins involved in outside-in signaling (Dedhar and Hannigan, 1996), but less is known about the molecules that regulate inside-out signaling.

Recent studies have identified molecules that bind to the cytoplasmic domains of both the α and β subunits and may play a role in inside-out signaling. The calcium-binding protein calreticulin has been shown to bind to the conserved KxGFFKR motif of α subunit cytoplasmic domains in vitro (Rojiani et al., 1991) and an in vivo study demonstrated an association with the active but not inactive form of the α2β1 integrin (Coppolino et al., 1995). Yeast two-hybrid screens have identified two novel molecules integrin-linked kinase (ILK; Hannigan et al., 1996) and β3 endonexin (Shattil et al., 1995). Whereas ILK has been shown to bind to the cytoplasmic tails of the β1A, β2 and β3 integrin subunits, β3 endonexin appears to be specific for the β3 subunit. A recent study has shown that the membrane distal NITY motif (see Figure 4) in the β3 cytoplasmic domain is required for β3 endonexin binding (Eigenthaler et al.. 1997). Furthermore. exchange of the β3 NITY residues for the corresponding β1 NPKY sequence enabled the β1 cytoplasmic tail to bind to β3 endonexin and conversely binding of the β3 subunit was abolished when NITY was replaced with NPKY. The physiological roles of calreticulin, ILK and β3 endonexin are unknown.

Members of the small GTP-binding protein family may also be involved in the regulation of inside-out signaling. Zhang and colleagues (1996) demonstrated that ectopic expression of R-ras induced cell attachment and spreading in nonadherent cells by activating the integrin, while expression of the dominant negative R-ras inhibited cell attachment. The Rho family of small GTPases have been shown to play an essential role in the regulation of integrin clustering and the assembly of focal adhesion complexes (Nobes and Hall, 1995; Hotchin and Hall, 1995). These proteins may also be involved in integrin activation possibly mediating their effects via the actin cytoskeleton.

In the future, a key area of investigation will be the link between integrin engagement by ligand, integrin activation and intracellular signaling responses. Elucidation of the molecular mechanisms involved will improve our understanding of the effects of adhesion on cell function and will also suggest strategies to control adhesion in disease.

ACKNOWLEDGMENTS

Work in the authors' laboratory was supported by grants from the Wellcome Trust to M.J.H.

REFERENCES

Akiyama, S. K., Yamada, S. S., Chen, W. T., & Yamada, K. M. (1989). Analysis of migration, matrix assembly, and cytoskeletal organization. J. Cell Biol. 109, 863-875.

Altruda, F., Cervella, P., Tarone, G., Botta, C., Balzac, F., Stefanuto, G., & Silengo, L. (1990). A human integrin beta 1 subunit with a unique cytoplasmic domain generated by alternative mRNA processing. Gene 95, 261-266.

Aota, S., Nomizu, M., & Yamada, K. M. (1994). The short amino acid sequence Pro-His-Ser-Arg-Asn in human fibronectin enhances cell-adhesive function. J. Biol. Chem. 269, 24756-24761.

Baker, E. K., Tozer, E. C., Pfaff, M., Shattil, S. J., Loftus, J., & Ginsberg, M. H. (1997). A genetic analysis of integrin function: Glanzmann thrombasthenia in vitro. Proc. Nat. Acad. Sci. U.S.A. 94, 1973-1978.

Bajt, M. L., Goodman, T., & McGuire, S. L. (1995). beta 2 (CD18) mutations abolish ligand recognition by I domain integrins LFA - 1 (alpha L beta 2, CD11a / CD18) and MAC -1 (alpha M beta 2, CD11b / CD18). J. Biol. Chem. 270, 94-98.

Bajt, M. L., & Loftus, J. C. (1994). Mutation of a ligand binding domain of beta 3 integrin . Integral role of oxygenated residues in alpha IIb beta 3 (GPIIb-IIIa) receptor function. J. Biol. Chem. 269, 20913-20919.

Balzac, F., Belkin, A. M., Koteliansky, V. E., Balabanov, Y. V., Altruda, F., Silengo, L., & Tarone, G. (1993). Expression and functional analysis of a cytoplasmic domain variant of the beta 1 integrin subunit. J. Cell Biol. 121, 171-178.

Bazzoni, G., Shih, Daw-T., Buck, C. A., & Hemler, M. E. (1995). Monoclonal antibody 9EG7 defines a novel beta 1 integrin epitope induced by soluble ligand and manganese, but inhibited by calcium. J. Biol. Chem. 270, 25570-25577.

Bowditch, R. D., Hariharan, M., Tominna, E. F., Smith, J. W., Yamada, K. M., Getzoff, E. D., & Ginsberg, M. H. (1994). Identification of a novel integrin binding site in fibronectin. Differential utilization by beta 3 integrins. J. Biol. Chem. 269, 10856-10863.

Calvete, J. J., Henschen, A., & Gonzalez-Rodriguez, J. (1991). Assignment of disulfide bonds in human platelet GPIIIa. A disulfide pattern for the beta.- subunits of the integrin family. Biochem. J. 274, 63-71.

Chen, Y. P., Djaffar, I., Pidard, D., Steiner, B., Cieutat, A. M., Caen, J. P., & Rosa, J. P. (1992). Ser-752 → Pro mutation in the cytoplasmic domain of integrin beta 3 subunit and defective activation of platelet integrin alpha IIb beta 3 (glycoprotein IIb - IIIa) in a variant of Glanzmann thrombasthenia. Proc. Natl. Acad. Sci. U. S. A. 89, 1016910173.

Chen, Yi-P., O'Toole, T. E., Shipley, T., Forsyth, J., LaFlamme, S. E., Yamada, K. M., Shattil, S. J., & Ginsberg, M. H. (1994). "Inside-out" signal transduction inhibited by isolated integrin cytoplasmic domains. J. Biol. Chem. 269, 18307-18310.

Cooper, H. M., Tamura, R. N., & Quaranta, V. (1991). The major laminin receptor of mouse embryonic stem cells is a novel isoform of the: alpha 6 beta 1 integrin. J.Cell Biol. 115, 843-850.

Coppolino, M., Leung-Hagesteijn, C., Dedhar, S., & Wilkins, J. (1995). Inducible interaction of integrin alpha 2 beta 1 with calreticulin. Dependence on the activation state of the integrin. J. Biol. Chem. 270, 23132-23138.

Dedhar, S. & Hannigan, G. E. (1996). Integrin cytoplasmic interactions and bidirectional transmembrane signaling. Curr. Opin. Cell Biol. 8, 657-669.

Diamond, M. S., Garcia-Aguilar, J., Bickford, J. K., Corbi, A. L., & Springer, T. A. (1993). The I domain is a major recognition site on the leukocyte integrin Mac - 1 (CD11b / CD18) for four distinct adhesion ligands. J. Cell Biol. 120, 1031-1043.

D'Souza, S. E., Ginsberg, M. H., Burke, T. A., & Plow, E. F. (1990). The ligand binding site of the platelet integrin receptor GPIIb-IIIa is proximal to the second calcium binding domain of its alpha subunit. J. Biol. Chem. 265, 3440-3446.

D'Souza, S. E., Ginsberg, M. H., Lam, S. C. T., & Plow, E. F. (1988). Chemical crosslinking of arginyl-glycyl-aspartic acid peptides to an adhesion receptor on platelets. J. Biol. Chem. 263, 3943-3951.

D'Souza, S. E., Haas, T. A., Piotrowicz, R. S., Byers-Ward, V., McGrath, D. E., Soule, H. R., Cierniewski, C., Plow, E. F., & Smith, J. W. (1994). Ligand and cation binding are dual functions of a discrete segment of the integrin beta 3 subunit: cation displacement is involved in ligand binding. Cell 79, 659-667.

Du, X., Gu, M., Weisel, J. W., Nagaswami, C., Bennett, J. S., Bowditch, R., & Ginsberg, M. H. (1993). Long range propagation of conformational changes in integrin alpha IIb beta 3. J. Biol. Chem. 268, 23087-23092.

Dufour, S., Duband, J. L., Humphries, M. J., Obara, M., Yamada, K. M., & Thiery, J. P. (1988). Attachment, spreading and locomotion of avian neural crest cells are mediated by multiple adhesion sites on fibronectin molecules. EMBO J. 7, 2661-2671.

Eble, J. A., Golbik, R., Mann, K., & Kühn, K. (1993). The alpha 1 beta 1 integrin recognition site of the basement membrane collagen molecule [alpha $1(IV)]_2$ alpha 2(IV). EMBO J. 12, 4795-4802.

Eigenthaler, M., Hofferer, L., Shattil, S. J., & Ginsberg, M. H. (1997). A conserved sequence motif in the integrin beta 3 cytoplasmic domain is required for its specific interaction with beta 3 endonexin. J. Biol. Chem. 272, 7693-7698.

Faull, R. J. & Ginsberg, M. H. (1995). Dynamic regulation of integrins. Stem Cells (Dayton) 13, 38-46.

Faull, R. J., Wang, J., Leavesley, D. I., Puzon, W., Russ, G. R., Vestweber, D., & Takada, Y. (1996). A novel activating anti-beta 1 integrin monoclonal antibody binds to the cysteine-rich repeats in the beta 1 chain. J. Biol. Chem. 271, 25099-25106.

Giancotti, F. G., Stepp, M. A., Suzuki, S., Engvall, E., & Ruoslahti, E. (1992). Proteolytic processing of endogenous and recombinant beta 4 integrin subunit. J. Cell Biol. 118, 951-959.

Ginsberg, M. H., Du, X., & Plow, E. F. (1992). Inside-out integrin signaling. Curr. Opin. Cell Biol. 4, 766-771.

Goodman, T. G. & Bajt, M. L. (1996). Identifying the putative metal ion-dependent adhesion site in the beta 2 (CD18) subunit required for alpha L beta 2 and . alpha M beta 2 ligand interactions. J. Biol. Chem. 271, 23729-23736.

Gulino, D., Boudignon, C., Zhang, L., Concord, E., Rabiet, M. J., & Marguerie, G. (1992). Calcium-binding properties of the platelet glycoprotein IIb ligand-interacting domain. J. Biol. Chem. 267, 1001-1007.

Hannigan, G. E., Leung-Hagesteijn, C., Fitz-Gibbon, L., Coppolino, M. G., Radeva, G., Filmus, J., Bell, J. C., & Dedhar, S. (1996). Regulation of cell adhesion and anchorage-dependent growth by a new beta-integrin-linked protein kinase. Nature (London) 379, 9196.

Hayashi, Y., Haimovich, B., Reszka, A., Boettiger, D., & Horwitz, A. (1990). Expression and function of chicken integrin beta 1 subunit and its cytoplasmic domain mutants in mouse NIH 3T3 cells. J. Cell Biol. 110, 175-184.

Hibbs, M. L., Xu, H., Stacker, S. A., & Springer, T. A. (1991). Regulation of adhesion to ICAM-1 by the cytoplasmic domain of LFA - 1 integrin beta subunit. Science 251, 1611-1613.

Honda, S., Tomiyama, Y., Pelletier, A. J., Anr is, D., Honda, Y., Orchekowski, R., Ruggeri, Z., & Kunicki, T. J. (1995). Topography of ligand-induced binding sites, including a novel cation-sensitive epitope (AP5) at the amino terminus. of the human integrin beta 3 subunit. J. Biol. Chem. 270, 11947-11954.

Hotchin, N. A. & Hall, A. (1995). The assembly of integrin adhesion complexes requires both extracellular matrix and intracellular rho/rac GTPases. J. Cell Biol. 131, 1857-1865.

Hughes, P. E., Diaz-Gonzalez, F., Leong, L., Wu, C., McDonald, J. A., Shattil, S. J., & Ginsberg, M. H. (1996). Breaking the integrin hinge. A defined structural constraint regulates integrin signaling. J. Biol. Chem. 271, 6571-6574.

Hughes, P. E., O'Toole, T. E., Ylanne, J., Shattil, S. J., & Gingsberg, M. H. (1995). The conserved membrane-proximal region of an integrin cytoplasmic domain specifies ligand binding affinity. J. Biol. Chem. 270, 12411-12417.

Humphries, M. J. (1990). The molecular basis and specificity of integrin -ligand interactions. J. Cell Sci. 97, 585-592.

Humphries, M. J. (1996). Integrin activation: the link between ligand binding and signal transduction. Curr. Opin. Cell Biol. 8' 632-640.

Hynes, R. O. (1992). Integrins: versatility, modulation, and signaling in cell adhesion. Cell 69, 11-25.

Irie, A., Kamata, T., Puzon-McLaughlin, W., & Takada, Y. (1995). Critical amino acid residues for ligand binding are clustered in a predicted beta-turn of the third N-terminal repeat in the integrin alpha 4 and alpha 5 subunits. Embo J. 14, 5550-5556.

Jones, E. Y., Harlos, K., Bottomley, M. J., Robinson, R. C., Driscoll, P. C., Edwards, R. M., Clements, J. M., Dudgeon, T. J., & Stuart, D. I. (1995). Crystal structure of an integrin -binding fragment of vascular cell adhesion molecule -1 at 1.8 .ANG. resolution. Nature (London) 373, 539-544.

Juliano, R. L. & Haskill, S. (1993). Signal transduction from the extracellular matrix. J. Cell Biol. 120, 577-585.

Kamata, T., Irie, A., Tokuhira, M., & Takada, Y. (1996). Critical residues of integrin alpha IIb subunit for binding of alpha IIb beta 3 (glycoprotein IIb -IIIa) to fibrinogen and ligand-mimetic antibodies (PAC-1, OP-G2, and LJ-CP3). J. Biol. Chem. 271, 1861018615.

Kamata, T., Puzon, W., & Takada, Y. (1995). Identification of putative ligand-binding sites of the integrin alpha 4 beta 1 (VLA -4, C])49d / CD29). Biochem. J. 305, 945951.

Kassner, P. D. & Hemler, M. E. (1993) . Interchangeable alpha chain cytoplasmic domains play a positive role in control of cell adhesion mediated by VLA -4, a .beta 1 integrin. J Exp. Med. 178, 649-660.

Kawaguchi, S. & Hemler, M. E. (1993). Role of the alpha subunit cytoplasmic domain in regulation of adhesive activity mediated by the integrin VLA -2. J. Biol. Chem. 268, 16279-16285.

Kern, A., Eble, J., Golbik, R., & Kühn, K. (1993). Interaction of type IV collagen with the isolated integrins alpha 1 beta 1 and alpha 2 beta 1. Eur. J. Biochem. 215, 151-159.

Komoriya, A., Green, L. J., Mervic, M., Yamada, S. S., Yamada, K. M., & Humphries, M. J. (1991). The minimal essential sequence for a major cell type-specific adhesion site (CS1) within the alternatively spliced type III connecting segment domain of fibronectin is leucine-aspartic acid-valine. J. Biol. Chem. 266, 15075-15079.

Kramer, R. H., Vu, M. P., Cheng, Yao-F., Ramos, D. M., Timp, R., & Waleh, N. (1991). Laminin-binding integrin alpha 7 beta 1: functional characterization and expression in normal and malignant melanocytes. Cell Regul. 2, 805-817.

Kretsinger, R. H . & Nockolds C. E. (1973) Carp muscle calcium-binding protein 11: Structure determination and general description. J. Biol. Chem. 248, 3313-3326.

Kühn, K. & Eble, J. (1994). The structural bases of integrin -ligand interactions. Trends Cell Biol. 4, 256-261.

Lee, Jie-O., Rieu, P., Arnaout, M. A., & Liddington, R. (1995a). Crystal structure of the A domain from the alpha subunit of integrin CR3 (CD11b / CD18). Cell (Cambridge, Mass.) 80, 631-638.

Lee, Jie-O., Bankston, L. A., Arnaout, M. A., & Liddington, R. C. (1995b). Two conformations of the integrin A-domain (I-domain): a pathway for activation? Structure (London) 3, 1333-1340.

Loftus, J. C., Halloran, C. E., Ginsberg, M. H., Feigen, L. P., Zablocki, J. A., & Smith, J. W. (1996). The amino-terminal one-third of alpha IIb defines the ligand recognition specificity of integrin alpha IIb beta 3. J. Biol. Chem. 271, 2033- 2039.

Loftus, J. C., O'Toole, T. E., Plow, E. F., Glass, A., Frelinger, A. L., III, & Ginsberg, M. H. (1990). A beta 3 integrin mutation abolishes ligand binding and alters divalent cation-dependent conformation. Science 249, 915-918.

Loftus, J. C., Smith, J. W., & Ginsberg, M. H. (1994) Integrin-mediated cell adhesion: The extracellular Face. J. Biol. Chem. 269, 25235-25238.

Main, A. L., Harvey, T. S., Baron, M., Boyd, J., & Campbell, I. D. (1992). The three-dimensional structure of the tenth type III module of fibronectin: an insight into RGD mediated interactions. Cell (Cambridge, Mass.) 71, 671-678.

Masumoto, A. & Hemler, M. E. (1993). Mutation of putative divalent cation sites in the .alpha 4 subunit of the integrin VLA -4: Distinct effects on adhesion to CS1 / fibronectin, VCAM -1, and Invasin. J. Cell Biol. 123, 245-253.

Michishita, M., Videm, V., & Arnaout, M. A. (1993). A novel divalent cation-binding site in the A domain of the beta 2 integrin CR3 (CD11b / CD18) is essential for ligand binding. Cell 72, 857-867.

Mould, A. P. (1996). Getting integrins into shape: recent insights into how integrin activity is regulated by conformational changes. J. Cell Sci. 109, 2613-2618.

Mould, A. P., Akiyama, S. K., & Humphries, M. J. (1995a). Regulation of integrin. alpha 5 beta1-fibronectin interactions by divalent cations. Evidence for distinct classes of binding sites for Mn^{2+}, Mg^{2+}, and Ca^{2+}. J. Biol. Chem. 270, 26270-26277.

Mould, A. P., Akiyama, S. K., & Humphries, M. J. (1996). The inhibitory anti-beta 1 integrin monoclonal antibody 13 recognizes an epitope that is attenuated by ligand occupancy. Evidence for allosteric inhibition of integrin function. J. Biol. Chem. 271, 20365-20374.

Mould, A. P., Askari, J. A., Aota, S., Yamada, K. M., Irie, A., Takada, Y., Mardon, H., and Humphries, M. J. (1997). Defining the topololgy of integrin alpha 5 beta 1-fibronectin interactions using inhibitory anti-alpha 5 and anti-beta 1 monoclonal antibodies. J. Biol. Chem. 272, 17283-17292.

Mould, A. P., Garratt, A. N., Askari, J. A., Akiyama, S. K., & Humphries, M. J. (1995b). Regulation of integrin alpha 5 beta 1 function by anti-integrin antibodies and divalent cations. Biochem. Soc. Trans. 23, 395S.

Mould, A. P., Garratt, A. N., Askari, J. A., Akiyama, S. K., & Humphries, M. J. (1995c). Identification of a novel anti- integrin monoclonal antibody that recognizes a ligand-induced binding site epitope on the beta 1 subunit. FEBS Lett. 363, 118-122.

Mould, A. P. & Humphries, M. J. (1991). Identification of a novel recognition sequence for the integrin alpha 4 beta 1 in the carboxy-terminal heparin- binding domain of fibronectin. Embo J. 10, 4089-4095.

Mould, A. P., Komoriya, A., Yamada, K. M., & Humphries, M. J. (1991). The CS5 eptide is a second site in the IIICS region of fibronectin recognized by the integrin alpha 4 beta 1. Inhibition of alpha 4 beta 1 function by RGD peptide homologs. J. Biol. Chem. 266, 35793585.

Nermut, M. V., Green, N. M., Eason, P., Yamada, S. S., & Yamada, K. M. (1988). Electron microscopy and structural model of human fibronectin receptor. Embo J. 7, 4093-4099.

Newham, P., Craig, S. E., Clark, K., Mould, A. P., & Humphries, M. J. (1998). Analysis of ligand-induced and ligand-attenuated epitopes on the leukocyte integrin alpha4 beta 1: VCAM-1, Mucosal Addressin Cell Adhesion Molecule-1, and Fibronectin induce distinct conformational changes. J. Immunol, 160, 4508-4517.

Newham, P., Craig, S. E., Seddon, G. N., Schofield, N. R., Rees, A., Edwards, R. M., Jones, E. Y., and Humphries, M. J. (1997). Alpha 4 integrin binding interfaces on VCAM-1 and MAdCAM-1: Integrin binding footprints identify accessory binding sites that play a role in integrin specificity. J. Biol. Chem. 272, 19429-19440

Newham, P. & Humphries, M. J. (1996). Integrin adhesion receptors: structure, function and implications for biomedicine. Mol. Med. Today 2, 304-313.

Nobes, C. D. & Hall, A. (1995). Rho, rac and cdc42 GTPases regulate the assembly of multimolecular focal complexes associated with actin stress fibres, lamellipodia, and filopodia. Cell 81, 53-62.

O'Toole, T. E., Katagiri, Y., Faull, R. J., Peter, K., Tamura, R., Quaranta, V., Loftus, J. C., Shattil, S. J., & Ginsberg, M. H. (1994). Integrin cytoplasmic domains mediate inside-out signal transduction. J. Cell Biol. 124, 1047- 1059.

O'Toole, T. E., Mandelman, D., Forsyth, J., Shattil, S. J., Plow, E. F., & Ginsberg, M. H. (1991). Modulation of the affinity of integrin alpha IIb beta 3 (GPIIb-IIIa) by the cytoplasmic domain of alpha IIb. Science 254, 845-847.

O'Toole, T. E., Ylanne, J., & Culley, B. M. (1995). Regulation of integrin affinity states through an NPXY motif in the beta subunit cytoplasmic domain. J. Biol. Chem. 270, 8553-8558.

Pierschbacher, M. D., & Ruoslahti, E. (1984). Variants of the cell recognition site of fibronectin that retain attachment-promoting activity. Proc. Natl. Acad. Sci. U.S.A. 81, 5985-5988.

Puzon-McLaughlin, W., & Takada, Y. (1996). Critical residues for ligand binding in an I domain-like structure of the integrin beta 1 subunit. J. Biol. Chem. 271, 2043820443.

Puzon-McLaughlin, W., Yednock, T. A., & Takada, Y. (1996). Regulation of conformation and ligand binding function of integrin alpha 5 beta 1 by the beta. 1 cytoplasmic domain. J. Biol. Chem. 271, 16580-16585.

Randi, A. M., & Hogg, N. (1994). I Domain of beta 2 Integrin Lymphocyte Function-associated Antigen-1 Contains a Binding Site for Ligand Intercellular Adhesion Molecule-1. J. Biol. Chem. 269, 12395-12398.

Rojiani, M. V., Finlay, B. B., Gray, V., & Dedhar, S. (1991). In vitro interaction of a polypeptide homologous to human Ro/SS-A antigen (calreticulin) with a highly conserved amino acid sequence in the cytoplasmic domain of integrin alpha subunits. Biochemistry 30, 9859-9866.

Sastry, S. K. & Horwitz, A. F. (1993). Integrin cytoplasmic domains: Mediators of cytoskeletal linkages and extra- and intracellular initiated transmembrane signaling. Curr. Opin. Cell Biol. 5, 819-831.

Shattil, S. J., O'Toole, T., Eigenthaler, M., Thon, V., Williams, M., Babior, B. M., & Ginsberg, M. H. (1995). Beta 3-Endonexin, a novel polypeptide that interacts specifically with the cytoplasmic tail of the integrin beta3 subunit. J. Cell Biol. 131, 807-816.

Smith, J. W. & Cheresh, D. A. (1991). Labeling of integrin alpha v beta 3 with cobalt-58(III). Evidence of metal ion coordination sphere involvement in ligand binding. J. Biol. Chem. 266, 11429-11432.

Smith, J. W. & Cheresh, D. A. (1990). Integrin (alpha v beta 3)-ligand interaction. Identification of a heterodimeric RGD binding site on the vitronectin receptor. J. Biol. Chem. 265, 2168-2172.

Smith, J. W. & Cheresh, D. A. (1988). The Arg - Gly - Asp binding domain of the vitronectin receptor. Photoaffinity cross-linking implicates amino acid residues 61-203 of the beta subunit. J. Biol. Chem. 263, 18726-18731.

Smith, J. W., Piotrowicz, R. S., & Mathis, D. (1994). A mechanism for divalent cation regulation of beta 3-integrins. J. Biol. Chem. 269, 960-967.

Sonnenberg, A. (1993). Integrins and their ligands. Curr. Top. Microbiol. Immunol. 184, 7-35.

Springer, T. A. (1997). Folding of the N-terminal, ligand-binding region of integrin alpha-subunits into a beta-propeller domain. Proc. Natl. Acad. Sci. U. S. A.94, 65-72.

Stanley, P., Bates, P. A., Harvey, J., Bennett, R. I., & Hogg, N. (1994). Integrin LFA-1.alpha subunit contains an ICAM -1 binding site in domains V and VI. Embo J. 13, 17901798.

Strynadka, N. C. J., & James, M. N. G. (1989). Crystal structures of the helix-loop-helix calcium binding proteins. Annul Rev. Biochem. 58, 951-998.

Takada, Y. & Puzon, W. (1993). Identification of a regulatory region of integrin beta 1 subunit using activating and inhibiting antibodies. J. Biol. Chem. 268, 17597-17601.

Tamura, R. N., Cooper, H. M., Collo, G., & Quaranta, V. (1991). Cell type-specific integrin variants with alternative alpha chain cytoplasmic domains. Proc. Natl. Acad. Sci. U. S. A. 88, 10183-10187.

Tamura, R. N., Rozzo, C., Starr, L., Chambers, J., Reichardt, L. F., Cooper, H. M., & Quaranta, V. (1990). Epithelial integrin alpha 6 beta 4: complete primary structure of alpha 6 and variant forms of beta 4. J. Cell Biol. 111, 1593-1604.

Teixido, J., Parker, C. M., Kassner, P. D., & Hemler, M. E. (1992). Functional and structural analysis of VLA -4 integrin alpha 4 subunit cleavage. J. Biol. Chem. 267, 17861791.

Tozer, E. C., Liddington, R. C., Sutcliffe, M. J., Smeeton, A. H., & Loftus, J. C. (1996). Ligand binding to integrin alpha IIb beta 3 is dependent on a MIDAS-like domain in the beta 3 subunit. J. Biol. Chem. 271, 21978-21984.

Tuckwell, D. S., Brass, A., & Humphries, M. J. (1992). Homology modeling of integrin EF-hands. Evidence for widespread use of a conserved cation-binding site. Biochem. J . 285, 325 331.

Tuckwell, D., Calderwood, D. A., Green, L. J., & Humphries, M. J. (1995). Integrin alpha 2 I-domain is a binding site for collagens. J. Cell Sci. 108, 1629-1637.

Tuckwell, D. S. & Humphries, M. J. (1997). A structure prediction for the ligand-binding region of the integrin beta subunit: evidence for the presence of a von Willebrand factor A domain. FEBS Lett. 400, 297-303.

Tuckwell, D. S., Humphries, M. J., & Brass, A. (1994). A secondary structure model of the integrin alpha subunit N-terminal domain based on analysis of multiple alignments. Cell Adhes. Commun. 2, 385-402.

Ueda, T., Rieu, P., Brayer, J., & Arnaout, M. A. (1994). Identification of the complement iC3b binding site in the beta 2 integrin CR3 (CD11b / CD18). Proc. Natl. Acad. Sci. U. S. A. 91, 10680-10684.

van Kuppervelt, T. H., Languino, L. R., Gailit, J. O., Suzuki, S., & Ruoslahti, E. (1989). An alternative cytoplasmic domain of the integrin beta 3 subunit. Proc. Nat. Acad. Sci. U.S.A. 86, 5415-5418.

Williams, M. J., Hughes, P. E., O'Toole, T. E., & Ginsberg, M. H. (1994). The inner world of cell adhesion: Integrin cytoplasmic domains. Trends Cell Biol. 4, 109-112.

Yamada, K. M. & Kennedy, D. W. (1985). Amino acid sequence specificities of an adhesive recognition signal. J. Cell. Biochem. 28. 99-104.

Yamada, K. M., & Kennedy, D. W. (1984). Dualistic nature of adhesive protein function: fibronectin and its biologically active peptide fragments can autoinhibit fibronectin function. J. Cell Biol. 99, 29-36.

Zhang, Z., Vuori, K., Wang, Hong-G., Reed, J. C., & Ruoslahti, E. (1996). Integrin activation by R-ras. Cell 85, 61-69.

Zhou, L., Lee, D. H. S., Plescia, J., Lau, C. Y., & Altieri, D. C. (1994). Differential ligand binding specificities of recombinant CD11b / CD18 integrin I-domain. J. Biol. Chem. 269, 17075-17079.

THE CADHERIN SUPERFAMILY

Jörg Stappert and Rolf Kemler

Advances in Molecular and Cell Biology
Volume 28, pages 27-63.

ISBN: 0-7623-0495-2

I. INTRODUCTION

Between 1939 and 1955, Johannes Holfreter investigated the behavior of amphibian embryonic cells and tissues in vitro (Townes and Holfreter, 1955). He was able to show that embryonic tissues expressed "preferences" in associating with other tissues. Generally, tissues would rearrange in a specific manner to finally adopt particular arrangements similar to those formed in the course of normal embryonic development. At the time Holfreter formulated the term "tissue affinities" to describe the forces responsible for the differential adhesion. To identify the molecules causing tissue affinities investigators have tried for decades since then to isolate and characterize them.The advent of genetic engineering finally made it possible to clone various cell adhesion molecules, generally called CAMs. These studies revealed that most CAMs belong to one of the following four protein families: cadherins, immunglobulins, integrins, or selecting (reviewed in Hynes and Lander, 1992)

Calcium-dependent adhesion is mediated by the cadherin superfamily. More than 30 different cadherins from a variety of different organisms and tissues have been described so far (reviewed in Kemler, 1992; Takeichi, 1995; Suzuki, 1996). Their pivotal roles in cell-cell adhesion and morphogenesis have been established, but there are structural differences among cadherins suggesting considerable diversity in this superfamily. In addition, recent experiments support the hypothesis of cadherins being actively involved in tissue formation (Larue et al., 1996).

In this chapter we will mainly focus on the structure and function of the so-called "classical cadherins" which can be regarded as the prototypes of all cadherins. For a detailed description of desmosomal cadherins as well as aspects of signaling via catenins the reader is referred to the corresponding chapters in this book.

II. CLASSIFICATION WITHIN THE CADHERIN SUPERFAMILY

Sequence similarities among the different members of the cadherin superfamily provide the basis for their assignment to four groups: classical cadherins (types I, II, and III), desmosomal cadherins, protocadherins, and other cadherin-related proteins (see Table 1; reviewed in Redies, 1995; Suzuki, 1996). Almost all cadherins consist of a cytoplasmic domain, a transmembrane domain, and an extracellular domain. Exceptions are T-cadherin and human cadherin-13/H-cadherin, which lack the cytoplasmic domain but are connected to the cell membrane via a glycosyl phosphatidylinositol anchor (Ranscht and Dours-Zimmermann, 1991; Vestal and

Table 1. Summary of the different cadherins cloned so far and their classification

Classical cadherins		
name	homolog	species
type I		
E-cadherin		human, mouse, Xenopus, canine
	L-CAM	chicken
N-cadherin		human, mouse, bovine, chicken, zebrafish
	N1/N2	Xenopus
P-cadherin		human, mouse, bovine
	B-CAM	chicken
	XB/U; EP/C	Xenopus
R-cadherin		mouse, rat, chicken
	cadherin-4	human
	XmN	Xenopus
M-cadherin		mouse
	cadherin-15	human
T-cadherin		mouse, chicken
	cadherin-13/ H-cadherin	human
type II		
cadherin 5-12		human, rat, chicken
	cadherin-5/ VE-caderin	mouse
	cadherin-6/ K-cadherin/	rat
	F-cadherin	Xenopus
	cadherin-11/ OB-cadherin/	mouse
	VN-cadherin	zebrafish
	cadherin-12/ BR-cadherin	human
type III		
DE-cadherin		Drosophila

Protocadherins		
name	homolog	species
Pcdh -1,-2,-3		human, rat
partial sequences		human, rat, mouse, Xenopus, C.elegans
fat		Drosophila, human
dachsous		Drosophila

Desmosomal cadherins		
name	homolog	species
desmogleins 1,2 &3		human, mouse, bovine
desmocollins 1,2 &3		human, mouse, bovine

st-cadherins		
name	homolog	species
HPT-1		human
LI		rat
KSP		rabbit

Ranscht, 1992; Tanihara et al., 1994, 1996). A truncated soluble form of N-cadherin arising in vivo by proteolysis and retaining its adhesive function has also been described (Paradies and Grunwald, 1993). The most characteristic feature of all cadherins is the architecture of their extracellular domain. It is made up of multiple repeats of a cadherin-specific motif or subdomain approximately 110 amino acid residues in length, (see Figure 1; Ringwald et al., 1987; Hatta et al., 1988; Sano et al., 1993; Tanihara et al., 1994a). Typical features of these subdomains include conserved amino acid residues and motifs such as DxNDNxPxF, as well as DxD and DRE/DYE/DFE motifs (amino acid one-letter code) implicated in Ca^{2+} coordination. The number of these cadherin repeats varies among different members of the cadherin superfamily from 5 (classical cadherins) to more than 30 (*Drosophila* cadherin Fat; Mahoney et al., 1991). While clearly conserved, a comparison of the primary structure of the cadherin repeats reveals a clear divergence between repeats from the different cadherin groups, and even between one repeat and another within one protein.

In contrast to the conservation of the extracellular structure of cadherins, which is the primary basis for the identification of a cadherin as such, the cytoplasmic domains show a much higher variability among the different cadherin classes. These differences may well reflect separate functions of the four cadherin classes.

A. Classical Cadherins

The classical cadherins such as E-, N- or P-cadherin, were the first cadherins described. The abbreviations E-, N- and P- reflect the tissues in which these cadherins were originally identified: Epithelial-, Neuronal and Placental (Nagafuchi et al., 1987; Nose et al., 1987; Ringwald et al., 1987; Hatta et al., 1988). Similiar nomenclature is still applicable for most of the other members of this group. As shown in Table 1, for most tissues members of this group can be found in a variety of different species.

The overall structural organization of all classical cadherins is essentially the same. The proteins are synthesized as a prepropeptide. The signal sequence at the N-terminus is followed by a prosequence that ends with a protease processing signal sequence, K/RRXKR. Proteolytic cleavage of the prosequence is absolutely essential for the activation of the classical cadherins. Total proteolytic inhibition or a mutated cleavage site leads to an adhesion-defective cadherin, suggesting that the very N-terminal-most amino acids of classical cadherins are directly involved in cell adhesion or are crucial for the proper folding of the first cadherin sub-domain (Ozawa and Kemler, 1990). The prosequence is followed by the extracellular domain, which consists of five cadherin repeats EC1–EC5 (extracellular sub-domains 1–5). Although highly conserved in sequence and position, each motif shows its own individual signature, suggesting a certain variability in the function of each sub-domain. Indeed, a sequence comparison among the different cadherins clearly shows that the intermolecular conservation of a given EC repeat is usually higher than the intramolecular conservation. For example, the intermolecular amino acid identity for EC1 through EC5 among E-,P- and N-cadherin varies between 43% and

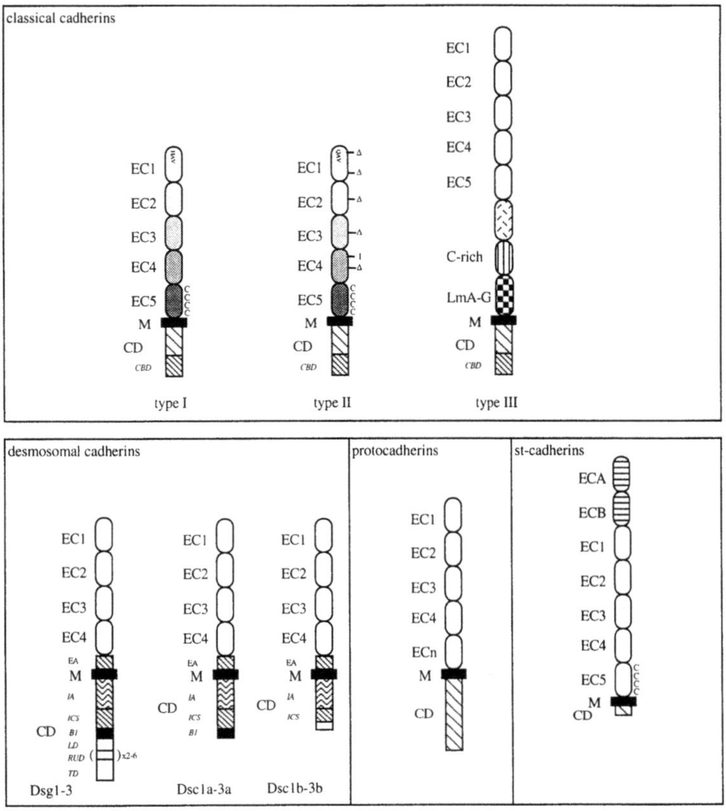

Figure 1. Schematic summary of the domain organization of the different cadherin classes. The domains and segments shown are as follows: EC1-EC5, extracellular cadherin subdomain 1–5; EC*n*, *n*-times repetition of EC subdomains; C-rich, cysteine rich domain in DE-cadherin; LmA-G, laminin A globular repeat; C, cysteine; EA, extracellular anchor repeat; M, transmembrane domain; CD, cytoplasmic domain, with CBD, catenin binding domain; IA, intracellular anchor domain; ICS, intracellular cadherin-typical segment; B1, binding site contributing to the binding of plakoglobin and assembly of the IF anchor; LD, linker domain specific for Dsgl-3; RUD, repeating unit domain; TD, terminal domain in Dsg1-3. HAV or QAV, amino acids presumably involved in homophilic interactions. Δ and I, short amino acid deletions or insertions, typical for type II classical cadherins. The different patterns of the extracellular domains of type I and type II classical cadherins reflect close amino acids similarity between these two classes. For a detailed description see text.

71%, whereas for E-cadherin the intramolecular conservation among EC1–EC5 is 24%–32%. It seems reasonable to think of classical cadherins as a family of proteins having diverged from an ancestral cadherin after the pentameric segmentation of the extracellular domain. The most N-terminal located sub-domain of classical type I cadherins, EC1, contains the highly conserved tripeptide HAV, necessary for cell adhesion (Blaschuk et al., 1990a, 1990b). The EC3 sub-domain contains a DY/F E sequence instead of a DRE sequence in the middle of the repeat and a one amino acid deletion near the end of the repeat. EC1–EC4 also show a characteristic AxDxGxP sequence near the C-terminal region. The EC5 sub-domain of classical cadherins shows another distinctive feature: the typical DRE and DxNDNxPxF sequences are missing and are replaced by four highly conserved cysteine residues.

The cytoplasmic domain of classical cadherins is their most conserved region. The amino acid identity of this 150 amino acid long domain can be as high as 89% among different classical cadherins. A site in this domain binds to molecules called catenins which provide a connection to the actin-based cytoskeleton (reviewed in Kemler, 1993; Ranscht, 1994; Aberle et al., 1996b).

Based on the amino acid alignment of classical cadherins, the family can be subdivided into three types. The differences between type I and type II cadherins are very subtle. For example, the amino acid identities among cadherin-5, cadherin-8, cadherin-11 and cadherin-12 (on average more than 40%) are higher than those between any of these cadherins and the other cadherins (mostly less than 35%). Furthermore, certain well-conserved aromatic amino acids as well as characteristic amino acid deletions or insertions are present in only one type or the other (Tanihara et al., 1994a, 1994b).

The *Drosophila* E-cadherin homolog DE-cadherin has some special features which clearly separate this protein from type I and type II cadherins (Ode et al., 1994). While its extracellular domain would classify this protein in the protocadherin sub-family, the cytoplasmic domain reveals a much higher homology to classical cadherins than to any other cadherin class. Additionally, DE-cadherin associates with catenins, while protocadherins do not. The extracellular part of DE-cadherin consists of the classical cadherin sub-domains EC1–EC4. EC1 is somewhat longer than in type I and type II cadherins. However, even characteristic features typical for type I cadherins such as the D Y/F E substitution in the EC3 sub-domain as well as the single amino acid deletion at the end of the same repeat, are evolutionary conserved. In contrast, the EC5 sub-domain with its conserved Cys residues is missing and seems to be replaced by an additional Ca^{2+}-binding sub-domain. Another unique feature of DE-cadherin is a large insertion at the proximal region of the extracellular domain with local sequence similarity to Fat, laminin A chain, Slit, and neurexin (Ode et al., 1994).Thus, DE-cadherin combines features of the protocadherin family with those of the classical cadherin family. It may be an archaic cadherin, combining multiple functions which in higher organisms became separated by the development of specialized cadherins. In other words, all vertebrate cadherins may have evolved from a single DE-cadherin-like protocadherin gene or subfamily.

B. Desmosomal Cadherins

To date, three desmoglein (Dsg) isoforms (Dsg 1, 2 and 3) and three desmocollin (Dsc) isoforms (Dsc 1, 2 and 3), each the product of a distinct gene, have been isolated from three different species, namely man, mouse and cow (for review see Koch and Franke, 1994; Kowalczyk et al., 1994; Schmidt et al., 1994; Amagai, 1995; Garrod et al., 1996). The overall structure of desmosomal cadherins is very similiar to that of classical cadherins. Like the classical cadherins, the extracellular domain of desmosomal cadherins is organized in to five subdomains. Desmoglein 1 however, lacks the EC5 subdomain, having four cadherin repeats in its extracellular domain (Koch et al., 1990). The main differences between classical cadherins and desmosomal cadherins reside in the cytoplasmic domain. The cytoplasmic domains of desmosomal cadherins vary in length, and a sequence similarity to classical cadherins can only been found in a small region within the cytoplasmic region. Interestingly, this small region is essential for the interaction with plakoglobin, one of the catenins, which also binds to the cytoplasmic domain of classical cadherins (Mathur et al., 1994; Troyanovsky et al., 1994a, 1994b; Roh and Stanley, 1995b; Troyanovsky et al., 1996). Despite this similarity, desmosomal cadherins are connected to the intermediate filaments instead to actin. Furthermore, desmocollins show an alternative splicing of the cytoplasmic domain which has not been observed in classical cadherins. The greater heterogeneity may serve to regulate their interaction with various cytoplasmic proteins.

C. Protocadherins

This recently described family of cadherins was identified using the polymerase chain reaction. Partial cDNA sequences were obtained from a number of species ranging from the worm *Caenorhabditis elegans* to humans (Sang et al., 1993; Suzuki, 1996). At present, it is still unclear how many members of this subgroup exist. So far, full-length cDNAs for three protocadherins, Pcdh1 and Pcdh2 from human (Sang et al., 1993; Obata et al., 1995) and Pcdh3 from rat (Sago et al., 1995) have been published.

Protocadherins are clearly distinguishable from other cadherins on the basis of two main features: (a) the extracellular domain of protocadherins usually consists of more than five of the typical EC repeats and (b) their cytoplasmic domain shows no apparent homology to those of other cadherins. The extracellular domain of protocadherins usually consists of 6 or 7 cadherin repeats, but can reach as many as 34 in the case of the product of the *Drosophila* tumor suppressor gene *fat*. The repeats are similiar to EC2 or EC4 of classical cadherins, whereas EC3 and EC5 repeats, a characteristic feature of all other cadherins, are missing.

A much higher diversity can be found in the cytoplasmic part of protocadherins. The cytoplasmic domains of the protocadherin members cloned so far differ not only from those of the classical cadherin family but also show a high divergence among themselves. Although still preliminary, these findings already point toward the need for sub-classification within this cadherin family.

D. ST-Cadherins

The last family of cadherins, here designated short-tail (ST) cadherins, consists of three homologous proteins, independently cloned in three different species: human proton-dependent peptide transporter HPT-1 (Dantzig et al., 1994), rat liver intestine LI-cadherin (Berndorff et al., 1994) and a kidney-specific KSP-cadherin (Thomson et al., 1995).

The overall structure of the extracellular domains of st-cadherins are well conserved. They possess definitive cadherin specific motifs such as LDRE, DxND, DxD and the characteristic cysteine residues found in the last sub-domain, EC5, of most other cadherins. The extracellular region of st-cadherins is built up of seven instead of the classical five EC sub-domains. Nevertheless, the high conservation of the Cys residues in the EC5 sub-domain, as well as the conservation of additional features mentioned above, clearly places this class in close proximity to classical cadherins.

A striking divergence between st-cadherins and classical cadherins can be found in the cytoplasmic region. In contrast to the classical cadherins which have a highly conserved cytoplasmic tail of about 150 amino acid residues, this new subfamily of cadherins has a very short cytoplasmic tail of only 18 amino acids, clearly missing the typical catenin binding site. Thus, st-cadherins do not interact with catenins. In addition, all members of this group cloned so far are not synthesized as propeptides. The conserved protease cleavage site for classical cadherin propeptides and the first LDRE-DxND couplet are located much closer to the putative signal sequence than the corresponding motifs in the classical cadherin propeptides.

III. STRUCTURE OF THE EXTRACELLULAR DOMAIN OF CLASSICAL CADHERINS

A. Structural Basis of Calcium-Induced Rigidification and Dimerization of Classical Cadherins

Even before the first members of the cadherin superfamily were cloned it became evident that this class of proteins strictly relies on extracellular calcium (Kemler et al., 1977; Takeichi, 1977; Takeichi et al., 1979; Urushihara et al., 1979; Grunwald et al., 1980, Grunwald, 1981; Magnani et al., 1981; Thomas and Steinberg, 1981; Thomas et al., 1981). Cadherins exert their adhesive function exclusively in the presence of calcium, thus leading to the name Cadherin (Ca^{2+} dependent adherin) (Yoshida-Noro et al., 1984). After the first members of the classical cadherin family were cloned, sequence analysis revealed the pentameric segmentation of the extracellular domain (see above) as well as amino acid motifs with clusters of acidic residues, such as PEN, LDRE, DxNDN and DxD, located in each sub-domain and hypothesized to be involved in Ca^{2+} binding. This idea was supported by the fact that a synthetic peptide containing the putative Ca^{2+} binding site DxD at amino acid positions 134–136 of the EC2 domain of mouse E-cadherin was found to complex Ca^{2+} (Ozawa et al., 1990). Indeed, substitution of Asp in amino acid position 134 to Lys abolished the Ca^{2+} binding. Furthermore, mutant E-cadherins bearing the same mutation as the peptide were found to be adhesion defective when transfected into mouse L-cells. In the presence of calcium the mutated E-cadherin protein also showed an increased susceptibility to trypsin digestion. Wild-type, classical cadherins are protected by Ca^{2+} from trypsin digestion. Indeed, this protective feature played a key role in the initial identification of classical cadherins as adhesion molecules (Takeichi et al., 1981). It is thought that binding of Ca^{2+} to the EC sub-domains protects trypsin-sensitive sites by physically covering putative cleavage sites. In fact, Ca^{2+} binding was found to fundamentally change the three-dimensional structures of cadherins (Pokutta et al., 1994).

The extent to which Ca^{2+} is involved in mediating the homophilic interaction of cadherins was unclear until recently. This has been changed profoundly by the crystallization of the first two repeats EC1–EC2 of mouse E-cadherin as well as the crystallization of a part of the EC1 sub-domain of mouse N-cadherin (Overduin et al., 1995; Shapiro et al., 1995a, 1995b; Nagar et al., 1996). These studies confirmed the existence of the theoretically proposed pentameric structure of the cadherin extracellular domain and concluded (a) that Ca^{2+} does not seem to be directly involved in mediating homophilic interactions but is essential for rigidifying the molecule and (b) that the active form of adhesive cadherins is dimeric and not monomeric (for reviews see Weis, 1995; Jones, 1996).

The cadherin sub-domain is approximately 45 Å x 25 Å x 25 Å and consists of seven antiparallel β-strands (designated A to G) folded up in a "Greek key" topology similar to that of the immunglobulin (Ig) fold. The domains apparently form a barrel-like structure, connected by a 10-residue linker (101-110), and bridged by an arrangement of three contiguous calcium ions (Cal, Ca2 and Ca3). Cal and Ca3, which are associated most closely with EC1 and EC2 respectively, are linked through Ca2 by bridging interactions involving the side-chain carboxylates of Glu 11 (PEN motif), Glu 69 (LDRE motif), Asp 103 (DxNDN motif) and Asp 136 (DxD motif of the second repeat EC2). A further stabilization between the EC1 and EC2 sub-domains is achieved by the adjacent linker residues Asn 102 and Asp 103, which bind Cal/Ca2 and Ca3, respectively. Interestingly, the mutation in amino acid position Asp 134 of E-cadherin described above removes the interaction with Ca^{2+} between the first and second EC sub-domains. Mutations in the PEN and DAD motifs of EC2/EC3 and EC4/EC5, respectively, have been found in two human carcinomas. Based on sequence homology among the five EC sub-domains of classical cadherins, a stoichiometry of 12 calcium ions per molecule is calculated. This network of interactions further suggests that calcium binding should be a highly cooperative process. Conclusive data on the different Ca^{2+} -binding sites are not yet available, but calcium-binding assays with the recombinantly expressed extracellular domain of E-cadherin point toward a cooperative calcium binding and an average Kd value of 150 μM (Pokutta et al., 1994).

The crystal structures of both E- and N-cadherin also reveal a propensity for a parallel paired arrangement of the β-barrels of two amino-terminal domains, but the models predicted for the entire extracellular domains of N-cadherin and E-cadherin differ in the details of dimerization and the overall architecture of the extracellular domain. While the EC1-EC5 subdomains of N-cadherin seem to be intimately entwined through interchain dimerization of corresponding subdomains, for E-cadherin dimerization of the extracellular sub-domains seems to be restricted to EC1. In contrast to N-cadherin, two E-cadherin protomers are thought to be arranged in a "V-shape." Each arm would be ~240 Å in length, very close to the 220 Å measured by electron microscopy. It is noteworthy that recent results with recombinant E-cadherin, fused to the assembly domain of rat cartilage oligomeric matrix protein (COMP), also limit E-cadherin dimerization to EC1 (Tomschy et al., 1996). Another difference between the N-cadherin and E-cadherin crystal lattices was found in the orientation of the EC1 or EC1–EC2 dimers. Besides forming parallel dimers, crystal lattices of EC1 of N-cadherin also showed antiparallel dimers, leading to a zipper-like model for cadherin-mediated cell adhesion. In contrast, no such antiparallel adhesive interactions were found in the EC1–EC2 crystals of E-cadherin, even though the domains could be accommodated in the zipper-like structure.

In any case, the crystal structures of N-cadherin and E-cadherin clearly show that the building block of any higher order organization is a parallel cadherin dimer whose structure is promoted by and dependent on the presence of bound calcium ions. Given the high similarity of EC sub-domains within the cadherin superfamily, this higher order principle should be a feature shared by all cadherins. The recent observation that lateral dimerization is required for the homophilic activity of C-cadherin strongly supports this hypothesis (Brieher et al., 1996). Another, highly speculative conclusion can be drawn from the different lateral arrangements of N-cadherin and E-cadherin dimers. The different shapes of the protomers, V-shaped in the case of E-cadherin versus I-shaped as described for N-cadherin, may have a profound impact on the overall structure of the membrane-based cytoskeleton. This may also lead to the recruitment of different proteins involved in different signaling cascades. In other words, the differential expression of members of the cadherin family may not only result in the separation of groups of cells but may also be directly involved in cell-fate decisions.

B. Specificities in Homophilic Cadherin Interactions

In most cases, classical cadherins interact homophilically with each other, preferentially binding to the same type of molecule on the neighboring cell. This has been found not only for type I classical cadherins but also for members of the type II group as well as for DE-cadherin. Weak heterophilic interactions have been observed in *in vitro* cell aggregation assays (Volk et al., 1987; Inuzuka et al., 1991; Matsunami et al., 1993; Redies et al., 1993; Nakagawa and Takeichi, 1995). In one case of a known heterophilic interaction, the interacting residues of N- and R-cadherin were almost identical at the adhesion interface (Weis, 1995). One unusual heterophilic heterotypic interaction was discovered in the lymphocyte adhesion system, where the interaction between epithelial cells of the mucosa and a subpopulation of T-lymphocytes was mediated by an E-cadherin-integrin interaction (Cepek et al., 1994; Karecla et al., 1995: Cepek et al., 1996; Karecla et al.. 1996). The integrin subtype involved was identified as aEβ7 in the human and aM290β7 in the mouse.

The molecular basis of the specificity in homophilic classical cadherin interactions still remains to be elucidated, but there is good evidence that the EC1 subdomain harbors the specificity. E-cadherin in which EC1 is replaced by the EC1 sub-domain of P-cadherin shows P-cadherin specificity and vice versa (Nose et al., 1990). In addition, the majority of monoclonal antibodies that inhibit cadherin-mediated cell adhesion are directed against epitopes localized to the first extracellular domain. Several sequences in the EC1 repeat are thought to be essential for the homophilic interaction and/or the specificity-determining site of cell adhesion. Of special interest is the amino acid triplet "HAV" located close to the C-terminal end of the EC1 sub-domain (Blaschuk et al., 1990a, 1990b). Synthetic peptides containing the sequence HAV have been shown to inhibit the compaction of eight-cell mouse embryos as well as neurite outgrowth on rat astrocytes, two classical cadherin-dependent processes. It should be noted, however, that the HAV motif is only conserved among various members of class I, while it is partially substituted by QAV in some class II cadherins and is absent in all other cadherin classes. Additionally, the results of the three-dimensional analysis of E-cadherin structure discussed above have shown that this motif is involved in a translational crystal lattice contact unlikely to be of physiological relevance. Thus the HAV sequence surely cannot be a key sequence for general interactions (Nagar et al., 1996). However, it is still possible that the HAV sequence constitutes the core structure of the binding site in conjunction with other sequences and plays an important role in strong cell-cell adhesion activity of type I classical cadherins. Indeed, substitutions of other amino acid residues immediately flanking the tripeptide HAV in mouse E-cadherin for those of P-cadherin were shown to partially switch the specificity of the chimeric molecule from E- to P-cadherin (Nose et al., 1990). Evidently, multiple portions of the amino-terminal EC1 sub-domain, including the HAV-flanking region, are involved in the adhesive interactions between the molecules. Even small changes in their sequences apparently alters their specificity.

Much less is known about the homophilic interaction and binding specificities of the other cadherin classes. Protocadherin-1 and protocadherin-2 transfectants show very weak cell aggregation activity in the conventional in vitro assay system (Sang et al., 1993; Obata et al., 1995). No significant cell adhesion activity was measurable for protocadherin-3 (Sago et al., 1995). One reason may be the kind of cytoplasmic anchorage of protocadherin-3, since its cytoplasmic domain differs from that of other members of this class. Indeed, a chimeric construct of protocadherin-2 in which the cytoplasmic domain was replaced by the cytoplasmic domain of E-cadherin showed stronger cell aggregation activity (Obata et al., 1995). On the other hand, a similiar chimeric protein consisting of the Dsc1 extracellular domain and the E-cadherin transmembrane and cytoplasmic domains was not adhesive in the same system (Chidgey et al., 1996). This experiment was done to understand why the expression of full-length Dsc1a or Dsc1b was not sufficient to confer adhesiveness to transfected mouse fibroblastic L-cells (Chidgey et al., 1996). It is still widely believed that Dsc's and Dsg's are involved in cell-cell adhesion, but little direct evidence for this is available at present. Interestingly, a chimeric protein with the extracellular domain of E-cadherin and the cytoplasmic part of Dsg3 mediates strong cell adhesiveness (Roh and Stanley, 1995a). One might assume that the extracellular domains of classical cadherins have a much higher intrinsic adhesive property than those of desmosomal cadherins. This would be in good agreement with the secondary status of desmosomal junctions. In general, desmosomal junctions are made after cell-cell adhesion has been established. In other words, in order for desmosomal cadherins to adhere and form tight desmosomal contacts it is essential to bring them in close contact via other cadherins (Marrs et al., 1995).

All these results clearly show that there is no common rule for how cadherins may specifically interact with each other. However, it is clear in most cases that the strength of adhesiveness depends on an interplay between the extracellular and the cytoplasmic domain although cytoplasmic anchorage of cadherins does not always seem to be necessary for cell adhesion. T-cadherin, as well as LI-cadherin, and later a member of the st-cadherins, were all shown to confer cadherin-dependent cell adhesiveness (Vestal and Ranscht, 1992; Berndorff et al., 1994).

IV. CYTOPLASMIC ANCHORAGE AND HIGHER ORDER STRUCTURE OF CLASSICAL CADHERINS IN ADHERENS JUNCTIONS

Classical cadherins and desmosomal cadherins have been shown to be connected to the cytoskeleton, classical cadherins to the actin-based cytoskeleton and desmosomal cadherins to the intermediate filament (IF) system (for review see Schmidt et al., 1994; Aberle et al., 1996b). The switch from the cytoskeleton-bound to the unbound form is usually reflected by a change in the solubility of the proteins. Whereas unbound cadherins are easily extractable with detergents like TX-100 or NP-40, cytoskeletal-connected cadherins are resistant to extraction (Nagafuchi and Takeichi, 1988; Ozawa et al., 1989; Hinck et al., 1994; Nathke et al., 1994; Oyama et al., 1994). A decrease in solubility coincides with an increase in cell-cell contact

(McNeill et al., 1993; Adams et al., 1996; Angres et al., 1996). For example, during initial formation of cell-cell contacts, E-cadherin is readily extracted from cells. However, 10-15 min after stabilization of the contact, E-cadherin becomes resistant to extraction, although there is little increase in the absolute amount of E-cadherin at the contact sites (Adams et al., 1996). Additionally, cell-cell adhesion is dependent on an intact actin cytoskeleton. E-cadherin mediated adhesion is disrupted if cells are treated with cytochalasin D (Angres et al., 1996). Extracellular interactions between cadherin molecules are generally not sufficient to provide strong adhesion but mainly serve in the initial recognition event. To make durable cell-cell contacts it is essential to link classical cadherins and desmosomal cadherins to the actin- and IF-cytoskeleton respectively, or to cross-link preformed cadherin complexes into higher order structures. Most if not all cadherins molecules are concentrated in such higher order structures, such as the "adherens junctions" of epithelial cells (for review see Schmidt et al., 1994).

In the following we will summarize present knowledge about the cytoplasmic anchorage of classical cadherins and briefly describe their higher order structure in adherens junctions. The cytoplasmic interactions of desmosomal cadherins are described elsewhere in the book.

A. Cytoplasmic Anchorage

Most if not all classical cadherins interact with a group of more or less tightly associated cytoplasmic proteins known as catenins (reviewed in Kemler, 1993; Aberle et al., 1996b). These proteins, α-, β-, and γ- (plakoglobin) catenins, were originally identified as proteins co-immunoprecipitating with E-cadherin from *35*S-Met labeled cells (Nagafuchi and Takeichi, 1989; Ozawa et al., 1989). Another catenin, p120*cas*, originally identified as a substrate of Src and several receptor tyrosine kinases, and the related protein p100 also interact with the cadherins (Reynolds et al., 1994; Aghib and McCrea, 1995; Shibamoto et al., 1995).

Biochemical studies on cultured cells as well as the use of recombinant proteins have provided a detailed picture of how these different components interact (Nagafuchi and Takeichi, 1988; Ozawa et al., 1989; Ozawa and Kemler, 1992; Hinck et al., 1994; Nathke et al., 1994). Pulse-chase experiments have shown a clear hierarchy in the assembly of the cadherin-catenin complex. While β-catenin or plakoglobin can be found in direct association with the cytoplasmic domain of the E-cadherin precursor, α-catenin first associates with the complex at a time concordant with arrival of the E-cadherin/β-catenin complex at the plasma membrane (Ozawa and Kemler, 1992). Here α-catenin binds directly to β-catenin or plakoglobin. This scheme has been confirmed in vitro using recombinant proteins (Aberle et al., 1994a). Interestingly, immunoprecipitations with anti-β-catenin or plakoglobin-specific antibodies show that cadherins form mutually exclusive complexes with either β-catenin or plakoglobin (Butz and Kemler, 1994; Nathke et al., 1994). At present, nothing is known about the necessity for two different complexes. However, plakoglobin also interacts with the cytoplasmic domain of desmosomal cadherins, suggesting that cross-talk between classical and desmosomal cadherins may occur (Cowin et al., 1986; Franke et al., 1987, 1989; Lewis et al., 1997 for review see Schmidt et al., 1994). Cross-talk between proteins of tight junctions and cadherins may also occur, since the tight junctional protein ZO-1 interacts with α-, β-, and γ-catenins (Rajasekaran et al., 1996).

In general, the significance of the choice of a cytoplasmic partner and the factors which govern this selection are poorly understood. However, for E-cadherin, phosphorylation of a cluster of eight serine residues within a 30 amino acid region of the cytoplasmic domain appears to regulate positively its association with β-catenin (Stappert and Kemler, 1994). Intriguingly, although only a small proportion of E-cadherin is bound to p120*cas* in normal cells, the latter appears to be the catenin of choice in a Ras-transformed mammary cell line (Kinch et al., 1995).

β-catenin, plakoglobin and p120*cas* are members of the so-called Armadillo repeat family (reviewed in Peifer et al., 1994; Aberle et al., 1996b; Cowin and Burke, 1996). Originally identified in the product of the *Drosophila* segment polarity gene *armadillo*, the 42 amino acid Arm motif is repeated 12–13 times in the central region of these homologous proteins (Peifer and Wieschaus, 1990). At least 14 different proteins, fulfilling different functions in the cell, have been identified as Arm-proteins. Among these is the tumor suppressor protein APC (adenomatous polyposis coli), which interacts directly with β-catenin and plakoglobin, leading to the degradation of the two catenins in a phosphorylation-dependent manner (Rubinfeld et al., 1993; Su et al., 1993; Shibata et al., 1994; Rubinfeld et al., 1996, 1997; for review see Gumbiner, 1995; Peifer, 1995). Classical cadherins bind to the central region of β-catenin and plakoglobin at Arm repeats 5–8, whereas α-catenin binds to the amino-terminal domain and the first Arm-repeat (Aberle et al., 1994; Hulsken et al., 1994; Sacco et al., 1995; Pai et al., 1996; Witcher et al., 1996). In vitro binding assays using variant β-catenins with point mutations in the α-catenin binding site have identified a stretch of hydrophobic amino acid residues crucial for binding with α-catenin (Aberle et al., 1996a). Although highly conserved among the more closely related members of the Arm family, the sequence motif of the α-catenin binding site is absent in distantly related members. Indeed, it has been shown that the tumor suppressor protein APC (another member of the Armadillo repeat family) and p120*cas*, which both lack this binding motif, do not interact directly with α-catenin (Hulsken et al., 1994; Daniel and Reynolds, 1995; Jou et al., 1995). Thus, p120*cas*-cadherin complexes are not bound to the cytoskeleton, which may explain the poor-adhesive phenotype of Ras-transformed cells, in which p120*cas*-E-cadherin complexes predominate (Kinch et al., 1995).

The cadherin/β-catenin or cadherin/plakoglobin membrane-associated complex is connected to the actin cytoskeleton either directly by α-catenin, via both its amino and carboxyl termini, or indirectly, by association with α-actinin, another actin-binding protein (Knudsen et al., 1995; Rimm et al., 1995; Obama and Ozawa,

1997). α-catenin resembles vinculin, a protein involved in the cytoplasmic anchorage of integrins to actin filaments in focal contacts (Herrenknecht et al., 1991; Nagafuchi et al., 1991; Oda et al., 1993b). Both α-catenin and vinculin consist of five domains, and the amino acid sequences of the first, third and fifth domains in a-catenin are homologous to those of the corresponding domains in vinculin (26%, 32% and 34 % identity, respectively). Domains 2 and 4 show no significant similarity between these proteins. Isoforms of α-catenin, designated αN-catenin I and II, have been characterized and found to be expressed almost exclusively in the nervous system (Hirano et al., 1992). Thus, there is good evidence for the existence of a vinculin superfamily. Expression of E-cadherin-α-catenin chimeras has been found to confer a strong and inflexible adhesive phenotype on cells (Nagafuchi et al., 1994). Although the adhesive properties of a β-catenin-deficient cadherin complex were indistinguishable from wild-type transfectants in this system, cell motility and down-regulation of adhesion during cytokinesis were significantly suppressed. These experiments, in addition to confirming the central role of α-catenin in cadherin-cytoskeleton interactions, also suggest that β-catenin may serve as a regulatory element. In support of this model, other studies show that β-catenin is a good target for various tyrosine and serine kinases (as discussed below).

Although many cadherin-catenin-cytoskeletal interactions have been elucidated, the mechanism of lateral clustering of cadherins into junctions remains obscure. Plakoglobin is known to form homodimers in the cytosol; however, it is not clear if it also does so at the membrane (Cowin et al., 1986). Co-immunoprecipitation assays of detergent lysates from cells have failed to demonstrate either dimerization of β-catenin or its association with more than one molecule of cadherin (Ozawa and Kemler, 1992; Aberle et al., 1994; Hulsken et al., 1994; Jou et al., 1995). Furthermore, despite the presence of a putative self-association domain in α-catenin, two-hybrid experiments in yeast provide no evidence that such a mechanism actually exists (Jou et al., 1995). It is therefore likely that lateral clustering requires either post-translational modifications of α-catenin and/or its interaction with other membrane-organizing proteins such as vinculin and spectrin.

B. Higher Order Structure

The local clustering of the cadherin-catenin complex at the membrane and its cytoplasmic crosslinking apparently represents the last step in generating tight cell-cell interactions. As has been demonstrated very recently, E-cadherin-expressing MDCK cells briefly show, after cell-cell contacts, the cadherin–catenin complex in spatially discrete micro-domains, designated "puncta" (Adams et al., 1996). As the time of contact increases, the number of puncta increases proportionally along the contact and each punctum is associated with a bundle of actin filaments. These puncta may be the "only" adhesion centers in various cells types but may develop into the adherers junctions of epithelial cells (reviewed in Schmidt et al., 1994; Cowin and Burke, 1996; Knust and Leptin, 1996).

Adherens junctions are usually localized at the baso-lateral plasma membrane, underlaid with a dense, 10 to 30 nm thick, cytoplasmic, submembranous plaque, at which bundles of cytoskeletal filaments attach in most but not all situations. The two major and most widespread plaque-bearing junctions, the E- and/or N-cadherin-containing adherers junctions and the desmosomes (*maculae adhaerentes*), usually coexist in the same cells, often in close proximity, as in the subapical region ("terminal web") of polar epithelial cells or at the "intercalated disks" of cardiomyocytes. Different kinds of adherers junctions have been described, mainly based on their electron microscopical appearance but also dependent on which kind of cadherin is concentrated in these junctions (reviewed in Schmidt et al., 1994). N-, P-, and the vascular endothelium-specific VE-cadherin have been found in plaque-bearing junctions of blood vascular endothelia and cultured cells derived therefrom (Lampugnani et al., 1992; Schmelz and Franke, 1993; Lampugnani et al., 1995). Another member of the classical cadherins, M-cadherin, is found in adherers junctions of the cerebellar glomerulus, where they connect the neurites of the granule cells (Rose et al., 1995).

As mentioned at the beginning, one can think of adherers junctions as the quaternary structure of cadherins. Here, the linear "zippers" of the cadherins, formed between neighboring cells and stabilized by catenins, are crosslinked by additional proteins. Many different proteins localized to the undercoat of adherers junctions have been identified and characterized (for review see, Geiger et al., 1995; Yamada and Geiger, 1997). A clear picture of the architecture of this complex structure is still missing. However, the undercoat of adherers junctions does appear somewhat similar to that found in erythrocytes (Tsukita et al., 1992). For example, radixin, a member of the so-called ERM (ezrin, radixin and moesin) family, is a major constituent of the undercoat of the isolated junctions (for review see Tsukita et al., 1997). The ERM family is included in the band 4.1 superfamily. The band 4.1 protein is one of the major constituents of the undercoat of erythrocyte membranes. In these cells, band 4.1 protein binds to spectrin, a key element in the association of short actin filaments with another plasma membrane protein, glycophorin C. In addition, fodrin, a spectrin homolog, has long been known to be localized to adherers junctions (Nelson and Veshnock, 1986, 1987).

As well as the many proteins presumably involved in the architecture of adherens junctions, members of different kinase families like c-src, c-yes and receptor tyrosine kinases such as the epidermal growth factor EGF receptor as well as protein kinase C (PKC) (Fukuyama and Shimizu, 1991; Tsukita et al., 1991; Blum et al., 1994; Lewis et al., 1994) are concentrated in the area of adherers junctions. Thus, adherers junctions seem to fulfill dual functions, being centers for cell adhesion as well as sending or receiving signals: the latter can result in a modulation of the junctional architecture (Volberg et al., 1991, 1992).

In summary, strong cadherin-mediated cell-cell adhesion is established in three steps. In the first step, loose contacts among the extracellular domains of cadherins are made. For classical and desmosomal cadherins these contacts are then stabilized in the second step by an anchorage of cadherins to different components of the cytoskeleton. The strength of adhesiveness may be an additive effect of the intrinsic adhesion properties of the extracellular domain, with clear differences across the cadherin superfamily and the cytoplasmic proteins such as catenins directly interacting with the cytoplasmic domain. In a third step, these secondary structures are then connected with a network of different proteins to give a complex adhesion and signaling center.

V. CADHERINS IN MORPHOGENESIS AND THEIR DISRUPTION BY DOMINANT-NEGATIVE AND LOSS-OF-FUNCTION MUTATIONS

It has been long known that the segregation and remodeling of embryonic tissues correlates well with the expression of different types of cadherins. Thus, great effort was put into the analysis of cadherin expression patterns during early and late developmental stages, mainly for *Xenopus*, chicken and mouse (reviewed in Ranscht, 1994; Amagai, 1995; Marrs et al., 1995; Redies, 1995; Takeichi, 1995; Kuhl and Wedlich, 1996; Redies and Takeichi, 1996). This rather descriptive analysis of cadherin expression patterns has now been extended by disrupting cadherins in morphogenesis, using dominant-negative as well as loss-of-function approaches (reviewed Takeichi, 1995; Huber et al., 1996a; Hynes, 1996). Since many excellent reviews already describe the expression patterns of cadherins in different organisms in great detail, we will only briefly describe a few classical examples of the spatio-temporal cadherin expression during mouse development, mainly focusing on the effects on morphogenesis of suppression or overexpression of cadherin function.

A classical example of how the spatio-temporal expression of different cadherins leads to the separation of different tissues is clearly provided by the formation of the neural tube and the migration of neural crest cells in mouse and chicken (for review see Takeichi, 1988; Nakagawa and Takeichi, 1995; Takeichi, 1995). During induction of the neural plate, ectodermal cells in this differentiating region switch from E-cadherin to N-cadherin expression. This may allow segregation of neural precursor cells from other ectodermal cells. In contrast, migrating crest cells down-regulate both E- and N-cadherin but may initiate expression of other cadherins to mediate cell sorting. Indeed, in the chicken embryo, expression patterns of cadherin-6B and -7 suggest that subpopulations of migrating neural crest cells find each other by the sequential and subgroup-specific expression of different cadherins. While the neural fold expresses cadherin-6B, a subpopulation of neural crest cells switches to the expression of cadherin-7 when starting to migrate. None of the cadherin-7 positive cells migrates toward the dermis, and only a few of them populate the dorsal root ganglia (Nakagawa and Takeichi, 1995). On the other hand, N-cadherin is expressed in forming dorsal root ganglia, and cadherin-11 is found in cells of the differentiating mesenchyme (Hoffmann and Balling, 1995; Kimura et al., 1995; Simonneau et al., 1995). Thus, neural crest cells seem to express cadherins in a subpopulation-specific manner, and this may result in the migration of segregated groups.

Cadherin-11 is interesting since it seems to be expressed exclusively in mesenchymal cells and their derivatives (Okazaki et al., 1994; Hoffmann and Balling, 1995; Simonneau et al., 1995). This is surprising since classical cadherins were thought to confer strong adhesiveness mainly to solid tissues. In early mouse embryos, cadherin-11 is most strongly expressed in the mesenchymes surrounding organ-anlages, in the cephalic mesoderm and during somite formation. At later developmental stages, cadherin-11 is expressed in a wide variety of mesenchymal tissues, in both mesodermal and neural crest derivatives. Although cadherin-11 is currently considered to be the major mesenchymal cadherin, M-cadherin and N-cadherin are expressed in other mesenchymal components such as muscle and condensing chondrocytes (Donalies et al., 1991; Moore and Walsh, 1993; Bornemann and Schmalbruch, 1994; Cifuentesdiaz et al., 1994; Irintchev et al., 1994; Soler and Knudsen, 1994; Cifuentesdiaz et al., 1996).

Most of the classical as well as recently identified cadherins are expressed in the developing nervous system (reviewed in Redies, 1995; Redies and Takeichi, 1996). However, their expression is always regional except for the rather ubiquitous expression of N-cadherin (Shimamura and Takeichi, 1992; Redies et al., 1993; Murphy-Erdosh et al., 1994; Ganzler and Redies, 1995). Here, the localized cadherin expression may lead to (1) regionalizing of the neural tube along the rostrocaudal or dorsoventral axis; (2) segmentation of the sublayers of the neuroepithelium or cortex; and (3) clustering or wiring of neurons in the visual network, based on expression in chicken. Interestingly, E-cadherin, thought to be exclusively involved in the adhesion of epithelial cells, is also found in the peripheral nervous system (Matsunami and Takeichi, 1995). It is expressed in a subset of sensory neurons whose afferent fibers terminate in a specific zone of the dorsal horn. At the electron microscopic level, E-cadherin is localized at lateral contacts between members of a subset of unmyelinated axons (Uchiyama et al., 1994). In addition, E-cadherin is also found in Schwann cells, mediating intracellular adhesion between membrane wraps within the cell (Fannon et al., 1995).

Originally, E-cadherin was identified as being essential in the compaction of the preimplantation mouse embryo. At the 8-cell stage, uniformly distributed E-cadherin molecules become concentrated at cell-cell contacts of blastomeres (reviewed in Collins and Fleming, 1995). Clustering of E-cadherin is accompanied by a reorganization of the cytoskeleton and changes in cell morphology. Loosely attached blastomeres flatten and adhere to each other with maximal strength, forming a compact morula. At the implantation stage, E-cadherin is expressed in all cells of the embryo. However, as cells differentiate into various types, this molecule disappears from some cell layers. The most prominent change in E-cadherin expression is seen at gastrulation. At this stage, the epithelial continuity of the ectodermal cell layer is lost in the streak region, and cells move or delaminate through the primitive streak to emerge between the ectoderm and visceral endoderm as a new intermediate layer of mesoderm. In parallel, these cells lose E-cadherin during migration but start to re-express the mesenchyme-specific cadherin-11 (Butz and Larue, 1995; Hoffmann and Balling, 1995; Ohsugi et al., 1996). Other regions of the ectoderm and all ectodermal cells maintain the expression of E-cadherin, and this expression persists as long as they differentiate into epithelial cells. The epithelial-mesenchymal transition at this time in development is of special interest since modifications of β-catenin and/or plakoglobin seem to orchestrate this step (reviewed in Gumbiner, 1995; Peifer, 1995; Miller and Moon, 1996).

In sharp contrast to the spatio-temporal expression patterns of cadherins, catenins are expressed rather ubiquitiously, although some distinct differences have been described (Butz and Larue, 1995; Ohsugi et al., 1996; Schneider et al., 1996; Larabell et al., 1997). During mouse preimplantation development, both α-catenin and β-catenin are maternally provided as protein and mRNA, and the embryonic genes become activated at the late two-cell stage (Ohsugi et al., 1996). Plakoglobin, however, exhibits a distinct expression profile, lacking the maternal component and appearing first at late morula stage.

Asymmetric expression of β-catenin has been described recently (Schneider et al., 1996; Larabell et al., 1997); transient accumulation of β-catenin in the nuclei of the dorsal side of the blastulae has been observed in *Xenopus* and zebrafish (Schneider et al., 1996). The *Drosophila* homolog Armadillo has long been known to be part of the Wg/Wnt pathway (reviewed in Orsulic and Peifer, 1996). The activation of this pathway leads to a stabilization of Armadillo, resulting in the accumulation of the protein in the cytoplasm and in the nucleus. Together with DNA binding factors of the HMG-box family, Armadillo seems to activate or deactivate genes which are involved in the determination of cell fate (Brunner et al., 1997; Vandewetering et al., 1997). There is compelling evidence showing that β-catenin and/or plakoglobin fulfills the same function in vertebrates (Behrens et al., 1996; Huber et al., 1996b; Molenaar et al., 1996; Korinek et al., 1997). Most intriguingly, the local accumulation of β-catenin in the nuclei of *Xenopus* and zebrafish, as described above, correlates well with the development of signaling centers that are crucial for the segmentation along the whole body axis.

A. Loss-of-Function Analysis

One way to gain insights into the developmental function of cadherins and catenins is to inhibit their expression either by targeted mutations or by repressing their translation in antisense experiments. In mice, knockout mutations have been described for E- and N-cadherin as well as for their cytoplasmic associates, β- and γ-catenin (reviewed Hynes, 1996). As mentioned above, E-cadherin was originally described as a molecule involved in compaction of the morula-stage embryo. E-cadherin null-mutant embryos are still able to compact because of maternal E-cadherin mRNA and protein, but they fail to form a functional trophectoderm epithelium or blastocyst cavity. As a consequence, mutant embryos die at the time of implantation (Larue et al., 1994). The *Drosophila* homolog of E-cadherin (DE-cadherin), is encoded by the *shotgun* gene locus. When zytogotically deleted, embryos exhibit no severe defects in epithelial organization because maternally derived DE-cadherin is sufficient (Uemura et al., 1996). Nevertheless, the formation of epithelia known to involve extensive morphogenetic movements (for example, forming of Malphigian tubules and tracheal ducts) is impaired. In contrast, the complete absence of maternal and zygotic DE-cadherin expression disturbs the formation of all epithelia (Tepass et al., 1996). N-cadherin null-mutant mouse embryos somewhat unexpectedly form neural tubes, somites and myocardium, structures thought to be dependent on the presence of N-cadherin (Radice et al., 1997). However, all these structures become defective later, leading to early embryonic lethality. It is tempting to speculate that other cadherins or even adhesion molecules of different families can partially compensate for N-cadherin loss.

Mouse embryos lacking β-catenin fail to gastrulate, and the embryonic ectoderm dissociates (Haegel et al., 1995). Mutant cells detach from the ectodermal cell layer and are dispersed in the proamniotic cavity, but, more strikingly, mutant embryos fail to form mesoderm. Lack of β-catenin in earlier stages of development may be partially compensated by plakoglobin, since plakoglobin appears to be upregulated in mutant mice. In Xenopus, depletion of maternal β-catenin by an antisense oligodeoxynucleotide complementary to β-catenin affects dorsal mesoderm induction, an effect presumably based on disruption of the Wnt signaling cascade (Heasman et al., 1994).

Plakoglobin null-mutant mouse embryos show severe heart defects, as well as skin blistering and subcorneal acantholysis (Bierkamp et al., 1996; Ruiz et al., 1996). In the skin, desmosomes were dramatically reduced and structurally altered. Obviously, lack of plakoglobin does not impair embryonic paterning or basic morphogenetic events during pre- and early post-implantation development. This may indicate that plakoglobin is not part of the Wg/Wnt signaling pathway. In addition, the fact that plakoglobin null-mutant embryos develop quite far suggests that β-catenin can partially compensate for plakoglobin loss. Thus, β-catenin seems to fulfill rather diverse functions, whereas plakoglobin may largely serve as a structural protein in desmosomal junctions.

B. Dominant-Negative Approach

An alternative way to disrupt cadherin function is the dominant-negative approach. For this purpose, two basic types of truncated cadherin molecules have been used. One encodes a cadherin with a deletion of the extracellular domain, while the second type encodes a protein with a partial or complete deletion of the cytoplasmic domain. Overexpression of this type of mutant cadherins is thought to compete with the endogenous cadherin-specific homophilic interaction mechanisms, resulting in developmental abnormalities restricted to tissues expressing the respective cadherin. This was proven by the ectopic expression of cytoplasmically truncated E- and N-cadherin in *Xenopus* embryos (Levine et al., 1994; Lee and Gumbiner, 1995; Kuhl et al., 1996). Whereas the N-cadherin mutant leads to perturbation of the neural tube, E-cadherin seems specifically required for maintaining the integrity of the ectoderm during epiboly gastrulation. However, the same mutants did not significantly affect cell adhesion when transfected into cell lines (Fujimori and Takeichi, 1993). The *Xenopus* system must, therefore, be rather sensitive to this kind of perturbation.

The second type of cadherin mutants, encoding the transmembrane and cytoplasmic domains but lacking the extracellular domain, compete for cytoplasmic components such as catenins which interact with all classical cadherins. Thus, overexpression of these mutant forms clearly has a much broader impact on cell adhesion and tissue formation than mutants with a deletion of the extracellular domain. This has been proven by injecting extracellularly deleted N-cadherin, again into early *Xenopus* embryos (Kintner, 1992). Expression of mutated N-cadherin results in a dramatic inhibition of cell adhesion and a perturbation of the integrity of the ectodermal cell layer at mid-gastrulation. Similar results have been obtained by injecting mutant constructs of N-cadherin or XB-cadherin or both E-cadherin and EP-cadherin into specific blastomeres of 32-cell stage *Xenopus* embryos (Dufour et al., 1994; Broders and Thiery, 1995). The local overproduction lead to a disruption of tissues descending from these cells. In addition, overexpression of the cytoplasmic domain of N-cadherin in the presumptive muscle anlage of early *Xenopus* embryos inhibits muscle differentiation, a process known to require cell-cell interactions (Holt et al., 1994). A similar construct of N-cadherin and E-cadherin has been used in transgenic mice under the control of various tissue- type specific promotors (Hermiston and Gordon, 1995a, 1995b; Dahl et al., 1996). With such constructs it is possible to study cadherin-mediated adhesion in specific cell lineages such as epithelial cells of the small intestine and pancreatic β-cells. In the mouse intestine, overexpression of mutated N-cadherin led to a disruption of cell-cell as well as cell-matrix contacts (Hermiston and Gordon, 1995a). Most surprisingly, mutated cells clearly entered apoptosis, indicating a linkage between cell adhesion, signaling and cell death. Another aspect of cadherin expression was revealed by the ectopic expression of E-cadherin in a retinal pigment epithelial cell line (Marrs and Nelson, 1995). E-cadherin expression induced epithelial cell polarity and even caused de novo assembly of desmosomes in these cells. Although other cadherins can be found in these cells, E-cadherin seems to overide their functions, resulting in a remodeling of the entire cell cytoskeleton. Thus, differences in the cytoskeletal architecture among cadherins clearly affect cell morphology.

Ectopic overexpression of β-catenin by injection of mRNA into the ventral blastomeres of *Xenopus* embryos leads to the formation of a second Spemann organizer and duplication of the anterior-posterior axis (Funayama et al., 1995; Guger and Gumbiner, 1995; Kelly et al., 1995). The same phenotype is generated by overexpressing plakoglobin (Karnovsky and Klymkowsky, 1995). This effect is mainly based on β-catenin (and/or plakoglobin) being a key protein in the Wg/Wnt signaling cascade. For a detailed description, see other chapters in this book.

VI. THE CADHERIN–CATENIN COMPLEX IN SIGNAL TRANSDUCTION AND PATHOGENESIS

The critical roles of cadherins in maintaining and organizing different kinds of cell-cell junctions as discussed above have been documented extensively in human cancer (for reviews see: Behrens, 1994; Birchmeier et al., 1995; Bussemakers and Schalken, 1996; Jiang, 1996; Mareel et al., 1996; Shiozaki et al., 1996). Various reports have indicated that the primary cause for the "scattering" of the cells in invasive carcinomas is a disturbance of the integrity of intercellular junctions, often correlating with the loss of E-cadherin. For various different cell lines derived from bladder, breast, lung and pancreas carcinomas, the correlation between cell invasion and E-cadherin expression is well documented. Cell lines with an epithelial morphology are generally non-invasive and express E-cadherin, whereas cell lines with a fibroblastic phenotype are invasive and show in most cases no expression of E-cadherin. Interestingly, invasiveness of de-differentiated breast carcinomas can be prevented by the ectopic expression of E-cadherin in these cells (Vleminckx et al., 1991). Thus, E-cadherin seems to act as an important suppressor of epithelial tumor cell invasiveness and metastasis. However, there is compelling evidence that loss of differentiation of epithelial cells in vitro and in vivo also occurs if E-cadherin is present. In various cases, mutations or truncations of catenins have been found instead. In the case of a poorly differentiated lung carcinoma cell line (PC9), E-cadherin expression is unchanged but cell-cell adhesion is disrupted due to a null mutation in the α-catenin gene (Ode et al., 1993a). Adhesion can be restored by transfecting these cells with wild-type α-catenin cDNA (Hirano et al., 1992; Shimoyama et al., 1992). A mutated form of plakoglobin, missing the fourth Armadillo repeat, was recently cloned from a human transitional carcinoma cell line, BOY (Ozawa et al., 1995a). The same research group showed that the fourth Armadillo repeat of plakoglobin is required for its high-affinity binding to the cytoplasmic domain of E-cadherin, Dsg2 or the tumor suppressor protein APC (Ozawa et al., 1995b). It would be of interest to see whether ectopic expression of plakoglobin can restore the adhesiveness of these cells. It was also shown that transfection of plakoglobin can suppress the tumorigenicity of cells in the presence of, or independently of, the cadherin–catenin complex. In addition, the human plakoglobin gene has been localized to chromosome 17_q21 and shown to be subject to a loss of heterozygosity in breast and ovarian cancer (Aberle et al., 1995).

There is increasing evidence that the accumulation of β-catenin may result in cell transformation. The first hint came from a genetic screen for oncogenes which cause a typical change in the morphology of NIH3T3 fibroblast cells upon expression (Whitehead et al., 1995). Along with classical oncogenes, β-catenin was found to be a putative candidate as well. This role is supported by recent findings showing mutations for β-catenin in certain colon carcinoma cell lines and in several melanoma cell lines (Morin et al., 1997; Rubinfeld et al., 1997). In these cells, β-catenin was stabilized, resulting in a cytoplasmic and nuclear localization of the protein. Point mutations at the N-terminus of β-catenin located within a putative phosphorylation site for glycogen synthase kinase (GSK3) or deletions comprising exons 2,3 and 4 of the β-catenin gene were found to be responsible for its stabilization. As a component of the Wg/Wnt signaling pathway, the interaction of β-catenin with members of the HMG-box family may activate genes which are necessary for the transformation of cells.

Desmogleins, desmoplakins and desmocollins are frequently down-regulated in oral squamous cell carcinomas where reduced expression is correlated with loss of differentiation and metastasis. Several autoimmune blistering diseases result from disruption of demosomes (reviewed in Stanley, 1995; Garrod et al., 1996). Pemphigus vulgaris, affecting cell-cell adhesion deep in the epidermis, is caused by autoantibodies to desmoglein-3. In another blistering disease, pemphigus foliaceous, the granular layer in the upper part of the skin is disturbed. In this case autoantibodies react with desmoglein-1. Thus, it seems to be rather clear that maintenance of cell-cell junctions is crucial for the integrity of cells within complex tissue structures. Mutations in junctional components may impair their function and eventually result in the transformation of cells.

Components known to be involved in different signaling cascades disrupt cell-cell junctions. β-catenin, plakoglobin and p120*cas* are known to be substrates for various tyrosine kinases (Matsuyoshi et al., 1992; Reynolds et al., 1992; Behrens et al., 1993; Hamaguchi et al., 1993; Reynolds et al., 1994; Shibamoto et al., 1994). Increased tyrosine phosphorylation of β-catenin, in particular, has been reported to occur in oncogene-transformed cells which show impaired intercellular junctions and enhanced invasiveness (Sommers et al., 1994; Nishimura et al., 1996). However, more recent data suggest that the tyrosine phosphorylation of β-catenin is not required for the strong to weak state shift of cadherin-mediated cell adhesion and that the phosphorylation of other junctional proteins like ERM, ZO-1 or other yet unidentified proteins may be involved in regulating cell–cell adhesion functions (Takeda et al., 1995). Growth factors, such as epidermal growth factor EGF and hepatocyte growth factor HGF, can change the cellular morphology and promote the mobility and/or invasiveness of cancer cells. Indeed, β-catenin was found to bind directly to the EGF receptor as well as to ErbB2, which encodes a kinase homologous to the EGF receptor (Hoschuetzky et al., 1994; Ochiai et al., 1994; Kanai et al., 1995). In addition, overexpression of ErbB2 results in the transcriptional inhibition of the E-cadherin gene and is associated with a reduced ability of a mammary epithelial cell line to form epithelial structures in vitro (Dsouza and Taylorpapadimitriou, 1994). The phenotype was reversible via antibodies blocking ErbB2 phosphorylation and signal transduction. Another example of a cadherin and receptor tyrosine kinase interaction is found in differentiating neurons. Among various CAMs, N-cadherin is thought to interact with the fibroblast growth factor receptor (FGFR), leading to receptor activation by autophosphorylation (reviewed in Doherty and Walsh, 1996; Hall et al., 1996). This results in the recruitment and activation of additional effector molecules positively stimulating axonal growth.

As well as the interaction with tyrosine kinase receptors, different protein tyrosine phosphatases (PTPs), such as PTPμ, PTPκ and a member of the leukocyte antigen-related protein (LAR) transmembrane tyrosine phosphatase family (LAR-PTP), interact with the cadherin–catenin complex. Interactions for E- and N-cadherin with PTPμ as well as an interaction between PTPκ or LAR and β-catenin have been described (Brady-Kalnay et al., 1995; Fuchs et al., 1996; Kypta et al., 1996). Dephosphorylation of β-catenin by PTPκ and LAR-PTP has been shown in vitro. It is tempting to speculate that PTPases like PTPκ and PTPμ may serve to regulate negatively the action of tyrosine kinase-induced signal events at intercellular junctions by dephosphorylating β-catenin and/or plakoglobin. If the model is correct, the antagonizing activity of PTPs should stabilize cell–cell contacts. Interestingly, protein levels of PTPμ and PTPκ are elevated in confluent cell layers. Additionally, LAR-PTPs are phosphorylated on tyrosine in a TrkA-dependent manner, and their association with the cadherin–catenin complex is reduced in cells treated with NGF. Thus phosphorylation and dephosphorylation of the cadherin–catenin complex seem to be crucial in the regulation of cell contact and adhesion-controlled events such as cell proliferation, tumor invasiveness, and metastatic spread.

In contrast to cadherin-associated proteins negatively influencing cadherin function, activation of G protein-coupled receptors increases cadherin-mediated cell adhesion (Williams et al., 1993). Activation of M3 muscarinic acetylcholine receptors induces E-cadherin-mediated cell adhesion and possibly involves cell signaling via protein kinase C. Protein kinase C localizes to the zonula adherens and was found to induce premature compaction of early mouse embryos upon stimulation with TPA (12-0-tetradecanoylphorbol 13-acetate; Winkel et al., 1990; Sefton et al., 1996).

A totally different aspect in cadherin-mediating signaling, namely cadherins directly stimulating the differentiation of cells into certain types of tissue, has been described recently (Larue et al., 1996). Mouse embryonic stem (ES) cells negative for E-cadherin are defective in cell aggregation. E-cadherin -/- ES cells were unable to form any organized structures when injected into syngenic hosts. As expected, adhesion of E-cadherin -/- ES cells was restored when cells were retransfected with either E- or N-cadherin cDNA . Interestingly, teratomas derived from the E- or N-cadherin-restored ES cells showed a preference for tissues formed: E-cadherin led exclusively to the formation of epithelia, whereas the expression of N-cadherin led to neuroepithelium and cartilage. Thus, expression of a specific cadherin subtype may recruit proteins of different signaling cascades into the junctional complexes, resulting in the activation of cell-type specific genes. One may speculate that differences in the architecture of the adhesion zippers (I-shaped for N-cadherin, V-shaped for E-cadherin as described above) may cause such differences. For this reason it will be most interesting to see which direction chimeric cadherin proteins force the cell to go, when the cytoplasmic or extracellular domain of E-cadherin is swapped with the analogous domain of N-cadherin. In any case, adherers junctions seem to play a central role in focusing different signaling cascades. It remains to be determined, to what degree cadherins are directly involved in these signaling events.

CONCLUSIONS

In summary, the structure and functions of cadherins have been found to be much more complex and diverse than previously assumed. Cadherins are involved in the regulation of morphogenesis in many organisms, including invertebrate species, thus indicating the general importance of this family in organizing multicellular structures. In the classical view cadherins function mainly in the adhesion of solid tissues, but they have been found also in mesenchymal tissues and on migrating cells as well, indicating that they may be involved in cell rearrangement of most if not all tissues. This seems to be achieved at least in part by the expression of a large number of cadherin subtypes with distinct adhesion specificities.

Crystal structures from parts of the extracellular domains of E- and N-cadherin reveal new insights in the interaction mechanisms of these proteins. However, it is still unclear whether each cadherin varies in its three-dimensional structure and what the consequences of its cytoplasmic anchorage are. Thus, a major area for future studies is to resolve the three-dimensional structure of other cadherins and the complex cytoplasmic network underlying cadherin clusters. The knowledge about the proteins being differentially recruited into junctions mediated by different cadherins may answer the question about the signals cadherins are able to transmit. However, it is already clear that modifications of catenins, release of the catenins from cadherins, and other types of cytoskeletal reorganization can modulate cadherin-mediated adhesion. An intriguing idea is that each cadherin may be directly involved in specification of tissue-type, as suggested by the effects of the expression of E- or N-cadherin in E-cadherin -/- ES cells. Future experiments such as looking at the effect of domain swapping between different cadherins should test this hypothesis. Independent of the outcome of such experiments, it is clear now: cadherins are more than just a glue!

REFERENCES

Aberle, H., Butz, S., Stappert, J., Weissig, H., Kemler, R. & Hoschuetzky, H. (1994). Assembly of the cadherin–catenin complex in vitro with recombinant proteins. J. Cell Sci. 107, 3655-3663.

Aberle, H., Bierkamp, C., Torchard, D., Serova, O., Wagner, T., Natt, E., Wirsching, J., Heidkamper, C., Montagna, M., Lynch, H. T., Lenoir, G. M., Scherer, G., Feunteun, J., & Kemler, R. (1995). The human plakoglobin gene localizes on chromosome 17q21 and is subjected to loss of heterozygosity in breast and ovarian cancers. Proc. Natl. Acad. Sci. U S A. 92, 6384-6388.

Aberle, H., Schwartz, H., Hoschuetzky, H., & Kemler, R. (1996a). Single Amino Acid Substitutions In Proteins Of the Armadillo Gene Family Abolish Their Binding to Alpha-Catenin. J. Biol. Chem. 271, 1520-1526.

Aberle, H., Schwartz, H., & Kemler, R. (1996b). Cadherin-catenin complex: protein interactions and their implications for cadherin function. J. Cell Biochem. 61, 514-523.

Adams, C. L., Nelson, W. J., & Smith, S. J. (1996). Quantitative analysis of cadherin-catenin-actin reorganization during development of cell-cell adhesion. J. Cell Biol. 135, 1899-1911.

Aghib, D. F. & McCrea, P. D. (1995). The E-Cadherin Complex Contains the Src Substrate P120. Exp. Cell Res. 218, 359-369.

Amagai, M. (1995). Adhesion molecules. I: Keratinocyte-keratinocyte interactions; cadherins and pemphigus. J. Invest. Dermatol. 104, 146-152.

Angres, B., Barth, A., & Nelson, W. J. (1996). Mechanism For Transition From Initial to Stable Cell-Cell Adhesion - Kinetic Analysis Of E-Cadherin-Mediated Adhesion Using a Quantitative Adhesion Assay. J. Cell Biol. 134, 549-557.

Behrens, J., Vakaet, L., Friis, R., Winterhager, E., Vauroy, F., Mareel, M. M., & Birchmeier, W. (1993). Loss Of epithelial differentiation and gain of invasiveness correlates with tyrosine phosphorylation of the E-cadherin beta-catenin complex in cells transformed with a temperature-sensitive v-src gene. J. Cell Biol. 120, 757-766.

Behrens, J. (1994). Cell contacts. differentiation, and invasiveness of epithelial cells. Invasion Metastasis. 14, 61-70.

Behrens, J., von Kries, J. P., Kuhl, M., Bruhn, L., Wedlich, D., Grosschedl, R., & Birchmeier, W. (1996). Functional interaction of beta-catenin with the transcription factor LEF-1. Nature. 382, 638-642.

Berndorff, D., Gessner, R., Kreft, B., Schnoy, N., Lajouspetter, A. M., Loch, N., Reutter, W., Hortsch, M., & Tauber, R. (1994). Liver-Intestine cadherin - molecular cloning and characterization of a novel Ca^{2+}-dependent cell adhesion molecule expressed in liver and intestine. J. Cell Biol. 125, 1353-1369.

Bierkamp, C., McLaughlin, K. J., Schwarz, H., Huber, O., & Kemler, R. (1996). embryonic heart and skin defects in mice lacking plakoglobin. Dev. Biol. 180, 780-785.

Birchmeier, W., Hulsken, J., & Behrens, J. (1995). Adherens junction proteins in tumour progression. Cancer Surveys 1995;24: 129-140.

Blaschuk, O. W., Pouliot, Y., & Holland, P. C. (1990a). Identification of a conserved region common to cadherins and influenza strain A hemagglutinins. J. Mol. Biol. 211, 679-682.

Blaschuk, O. W., Sullivan, R., David, S., & Pouliot, Y. (1990b). Identification of a cadherin cell adhesion recognition sequence. Dev. Biol. 139, 227-229.

Blum, S., Ness, W., Petrow, W., & Achenbach, F. (1994). Localization of protein kinase C in primary cultures of human keratinocytes in relation to cell contact proteins. Cell Signal. 6, 157-165.

Bornemann, A. & Schmalbruch, H. (1994). Immunocytochemistry of m-cadherin in mature and regenerating rat muscle. Anat. Rec. 239, 119-125.

Brady-Kalnay, S. M., Rimm, D. L., & Tonks, N. K. (1995). Receptor protein tyrosine phosphatase PTPmu associates with cadherins and catenins in vivo. J. Cell Biol. 130, 977-986.

Brieher, W. M., Yap, A. S., & Gumbiner, B. M. (1996). Lateral dimerization is required for the homophilic binding activity of C-cadherin. J. Cell Biol. 135, 487-496.

Broders, F. & Thiery, J. P. (1995). Contribution of cadherins to directional cell migration and histogenesis in Xenopus embryos. Cell Adhes. Commun. 3, 419-440.

Brunner, E., Peter, O., Schweizer, L., & Basler, K. (1997). Pangolin Encodes a Lef-1 Homologue That Acts Downstream Of Armadillo to Transduce the Wingless Signal In Drosophila. Nature. 385, 829-833.

Bussemakers, M. J. & Schalken, J. A. (1996). The role of cell adhesion molecules and proteases in tumor invasion and metastasis. World J. Urol. 14, 151-156.

Butz, S. & Kemler, R. (1994). Distinct cadherin-catenin complexes in Ca($^{2+}$) -dependent cell-cell adhesion. FEBS Lett. 355, 195-200.

Butz, S. & Larue, L. (1995). Expression of catenins during mouse embryonic development and in adult tissues. Cell Adhes. Commun. 3, 337-352.

Cepek, K. L., Shaw, S. K., Parker, C. M., Russell, G. J., Morrow, J. S., Rimm, D. L., & Brenner, M. B. (1994). Adhesion between epithelial cells and t lymphocytes mediated by E-cadherin and the alpha(E)beta(7) integrin. Nature. 372, 190-193.

Cepek, K. L., Rimm, D. L., & Brenner, M. B. (1996). Expression of a candidate cadherin in t-lymphocytes. Proc. Natl. Acad. Sci. U S A. 93, 6567-6571.

Chidgey, M. A., Clarke, J. P., & Garrod, D. R. (1996). Expression of full-length desmosomol glycoproteins (desmocollins) is not sufficient to confer strong adhesion on transfected L929 cells. J. Invest. Dermatol. 106, 689-695.

Cifuentesdiaz, C., Nicolet, M., Goudou, D., Rieger, F., & Mege, R. M. (1994). N-Cadherin expression in developing, adult and denervated chicken neuromuscular system - accumulations at both the neuromuscular junction and the node of ranvier. Development. 120, 1-11.

Cifuentesdiaz, C., Goudou, D., Padilla, F., Facchinetti, P., Nicolet, M., Mege, R. M., & Rieger, F. (1996). M-Cadherin distribution in the mouse adult neuromuscular system suggests a role in muscle innervation. Eur. J. Neurosci. 8, 1666-1676.

Collins, J. E. & Fleming, T. P. (1995). Epithelial differentiation in the mouse preimplantation embryo: making adhesive cell contacts for the first time. Trends Biochem. Sci. 20, 307-312.

Cowin, P., Kapprell, H. P., Franke, W. W., Tamkun, J., & Hynes, R. O. (1986). Plakoglobin: a protein common to different kinds of intercellular adhering junctions. Cell. 46, 1063-1073.

Cowin, P. & Burke, B. (1996). Cytoskeleton-membrane interactions. Curr. Opin. Cell Biol. 8, 56-65.

Dahl, U., Sjodin, A., & Semb, H. (1996). cadherins regulate aggregation of pancreatic beta-cells in vivo. Development. 122, 2895-2902.

Daniel, J. M. & Reynolds, A. B. (1995). The tyrosine kinase substrate p120*cas* binds directly to E-cadherin but not to the adenomatous polyposis cold protein or alpha-catenin. Mol. Cell. Biol. 15, 4819-4824.

Dantzig, A. H., Hoskins, J. A., Tabas, L. B., Bright, S., Shepard, R. L., Jenkins, I. L., Duckworth, D. C., Sportsman, J. R., Mackensen, D., Rosteck, P. R., Jr., & et al. (1994). Association of intestinal peptide transport with a protein related to the cadherin superfamily. Science. 264. 430-433.

Doherty, P. & Walsh, F. S. (1996). CAM-FGF receptor interactions - a model for axonal growth. Mol. Cell. Neurosci. 8, 99-111.

Donalies, M., Cramer, M., Ringwald, M., & Starzinski-Powitz, A. (1991). Expression of M-cadherin, a member of the cadherin multigene family, correlates with differentiation of skeletal muscle cells. Proc. Natl. Acad. Sci. U S A. 88, 8024-8028.

Dsouza, B. & Taylorpapadimitriou, J. (1994). overexpression of erbb2 in human mammary epithelial cells signals inhibition of transcription of the E-cadherin gene. Proc. Natl. Acad. Sci. U S A. 91, 7202-7206.

Dufour, S., Saint-Jeannet, J. P., Broders, F., Wedlich, D., & Thiery, J. P. (1994). Differential perturbations in the morphogenesis of anterior structures induced by overexpression of truncated XB- and N-cadherins in Xenopus embryos. J. Cell Biol. 127, 521-535.

Fannon, A. M., Sherman, D. L., Ilyina-Gragerova, G., Brophy, P. J., Friedrich, V. L., Jr., & Colman, D. R. (1995). Novel E-cadherin-mediated adhesion in peripheral nerve: Schwann cell architecture is stabilized by autotypic adherers junctions. J. Cell Biol. 129, 189-202.

Franke, W. W., Kapprell, H. P., & Cowin, P. (1987). Plakoglobin is a component of the filamentous subplasmalemmal coat of lens cells. Eur. J. Cell Biol. 43, 301-315.

Franke, W. W., Goldschmidt, M. D., Zimbelmann, R., Mueller, H. M., Schiller, D. L., & Cowin, P. (1989). Molecular cloning and amino acid sequence of human plakoglobin, the common junctional plaque protein. Proc. Natl. Acad. Sci. U S A. 86, 4027-4031.

Fuchs, M., Muller, T., Lerch, M. M., & Ullrich, A. (1996). Association of human protein-tyrosine phosphatase kappa with members of the armadillo family. J. Biol. Chem. 271, 16712-16719.

Fujimori, T. & Takeichi, M. (1993). Disruption of epithelial cell-cell adhesion by exogenous expression of a mutated nonfunctional N-cadherin. Mol. Biol. Cell. 4. 37-47.

Fukuyama, R., & Shimizu, N. (1991). Detection of epidermal growth factor receptors and E-cadherins in the basolateral membrane of A431 cells by laser scanning fluorescence microscopy. Jpn. J. Cancer Res. 82, 8-11.

Funayama, N., Fagotto, F., McCrea, P., & Gumbiner, B. M. (1995). Embryonic axis induction by the armadillo repeat domain of beta catenin: evidence for intracellular signaling. J. Cell Biol. 128, 959-968.

Ganzler, S. I. & Redies, C. (1995). R-cadherin expression during nucleus formation in chicken forebrain neuromeres. J. Neurosci. 15, 4157-4172.

Garrod, D., Chidgey, M., & North, A. (1996). Desmosomes - differentiation, development, dynamics and disease. Curr. Opin. Cell Biol. 8, 670-678.

Geiger, B., Yehuda-Levenberg, S., & Bershadsky, A. D. (1995). Molecular interactions in the submembrane plaque of cell-cell and cell-matrix adhesions. Acta Anat. 154, 46-62.

Grunwald, G. B., Geller, R. L., & Lilien, J. (1980). Enzymatic dissection of embryonic cell adhesive mechanisms. J. Cell Biol. 85, 766-776.

Grunwald, G. B., Bromberg, R. E., Crowley, N. J., & Lilien, J. (1981). Enzymatic dissection of embryonic cell adhesive mechanisms. II. Developmental regulation of an endogenous adhesive system in the chick neural retina. Dev. Biol. 86, 327-338.

Guger, K. A. & Gumbiner, B. M. (1995). beta-Catenin has Wnt-like activity and mimics the Nieuwkoop signaling center in Xenopus dorsal-ventral patterning. Dev. Biol. 172, 115-125.

Gumbiner, B. M. (1995) . Signal transduction of beta-catenin. Curr. Opin. Cell Biol. 7, 634-640.

Haegel, H., Larue, L., Ohsugi, M., Fedorov, L., Herrenknecht, K., & Kemler, R. (1995). Lack of beta-catenin affects mouse development at gastrulation. Development. 121, 3529-3537

Hall, H., Walsh, F. S., & Doherty, P. (1996). Review: a role for the FGF receptor in the axonal growth response stimulated by cell adhesion molecules? Cell Adhes. Commun. 3, 441-450.

Hamaguchi, M., Matsuyoshi, N., Ohnishi, Y., Gotoh, B., Takeiehi, M., & Nagai, Y. (1993). p60v-src causes tyrosine phosphorylation and inactivation of the N-cadherin-catenin cell adhesion system. EMBO J. 12, 307-314.

Hatta, K., Nose, A., Nagafuchi, A., & Takeichi, M. (1988). Cloning and expression of cDNA encoding a neural calcium-dependent cell adhesion molecule: its identity in the cadherin gene family. J. Cell Biol. 106, 873-881.

Heasman, J., Crawford, A., Goldstone, K., Garner-Hamriek, P., Gumbiner, B., MeCrea, P., Kintner, C., Noro, C. Y., & Wylie, C. (1994). Overexpression of cadherins and underexpression of beta-catenin inhibit dorsal mesoderm induction in early Xenopus embryos. Cell. 79, 791-803.

Hermiston, M. L., & Gordon, J. I. (1995a). In vivo analysis of cadherin function in the mouse intestinal epithelium: essential roles in adhesion, maintenance of differentiation, and regulation of programmed cell death. J. Cell Biol. 129, 489-506.

Hermiston, M. L., & Gordon, J. I. (1995b). Inflammatory bowel disease and adenomas in mice expressing a dominant negative N-cadherin. Science. 270, 1203-1207.

Herrenknecht, K., Ozawa, M., Eckerskorn, C., Lottspeich, F., Lenter, M., & Kemler, R. (1991). The uvomorulin-anchorage protein alpha catenin is a vinculin homologue. Proc. Natl. Acad. Sci. U S A 88, 9156-9160.

Hinck, L., Nathke, I. S., Papkoff, J., & Nelson, W. J. (1994). Dynamics of cadherin/catenin complex formation: novel protein interactions and pathways of complex assembly. J. Cell Biol. 125, 1327-1340.

Hirano, S., Kimoto, N., Shimoyama, Y., Hirohashi, S., & Takeichi, M. (1992). Identification of a neural alpha-catenin as a key regulator of cadherin function and multicellular organization. Cell. 70, 293-301.

Hoffmann, I. & Balling, R. (1995). Cloning and expression analysis of a novel mesodermally expressed cadherin. Dev. Biol. 169, 337-346.

Holt, C. E., Lemaire, P., & Gurdon, J. B. (1994). Cadherin-mediated cell interactions are necessary for the activation of myod in xenopus mesoderm. Proc. Natl. Acad. Sci. U S A. 91, 10844-10848.

Hosehuetzky, H., Aberle, H., & Kemler, R. (1994). Beta-catenin mediates the interaction of the cadherin-catenin complex with epidermal growth factor receptor. J. Cell Biol. 127. 1375-1380.

Huber, O., Bierkamp, C., & Kemler, R. (1996a). cadherins and catenins in development. Curr. Opin. Cell Biol. 8, 685-691.

Huber, O., Korn, R., McLaughlin, J., Ohsugi, M., Herrmann, B. G., & Kemler, R. (1996b). Nuclear localization of beta-catenin by interaction with transcription factor LEF-1. Mech. Dev. 59, 3-10.

Hulsken, J., Birchmeier, W., & Behrens, J. (1994). E-cadherin and APC compete for the interaction with beta-catenin and the cytoskeleton. J. Cell Biol. 127, 2061-2069.

Hynes, R. O. (1996). Targeted mutations in cell adhesion genes - what have we learned from them. Dev. Biol. 180, 402-412.

Hynes, R. O. & Lander, A. D. (1992). Contact and adhesive specificities in the associations, migrations, and targeting of cells and axons. Cell. 68, 303-322.

Inuzuka, H., Redies, C., & Takeichi, M. (1991). Differential expression of R- and N-cadherin in neural and mesodermal tissues during early chicken development. Development. 113, 959-967.

Irintchev, A., Zeschnigk, M., Starzinskipowitz, A., & Wernig, A. (1994). Expression pattern of M-cadherin in normal, denervated, and regenerating mouse muscles. Dev. Dyn. 199, 326-337.

Jiang, W. G. (1996). E-cadherin and its associated protein catenins, cancer invasion and metastasis. Br. J. Surg. 83, 437-446.

Jones, E. Y. (1996). Three-dimensional structure of cell adhesion molecules. Curr. Opin. Cell Biol. 8, 602-608.

Jou, T. S., Stewart, D. B., Stappert, J., Nelson, W. J., & Marrs, J. A. (1995). Genetic and biochemical dissection of protein linkages in the cadherin-catenin complex. Proc. Natl. Acad. U S A. 92, 5067-5071.

Kanai, Y., Oehiai, A., Shibata, T., Oyama, T., Ushijima, S., Akimoto, S., & Hirohashi, S. (1995).c-erbB-2 gene product directly associates with beta-catenin and plakoglobin. Biochem. Biophys. Res. Commun. 208, 1067-1072.

Karecla, P. I., Bowden, S. J., Green, S. J., & Kilshaw, P. J. (1995). Recognition of E-cadherin on epithelial cells by the mucosal T cell integrin alpha M290 beta 7 (alpha E beta 7). Eur. J. Immunol. 25, 852-856.

Karecla, P. I., Green, S. J., Bowden, S. J., Coadwell, J., & Kilshaw, P. J. (1996). Identification of a binding site for integrin alphaEbeta7 in the N-terminal domain of E-cadherin. J. Biol. Chem. 271, 30909-30915.

Karnovsky, A. & Klymkowsky, M. W. (1995). Anterior axis duplication in Xenopus induced by the over-expression of the cadherin-binding protein plakoglobin. Proc. Natl. Acad. Sci. U S A. 92, 4522-4526.

Kelly, G. M., Erezyilmaz, D. F., & Moon, R. T. (1995). Induction of a secondary embryonic axis in zebrafish occurs following the overexpression of beta-catenin. Mech. Dev. 53, 261-273.

Kemler, R. (1992). Classical cadherins. Semin. Cell Biol. 3, 149-155.

Kemler, R. (1993). From cadherins to catenins: cytoplasmic protein interactions and regulation of cell adhesion. Trends Genet. 9, 317-321.

Kemler, R., Babinet, C., Eisen, H., & Jacob, F. (1977). Surface antigen in early differentiation. Proc. Natl. Acad. Sci. U S A. 74, 4449-4452.

Kimura, Y., Matsunami, H., Inoue, T., Shimamura, K., Uchida, N., Ueno, T., Miyazaki, T., & Takeichi, M. (1995). Cadherin-11 expressed in association with mesenchymal morphogenesis in the head, somite, and limb bud of early mouse embryos. Dev. Biol. 169, 347-358.

Kinch, M. S., Clark, G. J., Der, C. J., & Burridge, K. (1995). Tyrosine phosphorylation regulates the adhesions of ras-transformed breast epithelia. J. Cell Biol. 130, 461-471.

Kintner, C. (1992). Regulation of embryonic cell adhesion by the cadherin cytoplasmic domain. Cell. 69, 225-236.

Knudsen, K. A., Soler, A. P., Johnson, K. R., & Wheelock, M. J. (1995). Interaction of alpha-actinin with the cadherin/catenin cell-cell adhesion complex via alpha-catenin. J. Cell Biol. 130, 67-77.

Knust, E. & Leptin, M. (1996). Adherens junctions in the drosophila embryo - the role of E-cadherin in their establishment and morphogenetic function. Bioessays. 18, 609-612.

Koch, P. J., Walsh, M. J., Schmelz, M., Goldschmidt, M. D., Zimbelmann, R., & Franke, W. W. (1990). Identification of desmoglein, a constitutive desmosomal glycoprotein, as a member of the cadherin family of cell adhesion molecules. Eur. J. Cell Biol. 53, 1-12.

Koch, P. J. & Franke, W. W. (1994). Desmosomal cadherins: another growing multigene family of adhesion molecules. Curr. Opin. Cell Biol. 6, 682-687.

Korinek, V., Barker, N., Morin, P. J., Vanwichen, D., Deweger, R., Kinzler, K. W., Vogelstein, B., & Clevers, H. (1997). constitutive transcriptional activation by a beta-catenin-TCF complex in APC(-/-) colon carcinoma. Science. 275, 1784-1787.

Kowalczyk, A. P., Stappenbeck, T. S., Parry, D. A., Palka, H. L., Virata, M. L., Bornslaeger, E. A., Nilles, L. A., & Green, K. J. (1994). Structure and function of desmosomal transmembrane core and plaque molecules. Biophys. Chem. 50, 97-112.

Kuhl, M., Finnemann, S., Binder, O., & Wedlich, D. (1996). Dominant negative expression of a cytoplasmically deleted mutant of XB/U-cadherin disturbs mesoderm migration during gastrulation in Xenopus laevis. Mech. Dev. 54, 71-82.

Kuhl, M. & Wedlich, D. (1996). Xenopus cadherins: sorting out types and functions in embryogenesis. Dev. Dyn. 207, 121-134.

Kypta, R. M., Su, H., & Reichardt, L. F. (1996). Association between a transmembrane protein tyrosine phosphatase and the cadherin-catenin complex. J. Cell Biol. 134, 1519-1529.

Lampugnani, M. G., Resnati, M., Raiteri, M., Pigott, R., Pisacane, A., Houen, G., Ruco, L. P., & Dejana, E. (1992). A novel endothelia -specific membrane protein is a marker of cell-cell contacts. J. Cell Biol. 118, 1511-1522.

Lampugnani, M. G., Corada, M., Caveda, L., Breviario, F., Ayalon, O., Geiger, B., & Dejana, E. (1995). The molecular organization of endothelial cell to cell junctions: differential association of plakoglobin, beta-catenin, and alpha-catenin with vascular endothelial cadherin (VE-cadherin). J. Cell Biol. 129, 203-217.

Larabell, C. A., Torres, M., Rowning, B. A., Yost, C., Miller, J. R., Wu, M., Kimelman, D., & Moon, R. T. (1997). Establishment of the dorso-ventral axis in xenopus embryos is presaged by early asymmetries in beta-catenin that are modulated by the wnt signaling pathway. J. Cell Biol. 136, 1123-1136.

Larue, L., Ohsugi, M., Hirchenhain, J., & Kemler, R. (1994). E-cadherin null mutant embryos fail to form a trophectoderm epithelium. Proc. Natl. Acad. Sci. U S A. 91, 8263-8267.

Larue, L., Antos, C., Butz, S., Huber, O., Delmas, V., Dominis, M., & Kemler, R. (1996). A role for cadherins in tissue formation. Development. 122, 3185-3194.

Lee, C. H. & Gumbiner, B. M. (1995). Disruption of gastrulation movements in Xenopus by a dominant-negative mutant for C-cadherin. Dev. Biol. 171, 363-373.

Lee, S. W. (1996). H-cadherin, a novel cadherin with growth inhibitory functions and diminished expression in human breast cancer. Nat. Med. 2, 776-782.

Levine, E., Lee, C. H., Kintner, C., & Gumbiner, B. M. (1994). Selective disruption of E-cadherin function in early Xenopus embryos by a dominant negative mutant. Development. 120, 901-909.

Lewis, J. E., Jensen, P. J., Johnson, K. R., & Wheelock, M. J. (1994). E-cadherin mediates adherers junction organization through protein kinase C. J. Cell Sci. 107, 3615-3621.

Lewis, J. E., Wahl, J. K., Sass, K. M., Jensen, P. J., Johnson, K. R., & Wheelock, M. J. (1997). Cross-talk between adherers junctions and desmosomes depends on plakoglobin. J. Cell Biol. 136, 919-934.

Magnani, J. L., Thomas, W. A., & Steinberg, M. S. (1981). Two distinct adhesion mechanisms in embryonic neural retina cells. I. A kinetic analysis. Dev. Biol. 81, 96-105.

Mahoney, P. A., Weber, U., Onofrechuk, P., Biessmann, H., Bryant, P. J., & Goodman, C. S. (1991). The fat tumor suppressor gene in Drosophila encodes a novel member of the cadherin gene superfamily. Cell. 67, 853-868.

Mareel, M., Berx, G., Vanroy, F., & Bracke, M. (1996). Cadherin/catenin complex - a target for antiinvasive therapy. J. Cell Biochem. 61, 524-530.

Marrs, J. A., Andersson-Fisone, C., Jeong, M. C., Cohen-Gould, L., Zurzolo, C., Nabi, I. R., Rodriguez-Boulan, E., & Nelson, W. J. (1995). Plasticity in epithelial cell phenotype: modulation by expression of different cadherin cell adhesion molecules. J. Cell Biol. 129, 507-519.

Marrs, J. A. & Nelson, W. J. (1995). Cadherin cell adhesion molecules in differentiation and embryogenesis. Int. Rev. Cytol. 165, 159-205.

Mathur, M., Goodwin, L., & Cowin, P. (1994). Interactions of the cytoplasmic domain of the desmosomal cadherin Dsg1 with plakoglobin. J. Biol. Chem. 269, 14075-14080.

Matsunami, H., Miyatani, S., Inoue, T., Copeland, N. G., Gilbert, D. J., Jenkins, N. A., & Takeichi, M. (1993). Cell binding specificity of mouse R-cadherin and chromosomal mapping of the gene. J. Cell Sci. 106, 401-409.

Matsunami, H. & Takeichi, M. (1995). Fetal brain subdivisions defined by R- and E-cadherin expressions: evidence for the role of cadherin activity in region-specific, cell-cell adhesion. Dev. Biol. 172, 466-478.

Matsuyoshi, N., Hamaguchi, M., Taniguchi, S., Nagafuchi, A., Tsukita, S., & Takeichi, M. (1992). Cadherin-mediated cell-cell adhesion is perturbed by v-src tyrosine phosphorylation in metastatic fibroblasts. J. Cell Biol. 118, 703-714.

McNeill, H., Ryan, T. A., Smith, S. J., & Nelson, W. J. (1993). Spatial and temporal dissection of immediate and early events following cadherin-mediated epithelial cell adhesion. J. Cell Biol. 120, 1217- 1226.

Miller, & Moon (1996). Confocal micrograph of four-cell-stage Xenopus embryo stained with anti-beta-catenin antibody . Genes Dev. 10, 3216.

Molenaar, M., Vandewetering, M., Oosterwegel, M., Petersonmaduro, J., Godsave, S., Korinek, V., Roose, J., Destree, O., & Clevers, H. (1996). Xtcf-3 transcription factor mediates beta-catenin-induced axis formation in xenopus embryos. Cell. 86, 391-399.

Moore, R. & Walsh, F. S. (1993). the cell adhesion molecule M-cadherin is specifically expressed in developing and regenerating, but not denervated skeletal muscle. Development. 117, 1409-1420.

Morin, P. J., Sparks, A. B., Korinek, V., Barker, N., Clevers, H., Vogelstein, B., & Kinzler, K. W. (1997). Activation of beta-catenin-Tcf signaling in colon cancer by mutations in beta-catenin or APC. Science. 275, 1787-1790.

Murphy-Erdosh, C., Napolitano, E. W., & Reichardt, L. F. (1994). The expression of B-cadherin during embryonic chick development. Dev. Biol. 161, 107-125.

Nagafuchi, A., Shirayoshi, Y., Okazaki, K., Yasuda, K., & Takeichi, M. (1987). Transformation of cell adhesion properties by exogenously introduced E-cadherin cDNA. Nature 329, 341-343.

Nagafuchi, A. & Takeichi, M. (1988). Cell binding function of E-cadherin is regulated by the cytoplasmic domain. EMBO J. 7, 3679-3684.

Nagafuchi, A. & Takeichi, M. (1989). Transmembrane control of cadherin-mediated cell adhesion: a 94 kDa protein functionally associated with a specific region of the cytoplasmic domain of E-cadherin. Cell Regulation. 1, 37-44.

Nagafuchi, A., Takeichi, M., & Tsukita, S. (1991). The 102 kd cadherin-associated protein: similarity to vinculin and posttranscriptional regulation of expression. Cell. 65, 849-857.

Nagafuchi, A., Ishihara, S., & Tsukita, S. (1994). The roles of catenins in the cadherin-mediated cell adhesion: functional analysis of E-cadherin-alpha catenin fusion molecules. J. Cell Biol. 127, 235-45.

Nagar, B., Overduin, M., Ikura, M., & Rini, J. M. (1996). Structural basis of calcium-induced E-cadherin rigidification and dimerization. Nature 380, 360-364.

Nakagawa, S., & Takeichi, M. (1995). Neural crest cell-cell adhesion controlled by sequential and subpopulation-specific expression of novel cadherins. Development. 121. 1321-1332.

Nathke, I. S., Hinck, L., Swedlow, J. R., Papkoff, J., & Nelson, W. J. (1994). Defining interactions and distributions of cadherin and catenin complexes in polarized epithelial cells. J. Cell Biol. 125, 1341-1352.

Nelson, W. J. & Veshnock, P. J. (1986). Dynamics of membrane-skeleton (fodrin) organization during development of polarity in Madin-Darby canine kidney epithelial cells. J Cell Biol. 103, 1751-1765.

Nelson, W. J. & Veshnock, P. J. (1987). Modulation of fodrin (membrane skeleton) stability by cell-cell contact in Madin-Darby canine kidney epithelial cells. J. Cell Biol. 104, 1527-1537.

Nishimura, M., Machida, K., Imaizumi, M., Abe, T., Umeda, T., Takeshima, E., Watanabe, T., Ohnishi, Y., Takagi, K., & Hamaguchi, M. (1996). Tyrosine phosphorylation of 100-130 KDa proteins in lung cancer correlates with poor prognosis. Br. J. Cancer. 74. 780-787.

Nose, A., Nagafuchi, A., & Takeichi, M. (1987). Isolation of placental cadherin cDNA: identification of a novel gene family of cell-cell adhesion molecules. EMBO J. 6, 3655-3661.

Nose, A., Tsuji, K., & Takeichi, M. (1990). Localization of specificity determining sites in cadherin cell adhesion molecules. Cell. 61, 147-155.

Obama, H. & Ozawa, M. (1997). Identification of the domain of α-catenin involved in its association with β-catenin and plakoglobin (γ-catenin). J. Biol. Chem. 272, 11017-11020.

Obata, S., Sago, H., Mori, N., Rochelle, J. M., Seldin, M. F., Davidson, M., John, T. S., Taketani, S., & Suzuki, S. T. (1995). Protocadherin Pcdh2 shows properties similar to, but distinct from, those of classical cadherins. J. Cell Sci. 108, 3765-3773.

Ochiai, A., Akimoto, S., Kanai, Y., Shibata, T., Oyama, T., & Hirohashi, S. (1994). c-erbB-2 gene product associates with catenins in human cancer cells. Biochem. Biophys. Res. Commun. 205, 73-78.

Oda, T., Kanai, Y., Shimoyama, Y., Nagafuchi, A., Tsukita, S., & Hirohashi, S. (1993a). Cloning of the human α-catenin cDNA and its aberrant mRNA in a human cancer cell line. Biochem. Biophys. Res. Commun. 193, 897-904.

Oda, H., Uemura, T., Shiomi, K., Nagafuchi, A., Tsukita, S., & Takeichi, M. (1993b). Identification of a Drosophila homologue of α-catenin and its association with the armadillo protein. J. Cell Biol. 121, 1133-1140.

Oda, H., Uemura, T., Harada, Y., Iwai, Y., & Takeichi, M. (1994). A Drosophila homolog of cadherin associated with armadillo and essential for embryonic cell-cell adhesion. Dev. Biol. 165, 716-726.

Ohsugi, M., Hwang, S. Y., Butz, S., Knowles, B. B., Solter, D., & Kemler, R. (1996). Expression and cell membrane localization of catenins during mouse preimplantation development. Dev. Dyn. 206, 391-402.

Okazaki, M., Takeshita, S., Kawai, S., Kikuno, R., Tsujimura, A., Kudo, A., & Amann, E. (1994). Molecular cloning and characterization of OB-cadherin, a new member of cadherin family expressed in osteoblasts. J. Biol. Chem. 269, 12092-12098.

Orsulic, S. & Peifer, M. (1996). Cell-cell signalling: Wingless lands at last. Curr. Biol. 6, 1363-1367.

Overduin, M., Harvey, T. S., Bagby, S., Tong, K. I., Yau, P., Takeichi, M., & Ikura, M. (1995). Solution structure of the epithelial cadherin domain responsible for selective cell adhesion [see comments]. Science. 267, 386-389.

Oyama, T., Kanai, Y., Ochiai, A., Akimoto, S., Oda, T., Yanagihara, K., Nagafuchi, A., Tsukita, S., Shibamoto, S., Ito, F., Takeichi, M., Matsuda, H., & Hirohashi, S. (1994). A truncated β-catenin disrupts the interaction between E-cadherin and α-catenin: a cause of loss of intercellular adhesiveness in human cancer cell lines. Cancer Res. 54, 6282-6287.

Ozawa, M., Baribault, H., & Kemler, R. (1989). The cytoplasmic domain of the cell adhesion molecule uvomorulin associates with three independent proteins structurally related in different species, EMBO J. 8, 1711-1717.

Ozawa, M., Engel, J., & Kemler, R. (1990). Single amino acid substitutions in one Ca^{2+} binding site of uvomorulin abolish the adhesive function. Cell. 63, 1033-1038.

Ozawa, M. & Kemler, R. (1990). Correct proteolytic cleavage is required for the cell adhesive function of uvomorulin. J. Cell Biol. 111, 1645-1650.

Ozawa, M. & Kemler, R. (1992). Molecular organization of the uvomorulin-catenin complex. J. Cell Biol. 116, 989-996.

Ozawa, M., Nuruki, K., Toyoyama, H., & Ohi, Y. (1995a). Cloning of an alternative form of plakoglobin (γ-catenin) lacking the fourth armadillo repeat. J. Biochem. 118, 836-840.

Ozawa, M., Terada, H., & Pedraza, C. (1995b). The fourth armadillo repeat of plakoglobin (γ-catenin) is required for its high affinity binding to the cytoplasmic domains of E-cadherin and desmosomal cadherin Dsg2, and the tumor suppressor APC protein. J Biochem. 118, 1077-1082.

Pai, L. M. Kirkpatrick, C., Blanton, J., Oda, H., Takeichi, M., & Peifer, M. (1996). *Drosophila* α-catenin and E-cadherin bind to distinct regions of *Drosophila* armadillo. J. Biol. Chem. 271, 32411-32420.

Paradies, N. E., & Grunwald, G. B. (1993). Purification and characterization of NCAD90, a soluble endogenous form of N-cadherin, which is generated by proteolysis during retinal development and retains adhesive and neurite-promoting function. J. Neurosci. Res. 36, 33-45.

Peifer, M. & Wieschaus, E. (1990). The segment polarity gene armadillo encodes a functionally modular protein that is the *Drosophila* homolog of human plakoglobin. Cell. 63, 1167-1176.

Peifer, M., Berg, S., & Reynolds, A. B. (1994). A repeating amino acid motif shared by proteins with diverse cellular roles [letter]. Cell. 76, 789-791.

Peifer, M. (1995). Cell adhesion and signal transduction—The armadillo connection. Trends Cell. Biol. 5, 224-229.

Pokutta, S., Herrenknecht, K., Kemler, R., & Engel, J. (1994). Conformational changes of the recombinant extracellular domain of E-cadherin upon calcium binding. Eur. J. Biochem. 223, 1019-1026.

Radice, G. L., Rayburn, H., Matsunami, H., Knudsen, K. A., Takeichi, M., & Hynes, R. O. (1997). Developmental defects in mouse embryos lacking N-cadherin. Dev. Biol. 181, 64-78.

Rajasekaran, A. K., Hojo, M., Huima, T., & Rodriguez-Boulan, E. (1996). Catenins and zonula occludens-1 form a complex during early stages in the assembly of tight junctions. J.
Cell Biol. 132, 451-463.

Ranscht, B. & Dours-Zimmermann, M. T. (1991). T-cadherin, a novel cadherin cell adhesion molecule in the nervous system lacks the conserved cytoplasmic region. Neuron. 7, 391-402.

Ranscht, B. (1994). Cadherins and catenins: Interactions and functions in embryonic development. Curr. Opin. Cell Biol. 6, 740-746.

Redies, C., Engelhart, K., & Takeichi, M. (1993). Differential expression of N- and R-cadherin in functional neuronal systems and other structures of the developing chicken brain. J. Comp. Neurol. 333, 398-416.

Redies, C. (1995). Cadherin expression in the developing vertebrate CNS: from neuromeres to brain nuclei and neural circuits. Exp. Cell Res. 220, 243-256.

Redies, C. & Takeichi, M. (1996). Cadherins in the developing central nervous system: An adhesive code for segmental and functional subdivisions. Dev. Biol. 180, 413-423.

Reynolds, A. B., Herbert, L., Cleveland, J. L., Berg, S. T., & Gaut, J. R. (1992). p120, a novel substrate of protein tyrosine kinase receptors and of p60v-src, is related to cadherin-binding factors β-catenin, plakoglobin and armadillo. Oncogene. 7, 2439-2445.

Reynolds, A. B., Daniel, J., McCrea, P. D., Wheelock, M. J., Wu, J., & Zhang, Z. (1994). Identification of a new catenin: The tyrosine kinase substrate p120cas associates with E-cadherin complexes. Mol. Cell. Biol. 14, 8333-8342.

Rimm, D. L., Koslov, E. R., Kebriaei, P., Cianci, C. D., & Morrow, J. S. (1995). α 1(E)-catenin is an actin-binding and -bundling protein mediating the attachment of F-actin to the membrane adhesion complex. Proc. Natl. Acad. Sci. USA 92, 8813-8817.

Ringwald, M., Schuh, R., Vestweber, D., Eistetter, H., Lottspeich, F., Engel, J., Dolz, R., Jahnig, F., Epplen, J., Mayer, S. Müller, C., & Kemler, R. (1987). The structure of cell adhesion molecule uvomorulin. Insights into the molecular mechanism of $Ca2^{+}$-dependent cell adhesion. EMBO J. 6, 3647-3653.

Roh, J. Y. & Stanley, J. R. (1995a). Intracellular domain of desmoglein 3 (*Pemphigus vulgaris* antigen) confers adhesive function on the extracellular domain of E-cadherin without binding catenins. J. Cell Biol. 128, 939-947.

Roh, J. Y. & Stanley, J. R. (1995b). Plakoglobin binding by human Dsg3 (*Pemphigus vulgaris* antigen) in keratinocytes requires the cadherin-like intracytoplasmic segment. J. Invest. Dermatol. 104, 720-724.

Rose, O., Grund, C., Reinhardt, S., Starzinski-Powitz, A., & Franke, W. W. (1995). Contactus adherens, a special type of plaque-bearing adhering junction containing M-cadherin, in the granule cell layer of the cerebellar glomerulus. Proc. Natl. Acad. Sci. USA 92, 6022-6026.

Rubinfeld, B., Souza, B., Albert, I., Muller, O., Chamberlain, S. H., Masiarz, F. R., Munemitsu, S., & Polakis, P. (1993). Association of the APC gene product with β-catenin. Science. 262, 1731-1734.

Rubinfeld, B., Albert, I., Porfiri, E., Fiol, C., Munemitsu, S., & Polakis, P. (1996). Binding of gGSK3-β to the APC-β-catenin complex and regulation of complex assembly. Science. 272, 1023-1026.

Rubinfeld, B., Robbins, P., Elgamil, M., Albert, I., Porfiri, E., & Polakis, P. (1997). Stabilization of β-catenin by genetic defects in melanoma cell lines. Science. 275, 1790-1792.

Ruiz, P., Brinkmann, V., Ledermann, B., Behrend, M., Grund, C., Thalhammer, C., Vogel, F., Birchmeier, C., Gunthert, U., Franke, W. W., & Birchmeier, W. (1996). Targeted mutation of plakoglobin in mice reveals essential functions of desmosomes in the embryonic heart. J. Cell Biol. 135, 215-225.

Sacco, P. A., McGranahan, T. M., Wheelock, M. J., & Johnson, K. R. (1995). Identification of plakoglobin domains required for association with N-cadherin and α-catenin. J. Biol. Chem. 270, 20201-20206.

Sago, H., Kitagawa, M., Obata, S., Mori, N., Taketani, S., Rochelle, J. M., Seldin, M. F., Davidson, M., Stjohn, T., & Suzuki, S. T. (1995). Cloning, expression, and chromosomal localization of a novel cadherin-related protein, protocadherin-3. Genomics. 29, 631-640.

Sano, K., Tanihara, H., Heimark, R. L., Obata, S., Davidson, M., St John, T., Taketani, S., & Suzuki, S. (1993). Protocadherins: A large family of cadherin-related molecules in central nervous system. EMBO J. 12, 2249-2256.

Schmelz, M. & Franke, W. W. (1993). Complexus adhaerentes, a new group of desmoplakin-containing junctions in endothelial cells: The syndesmos connecting retothelial cells of lymph nodes. Eur. J. Cell Biol. 61, 274-289.

Schmidt, A., Heid, H. W., Schafer, S., Nuber, U. A., Zimbelmann, R., & Franke, W. W. (1994). Desmosomes and cytoskeletal architecture in epithelial differentiation: Cell type-specific plaque components and intermediate filament anchorage. Eur. J. Cell Biol. 65, 229-245.

Schneider, S., Steinbeisser, H., Warga, R. M., & Hausen, P. (1996). β-Catenin translocation into nuclei demarcates the dorsalizing centers in frog and fish embryos. Mech. Dev. 57, 191-198.

Sefton, M., Johnson, M. H., Clayton, L., & McConnell, J. M. L. (1996). Experimental manipulations of compaction and their effects on the phosphorylation of uvomorulin. Mol. Reprod. Dev. 44, 77-87.

Shapiro, L., Fannon, A. M., Kwong, P. D., Thompson, A., Lehmann, M. S., Grubel, G., Legrand, J. F., Als-Nielsen, J., Colman, D. R., & Hendrickson, W. A. (1995a). Structural basis of cell–cell adhesion by cadherins (see comments). Nature. 374, 327-337.

Shapiro, L., Kwong, P. D., Fannon, A. M., Colman, D. R., & Hendrickson, W. A. (1995b). Considerations on the folding topology and evolutionary origin of cadherin domains. Proc. Natl. Acad. Sci. USA. 92, 6793-6797.

Shibamoto, S., Hayakawa, M., Takeuchi, K., Hori, T., Oku, N., Miyazawa, K., Kitamura, N., Takeichi, M., & Ito, F. (1994). Tyrosine phosphorylation of β-catenin and plakoglobin enhanced by hepatocyte growth factor and epidermal growth factor in human carcinoma cells. Cell Adhes. Commun. 1, 295-305.

Shibamoto, S., Hayakawa, M., Takeuchi, K., Hori, T., Miyazawa, K., Kitamura, N., Johnson, K. R., Wheelock, M. J., Matsuyoshi, N., Takeichi, M., & Ito, F. (1995). Association of p120, a tyrosine kinase substrate, with E-cadherin/catenin complexes. J. Cell Biol. 128, 949-957.

Shibata, T., Gotoh, M., Ochiai, A., & Hirohashi, S. (1994). Association of plakoglobin with APC, a tumor suppressor gene product, and its regulation by tyrosine phosphorylation. Biochem. Biophys. Res. Commun. 203, 519-522.

Shimamura, K. & Takeichi, M. (1992). Local and transient expression of E-cadherin involved in mouse embryonic brain morphogenesis. Development. 116, 1011-1019.

Shimoyama, Y., Nagafuchi, A., Fujita, S., Gotoh, M., Takeichi, M., Tsukita, S., & Hirohashi, S. (1992). Cadherin dysfunction in a human cancer cell line: Possible involvement of loss of α-catenin expression in reduced cell-cell adhesiveness. Cancer Res. 52, 5770-5774.

Shiozaki, H., Oka, H., Inoue, M., Tamura, S., & Monden, M. (1996). E-cadherin mediated adhesion system in cancer cells. Cancer. 77, 1605-1613.

Simonneau, L., Kitagawa, M., Suzuki, S., & Thiery, J. P. (1995). Cadherin 11 expression marks the mesenchymal phenotype: Towards new functions for cadherins? Cell Adhes. Commun. 3, 115-130.

Soler, A. P. & Knudsen, K. A. (1994). N-cadherin involvement in cardiac myocyte interaction and myofibrillogenesis. Dev. Biol. 162, 9-17.

Sommers, C. L., Gelmann, E. P., Kemler, R., Cowin, P., & Byers, S. W. (1994). Alterations in β-catenin phosphorylation and plakoglobin expression in human breast cancer cells. Cancer Res. 54, 3544-3552.

Stanley, J. R. (1995). Defective cell-cell adhesion in the epidermis. Ciba Foundation Symposium 1995. 189, 107-120.

Stappert, J., & Kemler, R. (1994). A short core region of E-cadherin is essential for catenin binding and is highly phosphorylated. Cell Adhes. Commun. 2, 319-27.

Su, L. K., Vogelstein, B., & Kinzler, K. W. (1993). Association of the APC tumor suppressor protein with catenins. Science. 262, 1734-1737.

Suzuki, S. T. (1996). Structural and functional diversity of cadherin superfamily—are new members of cadherin superfamily involved in signal transduction pathway. J. Cell Biochem. 61, 531-542.

Takeda, H., Nagafuchi, A., Yonemura, S., Tsukita, S., Behrens, J., Birchmeier, W., & Tsukita, S. (1995). V-src kinase shifts the cadherin-based cell adhesion from the strong to the weak state and β catenin is not required for the shift. J. Cell Biol. 131, 1839-1847.

Takeichi, M. (1977). Functional correlation between cell adhesive properties and some cell surface proteins. J. Cell Biol. 75, 464-474.

Takeichi, M., Ozaki, H. S., Tokunaga, K., & Okada, T. S. (1979). Experimental manipulation of cell surface to affect cellular recognition mechanisms. Dev. Biol. 70, 195-205.

Takeichi, M. (1988). The cadherins: Cell-cell adhesion molecules controlling animal morphogenesis. Development. 102, 639-655.

Takeichi, M., Atsumi, T., Yoshida, C., Uno, K., & Okada, T. S. (1981). Selective adhesion of embryonal carcinoma cells and differentiated cells by $Ca2^+$-dependent sites. Dev. Biol. 87, 340-350.

Takeichi, M. (1995). Morphogenetic roles of classic cadherins. Curr. Opin. Cell Biol. 7, 619-627.

Tanihara, H., Kido, M., Obata, S., Heimark, R. L., Davidson, M., St John, T., & Suzuki, S. (1994a). Characterization of cadherin-4 and cadherin-5 reveals new aspects of cadherins. J. Cell Sci. 107, 1697-1704.

Tanihara, H., Sano, K., Heimark, R. L., St John, T., & Suzuki, S. (1994b). Cloning of five human cadherins clarifies characteristic features of cadherin extracellular domain and provides further evidence for two structurally different types of cadherin. Cell Adhes. Commun. 2, 15-26.

Tepass, U., Gruszynski-DeFeo, E., Haag, T. A., Omatyar, L., Torok, T., & Hartenstein, V. (1996). *shotgun* encodes *Drosophila* E-cadherin and is preferentially required during cell rearrangement in the neurectoderm and other morphogenetically active epithelia. Genes Dev. 10, 672-685.

Thomas, W. A. & Steinberg, M. S. (1981). Two distinct adhesion mechanisms in embryonic neural retina cells. II. An immunological analysis. Dev. Biol. 81, 106-114.
Thomas, W. A., Thomson, J., Magnani, J. L., & Steinberg, M. S. (1981). Two distinct adhesion mechanisms in embryonic neural retina cells. III. Functional specificity. Dev. Biol. 81, 379-385.
Thomson, R. B., Igarashi, P., Biemesderfer, D., Kim, R., Abu-Alfa, A., Soleimani, M., & Aronson, P. S. (1995). Isolation and cDNA cloning of Ksp-cadherin, a novel kidney-specific member of the cadherin multigene family. J. Biol. Chem. 270, 17594-17601.
Tomschy, A., Fauser, C., Landwehr, R., & Engel, J. (1996). Homophilic adhesion of E-cadherin occurs by a co-operative two-step interaction of N-terminal domains. EMBO J. 15, 3507-3514.
Townes, P. L. & Holtfreter, J. (1955). Directed movements and selective adhesion of embryonic amphibian cells. J. Exp. Zool. 128, 53-120.
Troyanovsky, S. M., Eshkind, L. G., Troyanovsky, R. B., Leube, R. E., & Franke, W. W. (1993). Contributions of cytoplasmic domains of desmosomal cadherins to desmosome assembly and intermediate filament anchorage. Cell. 72, 561-574.
Troyanovsky, S. M., Troyanovsky, R. B., Eshkind, L. G., Krutovskikh, V. A., Leube, R. E., & Franke, W. W. (1994a). Identification of the plakoglobin-binding domain in desmoglein and its role in plaque assembly and intermediate filament anchorage. J. Cell Biol. 127, 151-160.
Troyanovsky, S. M., Troyanovsky, R. B., Eshkind, L. G., Leube, R. E., & Franke, W. W. (1994b). Identification of amino acid sequence motifs in desmocollin, a desmosomal glycoprotein, that are required for plakoglobin binding and plaque formation. Proc. Natl. Acad. Sci. USA 91, 10790-10794.
Troyanovsky, R. B., Chitaev, N. A., & Troyanovsky, S. M. (1996). Cadherin binding sites of plakoglobin—localization, specificity and role in targeting to adhering junctions. J. Cell Sci. 109, 3069-3078.
Tsukita, S., Oishi, K., Akiyama, T., Yamanashi, Y., Yamamoto, T., & Tsukita, S. (1991). Specific proto-oncogenic tyrosine kinases of src family are enriched in cell-to-cell adherens junctions where the level of tyrosine phosphorylation is elevated. J. Cell Biol. 113, 867-879.
Tsukita, S., Tsukita, S., Nagafuchi, A., & Yonemura, S. (1992). Molecular linkage between cadherins and actin filaments in cell-cell adherens junctions. Curr. Opin. Cell Biol. 4, 834-839.
Tsukita, S., Yonemura, S., & Tsukita, S. (1997). Erm (Ezrin/Radixin/Moesin) family—from cytoskeleton to signal transduction. Curr. Opin. Cell Biol. 9, 70-75.
Uchiyama, N., Hasegawa, M., Yamashima, T., Yamashita, J., Shimamura, K., & Takeichi, M. (1994). Immunoelectron microscopic localization of E-cadherin in dorsal root ganglia, dorsal root, and dorsal horn of postnatal mice. J. Neurocytol. 23, 460-468.
Uemura, T., Oda, H., Kraut, R., Hayashi, S., Kotaoka, Y., & Takeichi, M. (1996). Zygotic *Drosophila* E-cadherin expression is required for processes of dynamic epithelial cell rearrangement in the *Drosophila* embryo. Genes Dev. 10, 659-671.
Urushihara, H., Ozaki, H. S., & Takeichi, M. (1979). Immunological detection of cell surface components related with aggregation of Chinese hamster and chick embryonic cells. Dev. Biol. 70, 206-216.
Vandewetering, M., Cavallo, R., Dooijes, D., Vanbeest, M., Vanes, J., Loureiro, J., Ypma, A., Hursh, D., Jones, T., Bejsovec, A., Peifer, M., Mortin, M., & Clevers, H. (1997). Armadillo coactivates transcription driven by the product of the *Drosophila* segment polarity gene Dtcf. Cell. 88, 789-799.
Vestal, D. J. & Ranscht, B. (1992). Glycosyl phosphatidylinositol—anchored T-cadherin mediates calcium-dependent, homophilic cell adhesion. J. Cell Biol. 119, 451-461.
Vleminckx, K., Vakaet, L., Jr., Mareel, M., Fiers, W., & van Roy, F. (1991). Genetic manipulation of E-cadherin expression by epithelial tumor cells reveals an invasion suppressor role. Cell. 66, 107-119.
Volberg, T., Geiger, B., Dror, R., & Zick, Y. (1991). Modulation of intercellular adherens-type junctions and tyrosine phosphorylation of their components in RSV-transformed cultured chick lens cells. Cell Regulation. 2, 105-120.

Volberg, T., Zick, Y., Dror, R., Sabanay, I., Gilon, C., Levitzki, A., & Geiger, B. (1992). The effect of tyrosine-specific protein phosphorylation on the assembly of adherens-type junctions. EMBO J. 11, 1733-1742.

Volk, T., Cohen, O., & Geiger, B. (1987). Formation of heterotypic adherens-type junctions between L-CAM-containing liver cells and A-CAM-containing lens cells. Cell. 50, 987-994.

Weis, W. I. (1995). Cadherin structure: A revealing zipper. Structure 3, 425-427.

Whitehead, I., Kirk, H., & Kay, R. (1995). Expression cloning of oncogenes by retroviral transfer of cDNA libraries. Mol. Cell. Biol. 15, 704-710.

Williams, C. L., Hayes, V. Y., Hummel, A. M., Tarara, J. E., & Halsey, T. J. (1993). Regulation of E-cadherin mediated adhesion by muscarinic acetylcholine receptors in small cell lung carcinoma. J. Cell Biol. 121, 643-654.

Winkel, G. K., Ferguson, J. E., Takeichi, M., & Nuccitelli, R. (1990). Activation of protein kinase C triggers premature compaction in the four-cell stage mouse embryo. Dev. Biol. 138, 1-15.

Witcher, L. L., Collins, R., Puttagunta, S., Mechanic, S. E., Munson, M., Gumbiner, B., & Cowin, P. (1996). Desmosomal cadherin binding domains of plakoglobin. J. Biol. Chem. 271, 10904-10909.

Yamada, K. M. & Geiger, B. (1997). Molecular interactions in cell adhesion complexes. Curr. Opin. Cell Biol. 9, 76-85.

Yoshida-Noro, C., Suzuki, N., & Takeichi, M. (1984). Molecular nature of the calcium-dependent cell-cell adhesion system in mouse teratocarcinoma and embryonic cells studied with a monoclonal antibody. Dev. Biol. 101, 19-27.

THE SELECTINS AND THEIR LIGANDS
ADHESION MOLECULES OF THE VASCULATURE

Thomas F. Tedder, Xuan Li, and Douglas A. Steeber

Advances in Molecular and Cell Biology
Volume 28, pages 65-111.

ISBN: 0-7623-0495-2

I. INTRODUCTION

At sites of inflammation, leukocyte recruitment from the circulation into tissues requires leukocytes to interact with vascular endothelial cells. Leukocyte–endothelial cell interactions are mediated in part by cell-surface adhesion molecules including members of the selectin, integrin and immunoglobulin families. Through cooperative interactions, these receptors mediate leukocyte capture from the flowing blood, rolling, firm adhesion, and subsequent extravasation as outlined in Figure 1 (Ley and Tedder, 1995). The selectin family of adhesion molecules predominantly mediates the initial attachment of leukocytes to endothelial cells, which allows leukocytes to roll along the venular wall (Tedder et al., 1995b). Leukocyte rolling involves a complex series of interactions between multiple families of adhesion molecules (Steeber et al., 1998). During their interactions with endothelial cells, leukocytes encounter chemoattractants (Ben-Baruch et al., 1995). Signal transduction through chemoattractant receptors results in increased binding activity for L-selectin (Spertini et al., 1991b) and β_1- β_2-integrins that stabilize leukocyte interactions with endothelial cells (Butcher, 1993; Springer, 1995). Subsequently, leukocyte integrins such as LFA-1 (CDi1a/CD18), Mo-1/Mac-1 (CD11b/CD18), and VLA-4 (CD49d/CD29) arrest rolling and mediate firm adhesion between leukocytes and vascular endothelium. Leukocyte integrins interact with immunoglobulin superfamily members expressed by vascular endothelial cells such as Intercellular Adhesion Molecule-1 (ICAM-1, CD54), Vascular Cell Adhesion Molecule-1 (VCAM-1, CD106), and ICAM-2 (CD102). Firm adhesion precedes diapedesis of leukocytes between endothelial cells.

The selectin family consists of three closely related cell-surface molecules: L-selectin (MEL-14, LAM-1, CD62L), E-selectin (ELAM-1, CD62E), and P-selectin (PADGEM, GMP140, CD62P) (Tedder et al., 1995a; Tedder et al., 1995b). The selectins provide precise control of leukocyte interactions with vascular endothelium as either their expression or function is completely restricted to the vasculature. L-selectin expressed by leukocytes, binds to constitutively expressed ligands on the surface of high endothelial venules (HEV) of peripheral lymphoid tissues and to inducible ligands on endothelium at sites of inflammation. E-selectin, expressed transiently on cytokine-activated endothelial cells, binds to ligands on myeloid cells

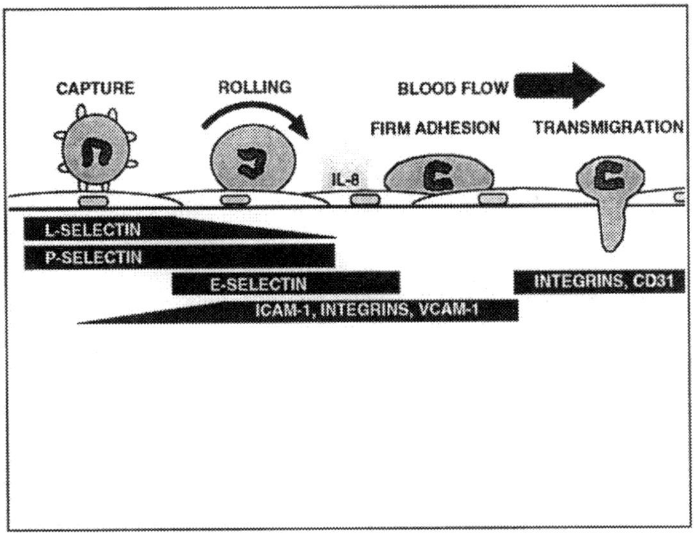

Figure 1. Involvement of the selectins in leukocyte capture, rolling, and firm adhesion. P- and L-selectin can mediate the initial capture of leukocytes from the flowing blood. PSGL-1 and L-selectin are concentrated on the microvilli projections of unstimulated neutrophils, which may be advantageous for early cell contacts. Subsequent rolling is optimally mediated by L-selectin in synergy with either P- or E-selectin function. E-selectin–mediated interactions facilitate firm adhesion by greatly decreasing the rolling velocity of leukocytes. Leukocyte integrins and their ligands also facilitate capture and/or rolling and are required for firm adhesion that proceeds diapedesis between endothelial cells.

and subsets of lymphocytes. P-selectin, stored in the membranes of secretory granules of platelets and endothelial cells, is rapidly redistributed to the cell surface where it binds to ligands on myeloid cells and subsets of lymphocytes. The differential expression and function of the selectins by various leukocyte subclasses and diverse beds of vascular endothelium allows for considerable specificity in leukocyte migration into tissues at sites of inflammation, wound healing, and immune responses.

II. SELECTIN STRUCTURE

Multiple factors influence adhesion molecule function during leukocyte-endothelial cell interactions, including the density and cell surface topology of adhesion molecule expression, molecular association and disassociation rates, and the lengths and lateral mobilities of the adhesion molecules. In these respects, the selectins are optimally engineered to carry out their specific functions. The selectins have a unique and characteristic extracellular region composed of an amino-terminal calcium-dependent lectin domain, an epidermal growth factor (EGF)-like domain, and two to nine short consensus repeat (SCR) units homologous to domains found in complement binding proteins (Figure 2). L-selectin contains two SCR domains (Lasky et al., 1989; Siegelman et al., 1989; Tedder et al., 1989), while E- and P-selectin contain six and nine SCR domains, respectively, although this depends on the species of origin (Bevilacqua et al., 1989; Johnston et al., 1989a). The selectins are the only known proteins in which these three domains are found in immediate juxtaposition.

Human L-, E-, and P-selectin are closely related in amino acid sequence, ranging from ~65% identity in the lectin and EGF domains to ~40% identity in the SCR domains. This high degree of sequence conservation between the lectin domains is consistent with these domains mediating essential interactions with similar, if not identical, carbohydrate determinants displayed on diverse ligands (Kansas et al., 1991; Erbe et al., 1992). By contrast, the cytoplasmic domains are unique to each selectin. During recent mammalian evolution, each selectin has been well conserved with human, mouse, rat, and bovine receptors sharing considerable amino acid sequence identity. Each structural unit is also encoded by individual exons which are organized identically within the selectin genes (Johnston et al., 1990; Ord

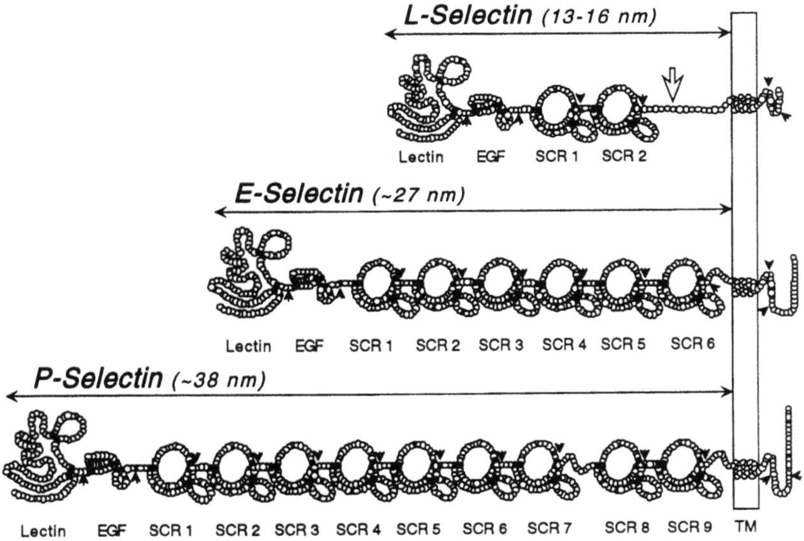

Figure 2. Human selectin structure. The lectin, EGF, SCR, and transmembrane (TM) domains are shown. Circles represent amino acids, with filled circles indicating conserved cysteine residues. Filled arrows indicate exon boundaries for the human and mouse receptors and the open arrow indicates the primary site for endoproteolytic cleavage of L-selectin from the cell surface.

et al., 1990; Collins et al., 1991). Further evidence for a close evolutionary relationship between the selectins includes the tandem clustering of the selectin genes along a 300 kb fragment of chromosome 1 in mice and humans (Tedder et al., 1989; Watson et al., 1990)

A. Molecular Features

L-selectin is heavily glycosylated and when expressed by human lymphocytes, has an ~74,000 M_r while neutrophil L-selectin is a more heterogeneous protein of 90-110,000 M_r (Griffin et al., 1990; Tedder et al., 1990a). Size variations result from glycosylation differences rather than changes in the protein core (Ord et al., 1990). The short cytoplasmic domain of L-selectin is highly conserved among species, particularly numerous basic residues surrounding two conserved serine residues which are sites of phosphorylation (Haribabu et al., 1997).

P-selectin is a 140,000 M_r protein in humans (Johnston et al., 1989b). Almost one-third of its apparent mass on SDS-PAGE analysis represents complex N-linked oligosaccharides. P-selectin is a rigid rod-shaped molecule of about 38 nm with a globular end that is likely to correspond to the lectin and EGF domains (Ushiyama et al., 1993). Length appears important for P-selectin function since its lectin domain must extend a sufficient length from the plasma membrane to mediate optimal rolling of neutrophils (Patel et al., 1995b). Three forms of P-selectin are predicted, two differ in numbers of SCR domains, while a third form lacks a transmembrane domain (Johnston et al., 1989a). Human platelets contain approximately equal amounts of mRNA encoding cell surface and soluble P-selectin (Johnston et al., 1990). P-selectin secreted by activated platelets is functionally active and is found in normal plasma at ~213 ng/ml (Dunlop et al., 1992). Soluble P-selectin may serve an anti-inflammatory function (Gearing and Newman, 1994) since it can prevent the adhesion of neutrophils to endothelium (Gamble et al., 1990) and inhibit superoxide release by neutrophils (Wong et al., 1991).

Human cytokine-activated endothelial cells express E-selectin as a 115,000 M_r glycoprotein, although a minor 97,000 isoform is also expressed (Bevilacqua et al., 1987). E-selectin is also a rigid, asymmetric molecule that extends about 27 nm from the cell surface (Hensley et al., 1994).

B. Functional Domains

The lectin domain is the dominant factor in ligand binding, while the EGF and SCR domains of the selectins modify lectin domain activity and/or ligand specificity. In fact, most monoclonal antibodies (mAbs) that block selectin function bind epitopes localized within the lectin domain (Bowen et al., 1990; Kansas et al., 1991; Spertini et al., 1991c; Tu et al., 1996; Steeber et al., 1997). Some adhesion-blocking mAbs define epitopes within the EGF domain of L-selectin (Siegelman et al., 1990; Kansas et al., 1991; Spertini et al., 1991c), but not within SCR domains. However, most mAbs that identify EGF domain epitopes also require the presence of either the appropriate lectin or SCR domains (Tu et al., 1996; Steeber et al., 1997). Thus, the EGF domain closely associates with its two adjacent domains such that most anti-EGF mAbs bind conformational determinants that are generated by combinations of domains rather than the EGF domain alone. A mAb that blocks selectin function has been reported to bind L- and E-selectin SCR domains (Jutila et al., 1992), but this mAb binds a lectin domain epitope (Steeber, D.A. & Teddler, T.F., unpublished observations). Although the crystal structure for the lectin-EGF domains of human E-selectin suggests that lectin and EGF domains are distinct modules with little apparent association (Drickamer, 1994; Graves et al., 1994), the structure of intact selectins is likely to be more complex.

Functional studies using chimeric selectins also demonstrate that both the lectin and EGF domains are directly involved in cell adhesion (Kansas et al., 1994; Gibson et al., 1995; Tu et al., 1996). In fact, the specific combination of lectin, EGF, and SCR domains is critical for appropriate ligand binding and cell adhesion by each of the selectins. Remarkably, chimeric selectins, which contain both the lectin domain of L-selectin and the EGF domain of P-selectin, have the ligand-binding properties/specificities of both L- and P-selectin (Kansas et al., 1994; Tu et al., 1996). Consistent with this, deletion of either the EGF or SCR domains from a soluble form of L-selectin greatly reduces HEV recognition (Bowen et al., 1990; Watson et al., 1991b). The cooperative action of multiple adhesive domains helps to explain why the selectins have overlapping carbohydrate binding specificities yet also identify functionally distinct ligands in vivo.

Although the levels of homology between selectin SCR domains are lower than for the other extracellular domains, the SCR domains are critical for optimal selectin function. SCR domains may mediate the proper structural presentation of the lectin-EGF domains (Kansas et al., 1991; Ley et al., 1993; Kansas et al., 1994). Extending the lectin and EGF domains of E- and P-selectin from the cell surface could also facilitate relevant ligand interactions which may not be accessible to membrane proximal receptors such as L-selectin, particularly under conditions of shear stress (Li et al., 1994; Patel et al., 1995b). However, the lectin, EGF, and first SCR domains may form the appropriate structural unit necessary for ligand binding since swapping of SCR domains between selectins has major effects on ligand binding activity and specificity (Tu et al., 1996). The SCR and cytoplasmic domains may also contribute indirectly to adhesion by serving as structural elements necessary for receptor oligomerization (Watson et al., 1991b; Li et al., 1998).

The cytoplasmic domain of L-selectin is essential for cell surface receptor function. When expressed in ectopic cells, L-selectin lacking its cytoplasmic domain is functionally inactive and does not mediate leukocyte–endothelial cell interactions (Kansas et al., 1993). This observation may relate to the finding that the affinity of L-selectin for ligand is rapidly upregulated by exposing leukocytes to a variety of pro-inflammatory agents that activate protein kinase C (PKC), including chemoattractants (Spertini et al., 1991b). The increase in ligand binding activity peaks, from within seconds to minutes, and rapidly declines to baseline levels thereafter (Spertini et al., 1991b). Within seconds of chemoattractant receptor activation, L-selectin is phosphorylated on cytoplasmic serine residues through a PKC-dependent pathway (Haribabu et al., 1997). In addition, specific PKC inhibitors completely block the increase in L-selectin ligand binding activity observed after leukocyte activation. Phosphorylation of the L-selectin cytoplasmic domain may therefore be a physiologically relevant mechanism regulating the binding activity of L-selectin. Rapid tyrosine phosphorylation of L-selectin following crosslinking by mAbs has also been reported (Brenner et al., 1996), although the functional significance of this event is unknown.

The cytoplasmic domain of L-selectin interacts constitutively with cytoskeletal proteins via α-actinin and vinculin (Pavalko et al., 1995). By contrast, the cytoplasmic domains of E- and P-selectin do not constitutively interact with α-actinin and are not essential for leukocyte adhesion when expressed by COS cells (Kansas and Pavalko, 1996). However, after leukocyte adhesion or crosslinking, E-selectin does associate with α-actinin and other cytoskeletal proteins including the nonreceptor tyrosine kinase FAK (Yoshida et al., 1996). Moreover, activated human umbilical vein endothelial cells (HUVEC) express E-selectin that is constitutively serine phosphorylated, but time-dependent dephosphorylation is induced following leukocyte adhesion or E-selectin crosslinking (Yoshida et al., 1998) suggesting an outside to inside signaling activity for E-selectin. The cytoplasmic domain of P-selectin is also transiently phosphorylated in a complex pattern on tyrosine, serine, threonine, and histidine residues following platelet activation (Fujimoto and McEver, 1993; Crovello et al., 1995). In addition, the P-selectin cytoplasmic domain directs its sorting into α-granules during synthesis and contains signals involved in receptor recycling from the cell surface (Disdier et al., 1992; Green et al., 1994).

III. SELECTIN EXPRESSION

The patterns of differential selectin expression by leukocytes, platelets and endothelial cells play controlling roles in leukocyte/endothelial cell interactions and in the kinetics of immune and inflammatory responses.

A. L-Selectin Expression

L-selectin is expressed by all classes of leukocytes at most stages of differentiation including most lymphocytes, and some circulating NK cells (Gallatin et al., 1983; Jutila et al., 1989; Kansas and Dailey, 1989; Griffin et al., 1990; Tedder et al., 1990a; Tedder et al., 1990b). Although subpopulations of immature and mature thymocytes express L-selectin, its role in their development is unknown (Tedder et al., 1990b). Interestingly, γδ T cells recently emigrating from the thymus express far more L-selectin than recently emigrating αβ T cells and contain the highest proportion of L-selectin expressing cells, but do not have a propensity for migrating to lymph nodes (Witherden et al., 1994). There is considerable heterogeneity in L-selectin expression among animal species. Although the majority of lymphocytes in tissues express L-selectin with *in situ* staining, most tissue lymphocytes in humans are L-selectin low or negative (Tedder et al., 1990b; Wallace and Beverley, 1993). However, since mouse tissue lymphocytes predominantly express L-selectin (Tang et al., 1998), human lymphocytes may downregulate L-selectin expression due to localized inflammation or other factors. Nonetheless, activation of lymphocytes with either mitogens or phorbol esters (PMA) causes a rapid release of L-selectin from the cell surface with a concomitant increase in expression of other adhesion molecules including CD2, LFA- 1, VLA-4, and LFA-3 (Tedder et al., 1990b). Therefore, L-selectin expression is a dynamic indicator of lymphocyte differentiation, activation, and function.

L-selectin expression is coordinately regulated during B cell development with pre-B cells increasing cell-surface L-selectin levels as they mature (Kansas and Dailey, 1989; Tedder et al., 1993a). Virgin, immunocompetent B cells and the majority of circulating B cells also uniformly express L-selectin (Kansas et al., 1985a; Tedder et al., 1990b; Tang et al., 1998). Regulation of L-selectin expression also occurs in B lymphocytes undergoing antigen-dependent maturation with germinal center cells being uniformly L-selectin negative, whereas cells in the surrounding mantle zone and paracortical areas express L-selectin (Kansas et al., 1985b; Kansas et al., 1989).

L-selectin directs the migration of lymphocytes to peripheral lymph nodes, mesenteric lymph nodes, and Peyer's patches by mediating lymphocyte interactions with HEV (Tedder et al., 1993b; Arbones et al., 1994). The majority of circulating virgin/naive T cells express L-selectin while distinct subpopulations of both $CD4^+$ and $CD8^+$ memory cells lacking L-selectin can be demonstrated in the circulation of humans and mice (Tedder et al., 1990a; Tedder et al., 1990b; Lee and Vitetta. 1991; Tang et al., 1998). Recently activated helper T cells with a memory phenotype generally lack L-selectin expression but reacquire surface receptor expression during further maturation into fully competent helper cells (Tedder et al., 1990a; Steeber et al., 1996b). Thus, both naive and mature memory cells utilize L-selectin to enter lymphoid tissues where effective helper function is provided. This fact has been borne out in studies of L-selectin-deficient mice where immune responses are shifted from the peripheral lymphoid tissues to the spleen (Catalina et al., 1996; Steeber et al., 1996a; Tang et al., 1997). Therefore, L-selectin expression influences the site of lymphocyte migration and whether immune responses can develop in particular tissue environments during antigen encounter.

Levels of L-selectin expression are also critical for regulating subset-specific patterns of lymphocyte migration into secondary lymphoid tissues in vivo. In mice, $CD4^+$ and $CD8^+$ T cells preferentially depended on L-selectin for migration, and thereby migrate in different patterns from B cells (Tang et al., 1998). This results from two-fold higher level of L-selectin expression by T cells, relative to B cells. Although other factors are also likely to regulate lymphocyte subset-specific migration, this establishes an important role for adhesion receptor expression levels. Therefore, while leukocyte-endothelial cell interactions are dictated by the presence or absence of specific adhesion receptors on lymphocytes, a second level of regulation exists whereby expression levels of adhesion receptors dramatically influence the dynamics of leukocyte–endothelial cell interactions.

L-selectin is expressed by most circulating neutrophils, monocytes, eosinophils, and their bone marrow precursors (Griffin et al., 1990; Tedder et al., 1990b; Knol et al., 1994). L-selectin is also expressed by human $CD34^+$ hematopoietic progenitor cells (Saeland et al., 1992; Lund-Johansen and Terstappen, 1993). Myeloid progenitor cells express L-selectin almost continuously throughout myeloid differentiation in the bone marrow (Griffin et al., 1990; Saeland et al., 1992; Lund-Johansen and Terstappen, 1993). Early erythroid progenitor cells (BFU-E) also express L-selectin, but mature erythrocytes do not. L-selectin expression decreases during monocyte differentiation into macrophages as alveolar macrophages express decreased levels of L-selectin compared to circulating monocytes (Prieto et al., 1994). This may reflect a change in cell activation status since monocytes lose L-selectin upon activation (Griffin et al., 1990). Similarly, neutrophil activation by either cytokines or PMA causes a rapid release of L-selectin from the cell surface (Griffin et al., 1990; Tedder et al., 1990b).

B. L-Selectin is Shed from the Cell Surface

L-selectin is rapidly lost from the surface of leukocytes following cellular activation, treatment with PMA, or overnight incubation of lymphocytes at 4°C (Tedder et al., 1985; Jung et al., 1988; Kishimoto et al., 1989; Griffin et al., 1990; Jung and Dailey, 1990; Tedder et al., 1990b; Tedder, 1991). The 69,000 M_r extracellular domain of L-selectin is released from the cell surface by endoproteolytic cleavage (Kishimoto et al., 1990; Spertini et al., 1991a). Release of L-selectin results from activation-induced changes in the tertiary conformation of the L-selectin protein which exposes nascent sites which are susceptible to cleavage by endogenous membrane-bound endoproteases that are ubiquitous in distribution (Spertini et al., 1991a; Chen et al., 1995). Although the endoprotease has a relaxed sequence specificity, it cleaves within the membrane proximal region of L-selectin (Figure 2) at a specific distance from the plasma membrane (Kahn et al., 1994; Chen et al., 1995; Migaki et al., 1995). A cell-surface metalloproteinase is likely to be responsible for L-selectin release (Feehan et al., 1996; Preece et al., 1996). Cross-linking of cell-surface L-selectin by mAb is reported to induce rapid activation-independent shedding of the receptor (Palecanda et al., 1992), but this has not been a generally reproducible observation (Steeber et al., 1997). Although L-selectin endoproteolytic release has been purported to regulate leukocyte rolling velocities (Walcheck et al.,

1996a), the functional significance of L-selectin endoproteolytic release remains controversial (Allport et al., 1997).

Soluble forms of L-selectin (sL-selectin) derived from leukocytes are present at high levels (1.6 ± 0.8 μg/ml) in human plasma (Schleiffenbaum et al., 1992; Spertini et al., 1992b). sL-selectin completely inhibits the attachment of lymphocytes to endothelium at concentrations of 8 to 15 μg/ml. Most likely, L-selectin is constitutively released from the cell-surface of leukocytes at a slow rate with its expression kept constant by continuous synthesis. During bone marrow transplantation, the levels of sL-selectin correlate closely with leukocyte counts and no detectable sL-selectin is present during periods of severe leukopenia (Zetterberg and Richter, 1993). In patients with lymphoid malignancies, sL-selectin levels correlate closely with disease activity and cell counts (Zetterberg and Richter, 1993; Spertini et al., 1994). sL-selectin is therefore principally, if not exclusively, generated by leukocytes. There is also a significant increase in serum concentrations of sL-selectin after coronary angioplasty (Kurz et al., 1994) and during hemodialysis (Rabb et al., 1995). In contrast, significantly decreased levels of sL-selectin have been observed in the serum of patients who progress to develop adult respiratory distress syndrome (Donnelly et al., 1994). In fact, reduced levels of sL-selectin is the best available prognostic indicator for the development of adult respiratory distress syndrome in at-risk patient groups. Soluble L-selectin is also detected in synovial fluid (Humbria et al., 1994). Thus, there are multiple situations where sL-selectin levels may be physiologically relevant. We suspect that sL-selectin normally acts as a buffering system which prevents unwanted leukocyte/endothelial cell interactions during noninflammatory situations.

C. P-Selectin Expression

P-selectin is constitutively found in the membranes of Weibel-Palade bodies within endothelial cells and in α-granules of platelets (Hsu-Lin et al., 1984; Hattori et al., 1989; Larsen et al., 1989; McEver et al., 1989). Within minutes following activation by thrombogenic and inflammatory mediators, granules containing P-selectin are mobilized to the cell surface, fuse with the plasma membrane and release P-selectin to the cell surface (Steinberg et al., 1985; Berman et al., 1986). Inducing agents include thrombin, histamine, complement fragments, oxygen-derived free radicals, and cytokines. In vivo, TNF-α treatment increases expression of P-selectin on endothelial cells (Sanders et al., 1992; Gotsch et al., 1994). Cell-surface expression of P-selectin is generally short-lived (minutes) which makes it an ideal candidate for mediating early leukocyte-endothelial cell interactions. However, in vivo studies reveal P-selectin to be also functionally important at later time points as a cytokine-induced adhesion molecule. Levels of P-selectin mRNA are increased in mice after treatment with cytokines or lipopolysaccharide (LPS), with maximal P-selectin expression at 4 hours following TNF-α stimulation. Nuclear factor (NF)-κB (NF-kB) binding sites have been identified in the promoter region of the P-selectin gene (Pan and McEver, 1994), which may account for its activation-induced transcription.

D. E-Selectin Expression

E-selectin is not normally expressed by resting endothelial cells and is not present in other cell types (Bevilacqua et al., 1987; Luscinskas et al., 1991). E-selectin protein production is strongly and rapidly induced by a variety of inflammatory mediators, including IL-1β, TNF-α, interferon-γ, substance P, and LPS (Bevilacqua et al., 1987). E-selectin is present on the surface of HUVEC 4-6 hours following activation and declines to basal levels by 24–48 hours (Bevilacqua et al., 1987; Bevilacqua et al., 1989). Activation of NF-κB correlates with enhanced E-selectin expression in endothelial cells (Montgomery et al., 1991; Whelan et al., 1991). Upstream regulatory elements of the human E-selectin gene include a cytokine-responsive region composed of three functional binding sites for NF-kB and one element (NF-ELAM-1) constitutively occupied by c-Jun and cAMP-independent activating transcription factor (ATF) family members (ATF-a, ATF-2, and ATF-3) (Whelan et al., 1991; Kaszubska et al., 1993; Schindler and Baichwal, 1994; Whitley et al., 1994; Collins et al., 1995). NF-κB activation alone is insufficient to induce E-selectin gene transcription (Essani et al., 1995), but maximal E-selectin gene transcription is induced following heterodimer formation of ATF members and the NF-κB subunits, p50 (NF-κB1), and p65 (RelA) (van Huijsduijnen et al., 1992; Kaszubska et al., 1993). Although ATF homodimers and ATF/c-Jun heterodimers interact with NF-ELAM-1 constitutively, TNF-α, increases binding of c-Jun-containing complexes (De Luca et al., 1994). NF-κB binding is a target of anti-inflammatory corticosteroids such as dexamethasone which downregulate E-selectin expression and leukocyte binding to activated endothelial cells (Cronstein et al., 1992; Brostjan et al., 1997).

The rapid decline in E-selectin expression by activated HUVEC may signify the importance of preventing inappropriate or prolonged expression of E-selectin after the induction of inflammation. As a potential mechanism, the E-selectin gene undergoes dramatic postinduction transcriptional repression in HUVEC transiently exposed to cytokines (Read et al., 1996). E-selectin is also rapidly endocytosed from the cell surface (von Asmuth et al., 1992; Smeets et al., 1993; Subramaniam et al., 1993) into lysosomal compartments of vesiculotubular shape which may involve interactions with cytoskeletal proteins such as tubulin (Kuijpers et al., 1994) or actin (Yoshida et al., 1996). Decreased E-selectin expression may also result from message instability since E-selectin mRNA contains a number of destabilizing AUUUA sites distributed throughout the 3' untranslated region. However, E-selectin expression in HUVEC may not reflect its temporal and spatial pattern of expression in vivo or in microvascular endothelial cells derived from different tissues. Sustained expression of E-selectin has been observed in endothelial cells isolated from different vascular beds (Keelan et al., 1994; Silber et al., 1994a;

Haraldsen et al., 1996). For example, cultured human dermal, lung, and subcutaneous fat microvascular endothelial cells show relatively persistent expression of E-selectin in response to cytokines through 24 hours of stimulation (Petzelbauer et al., 1993). It has also been proposed that persistent E-selectin expression in chronically inflamed skin results from termination of E-selectin mRNA transcripts at alternative polyadenylation sites, producing mRNAs with fewer destabilizing elements (Chu et al., 1994). However, shorter E-selectin transcripts were not found in microvascular cells in a subsequent study (Kluger et al., 1997), although E-selectin protein endocytosis from the plasma membrane and degradation were found to be slower in microvascular endothelial cells compared with large vessel endothelial cells (Kruger et al., 1997). Therefore, functionally significant differences in E-selectin expression by various vascular beds are likely to exist.

IV. SELECTIN FUNCTION

A. The Selectins Mediate Leukocyte Attachment to Endothelium

Prior to the identification of the selectin family, studies demonstrated that mouse L-selectin, termed the MEL-14 antigen, was critical for lymphocyte migration into some lymphoid tissues (Butcher et al., 1982; Gallatin et al., 1983). In subsequent studies, activated mouse neutrophils that had lost L-selectin expression failed to migrate into inflammatory sites in vivo (Lewinsohn et al., 1987; Jutila et al., 1989). Similarly, the intravenous administration of a function-blocking L-selectin mAb inhibited the accumulation of neutrophils in inflammatory lesions (Lewinsohn et al., 1987). Antibodies that bound E-selectin were also found to inhibit the adhesion of neutrophils to cytokine-activated vascular endothelial cells during in vitro binding assays (Bevilacqua et al., 1987). The simultaneous cloning of cDNA encoding each of the selectins then revealed a structurally related family of molecules with the potential for similar functional activities.

After the structural characterization of the selectins, each was shown to mediate leukocyte or platelet interactions with vascular endothelial cells. P-selectin mediated the adhesion of activated platelets to monocytes and neutrophils, the adhesion of myeloid cells to activated endothelium (Larsen et al., 1989; Geng et al., 1990; Larsen et al., 1990), and the adhesion of neutrophils within thrombi during fibrin deposition (Palabrica et al., 1992). E-selectin expressed by cytokine-activated HUVEC monolayers was found to support enhanced neutrophil transmigration (Luscinskas et al., 1991). E-selectin was also proposed to function as a tissue-specific receptor for T cell subsets (Shimizu et al., 1991) with functional ligands for the endothelial selectins found on 5–20% of circulating $\alpha\beta$ memory T cells (Picker et al., 1993; Luscinskas et al., 1995), a large percentage of $\gamma\delta$ T cells (Picker et al., 1990b; Jutila et al., 1994), natural killer cells (Moore and Thompson, 1992), and activated B cells (Postigo et al., 1994). L-selectin-mediated binding of leukocytes to cytokine-activated vascular endothelial cells was found to be completely dependent on the expression of an induced ligand (Spertini et al., 1991d) with L-selectin function-blocking mAbs inhibiting neutrophil, monocyte, lymphocyte, and eosinophil binding to cytokine-treated endothelial cells when assessed in vitro under conditions which mimic blood flow (Smith et al., 1991; Spertini et al., 1991d; Brady et al., 1992; Spertini et al., 1992a; Knol et al., 1994). L-selectin could mediate these interactions independent of E- and P-selectin since transfection of non-binding cells with the L-selectin cDNA conferred the ability to interact with activated endothelial cells in vitro and in vivo (Brady et al., 1992; Ley et al., 1993). L-selectin was also found to mediate monocyte adhesion to cytokine activated endothelial cells (Spertini et al., 1992a). Therefore, a specialized role for the selectins in mediating leukocyte interactions with endothelial cells was rapidly established.

B. The Selectins Initiate Leukocyte Rolling

Circulating leukocytes interact with activated vascular endothelium by rolling along the vessel wall, a phenomenon that was described more than 150 years ago in amphibian tissues. Studies using intravital microscopy revealed that, even in the absence of an inflammatory stimulus, rolling can occur in venules of the skin and associated tissues, but not internal organs (Janssen et al., 1994; Ley, 1994; Nolte et al., 1994). However, the exteriorization of internal tissues initiates a mild inflammatory response triggered largely by mast cell degranulation (Ley, 1994; Thorlacius et al., 1994; Hou et al., 1995). This results in histamine release, which induces the redistribution of P-selectin to the endothelial cell membrane and subsequent spontaneous leukocyte rolling (Asako et al., 1994; Kubes and Kanwar, 1994; Ley, 1994). Most rolling leukocytes are granulocytes (Atherton and Born, 1972; Fiebig et al., 1991), although histological investigation has established that mononuclear cells can also roll in inflamed venules (Westermann et al., 1994).

Involvement of the selectins in leukocyte rolling was first demonstrated in vivo when rolling was reduced by intravascular infusion of an antibody against L-selectin or an L-selectin-IgG fusion protein (Abbassi et al., 1991; Ley et al., 1991; von Andrian et al., 1991). Subsequently, transfection of cells that did not roll with L-selectin cDNA initiated their rolling in vivo (Ley et al., 1993; von Andrian et al., 1993). L-selectin also mediated monocyte rolling on IL-4 activated endothelial cell monolayers in vitro (Luscinskas et al., 1994) and promoted eosinophil rolling at physiological shear rates in vivo (Sriramarao et al., 1994). P-selectin incorporated into planar lipid bilayers or coated onto plastic also supported neutrophil rolling in flow chamber systems that mimic shear forces encountered by cells in the circulation (Lawrence and Springer, 1991). Endothelial cell stimulation with secretagogues also induced P-selectin dependent leukocyte rolling both in vitro (Jones et al., 1993) and in vivo (Kubes and Kanwar, 1994). E-selectin was found to support neutrophil attachment and rolling under conditions of physiological shear stress when it was immobilized on a surface, expressed by L cells, or displayed by IL-1 stimulated HUVEC (Abbassi et al., 1993; Lawrence and Springer, 1993). These studies clearly demonstrated a major role for the selectins in leukocyte rolling.

In mesenteric venules of mice and rats, surgery-induced inflammation initiates leukocyte rolling within minutes after exteriorization of the tissue, which reaches a peak 20–40 min and remains fairly constant over at least 2 h (Atherton and Born, 1972). Blockage of P-selectin function in this model system substantially reduces rolling (Dore et al., 1993). Nonetheless, rolling is progressively restored after a period of 10–15 minutes, which reflects the participation of L-selectin during later stages of the process (Ley et al., 1993; Ley and Tedder, 1995). Therefore, leukocyte rolling at the earliest time points is mediated by P-selectin with L-selectin serving as a major component of rolling at intermediate and later time points (20–120 min) (Ley et al., 1993; Ley and Tedder, 1995). Blockage of P-selectin function in vivo also reduces constitutive leukocyte rolling in the skin (Nolte et al., 1994). While considerable progress has been made in understanding the molecular basis of leukocyte rolling using in vitro flow chambers and by blocking adhesion molecule function in vivo with antibodies, the generation of adhesion molecule-deficient mice has provided considerable insight into the molecular events which regulate this process.

C. Leukocyte Rolling on Leukocytes

A recent observation under in vitro flow conditions is that neutrophils can roll on adherent neutrophils bound to cytokine-activated endothelial cells. This interaction is mediated via L-selectin on the rolling neutrophils (Bargatze et al., 1994). In addition, bovine $\gamma\delta$ T cells bound to endothelial cell monolayers can support continued rolling of newly arriving $\gamma\delta$ T cells through an L-selectin-dependent process (Jutila and Kurk, 1996). Bovine $\gamma\delta$ T cells can also roll on immobilized platelets (Jutila and Kurk, 1996). Furthermore, L-selectin mAbs can inhibit formyl peptide-induced neutrophil aggregation (Simon et al., 1992), and monolayers of KG1a hematopoietic progenitor cells can support L-selectin-dependent adhesion of human peripheral blood lymphocytes (Oxley and Sackstein, 1994). Therefore, L-selectin interacting with ligands on adherent leukocytes may help amplify leukocyte rolling at sites of inflammation and extravasation.

V. SELECTIN-DEFICIENT MICE

The generation of selectin-deficient mice by gene targeting in embryonic stem cells has provided new insights into the in vivo functions of the selectins and their interactions with other molecules during inflammation (Mayadas et al., 1993; Arbones et al., 1994; Labow et al., 1994). Mice deficient in individual selectins are viable, have no developmental defects, and do not succumb to multifocal infections.

A. L-Selectin Deficient Mice

Lymphocytes from L-selectin-deficient mice are completely blocked from entering resting or antigen-stimulated peripheral lymph nodes across HEVs, which results in a 70–90% reduction in cellularity of this tissue (Arbones et al., 1994; Steeber et al., 1996b). Since naive lymphocytes represent the largest proportion of L-selectin^{+} cells, blocking their entry results in a higher percentage of memory ($CD18^{hi}$ $CD44^{hi}$) lymphocytes in peripheral lymph nodes, but also alters the distribution of these lymphocyte subpopulations within other tissues (Steeber et al., 1996b). Specifically, a 30–55% increase in splenic cellularity results from increases in both naive and memory lymphocytes. Circulating lymphocyte numbers or subpopulations are not altered in young L-selectin-deficient mice, but circulating monocyte numbers are increased nearly three-fold. Older L-selectin-deficient mice have a disproportionate increase of both naive and memory $CD4^{+}$ T cells within the blood and spleen. These results—and the finding that memory lymphocytes in wild-type mice expressed L-selectin—demonstrate a requirement for L-selectin in the regulation of both naive and memory lymphocyte migration (Steeber et al., 1996b). Otherwise, leukocyte numbers and their distribution are generally normal in L-selectin-deficient mice (Arbones et al., 1994).

Overall, the frequency of rolling leukocytes in the exteriorized mesentery of L-selectin-deficient mice is reduced by ~70% (Arbones et al., 1994). L-selectin-deficient mice initially have normal levels of trauma-induced leukocyte rolling, but show a significant decline in rolling during the first hour following exteriorization of the mesentery (Arbones et al., 1994; Ley et al., 1995). Trauma-induced rolling in L-selectin-deficient mice is predominantly P-selectin dependent (Ley et al., 1995), with a rolling velocity identical to that seen in wild-type mice at early time points (Jung et al., 1996). L-selectin-deficient mice also have decreased rolling in cremaster muscle venules after TNF-α treatment (Ley et al., 1995; Kunkel and Ley, 1996).

Consistent with a decrease in leukocyte rolling in L-selectin-deficient mice is a significant reduction in the ability of leukocytes to migrate into inflamed tissues. L-selectin-deficient mice demonstrate decreased leukocyte recruitment into the peritoneal cavity at early (< 4 h) and late (4–72 h) time points after thioglycollate instillation (Arbones et al., 1994; Tedder et al., 1995c). Neutrophil, monocyte, and lymphocyte influx in L-selectin-deficient mice is impaired by 56–72%, 72–78%, 70–75%, respectively, when compared to control mice at 24–48 hours (Tedder et al., 1995c). Similar results have also been obtained with function-blocking L-selectin mAbs or a soluble L-selectin-IgG chimera in normal mice (Watson et al., 1991a; Pizcueta and Luscinskas, 1994). In other models of inflammation, swelling is reduced by 69–90% compared with swelling in control mice undergoing delayed contact hypersensitivity responses (Tedder et al., 1995c). Extreme resistance to LPS-induced septic shock is also a feature of L-selectin-deficient mice (Tedder et al., 1995c). L-selectin is also involved in lymphocyte migration to sites of inflammation in the skin since there is significantly delayed rejection of skin allografts in

L-selectin-deficient mice (Tang et al., 1997). Therefore, L-selectin plays a major role in leukocyte entry into a variety of tissues during inflammation.

Despite the redistribution of lymphocytes from peripheral lymph nodes to the spleen, L-selectin-deficient mice have relatively intact immune responses. In fact, L-selectin-deficient mice have significantly elevated levels of serum IgM and IgG_1, and generate augmented humoral immune responses following intraperitoneal immunizations (Steeber et al., 1996a). By contrast, subcutaneous immunization of L-selectin-deficient mice results in 40% lower serum IgM responses and essentially absent IgG responses on day 7. However, humoral responses are normal by day 14 and secondary responses are higher in L-selectin-deficient mice. Since lymphocytes still enter peripheral lymph nodes through the afferent lymphatics, the cellularity of draining lymph nodes can increase significantly during inflammatory responses (Steeber et al., 1996a). Germinal centers developed rapidly following immunization, although peripheral lymph nodes of L-selectin-deficient mice contain only a few follicles. Germinal centers within peripheral lymph nodes and the spleen are also consistently much larger and more well defined in L-selectin-deficient mice. Cellular immune responses are also relatively normal in L-selectin-deficient mice (Xu et al., 1996; Tang et al., 1997). These results confirm that lymphocyte migration plays an important role in the initiation of immune responses with complementary and overlapping roles for the spleen and other peripheral lymphoid tissues.

B. P-Selectin Deficient Mice

The most outward characteristic of P-selectin-deficient mice is their two- to three-fold higher numbers of circulating neutrophils (Mayadas et al., 1993). Leukocytosis results from a longer half-life of circulating neutrophils (Johnson et al., 1995), suggesting that P-selectin regulates steady state neutrophil numbers in the blood stream by controlling their entry into tissues where senescent neutrophils are cleared. Surprisingly, P-selectin expression is not required for platelet production, survival, or routine hemostasis.

Mice deficient in P-selectin expression have a complete absence of trauma-induced leukocyte rolling immediately following exteriorization of the mesentery (Mayadas et al., 1993) or the cremaster muscle (Ley et al., 1995). At time points beyond an hour, rolling can be seen in P-selectin-deficient mice but the frequency of rolling leukocytes remains very low (10–15% of wild-type values) (Johnson et al., 1995; Ley et al., 1995; Kunkel and Ley, 1996). Trauma-induced rolling in P-selectin-deficient mice is almost exclusively L-selectin dependent (Ley et al., 1995), with an average leukocyte rolling velocity that is three to five times faster than rolling in wild-type mice (Jung et al., 1996). Treatment of P-selectin-deficient animals with thioglycollate several hours prior to exteriorization of the mesentery also leads to leukocyte rolling, but at levels several fold lower than those observed in wild-type mice (Johnson et al., 1995). By contrast, leukocyte rolling is not significantly reduced in the venules of TNF-α-treated P-selectin-deficient mice (Ley et al., 1995). Therefore, P-selectin contributes significantly to rolling during the early phases of inflammatory responses, but is not strictly required during inflammation at later times.

P-selectin-deficient mice manifest a significant reduction in the ability of leukocytes to migrate into the peritoneal cavity subsequent to thioglycollate instillation at time points less than four hours (Mayadas et al., 1993). At later time points (4–48 hours), P-selectin-deficient mice have a greater than 50% reduction in monocyte infiltration, but near normal numbers of neutrophils and lymphocytes in the inflamed peritoneal cavity (Johnson et al., 1995). Delayed contact hypersensitivity responses are not inhibited in mice treated with P-selectin mAb (Labow et al., 1994), while P-selectin-deficient mice have been reported to have either normal (Staite et al., 1996) or reduced (Subramaniam et al., 1995) hypersensitivity responses. Rejection of allogeneic skin grafts is normal in P-selectin-deficient mice (Tang et al., 1997).

C. E-Selectin Deficient Mice

E-selectin-deficient mice are remarkably normal and generate normal inflammatory responses to peritonitis or delayed contact hypersensitivity responses (Labow et al., 1994). These studies contrast with studies in rats where neutrophil recruitment into an inflamed peritoneum is partially blocked by E-selectin neutralizing mAbs (Mulligan et al., 1991). Other studies in primates have suggested E-selectin involvement during cutaneous contact hypersensitivity reactions (Silber et al., 1994b). Although most inflammatory responses are normal in E-selectin-deficient mice, both the number of rolling leukocytes and rolling velocities are increased in cremaster venules of these mice after TNF-α treatment (Kunkel and Ley, 1996). In dermal microvessels, ~60% of circulating leukocytes roll in normal and E-selectin-deficient mice with equivalent increases after TNF-α treatment (Milstone et al., 1998). However, TNF-α treatment results in a dramatic increase in leukocyte stable adhesion to dermal endothelium which is significantly reduced in E-selectin-deficient mice (Milstone et al., 1998). A reduction in the frequency of microvessels supporting higher densities of adherent leukocytes is also observed in mesenteric microvessels of E-selectin-deficient mice. These results suggest that neutrophil influx into sites of inflammation may be subtly influenced by E-selectin expression.

In E-selectin-deficient mice, administration of anti-P-selectin mAbs significantly reduces inflammation (~80%) and neutrophil accumulation during delayed contact hypersensitivity responses while the same P-selectin mAbs have no effect in wild-type mice (Labow et al., 1994). Correspondingly, E-selectin antibodies block cytokine-induced leukocyte rolling in P-selectin-deficient mice, but have little effect in wild-type controls (Kunkel and Ley, 1996). Mice which are deficient in both E- and P-selectin display a virtual absence of leukocyte rolling and low extravasation of leukocytes during peritonitis (Bullard et al., 1996; Frenette et al., 1996).

However, these mice suffer from down-regulated L-selectin expression, chronic ulcerative cutaneous infections, dramatic leukocytosis, and altered hematopoiesis, which complicates the determination of primary or secondary defects in leukocyte interactions with endothelial cells. Nonetheless, these findings suggest that P-selectin and E-selectin are at least partially redundant and can replace each other as mediators of neutrophil rolling at inflammatory sites in vivo. These results further demonstrate that L-selectin function alone is insufficient to mediate neutrophil rolling in vivo in the absence of endothelial selectins.

VI. MOLECULAR BASIS OF LEUKOCYTE ROLLING

Alterations in leukocyte rolling observed in selectin-deficient mice clearly support the concept that each of the selectins have specialized functions during leukocyte interactions with endothelial cells (Ley and Tedder, 1995). Since early rolling is normal in L-selectin-deficient mice, P-selectin expressed alone can capture leukocytes from the flowing blood (see Figure 1). L-selectin can also capture leukocytes as seen in P-selectin-deficient mice, although the initiation of rolling is delayed. L-selectin can mediate leukocyte capture and rolling independent of P- or E-selectin, although cells that do not express P- or E-selectin ligands roll less effectively in mesenteric venules than neutrophils (Ley et al., 1993; von Andrian et al., 1993). Thus, although L- and P-selectin can mediate leukocyte capture, L-selectin alone does not appear sufficient to mediate leukocyte rolling at velocities typical of those seen in vivo. Therefore, interactions between leukocyte L-selectin-mediated and vascular selectin-mediated functions are required for optimal rolling.

A. Synergy Between L-Selectin and the Endothelial Selectins

Selectin-deficient mice provide sufficient information to sort out the overlapping functions of the selectins during leukocyte rolling in vivo. Spontaneous leukocyte rolling shortly after surgical trauma is initially P-selectin-dependent, but shows a prominent L-selectin-dependent component at later time points. After surgical trauma, ~13% of leukocytes in wild-type mice roll along the venular endothelium of the cremaster muscle during the first 20 minutes (Ley et al., 1995). The frequency of rolling leukocytes doubles between 20–60 min and returns to ~13% at 80–120 min. Rolling is initially absent in P-selectin-deficient mice, but reaches ~5% at 80–120 min (Ley et al., 1995). By contrast, rolling in L-selectin-deficient mice is initially similar to wild-type rolling, but declines to 5% by 80–120 min. In both wild-type and L-selectin-deficient mice, initial leukocyte rolling (0–60 minutes) is eliminated by P-selectin mAb treatment, but is unaffected by L-selectin mAb. Conversely, rolling at later time points (60–120 min) is inhibited by L-selectin mAb, but not by P-selectin mAb. Therefore, P-selectin predominantly mediates early trauma-induced rolling while L-selectin interactions with its inducible endothelial ligand (Spertini et al., 1991d) predominantly mediates rolling at later time points.

Leukocyte rolling in cytokine-treated venules is predominantly dependent on L-selectin function (Ley et al., 1995). Treatment of wild-type and P-selectin-deficient mice with TNF-α for 2 h results in rolling of ~25% of leukocytes in cremaster venules, a response which is completely blocked by L-selectin blocking mAbs. Nonetheless, E-selectin expression is also required during rolling to mediate leukocyte-endothelial interactions that result in slow rolling velocities (Kunkel and Ley, 1996). Consistent with this, the velocity of rolling leukocytes in P-selectin-deficient mice with peritonitis is much lower than in mesenteric venules of wild-type mice (Johnson et al., 1995). This is likely to result from cytokine-induced expression of E-selectin which supports leukocyte rolling at a typical velocity almost an order of magnitude lower than P-selectin at the site density prevailing in vivo (Kunkel and Ley, 1996). TNF-α treatment of the cremaster muscle induces P-selectin-dependent rolling of leukocytes in arterioles that also requires E-selectin for rolling at normal velocities (Kunkel et al., 1997). Therefore, E-selectin functions to retard the velocity of rolling leukocytes in vivo which increases the transit time of leukocyte rolling through a venule in inflamed tissue (Ley et al., 1998). Thus, L- and P-selectin serve as the initial anchors which mediate efficient capture of free-flowing leukocytes before they can then roll on P-, L-, and E-selectin in various combinations.

Maximal neutrophil accumulation in a given vascular bed or disease is dependent on L-selectin function in conjunction with the endothelial selectins. This helps explain why inhibiting either L- or P- selectin function has an almost complete blocking effect on inflammatory cell recruitment in some animal models of inflammation (Wine et al., 1993; Mihelcic et al., 1994). Distinct functional activities may also explain why neutrophil recruitment into the peritoneum of L-selectin– (Arbones et al., 1994; Ley et al., 1995; Tedder et al., 1995c) as well as P-selectin–deficient mice (Mayadas et al., 1993; Ley et al., 1995; Subramaniam et al., 1995) is inhibited to a similar level. This also explains why the simultaneous blockage of L-selectin and a vascular selectin is required to completely inhibit inflammation in some model systems. For example, acute lung injury induced by infusion of cobra venom factor in rats depends on both L- and P-selectin (Mulligan et al., 1992; Mulligan et al., 1993). Similarly, IgG-immune complex-mediated disease is dependent on both L- and E-selectin (Mulligan et al., 1991; Mulligan et al., 1993). Also, only a combination of L- and P-selectin mAbs is fully effective at blocking neutrophil recruitment into the peritoneum of wild-type mice (Bosse and Vestweber. 1994). In all cases, inhibitory reagents directed against single selectins give only partial protection. Thus, optimal leukocyte rolling is dependent on leukocyte L-selectin function in conjunction with either of the endothelial selectins to ensure sufficient adhesive interactions to resist shear stress.

B. P-selectin/ICAM-1– and L-selectin/ICAM-1–Deficient Mice

The dogma that only selectins mediate leukocyte rolling has been challenged recently by findings that some integrins support leukocyte rolling under reduced physiologic shear rates (Gaboury and Kubes, 1994). ICAM-1, a ligand for leukocyte LFA-1, is constitutively expressed by endothelial cells and is rapidly up-regulated during inflammation, resulting in increased leukocyte-endothelial cell adhesion (Dustin et al., 1986). Although LFA-1/ICAM-1 interactions do not support leukocyte rolling in vitro (Lawrence et al., 1990; Lawrence et al., 1995), recent studies have shown that under in vitro conditions of low shear flow, α_4,β_1 (CD49d/CD29) and α4β7 integrins, as well as hyaluronan receptors (CD44) expressed by leukocytes, can mediate rolling (Aloe et al., 1995b; Berlin et al., 1995; Clark et al., 1996; DeGrendele et al., 1996). Since integrin-dependent rolling is less efficient and is limited to lower shear rates than selectin-mediated rolling, the in vivo relevance of these observations has remained unknown until recently.

Recent studies in ICAM-1-, P-selectin/ICAM-1- and L-selectin/ICAM-1-deficient mice clearly demonstrate a role for ICAM-1 in leukocyte rolling at sites of inflammation. Leukocytes roll at increased velocities in ICAM-1-deficient mice (Steeber et al., 1998), although the frequency of rolling leukocytes is normal (Kunkel et al., 1996). P-selectin/ICAM-1-deficient mice display a profound decrease in trauma-induced leukocyte rolling that persists much longer than in mice deficient in P-selectin alone (Kunkel et al., 1996). As a result, P-selectin/ICAM-1-deficient mice show an almost complete lack of neutrophil emigration into an inflamed peritoneum at early time points (Bullard et al., 1995). Elimination of L-selectin expression in ICAM-1-deficient mice results in decreased leukocyte rolling in TNF-α treated cremaster venules, beyond what is observed with L-selectin loss alone (Steeber et al., 1998). Furthermore, the phenotype of L-selectin/ICAM-1-deficient mice reveals a physiologically significant role for ICAM-1 in regulating leukocyte rolling velocities mediated by P-selectin (Steeber et al., 1998). Consistent with this, circulating neutrophil, monocyte, and lymphocyte numbers are dramatically increased in L-selectin/ICAM-1-deficient mice. The loss of L-selectin and ICAM-1 also dramatically reduces leukocyte migration into sites of inflammation beyond what is observed with loss of either receptor alone. Thus, members of the selectin and immunoglobulin families function synergistically to mediate optimal leukocyte rolling in vivo, which is essential for the generation of effective inflammatory responses.

A specific role for ICAM-1 during selectin-mediated rolling is clarified by comparing the phenotypes of ICAM-1-deficient mice that also lack expression of either L- or P-selectin. Since the frequency of rolling leukocytes is normal in ICAM-1-deficient mice (Kunkel et al., 1996), selectin expression is sufficient to initiate leukocyte capture from the flowing blood (Ley and Tedder, 1995). Likewise, the findings in L-selectin/ICAM-1-deficient mice indicate that leukocyte rolling through P-selectin does not require ICAM-1 expression (Steeber et al., 1998). However,

ICAM-1 expression is required for P-selectin to mediate optimal rolling and to mediate rolling at its characteristic slow velocities. The absence of rolling in P-selectin/ICAM-1-deficient mice indicates that ICAM-1 expression is required for leukocyte capture and/or rolling through L-selectin (see Figure 1). These conclusions are further reinforced by previous findings that P-selectin function-blocking antibodies inhibit most rolling in ICAM-1-deficient mice (Kunkel et al., 1996) but only moderately inhibit rolling in wild-type mice (Ley et al., 1995). Conversely, blocking L-selectin function significantly reduces the frequency of rolling leukocytes in wild-type mice (Ley et al., 1995) but has modest effects in ICAM-1-deficient mice (Kunkel et al., 1996). Thus, L-selectin can only mediate leukocyte rolling in vivo when either ICAM-1, P-selectin, or both, are expressed, while P-selectin can mediate leukocyte rolling in the absence of L-selectin and ICAM-1 expression, albeit at significantly faster velocities. Thus, synergistic overlap between selectin and integrin functions are essential for optimal leukocyte entry into sites of inflammation.

The requirement for ICAM-1 expression and function for optimal selectin-mediated rolling in vivo may relate to the relative densities of adhesion molecules and their ligands. Under physiologic conditions in vivo, relative receptor and/or ligand densities appropriately displayed on leukocytes or endothelial cells may be limiting for individual receptor/ligand pairs, thereby requiring cooperative interactions between groups of adhesion molecules. Thus, selectin interactions with their ligand(s) may be insufficient to mediate stable leukocyte rolling under shear flow unless leukocyte interactions with vascular endothelial cells are supported by other adhesion molecules such as ICAM-1.

C. Molecular Interactions during Leukocyte Rolling

The selectins are well documented to arrest free flowing leukocytes and to mediate their rolling along the endothelium of blood vessels by the rapid formation and subsequent breakage of molecular bonds (Ley and Gaehtgens, 1991; Hammer and Apte, 1992; Tözeren and Ley, 1992). Cell adhesion under stress also requires selectin-ligand bonds to have high tensile strength (Aloe et al., 1995a). Nonetheless, the selectins also demonstrate individual characteristics. In vitro findings with flow chambers and reconstituted lipid bilayers containing the selectins at comparable site densities suggest that P- or E-selectin support leukocyte rolling at slower velocities than L-selectin (Lawrence and Springer, 1991; Lawrence et al., 1994; Lawrence et al., 1995; Patel et al., 1995a). This is consistent with the recent finding that leukocyte rolling velocities are significantly increased in P-selectin-deficient mice under conditions in which rolling is exclusively L-selectin dependent (Jung et al., 1996). E-selectin-mediated rolling in vivo also appears slow relative to P- and L-selectin-mediated rolling (Kunkel and Ley, 1996; Kunkel et al., 1997). Ultimately, the site density of the selectins and their ligands expressed during vascular inflammation will play a major role in determining which adhesion molecules dominate during leukocyte-endothelial cell interactions and which classes of leukocytes are retained at sites of inflammation (Jung and Ley, 1997).

The concept of a distinct role for L- and P-selectin in the capture of leukocytes from the blood is reinforced by the finding that both L-selectin (Picker et al., 1991; Erlandsen et al., 1993) and the P-selectin Glycoprotein Ligand (PSGL-1) (Moore et al., 1995) are localized at the tips of microvilli on neutrophils, where they are likely to support leukocyte capture during initial contact with the vessel wall under physiologic flow. Localization of L-selectin to the tips of micro-folds on the neutrophil surface is also required for effective attachment in a flow chamber, while the topographical position of L-selectin does not influence rolling velocity or detachment rate once rolling is established (von Andrian et al., 1995). Correct positioning on the cell surface is not sufficient for leukocyte adhesion mediated by L-selectin, as appropriate interactions with the cytoskeleton may also be required (Pavalko et al., 1995). Furthermore, L-selectin redistributes to the leading edge of leukocytes during interactions with endothelial cells, while leukocyte integrins do not (Rosenman et al., 1993). This is likely to be functionally important since leukocytes rolling along a vessel wall under shear stress are subjected to strong hydrodynamic forces and substantial cell deformations during flow (Damiano et al., 1996). These deformations would allow adhesion molecules expressed on the body of the leukocyte, such as integrins, to come into contact with adhesion molecules expressed by endothelial cells. Since selectin interactions with their ligands have rapid association and dissociation rate constants, interactions between integrins and their endothelial ligands may allow the leukocyte to remain tethered to the endothelium as selectin bonds dissociate. Also consistent with this idea is the finding that leukocyte attachment to endothelial cells, through each of the selectins, requires hydrodynamic shear above a threshold value for the promotion and maintenance of rolling during in vitro flow chamber assays and in vivo (Finger et al., 1996; Lawrence et al., 1997). ICAM-1/integrin and other adhesion molecule interactions may thereby stabilize selectin-mediated interactions between leukocytes and endothelial cells in vivo under conditions of both high and low shear, allowing effective rolling and leukocyte entry into sites of inflammation.

VII. SELECTIN LIGANDS

A. Carbohydrate Recognition

The molecular basis of selectin adhesion principally involves carbohydrate recognition (Feizi, 1993; Rosen and Bertozzi, 1994). The selectins bind a variety of carbohydrate structures in vitro, which has made identification of physiological ligands challenging. Lactosaminoglycans and related carbohydrates such as the tetrasaccharide sialyl Lewis^{X} (sLe^{X}) have been identified as prototype ligands for both P- and E-selectin (Phillips et al., 1990; Walz et al., 1990; Polley et al., 1991). Although expression of sLe^X alone is not generally sufficient for L-selectin binding (Sueyoshi et al., 1994), all three selectins can bind sLe^X, sialyl $Lewis^a$, and other carbohydrates under the appropriate conditions (Berg et al., 1991a; Handa et al., 1991; Tyrrell et al., 1991; Berg et al., 1992; Foxall et al., 1992; Imai et al., 1992). In addition, L-selectin adhesion can be specifically blocked by simple or complex carbohydrates (Stoolman et al., 1987; Yednock et al., 1987a; Imai et al., 1990). L- and P-selectin, but not E-selectin, also bind complex sulfated carbohydrates that do not contain sialic acid or fucose residues, for example, heparin, heparin sulfate proteoglycans, fucoidin, dextran sulfate, sulfatides, and the HNK-1-reactive sulfoglucuronyl glycolipids (Aruffo et al., 1991; Needham and Schnaar, 1993; Nelson et al., 1993; Varki, 1994; Giuffe et al., 1997). A polyphosphomannan (PPME) isolated from yeast is recognized specifically by L-selectin (Stoolman et al., 1984; Yednock et al., 1987b). Fucoidin, a sulfated fucose polymer, inhibits L-selectin-mediated lymphocyte attachment to HEV and also blocks leukocyte rolling in vivo (Ley et al., 1992). However, in most cases, the physiological significance of these various putative ligands remains to be demonstrated.

Sialylation is an essential component of the HEV ligand recognized by L-selectin (Imai et al., 1991, 1992). Desialylation of HEV in vivo or in vitro markedly reduces lymphocyte attachment (Rosen et al., 1985, 1989). However, a L-selectin-Ig fusion protein is able to bind endothelial heparin-like chains through a sialic acid-independent mechanism (Norgard-Sumnicht et al., 1993). Modified carbohydrate moieties such as $Le^{x/a}$ derivatives in which the sialic acid moiety is replaced with a SO_3^--group can also serve as higher affinity selectin ligands (Green et al., 1992; Yuen et al., 1992; Brandley et al., 1993; Yuen et al., 1994). In addition, inositol polyanions, simple 6-carbon ring structures that have multiple ester-linked phosphate or sulfate groups, can block L- and P-selectin function (Cecconi et al., 1994). Thus, under some yet undetermined circumstances, sulfation may replace sialylation as a critical determinant of selectin ligands.

Fucosylation of appropriate carbohydrate determinants is also critical for selectin ligands. Lymph node HEV express sLe^X (Paavonen and Renkonen, 1992; Sawada et al., 1993). An essential role for fucose has also been established for the myeloid ligands of both P- and E-selectin (Larsen et al., 1992). The addition of fucose alone to the appropriate sialylated precursor carbohydrate moieties expressed by COS and CHO cells results in the synthesis of sLe^X and generates functional E-selectin ligands (Lowe et al., 1990, 1991a). Cloning of the "ELAM ligand fucosyltransferase" (ELFT, fucosyltransferase-IV or FucT-IV) from the HL-60 promyelocytic cell line suggested that this enzyme regulates the synthesis of E-selectin ligands on leukocytes (Goelz et al., 1990; Meier et al., 1993). However, other fucosyltransferases are more likely to be involved since FucT-IV expressed at very high levels in CHO cells only synthesizes sLe^X at a low efficiency and only generates low avidity E-selectin binding sites (Lowe et al., 1991b; Sasaki et al., 1994). In human T lymphoblasts, FucT-VII is a principal regulator of E-selectin ligand synthesis while both FucT-VII and FucT-IV may direct ligand synthesis in some myeloid cells

(Knibbs et al., 1996). The dominance of FucT-VII in generating selectin ligands is illustrated in FucT-VII-deficient mice (Maly et al., 1996). FucT-VII-deficient mice are characterized by a severe selectin ligand deficiency resulting in blood leukocytosis, impaired leukocyte extravasation during inflammation, and faulty lymphocyte migration. FucT-VII is also likely to play a major role in the generation of L-selectin ligands on HEV since transfection of an endothelial cell line with FucT-VII results in the production of appropriate sLe^X determinants while transfection of the same cell line with other $\alpha(1,3)$ fucosyltransferases fails to generate appropriate sLe^X epitopes (Mitsuoka et al., 1997). The specific localization of FucT-VII expression to HEVs also correlates with expression of L-selectin ligands (Smith et al., 1996). Thus, $\alpha(1,3)$ fucosylated oligosaccharides generated by FucT-VII are likely to be a critical component of the selectin ligands.

Sulfation also appears important for L-selectin ligands expressed by HEV (Imai et al., 1993; Hemmerich et al., 1994a; Hemmerich et al., 1994b; Hemmerich and Rosen, 1994) and the P-selectin ligand, PSGL-1 (Pouyani and Seed, 1995; Sako et al., 1995; Wilkins et al., 1995). For L-selectin ligands, the critical sulfates may be present on modified carbohydrates. Specifically, a subclass of anti-sLe^X mAbs are able to block L-selectin-dependent adhesion of lymphocytes to HEV (Sawada et al., 1993; Mitsuoka et al., 1997). Significantly, this subset of anti-sLe^X mAbs reacts well with a distinct type of sLeX on HEV, while other anti-sLe^X mAbs do not react well with HEV of lymph nodes. These anti-sLe^X mAbs bind 6-sulfo and 6,6'-*bis*-sulfo sLe^X determinants which are found only on glycoproteins of HEV, not in glycolipid fractions. 6-Sulfo sLe^X binds to L-selectin with higher affinity than does sLex or 6'-sulfo Le^X (Asa et al., 1995). Therefore the L-selectin ligands on HEV may be synthesized through the concerted action of FucT-VII and sulfotransferase(s) (Mitsuoka et al., 1997).

L-selectin HEV ligands are also identified by the MECA-79 mAb which identifies a specific sulfation-dependent epitope on some carbohydrate side chains (Streeter et al., 1988; Berg et al., 1991b; Hemmerich et al., 1994b; Smith et al., 1996). The carbohydrates identified by the MECA-79 mAb are also found on human venular endothelium and are expressed at sites of chronic inflammation (Hanninen et al., 1993; Michie et al., 1993). Despite the association between MECA-79 epitope expression and L-selectin ligands on HEV, this determinant is expressed at normal levels on HEV of FucT-VII-deficient mice that do not support L-selectin binding (Maly et al., 1996). While it is unknown whether sulfation is an absolute requirement for L-selectin ligand binding, sulfation of amino-terminal tyrosine residues of PSGL-1 is critical for P-selectin binding (Pouyani and Seed, 1995; Sako et al.. 1995; Wilkins et al., 1995).

B. Mucins as Selectin Ligands

Although the selectins bind weakly with small, sialylated, fucosylated oligosaccharides, cell-surface expression of sLe^X alone does not necessarily result in cell adhesion. The selectins appear to bind with a higher avidity to carbohydrate determinants displayed in the proper context on a limited number of glycoproteins or proteoglycans (Shimizu and Shaw, 1993; Rosen and Bertozzi, 1994). Thus far, the selectins appear to bind preferentially to members of the sialomucin family of cell-surface molecules which serve as specific scaffolds for the display of oligosaccharides recognized by the selectin lectin/EGF domains

P-Selectin Glycoprotein Ligand-1

The only direct demonstration of a specific physiologic interaction between a selectin and a defined cell-surface glycoprotein under appropriate shear forces is for P-selectin and PSGL-1 (Moore et al., 1995; Norman et al., 1996). PSGL-1 was identified originally in human neutrophils and HL-60 cells as a P-and E-selectin ligand (Moore et al., 1992; Sako et al., 1993). A mAb specific for PSGL-1 inhibits the majority of P-selectin binding to neutrophils, lymphocytes, and HL-60 cells during both static conditions and over a range of physiologic shear stresses (Moore et al., 1995). All P-selectin-dependent leukocyte rolling in vivo is mediated by PSGL-1 (Norman et al., 1996). In addition, PSGL-1, P-selectin, and L-selectin function cooperatively to mediate optimal attachment of flowing neutrophils to E-selectin *in vitro* (Patel et al., 1995a). The heat-stable antigen (mouse CD24) also supports myeloid cell binding to endothelial and platelet P-selectin (Aigner et al., 1995), although the physiological significance of this interaction is yet to be demonstrated. Therefore, it appears that PSGL-1 is the dominant physiologic ligand for P-selectin.

PSGL-1 is expressed as a homodimer of disulfide-linked 120,000 M_r subunits (Moore et al., 1992). PSGL-1 expressed by human neutrophils displays *N*-linked glycans and numerous sialylated *O*-linked glycans, including *O*-linked polylactosamine determinants that carry sLe^X (Zhou et al., 1991; Moore et al., 1994) and the HECA-452 (CLA) determinant (Fuhlbrigge et al., 1997). *O*-linked glycans are required for P-selectin recognition, whereas the *N*-linked glycans are not required. Molecular cloning of the PSGL-1 cDNA and gene predicts a 402–412 residue type I membrane protein with a Ser/Thr-rich extracellular domain that contains 3 potential N-linked glycosylation sites and a single Cys residue that promotes dimerization (Sako et al., 1993; Veldman et al., 1995). Several of the structural features that define the P-selectin-binding region on PSGL-1 have been recently elucidated, particularly the presence of several amino-terminal sulfated tyrosine residues that are required for Ca^{2+}-dependent binding by P-selectin (Pouyani and Seed, 1995; Sako et al., 1995; Wilkins et al., 1995). P-selectin binding through the amino-terminal region of PSGL-1 is blocked by a specific anti-PSGL-1 mAb (PL1) and by O-sialoglycoprotein endopeptidase (OSGE)-mediated cleavage of PSGL-1 from the cell surface. Therefore, PSGL-1 functions as a ligand for P-selectin because it appropriately displays specific oligosaccharides in addition to its sulfated tyrosine residues. Whether specific protein-protein interactions are also critical for P-selectin/PSGL-1 interactions remains an area for speculation.

PSGL-1 expression is primarily restricted to cells of hematopoietic origin in normal human tissues. PSGL-1 is broadly expressed by immature and mature myeloid cells, eosinophils, thymocytes, lymphocytes, NK cells, follicular dendritic cells, and non-follicular circulating dendritic cells (Laszik et al., 1996). Immunofluorescence staining of blood leukocytes with anti-PSGL-1 mAbs indicates that all neutrophils, monocytes, and lymphocytes express PSGL-1 (Moore et al., 1995; Laszik et al., 1996). However, only a small fraction (~10–20%) of lymphocytes bind P-selectin with high avidity since specific glycosylation is required for ligand function (Moore and Thompson, 1992; Moore et al., 1995; Vachino et al., 1995). Thus, most blood lymphocytes do not express PSGL-1 in a form that can bind P-selectin. Nonetheless, P-selectin binding to lymphocytes is abolished by the PL1 mAb, indicating that PSGL-1 is the predominant P-selectin ligand on these cells (Moore et al., 1995). The functional heterogeneity of PSGL-1 on lymphocytes may be due to differential expression of glycosyltransferases and/or sulfotransferases that direct the post-translational modifications required for P-selectin recognition. PSGL-1 is also detected on the epithelial lining of the fallopian tube and on microvascular endothelium in some pathologic tissues, but not on most other non-hematopoietic tissues (Laszik et al., 1996).

L-Selectin Ligands

In vitro, mouse L-selectin binds to at least four different heavily glycosylated mucin-like proteins expressed by HEV: GlyCAM-1 (Imai et al., 1991; Lasky et al., 1992), CD34 (Imai et al., 1991; Baumhueter et al., 1993), MAdCAM-1 (Berg et al., 1993; Briskin et al., 1993), and a 200,000–M_r HEV ligand (sgp200) (Hemmerich et al., 1994b; Hoke et al., 1995). L-selectin also binds PSGL-1 expressed by leukocytes (Guyer et al., 1996; Tu et al., 1996; Walcheck et al., 1996b). Each of these molecules bear sulfated, sialylated and fucosylated *O*-linked carbohydrate side chains which appear to be essential for L-selectin binding activity (Imai et al., 1991; Imai et al., 1993; Imai and Rosen, 1993). Although GlyCAM-1 is constitutively expressed by peripheral lymph node HEV (Lasky et al., 1992; Dowgenko et al., 1993), it lacks a transmembrane domain and is likely to be a secretory product found primarily in serum (Kikita and Rosen, 1994). While GlyCAM-1 is expressed specifically by the endothelial cells of peripheral and mesenteric lymph nodes and in an unknown site in the lung, this protein is also expressed during lactation by mammary epithelial cells with the protein released into milk (Tiemeyer et al., 1991). GlyCAM-1 is released by endothelial cells into the circulation at high levels (1-2 μg/ml) (Brustein et al., 1992). Like GlyCAM-1, the sgp200 HEV ligand for L-selectin also bears the MECA-79 sulfate-dependent carbohydrate epitope (Hemmerich et al., 1994b). Sgp200 is endothelial cell-associated, but is also a secreted protein like GlyCAM-1 (Hoke et al., 1995). Therefore, GlyCAM-1 and sgp200 may also competitively inhibit, rather than promote L-selectin-mediated attachment to HEV

CD34 isolated from murine lymph nodes also binds L-selectin *in vitro* (Imai et al., 1991; Baumhueter et al., 1994). CD34 is a type I transmembrane glycoprotein of 140 amino acids of which, 40% are serines or threonines (Simmons et al., 1992). At least half of the mass of CD34 is predicted to result from *O*-linked carbohydrate chains. Because CD34 is constitutively expressed on most endothelial cells (Fina et al., 1990; Baumhueter et al., 1994) and hematopoietic stem cells, tissue-specific glycosylation is hypothesized to regulate its selectin binding activity. Interestingly, $CD34^+$ hematopoietic stem cells express a ligand for L-selectin (Oxley and Sackstein, 1994), although it is predominantly PSGL-1 (Spertini et al., 1996). Nonetheless, CD34 is reported to be the major L-selectin ligand in human tonsil endothelial venules (Purl et al., 1995). However, despite subtle defects in the hematopoietic progenitor cell compartment, mice lacking CD34 are generally normal (Majuri et al., 1994). Given this, the overall contribution of CD34 to L-selectin function is uncertain.

MAdCAM-1 purified from mesenteric lymph nodes of mice supports rolling of L-selectin transfected cells in vitro suggesting that it may also serve as a L-selectin ligand (Berg et al., 1993). MAdCAM-1 contains a putative mucin-like domain (Briskin et al., 1993) and immunoglobulin-like domains which serve as a ligand for the $\alpha_4\beta_7$ integrin (Berlin et al., 1993). However, the mucin-like recognition region for L-selectin on MAdCAM-1 is likely to reside relatively close to the membrane of intact cells where it may be sequestered from interactions with L-selectin on flowing leukocytes in vivo (Briskin et al., 1993). Nonetheless, it may contribute to L-selectin-mediated adhesion to Peyer's patch HEV in vivo (Bargatze et al., 1995). Murine MAdCAM-1 is expressed in mesenteric lymph nodes, lamina propria of the small and large intestine, and the lactating mammary gland. Human MAdCAM-1 is constitutively expressed on endothelium of venules of intestinal lamina propria with increased expression at inflammatory foci associated with ulcerative colitis and Crohn's disease (Briskin et al., 1997).

PSGL-1 also serves as an L-selectin ligand (Guyer et al., 1996; Spertini et al., 1996; Tu et al., 1996; Walcheck et al., 1996b). L- and P-selectin bind through similar, if not identical, regions of PSGL-1. However, L-selectin appears to be much less efficient than P-selectin in binding PSGL-1 (Tu et al., 1996). The lectin domains of L- and P-selectin are the dominant factors for PSGL-1 binding, although cooperative interactions between the lectin and EGF domains of the selectins also significantly modify ligand binding activity (Tu et al., 1996). Although L-selectin binding to PSGL-1 may contribute to leukocytes rolling on leukocytes, and neutrophil aggregation (Guyer et al., 1996), PSGL-1 does not appear to be required for L-selectin adhesion in some assay systems such as neutrophil-neutrophil interactions in flow (Aloe et al., 1996). Also, a non-mucin-like component of neutrophil adhesion to a L-selectin-IgG fusion protein has been demonstrated under flow conditions (Fuhlbrigge et al., 1996). Therefore, the contribution of L-selectin/PSGL-1 interactions to inflammation are unknown.

A lingering issue is the identity of the vascular endothelial L-selectin ligand expressed at sites of inflammation. A cytokine-inducible ligand for L-selectin has been described for cultured HUVEC and microvascular endothelial cells (Spertini et al., 1991d; Brady et al., 1992; Spertini et al., 1992a). This ligand is not CD34 (Wagers et al., 1997) and HUVEC do not express PSGL-1 or MAdCAM-1, and do not express the MECA-79 antigenic determinant (Tu et al., 1999). However, cultured HUVEC express unique Le^X and sLe^X epitopes on their surface, express sialyltransferases and fucosyltransferases and upregulate fucosyltransferase activity after cytokine activation (Majuri et al.. 1994). Most importantly, HEV and endothelial cells of skin venules express a sLe^X-like epitope defined by the 2H5 mAb (Sawada et al., 1993; Akahori et al., 1997; Mitsuoka et al., 1997). The 2H5-sLe^X determinant is induced on vascular endothelial cells with expression increased as inflammation progresses. Moreover, the 2H5 mAb blocks L-selectin-dependent binding of leukocytes to HEV and inflamed skin venules in vitro and blocks L-selectin-dependent migration of leukocytes to in vivo sites of skin inflammation. Therefore, it is likely that post-translational decoration of constitutively expressed cell-surface protein(s) with 2H5 sLe^X-like determinants represents functional L-selectin ligands. This would agree with enhanced endothelial sLe^X and 2H5-sLe^X expression at sites of inflammation (Turunen et al., 1994; Akahori et al., 1997). Since the 2H5 mAb identifies a 6-sulfo-sLe^X determinant (Mitsuoka et al., 1997), this is likely to be a critical determinant of the vascular L-selectin ligand.

Although there is considerable in vitro evidence that L-selectin interacts with multiple sialomucins, a physiological role for any of these molecules in the mediation of lymphocyte adhesion to endothelium has not yet been proven in vivo. Additional L-selectin ligands may be broadly expressed by a variety of tissues as leukocytes can also bind to the myelin sheaths of central nervous system axons in a L-selectin-dependent manner (Huang et al., 1991; Huang et al., 1994). While L-selectin appears to bind a variety of glycoproteins that express the appropriate carbohydrate determinants under in vitro conditions, it is likely that there are relatively few high affinity ligands which mediate physiological function. Therefore, experiments similar to those carried out with PSGL-1 will be required to establish whether all of the ligands which bind L-selectin *in vitro* mediate leukocyte adhesion in vivo, particularly under physiological shear stresses. Since L-selectin binding to non-HEV vascular endothelium is entirely dependent on the prior activation of the endothelial cells with proinflammatory mediators (Spertini et al., 1991d), the mechanism of ligand induction also needs to be determined.

E-Selectin Ligands

Although E- and P-selectin bind to similar if not identical carbohydrate moieties, glycoprotein ligands unique to P- and E-selectin have been identified. P- and E-selectin bind appropriately glycosylated PSGL-1 on myeloid cells, although each binds to different regions of PSGL-1 (Moore et al., 1992; Norgard et al., 1993; Sako et al., 1993; Moore et al., 1994; Asa et al., 1995). E-selectin also binds a carbohydrate determinant related to sLe^X that is recognized by the HECA-452

(CLA) mAb which is associated with tissue selective homing of T cells to skin at sites of chronic cutaneous inflammation (Picker et al., 1990a; Picker et al., 1993). Lymphocytes express CLA as a broad 260,000 M_r glycoprotein (Rossiter et al., 1994) that is actually PSGL-1 (Fuhlbrigge et al., 1997). E-selectin also binds a 260,000 M_r glycoprotein on bovine $\gamma\delta$ T cells (Walcheck et al., 1993) and a predominant 150,000 M_r glycoprotein and minor 250,000 M_r protein on mouse neutrophils (Levinovitz et al., 1993). The 150,000 M_r glycoprotein, termed the E-selectin ligand (ESL-1), is a variant of a receptor for fibroblast growth factor (Steegmaler et al., 1995). P-selectin has also been reported to selectively bind a predominant 160,000 M_r glycoprotein on mouse myeloid cells (Lenter et al., 1994). Although L-selectin has been proposed as an E-selectin ligand (Picker et al., 1991), an overwhelming number of experimental results in L- and E-selectin-deficient mice suggest that this interaction may not be physiologically relevant. E-selectin is also unique among the selectins since most cells that express FucT-VII bear E-selectin ligands (Wagers et al., 1997). Therefore, E-selectin may be capable of binding appropriate carbohydrate determinants without regard to the molecule bearing the carbohydrate. These observations have lead to the suggestion that E- and P-selectin recognize two categories of glycoprotein ligands—one class being monospecific and the second being common for both endothelial selectins (Lenter et al., 1994).

VIII. CONCLUSIONS

Leukocyte rolling and adhesion to endothelium are dynamic processes that involve multiple adhesion receptors and the active participation of the cells involved. Current data obtained from in vitro and in vivo assays, gene-targeted mice, and antibody blocking studies demonstrate that leukocyte rolling is likely to result from cumulative additive interactions with several molecules contributing in both specialized and redundant ways. Selectin expression is critical for capturing leukocytes from the streaming blood, while the synergistic action of the selectins and some integrins contribute to selectin-mediated rolling. Regulatory mechanisms which control these events include the rapid mobilization of presynthesized P-selectin to the cell surface, increased rates of selectin protein synthesis, induced transcription of the E-selectin gene, changes in cycling of E- and P-selectin from the cell surface to intracellular compartments, rapid shedding of L-selectin from the cell surface, activation-induced changes in L-selectin avidity for ligands, and alterations in cytoskeletal associations. Control of selectin ligand function through induction of their synthesis, differential glycosylation, and release from the cell surface are also important regulatory events. Therefore, although a decade of exciting progress has followed the identification of the selectins, considerably more knowledge is yet to be harvested in our understanding of the molecular basis of inflammation.

ACKNOWLEDGMENTS

This work was supported by grants IL-50985, AI-26872 and CA-54464 from the National Institutes of Health.

REFERENCES

Abbassi, O., Kishimoto, T. K., McIntire, L. V., Anderson, D. C., & Smith, C. W. (1993). E-selectin supports neutrophil rolling in vitro under conditions of flow. J. Clin. Invest. 92, 2719-2730.

Abbassi, O., Lane, C. L., Krater, S., Kishimoto, T. K., Anderson, D. C., McIntire, L. V., & Smith, C. W. (1991). Canine neutrophil margination mediated by lectin adhesion molecule-1 in vitro J. Immunol. 147, 2107-2115.

Aigner, S., Ruppert, M., Hubbe, M., Sammar, M., Sthooger, Z., Butcher, E. C., Vestweber, D., & Altevogt, P. (1995). Heat-stable antigen (mouse CD24) supports myeloid cell binding to endothelial and platelet P-selectin. Int. Immunol. 7, 1557-1565.

Akahori, T., Yuzawa, Y., Nishikawa, K., Tamatani, T., Kannagi, R., Miyasaka, M., Okada, H., Hotta, N., & Matsuo, S. (1997). Role of a sialyl Lewis^{X}-like epitope selectively expressed on vascular endothelial cells in local skin inflammation of the rat. J. Immunol. 158, 5384-5392.

Allport, J. R., Ding, H. T., Ager, A., Steeber, D. A., Tedder, T. F., & Luscinskas, F. W. (1997). L-selectin shedding does not regulate human neutrophil attachment, rolling or transmigration across human vascular endothelium in vitro. J. Immunol. 158, 4365-4372.

Alon, R., Fuhlbrigge, R. C., Finger, E. B., & Springer, T. A. (1996). Interactions through L-selectin between leukocytes and adherent leukocytes nucleate rolling adhesions on selectins and VCAM-1 in shear flow. J. Cell Biol. 135, 849-865.

Alon, R., Hammer, D. A., & Springer, T. A. (1995a). Lifetime of the P-selectin-carbohydrate bond and its response to tensile force in hydrodynamic flow. Nature 374, 539-542.

Alon, R., Kassner, P. D., Carr, M. C., Finger, E. B.. Hemler. M. E., & Springer, T. A. (1995b). The integrin VLA-4 supports tethering and rolling in flow on VCAM-1. J. Cell Biol. 128, 12431253.

Arbones, M. L., Ord, D. C., Ley, K., Radich, H., Maynard-Curry, C., Capon, D J., & Tedder, T. F. (1994). Lymphocyte homing and leukocyte rolling and migration are impaired in L-selectin (CD62L) deficient mice. Immunity 1, 247-260.

Aruffo, A., Kolanus, W., Walz, G., Freedman, P., & Seed, B. (1991). CD62/P-selectin recognition of myeloid and tumor cell sulfitides. Cell 67, 35-44.

Asa, D., Raycroft, L., Ma, L., Aeed, P. A., Kaytes. P. S.. Elhammer, A. P., & Geng, J. G. (1995). The P-selectin glycoprotein ligand functions as a common human leukocyte ligand for P- and E-selectins. J. Biol. Chem. 27O, 11662-11672.

Asako, H., Kurose, I., Wolf, R., DeFrees, S., Zheng, Z.-L., Phillips, M. L., Paulson, J. C., & Granger, D. N. (1994). Role of H1 receptors and P-selectin in histamine-induced leukocyte rolling and adhesion in postcapillary venules. J. Clin. Invest. 93, 1508-1515.

Atherton, A. & Born, G. V. R. (1972). Quantitative investigation of the adhesiveness of circulating polymorphonuclear leukocytes to blood vessel walls. J. Physiol. 222, 447-474.

Bargatze, R. F., Jutila, M. A., & Butcher, E. C. (1995). Distinct roles of L-selectin and integrins $\alpha4\beta7$ and LFA-1 in lymphocyte homing to Peyer's patch-HEV in situ: the multistep hypothesis confirmed and refined. Immunity 3, 99-108.

Bargatze, R. F., Kurk, S., Butcher, E. C., & Jutila, M. A. (1994). Neutrophils roll on adherent neutrophils bound to cytokine-induced endothelial cells via L-selectin on the rolling cells. J. Exp. Med. 180, 1785-1792.

Baumhueter, S., Dybdal, N., Klye, C., & Lasky, L. A. (1994). Global vascular expression of murine CD34, a sialomucin-like endothelial ligand for L-selectin. Blood 84, 2554-2565.

Baumhueter, S., Singer, M. S., Henzel, W., Hemmerich, S., Renz, M., Rosen, S. D., & Lasky, L. A. (1993). Binding of L-selectin to the vascular sialomucin CD34. Science 262, 436-438.

Ben-Baruch, A.. Michiel. D. F., & Oppenheim, J. J. (1995). Signals and receptors involved in recruitment of inflammatory cells. J. Biol. Chem. 270, 11703-11706.

Berg, E. L., Magnani, J., Warnock, R. A., Robinson, M. K., & Butcher, E. C. (1992). Comparison of L-selectin and E-selectin ligand specificities: the L-selectin can bind the E-selectin ligands sialyl Le^X and sialyl Le^a. Biochem. Biophys. Res. Commun. 184, 1048-1055.

Berg, E. L., McEvoy, L. M., Berlin, C., Bargatze, R. F., & Butcher, E. C. (1993). L-selectin-mediated lymphocyte rolling on MAdCAM-1. Nature 366, 695-698

Berg, E. L., Robinson, M. K., Mansson, O., Butcher, E. C., & Magnani, J. L. (1991a). A carbohydrate domain common to both sialyl Le^a and sialyl Le^X is recognized by the endothelial cell leukocyte adhesion molecule ELAM-1. J. Biol. Chem. 266, 14869-14872.

Berg, E. L., Robinson, M. K., Warnock, R. A., & Butcher, E. C. (1991b). The human peripheral lymph node vascular addressin is a ligand for LECAM-1, the peripheral lymph node homing receptor. J. Cell. Biol. 114, 343-349.

Berlin, C., Bargatze, R. F., Campbell, J. J., von Andrian, U. H., Szabo, M. C., Hasslen, S. R., Nelson, R. D., Berg, E. L., Erlandsen, S. L., & Butcher, E. C. (1995). α4 integrins mediate lymphocyte attachment and rolling under physiologic flow. Cell 80, 413-422.

Berlin, C.. Berg, E. L., Briskin, M. J., Andrew, D. P., Kilshaw, P. J., Holzmann, B., Weissman, I. L., Hamann, A., & Butcher, E. C. (1993). α4β7 integrin mediates lymphocyte binding to the mucosal vascular addressin MAdCAM-1. Cell 74, 185-195.

Berman, C. L., Yeo, E. L., Wencel-Drake, J. D., Furie, B. C., Ginsberg, M. H., & Furie, B. (1986). A platelet alpha granule membrane protein that is associated with the plasma membrane after activation. J. Clin. Invest. 78, 130-137.

Bevilacqua, M. P., Pober, J. S., Mendrick, D. L., Cotran, R. S., & Gimbrone Jr., M. A. (1987). Identification of an inducible endothelial-leukocyte adhesion molecule. Proc. Natl. Acad. Sci. USA 84, 9238-9243.

Bevilacqua. M. P. Stengelin, S., Gimbrone Jr., M. A., & Seed, B. (1989). Endothelial leukocyte adhesion molecule 1: an inducible receptor for neutrophils related to complement regulatory proteins and lectins. Science 243, 1 160-1 164.

Bosse, R. & Vestweber, D. (1994). Only simultaneous blocking of the L- and P-selectin completely inhibits neutrophil migration into mouse peritoneum. Eur. J. Immunol. 24, 30193024.

Bowen, B. R., Fennie, C., & Lasky, L. A. (1990). The MEL-14 antibody binds to the lectin domain of the murine peripheral lymph node homing receptor. J. Cell Biol. 110, 147-153.

Brady, H. R., Spertini, O., Jimenez, W., Brenner, B. M., Marsden, P. A., & Tedder, T. F. (1992). Neutrophils, monocytes and lymphocytes bind to cytokine-activated kidney glomerular endothelial cells through L-selectin (LAM-1) in vitro. J. Immunol. 149, 2437-2444.

Brandley, B. K., Kiso, M., Abbas, S., Nikrad, P., Srivasatava, O., Foxall, C., Oda, Y., & Hasegawa, A. (1993). Structure-function studies on selectin carbohydrate ligands. Modifications to fucose, sialic acid and sulphate as a sialic acid replacement. Glycobiology 3, 633-639.

Brenner, B., Gulbins, E., Schlottmann, K., Koppenhoefer, U., Busch, G. L., Walzog, B., Steinhausen, M., Coggeshall, K. M., Linderkamp, O., & Lang, F. (1996). L-selectin activates the Ras pathway via the tyrosine kinase $p56^{lck}$. Proc. Natl. Acad. Sci. USA 93, 15376- 15381.

Briskin, M., Winsor-Hines, D., Shyjan, A., Cochran, N., Bloom, S., Wilson, J., McEvoy, L. M., Butcher, E. C., Kassam, N., Mackay, C. R., Newman, W., & Ringler, D. J. (1997). Human mucosal addressin cell adhesion molecule-1 is preferentially expressed in intestinal tract and associated lymphoid tissue. Am. J. Pathol. 151, 97-110

Briskin, M. J., McEvoy, L. M., & Butcher, E. C. (1993). MAdCAM-1 has homology to immunoglobulin and mucin-like adhesion receptors and to IgA1. Nature 363, 461-464.

Brostjan, C., J., A., Csizmadia, V., Natarajan, G., & Winkler, H. (1997). Glucocorticoids inhibit E-selectin expression by targeting NF-kB and not ATF/c-Jun. J. Immunol. 158, 3836-3844.

Brustein, M., Kraal, G., Mebius, R. E., & Watson, S. R. (1992). Identification of a soluble form of a ligand for the lymphoctye homing receptor. J. Exp. Med. 176, 1415- 1419.

Bullard, D. C., Kunkel, E. J., Kubo, H., Hicks, M. J., Lorenzo. I., Doyle, N. A., Koerschuk, C. M., Ley, K., & Beaudet, A. L. (1996). Infectious susceptibility and severe deficiency of leukocyte rolling and recruitment in E-selectin and P-selectin double mutant mice. J. Exp. Med. 183, 2329-2336.

Bullard, D. C., Qin, L., Lorenzo, I., Quinlin, W. M., Doyle, N. A., Bosse, R., Vestweber, D., Doerschuk, C. M., & Beaudet, A. L. (1995). P-selectin/ICAM-1 double mutant mice: acute emigration of neutrophils into the peritoneum is completely absent but is normal into pulmonary alveoli. J. Clin. Invest. 95, 1782-1788.

Butcher, E. C. (1993). Specificity of leukocyte-endothelial interactions and diapedesis: physiologic and therapeutic implications of an active decision making process. Res. Immunol. 144, 695-698.

Butcher, E. C., Kraal, G., Stevens, S.K ., & Weissman, I. L. (1982). Selective migration of murine lymphocyte and lymphoblast populations and the role of endothelial cell recognition. Adv. Exp. Med. Biol. 149, 199-206.

Catalina, M. D., Carroll, M. C., Arizpe, H., Takashima, A., Estess, P., & Siegelman, M. H. (1996). The route of antigen entry determines the requirement for L-selectin during immune responses. J. Exp. Med. 184, 2341-2351.

Cecconi, O., Nelson, R. M., Roberts, W. G., Hanasaki, K., Mannori. G., Schultz, C., Ulich. T. R., Aruffo, A., & Bevilacqua, M. P. (1994). Inositol polyanions. Noncarbohydrate inhibitors of L- and P-selectin that block inflammation. J. Biol. Chem. 269, 15060-15066.

Chen, A., Engel, P., & Tedder, T. F. (1995). Structural requirements regulate endoproteolytic release of the L-selectin (CD62L) adhesion receptor from the cell surface of leukocytes. J. Exp. Med. 182, 519-530.

Chu, W., Presky, D. H., Swerlick, R. A., & Burns, D. K. (1994). Alternatively processed human E-selectin transcripts linked to chronic expression of E-selectin *in vivo*. J. Immunol. 153, 4179-4189.

Clark, R. A., Alon, R., & Springer, T. A. (1996). CD44 and hyaluronan-dependent rolling interactions of lymphocytes on tonsillar stroma. J. Cell Biol. 134, 1075-1087.

Collins, T., Read, M. A., Neish, A. S., Whitley, M. Z., Thanos, D., & Maniatis, T. (1995). Transcriptional regulation of endothelial cell adhesion molecules: NF-κB and cytokine-inducible enhancers. FASEB J. 9, 899-909.

Collins, T., Williams, A., Johnston, G. I., Kim, J., Eddy, R., Shows, T., Gimbrone Jr., M. A., & Bevilacqua, M. P. (1991). Structure and chromosomal location of the gene for endothelial leukocyte adhesion molecule 1. J Biol. Chem. 266, 2466-2473.

Cronstein, B. N., Kimmel, S. C., Levin, R. I., Martiniuk, F., & Weissmann, G. (1992). A mechanism for the anti-inflammatory effects of corticosterioids: the gulcocorticoid receptor regulates leukocyte adhesion to endothelial cells and expression of ELAM-1 and ICAM-1. Proc. Natl. Acad. Sci. USA 89, 9991-9995.

Crovello, C. S., Furie, B. C., & Furie, B. (1995). Histidine phosphorylation of P-selectin upon stimulation of human platelets: a novel pathway for activation-dependent signal transduction. Cell 82, 279-286.

Damiano, E. R., Westheider, J., Tozeren, A., & Ley, K. (1996). Variation in the velocity, deformation, and adhesion energy density of leukocytes rolling within venules. Circ. Res. 79, 1122-1130.

De Luca, L. G., Johnson, D. R., Whitley, M. Z., Collins, T., & Pober, J. S. (1994) cAMP and TNF competitively regulate transcriptional activation through and nuclear factor binding to the CRE-ATF element of the E-selectin promoter. J. Biol. Chem. 269, 19193-19196.

DeGrendele, H. C., Estess, P., Picker, L. J., & Siegelman, M. H. (1996). CD44 and its ligand hyaluronate mediate rolling under physiologic flow: a novel lymphocyte-endothelial cell primary adhesion pathway. J. Exp. Med. 183, 1119-1130.

Disdier, M., Morrissey, J. H., Fugate, R. D., Bainton, D. F., & McEver, R. P. (1992). Cytoplasmic domain of P-selectin (CD62) contains the signal for sorting into the regulated secretory pathway. Mol. Biol. Cell 3, 309-321.

Donnelly, S. C., Haslett, C., Dransfield, I., Robertson, C. E., Carter, D. C., Grant, I. S., & Tedder, T.F. (1994). Altered levels of soluble L-selectin adhesion receptor in plasma is correlated with development of the adult respiratory distress syndrome (ARDS) in at-risk patient groups. Lancet 344, 215-219.

Dore, M., Korthuis, R. J., Granger, D. N., Entman, M. L., & Smith, C. W. (1993). P-selectin mediates spontaneous leukocyte rolling in vivo. Blood 82, 1308-1316.

Dowgenko, D., Andalibi, A., Young, P. E., Lusis, A. J., & Lasky, L. A. (1993). Structure and chromosomal localization of the murine gene encoding GLYCAM 1. A mucin-like endothelial ligand for L selectin. J. Biol. Chem. 268, 4525-4529.

Drickamer, K. (1994). Three-dimensional view of a selectin cell adhesion molecule. Glycobiology 4, 245-248.

Dunlop, L. C., Skinner, M. P., Bendall, L. J., Favaloro, E. J., Castaldi, P. A., Gorman, J. J., Gamble, J. R., Vadas, M. A., & Berndt, M. C. (1992). Characterization of GMP-140 (P-selectin) as a circulating plasma protein. J. Exp. Med. 175, 1147-1150.

Dustin, M. L., Rothlein, R., Bhan, A. K., Dinarello, C. A., & Springer, T. A. (1986). Induction by IL 1 and interferon-gamma: tissue distribution, biochemistry, and function of a natural adherence molecule (ICAM-1). J. Immunol. 137, 245-253.

Erbe, D. V., Wolitzky, B. A., Presta, L. G., Norton, C. R., Ramos, R. J., Burns, D. K., Rumberger, J. M., Rao, B. N. N., Foxall, C., Brandley, B. K., & Lasky, L. A. (1992). Identification of an E-selectin region critical for carbohydrate recognition and cell adhesion. J. Cell Biol. 119, 215227.

Erlandsen, S. L., Hasslen, S. R., & Nelson, R. D. (1993). Detection and spatial distribution of the beta 2 integrin (Mac-1) and L-selectin (LECAM-1) adherence receptors on human neutrophils by high-resolution field emission SEM. J. Histochem. Cytochem. 41, 327-333.

Essani, N. A., McGuire, G. M., Manning, A. M., & Jaeschke, H. (1995). Differential induction of mRNA for ICAM-1 and selectins in hepatocytes, Kupffer cells, and endothelial cells during endotoxemia. Biochem. Biophys. Res. Commun. 211, 74-82.

Feehan, C., Darlak, K., Kahn, J., Walcheck, B., Spatola, A. F., & Kishimoto, T. K. (1996). Shedding of the lymphocyte L-selectin adhesion molecule is inhibited by a hydroxamic acid-based protease inhibitor. J. Biol. Chem. 271, 7019-7024.

Feizi, T. (1993). Oligosaccharides that mediate mammalian cell-cell adhesion. Curr. Opin. Struct. Biol. 3, 701-710.

Fiebig, E., Ley, K., & Arfors, K.-E. (1991). Rapid leukocyte accumulation by "spontaneous" rolling and adhesion in the exteriorized rabbit mesentery. Int. J. Microcirc. Clin. Exp. 1O, 127144.

Fina, L., Molgaard, H. V., Robertson, D., Bradley, N. J., Monaghan, P., Delia, D., Sutherland, D. R., Baker, M. A., & Greaves, M. F. (1990). Expression of CD34 gene in vascular endothelial cells. Blood 75, 2417-2426.

Finger, E. B., Puri, K. D., Alon, R., Lawrence, M. B., von Andrian, U. H., & Springer, T. A. (1996). Adhesion through L-selectin requires a threshold hydrodynamic shear. Nature 379, 266-269.

Foxall, C., Watson, S. R., Dowbenko, D., Fennie, C., Lasky, L. A., Kiso, M., Hasegawa, A., Asa, D., & Brandley, B. K. (1992). The three members of the selectin receptor family recognize a common carbohydrate epitope, the sialyl lewis^x oligosaccharide. J. Cell Biol. 117, 895-902.

Frenette, P. S., Mayadas, T. N., Rayburn, H., Hynes, R. O., & Wagner, D. D. (1996). Susceptibility to infection and altered hematopoiesis in mice deficient in both P- and E-selectins. Cell 84, 563-574.

Fuhlbrigge, R. C., Alon, R., Puri, K. D., Lowe, J. B., & Springer, T. A. (1996). Sialylated, fucosylated ligands for L-selectin expressed on leukocytes mediate tethering and rolling adhesions in physiologic flow conditions. J. Cell Biol. 135, 837-848.

Fuhlbrigge, R. C., Kieffer, J. D., Armerding, D., & Kupper, T. S. (1997). Cutaneous lymphocyte antigen is a specialized form of PSGL-1 expressed on skin-homing T cells. Nature 389, 978-981.

Fujimoto, T. & McEver, R. P. (1993). The cytoplasmic domain of P-selectin is phosphorylated on serine and threonine residues. Blood 82, 1758-1766.

Gaboury, J. P. & Kubes, P. (1994). Reductions in physiologic shear rates lead to CD11/CD18-dependent, selectin-independent leukocyte rolling in vivo. Blood 83, 345-350.

Gallatin. W. M., Weissman, I. L., & Butcher, E. C. (1983). A cell-surface molecule involved in organ-specific homing of lymphocytes. Nature 304, 30-34.

Gamble, J. R., Skinner, M. P., Berndt, M. C., & Vadas, M. A. (1990). Prevention of activated neutrophil adhesion to endothelium by soluble adhesion protein GMP-140. Science 249, 414-417.

Gearing, A. J. & Newman, W. (1994). Circulating adhesion molecules in disease. Immunol. Today 14, 506-512.

Geng, J. G., Bevilacqua, M. P., Moore, K. L., McIntyre, T. M., Prescott, S. M., Kim, J. M., Bliss, G. A., Zimmerman, G. A., & McEver, R. P. (1990). Rapid neutrophil adhesion to activated endothelium mediated by GMP-140. Nature 343, 757-760.

Gibson, R. M., Kansas, G. S., Tedder, T. F., Furie, B., & Furie, B. C. (1995). Lectin and epidermal growth factor domains of P-selectin at physiological density are the recognition unit for leukocyte binding. Blood 85, 151-158.

Giuffe, L., Cordey, A.-S., Monai, N., Tardy, Y., Schapira, M., & Spertini, O. (1997). Monocyte adhesion to activated aortic endothelium: role of L-selectin and heparan sulfate proteoglycans. J. Cell Biol. 136, 945-956.

Goelz, S. E., Hession, C., Goff, D., Griffiths, B., Tizard, R., Newman, B., Chi-Rosso, G., & Lobb, R. (1990). ELFT: A gene that directs the expression of an ELAM-1 ligand. Cell 63, 1349-1356.

Gotsch, U., Jager, U., Dominis, M., & Vestweber, D. (1994). Expression of P-selectin on endothelial cells is upregulated by LPS and TNF-alpha in vivo. Cell Adhed. Comm. 2, 7-14.

Graves, B. J., Crowther, R. L., Chandran, C., Rumberger, J. M., Li, S., Huang, K.-S., Presky, D. H., Familletti, P. C., Wolitzky, B. A., & Burns, D. K. (1994). Insight into E-selectin/ligand interaction from the crystal structure and mutagenesis of the lec/EGF domains. Nature 367, 532-538.

Green, P. J., Tamatani, T., Watanabe, T., Miyasaka, M., Hasegawa, A., Kiso, M., Yuen, C.-T., Stoll, M. S., & Feizi, T. (1992). High affinity binding of the leukocyte adhesion molecule L-selectin to 3'-sulfated-Le^a and -Le^X oligosaccharides and the predominance of sulphate in this interaction demonstrated by binding studies with a series of lipid-linked oligosaccharides. Biochem. Biophys. Res. Commun. 188, 244-251.

Green, S. A., Setiadi, H., McEver, R. P., & Kelly, R. B. (1994). The cytoplasmic domain of P-selectin contains a sorting determinant that mediates rapid degradation in lysosomes. J. Cell Biol. 124, 435-448.

Griffin, J. D., Spertini, O., Ernst, T. J., Belvin, M. P., Levine, H. B., Kanakura, Y., & Tedder, T. F. (1990). GM-CSF and other cytokines regulate surface expression of the leukocyte adhesion molecule-1 on human neutrophils, monocytes, and their precursors. J. Immunol. 145, 576-584.

Guyer, D. A., Moore, K. L., Lynam, E. B., Schammel, C. M. G., Rogeli, S., McEver, R. P., & Sklar, L. A. (1996). P-selectin glycoprotein ligand-1 (PSGL-1) is a ligand for L-selectin in neutrophil aggregation. Blood 88, 2415-2421.

Hammer, D. A. & Apte, S. M. (1992). Simulation of cell rolling and adhesion on surfaces in shear flow: general results and analysis of selectin-mediated neutrophil adhesion. Biophys. J. 63, 3557.

Handa, K., Nudelman, E. D., Stroud, M. R., Shiozawa, T., & Hakomori, S.-I. (1991). Selectin GMP-140 (CD62; PADGEM) binds to sialosyl-Le^a and sialosyl-Le^X, and sulfated glycans modulate this binding. Biochem. Biophys. Res. Commun. 181, 1223-1230.

Hanninen, A., Taylor, C., Streeter, P. R., Stark, L. S., Sarte, J. M., Shizuru, J. A., Simell, O., & Michie, S. A. (1993). Vascular addressins are induced on islet vessels during insulitis in nonobese diabetic mice and are involved in lymphoid binding to islet endothelium. J. Clin. Invest. 92, 2509-2515.

Haraldsen, G., Kvale, D., Lien, B., Farstad, I. N., & Brandtzaeg, P. (1996). Cytokine-regulated expression of E-selectin, intercellular adhesion molecule-1 (ICAM-1), and vascular cell adhesion molecule-1 (VCAM-1) in human intestinal microvascular endothelial cells. J. Immunol. 156, 2448-2564

Haribabu, B., Steeber, D. A., Ali. H., Richardson, R. M., Snyderman, R., & Tedder, T. F. (1997). Chemoattractant receptor-induced phosphorylation of L-selectin. J. Biol. Chem. 272, 13961-13965.

Hattori, R., Hamilton, K. K., Fugate, R. D., McEver, R. P., & Sims, P. J. (1989). Stimulated secretion of endothelial von Willebrand factor is accompanied by rapid redistribution to the cell surface of the intracellular granule membrane protein GMP-140. J. Biol. Chem. 264, 7768-7771.

Hemmerich, S., Bertozzi, C. R., Leffler, H., & Rosen, S. D. (1994a). Identification of the sulfated monosaccharides of GlyCAM-1, an endothelial-derived ligand for L-selectin. Biochemistry 33. 4820-4829.

Hemmerich, S., Butcher, E. C., & Rosen, S.D. (1994b). Sulfation-dependent recognition of HEV-ligands by L-selectin and MECA 79, an adhesion-blocking mAb. J. Exp. Med. 180, 2219-2226.

Hemmerich, S. & Rosen, S. D. (1994). 6'-Sulfated sialyl Lewis x is a major capping group for GlyCAM-1. Biochemistry 33, 4830-4835.

Hensley, P., McDevitt, P. J., Brooks, I., Trill, J. J., Field, J. A., McNulty, D. E., Connor, J. R., Griswold, D. E., Kumar, N. V., Kopple, K. D., Kalton, B. J., & Johanson, K. (1994). The soluble form of E-selectin is an asymmetric monomer. Expression, purification, and characterization of the recombinant protein. J. Biol. Chem. 269, 23949-23958.

Hoke, D., Mebius, R. E., Dybdal, N., Dowbenko, D., Gribling, P., Kyle, C., Baumhueter, S., & Watson, S. R. (1995). Selective modulation of the expression of L-selectin ligands by an immune response. Current Biol. 5, 670-678.

Hou, S., Hyland, L., Bradley, L. M., Watson, S. R., & Doherty, P. C. (1995). Subverting lymph node trafficking by treatment with the Mel-14 monoclonal antibody to L-selectin does not prevent an effective host response to Sendai Virus. J. Immunol. 155, 252-258.

Hsu-Lin, S.-C., Berman, C. L., Furie, B. C., August, D., & Furie, B. (1984). A platelet membrane protein expressed during platelet activation and secretion. Studies using a monoclonal antibody specific for thrombin-activated platelets. J. Biol. Chem. 259, 9121-9126.

Huang, K., Geoffroy, J. S., Singer, M. S., & Rosen, S. D. (1991). A lymphocyte homing receptor (L-selectin) mediates the in vitro attachment of lymphocytes to myelinated tracts of the central nervous system. J. Clin. Invest. 88, 1778-1783.

Huang, K., Kikuta, A., & Rosen, S. D. (1994). Myelin localization of a central nervous system ligand for L-selectin. J. Neuroimmunol. 52, 133-141.

Humbria, A., Diaz-Gonzalez, F., Campanero, M. R., Arroyo, A.G., Laffon, A., Gonzalez-Amaro, R., & Sanchez-Madrid, F. (1994). Expression of L-selectin, CD43, and CD44 in synovial fluid neutrophils from patients with inflammarory joint diseases. Evidence for a soluble form of L-selectin in synovial fluid. Arthritis Rheum. 37, 342-348.

Imai, Y., Lasky, L. A., & Rosen, S. D. (1992). Further characterization of the interaction between L-selectin and its endothelial ligands. Glycobiology 4, 373-381

Imai, Y., Lasky, L. A., & Rosen, S. D. (1993). Sulphation requirement for GlyCAM-1, an endothelial ligand for L-selectin. Nature 361, 555-557.

Imai, Y. & Rosen, S. D. (1993). Direct demonstration of hererogeneous, sulfated O-linked carbohydrate chains on an endothelial ligand for L-selectin. Glycoconj. J. 10, 34-39.

Imai, Y., Singer, M. S., Fennie, C., Lasky, L. A., & Rosen, S. D. (1991). Identification of a carbohydrate-based endothelial ligand for a lymphocyte homing receptor. J. Cell Biol. 113, 1213-1221.

Imai, Y., True, D. D., Singer, M. S., & Rosen, S. D. (1990). Direct demonstration of the lectin activity of $gp90^{mel}$, a lymphocyte homing receptor. J. Cell Biol. 111, 1225-1232

Janssen, G. H. G. W., Tangelder, G. J., oude Egbrink, M. G. A., & Reneman, R. S. (1994). Spontaneous leukocyte rolling in venules in untraumatized skin of conscious and anesthetized animals. Am. J. Physiol. 267, H1199-1204.

Johnson, R. C., Mayadas, T. N., Frenette, P. S., Mebius, R. E., Subramaniam, M., Lacasce, A., Hynes, R. O., & Wagner, D. D. (1995). Blood cell dynamics in P-selectin-deficient mice. Blood 86, 1106-1114.

Johnston, G. I., Bliss, G. A., Newman, P. J., & McEver, R. P. (1990). Structure of the human gene encoding granule membrane protein-140, a member of the selectin family of adhesion receptors for leukocytes. J. Biol. Chem. 265, 21381-21385.

Johnston, G. I., Cook, R. G., & McEver, R. P. (1989a). Cloning of GMP-140, a granule membrane protein of platelets and endothelium: sequence similarity to proteins involved in cell adhesion and inflammation. Cell 56, 1033-1044.

Johnston, G. I., Kurosky, A., & McEver, R. P. (1989b). Structural and biosynthetic studies of the granule membrane protein, GMP-140, from human platelets and endothelial cells. J. Biol. Chem. 264, 1816-1823.

Jones, D. A., Abbassi, O., V., M. L., McEver, R. P., & Smith, C. W. (1993). P-selectin mediates neutrophil rolling on histamine-stimulated endothelial cells. Biophys. J. 65, 1560-1569.

Jung, T. M. & Dailey, M. O. (1990). Rapid modulation of homing receptors ($gp90^{Mel-14}$) induced by activators of protein kinase C. Receptor shedding due to accelerated proteolytic cleavage at the cell surface. J. Immunol. 144, 3130-3136.

Jung, T. M., Gallatin, W. M., Weissman, I. L., & Dailey, M. O. (1988). Down-regulation of homing receptors after T cell activation. J. Immunol. 141, 4110-4117.

Jung, U., Bullard, D. C., Tedder, T. F., & Ley, K. (1996). Velocity differences between L- and P-selectin-dependent neutrophil rolling in venules of mouse cremaster muscle in vivo. Am. J. Physiol. 271, H2740-H2747.

Jung, U. & Ley, K. (1997). Regulation of E-selectin, P-selectin and ICAM-1 expression in mouse cremaster muscle vasculature. Microcirculation 4, 311-319.

Jutila, M. A., Bargatze, R. F., Kurk, S., Warnock, R. A., Ehsani, N., Watson, S. R., & Walcheck, B. (1994). Cell surface P- and E-selectin support shear-dependent rolling of bovine gamma/delta T cells. J. Immunol. 152, 3917-3928.

Jutila, M. A. & Kurk, S. (1996). Analysis of bovine $\gamma\delta$ T cell interactions with E-, P-, and L-selectin. Characterization of lymphocyte on lymphocyte rolling and the effects of *O*-glycoprotease. J. Immunol. 156, 289-296.

Jutila, M. A., Rott, L., Berg, E. L., & Butcher, E. C. (1989). Function and regulation of the neutrophil MEL-14 antigen in vivo: comparison with LFA-1 and MAC-1. J. Immunol. 143, 3318-3324

Jutila, M. A., Watts, G., Walcheck, B., & Kansas, G. S. (1992). Characterization of a functionally important and evolutionarily well-conserved epitope mapped to the short consensus repeats of E-selectin and L-selectin. J. Exp. Med. 175, 1565-1573.

Kahn, J., Ingraham, R. H., Shirley, F., Magaki, G. I., & Kishimoto, T. K. (1994). Membrane proximal clevage of L-selectin: identification of the cleavage site and a 6-kD transmembrane peptide fragment of L-selectin. J. Cell Biol. 125, 461-470.

Kansas, G. S. & Dailey, M. O. (1989). Expression of adhesion structures during B cell development in man. J. Immunol. 142, 3058-3062.

Kansas, G. S., Ley, K., Munro, J. M., & Tedder, T. F. (1993). Regulation of leukocyte rolling and adhesion to HEV through the cytoplasm domain of L-selectin. J. Exp. Med. 177, 833-838.

Kansas, G. S. & Pavalko, F. M. (1996). The cytoplasmic domain of E- and P-selectin do not constitutively interact with α-actinin and are not essential for leukocyte adhesion. J. Immunol. 157, 321-325.

Kansas, G. S., Saunders, K. B., Ley, K., Zakrzewicz, A., Gibson, R. M., Furie, B. C., Furie, B., & Tedder, T. F. (1994). A role for the epidermal growth factor-like domain of P-selectin in ligand recognition and cell adhesion. J. Cell Biol. 124, 609-618.

Kansas, G. S., Spertini, O., Stoolman, L. M., & Tedder, T. F. (1991). Molecular mapping of functional domains of the leukocyte receptor for endothelium, LAM-1. J. Cell Biol. 114, 351-358.

Kansas, G. S., Wood, G. S., & Dailey, M. O. (1989). A family of cell-surface glycoproteins defined by a putative anti-endothelial cell receptor antibody in man. J. Immunol. 142, 3050-3057.

Kansas, G.S.. Wood. G.S., & Engleman, E.G. (1985a). Maturational and functional diversity of human B lymphocytes delineated with anti-Leu-8. J. Immunol. 134, 3003-3006.

Kansas, G. S., Wood, G. S., Fishwild, D. M., & Engleman, E. G. (1985b). Functional characterization of human T lymphocyte subsets distinguished by monoclonal anti-Leu-8. J. Immunol. 134, 2995-3002

Kaszubska, W., van Huijsduijnen, R. H., Ghersa, P., Deraemyschenk, A. M., Chen, B. P. C., Hai, T., DeLamarter, J. F., & Whelan, J. (1993). cAMP-independent ATF family members interact with NF-κB and function in the activation of the E-selectin promoter in response to cytokines. Mol. Cell. Biol. 13, 7180-7190.

Keelan, E. T., Licence, S. T., Peters, A. M., Binns, R. M., & Haskard, D. O. (1994). Characterization of E-selectin expression in vivo with use of a radiolabeled monoclonal antibody. Am. J. Physiol. 266, H278-H290.

Kikita, A. & Rosen, S. D. (1994). Localization of ligands for L-selectin in mouse peripheral node high endothelial cells by colloidal gold conjugates. Blood 84, 3766-3775.

Kishimoto, T. K., Julita, M. A., Berg, E. L., & Butcher, E. C. (1989). Neutrophil Mac-1 and MEL-14 adhesion proteins inversely regulated by chemotactic factors. Science 245, 1238-1241.

Kishimoto, T. K., Jutila, M. A., & Butcher, E. C. (1990). Identification of a human peripheral lymph node homing receptor: a rapidly down-regulated adhesion molecule. Proc. Natl. Acad. Sci. USA 87, 2244-2248.

Kluger, M. S., Johnson, D. R., & Pober, J. S. (1997). Mechanism of sustained E-selectin expression in cultured human dermal microvascular endothelial cells. J. Immunol. 158, 887-896.

Knibbs, R. N., Craig, R. A., Natsuka, S., Chang, A., Cameron, M., Lowe, J. B., & Stoolman, L. M. (1996). The fucosyltransferase FucT-VII regulates E-selectin ligand synthesis in human T cells. J. Cell Biol. 133, 911-920.

Knol, E. F., Tackey, F., Tedder, T. F., Klunk, D. A., Bickel, C. A., & Bochner, B. S. (1994). Comparison of human eosinophil and neutrophil adhesion to endothelial cells under non-static conditions: the role of L-selectin. J. Immunol. 153, 2161-2167.

Kubes, P. & Kanwar, S. (1994). Histamine induces leukocyte rolling in post-capillary venules: a P-selectin-mediated event. J. Immunol. 152, 3570-3577.

Kuijpers, T. W., Raleigh, M., Kavanagh, T., Janssen, H., Calafat, J., Roos, D., & Harlan, J. M (1994). Cytokine-activated endothelial cells internalize E-selectin into a lysosomal compartment of vesiculotubular shape. J. Immunol. 152, 5060-5069.

Kunkel, E. J., Jung, U., Bullard, D. C., Norman, K. E., Wolitzky, B. A., Vestweber, D., Beaudet, A. L., & Ley, K. (1996). Absence of trauma-induced leukocyte rolling in mice deficient in both P-selectin and Intercellular adhesion molecule-1 (ICAM-1). J. Exp. Med. 183, 57-65.

Kunkel, E. J., Jung, U., & Ley, K. (1997). TNF-α induces selectin-mediated leukocyte rolling in mouse cremaster muscle arterioles. Am. J. Physiol. 272, H1391-H1400.

Kunkel, E. J. & Ley, K. (1996). Distinct phenotype of E-selectin-deficient mice. E-selectin is required for slow leukocyte rolling in vivo. Circ. Res. 79, 1196-1204

Kurz, R. W., Graf, B., Gremmel, F., Wurnig, C., & Stockenhuber, F. (1994). Increased serum concentrations of adhesion molecules after coronary angioplasty. Clin. Sci. 87, 627-633.

Labow, M. A., Norton, C. R., Rumberger, J. M., Lombard-Gillooly, K. M., Shuster, D. J., Hubbard, J., Bertko, R., Knaack, P. A., Terry, R. W., Harbison, M. L., Kontgen, F., Stewart, C. L., McIntyre, K. W., Will, P. C., Burns, D. K., & Wolitzky, B. A. (1994). Characterization of E-selectin-deficient mice: demonstration of overlapping function of the endothelial selectins. Immunity 1, 709-720.

Larsen, E., Celi, A., Gilbert, G. E., Furie, B. C., Erban, J. K., Bonfanti, R., Wagner, D. D., & Furie, B. (1989). PADGEM protein: a receptor that mediates the interaction of activated platelets with neutrophils and monocytes. Cell 59, 305-312.

Larsen, E., Palabrica, T., Sajer, S., Gilbert, G. E., Wagner, D. D., Furie, B. C., & Furie, B. (1990). PADGEM-dependent adhesion of platelets to monocytes and neutrophils is mediated by a lineage-specific carbohydrate, LNF III (CD 15). Cell 63, 467-474.

Larsen, G. R., Sako, D., Ahern, T. J., Shaffer, M.. Erban, J., Sajer, S. A., Gibson, R. M., Wagner, D. D., Furie, B. C., & Furie, B. (1992). P-selectin and E-selectin. Distinct but overlapping leukocyte ligand specificities. J. Biol. Chem. 267, 11104- 11110.

Lasky, L. A., Singer, M. S., Dowbenko, D., Imai, Y., Henzel, W. J., Grimley, C., Fennie, C., Gillett, N., Watson, S. R., & Rosen, S. D. (1992). An endothelial ligand for L-selectin is a novel mucin-like molecule. Cell 69, 927-938.

Lasky, L. A., Singer, M. S., Yednock, T. A., Dowbenko, D., Fennie, C., Rodriguez, H., Nguyen, T., Stachel, S., & Rosen, S. D. (1989). Cloning of a lymphocyte homing receptor reveals a lectin domain. Cell 56, 1045-1055.

Laszik, Z., Jansen, P. J., Cummings, R. D., Tedder, T. F., McEver, R. P., & Moore, K. L. (1996). P-selectin glycoprotein ligand-1 is broadly expressed in cells of myeloid, lymphoid, and dendritic lineage and in some non-hematopietic cells. Blood 88, 3010-3021.

Lawrence, M. B., Bainton, D. F., & Springer, T. A. (1994). Neutrophil tethering to and rolling on E-selectin are separable by requirement for L-selectin. Immunity 1, 137-145.

Lawrence, M. B., Berg, E. L., Butcher, E. C., & Springer, T. A. (1995). Rolling of lymphocytes and neutrophils on peripheral node addressin and subsequent arrest on ICAM-1 in shear flow. Eur. J. Immunol. 25, 1025-1031.

Lawrence, M. B., Kansas, G. S., Kunkel, E. J., & Ley, K. (1997). Threshold levels of fluid shear promote leukocyte adhesion through selectins (CD62L, P, E). J. Cell Biol. 136, 717-727.

Lawrence, M. B., Smith, C. W., Eskin, S. G., & McIntire, L. V. (1990). Effect of venous shear stress on CD18-mediated neutrophil adhesion to cultured endothelium. Blood 75, 227-237.

Lawrence, M. B. & Springer, T. A. (1991). Leukocytes roll on a selectin at physiologic flow rates: distinction from and prerequisite for adhesion through integrins. Cell 65, 859-873.

Lawrence, M. B. & Springer, T. A. (1993). Neutrophils roll on E-selectin. J. Immunol. 151, 6338-6346.

Lee, W. T. & Vitetta, E. S. (1991). The differential expression of homing and adhesion molecules on virgin and memory T cells in the mouse. Cell. Immunol. 132, 215-222.

Lenter, M., Levinovitz, S., Isenmann, S., & Vestweber, D. (1994). Monospecific and common glycoprotein ligands for E- and P-selectin on myeloid cells. J. Cell Biol. 125, 471-481.

Levinovitz, A., Muhlhoff, J., Isenmann, S., & Vestweber, D. (1993). Identification of a glycoprotein ligand for E-selectin on mouse myeloid cells. J. Cell Biol. 121, 449-459.

Lewinsohn, D. M., Bargatze, R. F., & Butcher, E. C. (1987). Leukocyte-endothelial cell recognition: evidence of a common molecular mechanism shared by neutrophils, lymphocytes, and other leukocytes. J. Immunol. 138, 4313-4321.

Ley, K. (1994). Histamine can induce leukocyte rolling in rat mesenteric venules. Am. J. Physiol. 267, H1017-H1023.

Ley, K., Allietta, M., Bullard, D. C., & Morgan, S. (1998). Importance of E-selectin for firm leukocyte adhesion in vivo. Circ. Res. 83, 287-294.

Ley, K. & Gaehtgens, P. (1991). Endothelial, not hemodynamic differences are responsible for preferential leukocyte rolling in rat mesenteric venules. Circ. Res. 69, 1034-1041.

Ley, K., Gachtgens. P., Fennie, C., Singer. M. S., Lasky, L. A., & Rosen, S. D. (1991). Lectin-like cell adhesion molecule 1 mediates leukocyte rolling in mesenteric venules in vivo. Blood 77, 2553-2555.

Ley, K., Linnemann, G., Meinen, M., Stoolman, L. M., & Gaehtgens, P. (1992). Fucoidin, but not yeast polyphosphomannan PPME inhibits leukocyte rolling in venules of the rat mesentery. Blood 81, 177-185.

Ley, K. & Tedder, T. F. (1995). Leukocyte interactions with vascular endothelium: new insights into selectin-mediated attachment and rolling. J. Immunol. 155, 525-528.

Ley, K., Tedder, T. F., & Kansas, G. S. (1993). L-selectin can mediate leukocyte rolling in untreated mesenteric venules in vivo independent of E- or P-selectin. Blood 82, 1632-1638.

Ley, K. E., Bullard, D., Arbones, M. L., Bosse, R., Vestweber, D., Tedder, T. F., & Beaudet, A. L. (1995). Sequential contribution of L- and P-selectin to leukocyte rolling in vivo. J. Exp. Med. 181, 669-675.

Li, X., Steeber, D.A., Tang, M.L.K., Farrar, M.A., Perlmutter, R.M., & Tedder, T.F. (1998). Regulation of L.-selectin-mediated rolling through receptor dimerization. J. Exp. Med. 188, 1385-1390.

Li, S. H., Burns, D. K., Rumberger, J. M., Presky, D. H., Wilkinson, V. L., Anosterio Jr., M., Wolitzky, B. A., Norton, C. R., Familletti, P. C., Kim, K. J., Goldstein, A. L., Cox, D. C., & Huang, K.-S. (1994). Consensus repeat domains of E-selectin enhance ligand binding. J. Biol. Chem. 269, 4431-4437.

Lowe, B., Stoolman, L. M., Nair, R. P., Larsen, R. D., Berhend, T. L., & Marks, R. M. (1991a). A transfected human fucosyltransferase cDNA determines biosynthesis of oligosaccharide ligand(s) for endothelial-leukocyte adhesion molecule 1. Biochem. Soc. Trans. 19, 649-653.

Lowe, B., Stoolman, L. M., Rajan, P. N., Larsen, R. D., Berhend, T. L., & Marks, R. M. (1990). ELAM-1-dependent cell adhesion to vascular endothelium determined by transfected human fucosyltransferase cDNA. Cell 63, 475-484.

Lowe, J. B., Kukowska-Latallo, J., Nair, R. P., Larsen, R. D., Marks, R. M., Macher, B. A., Kelly, R. J., & Ernst, L. K. (1991b). Molecular cloning of a human fucosyltransferase gene that determines expression of the Lewis x and VIM-2 epitopes but not ELAM-1-dependent cell adhesion. J. Biol. Chem. 266, 17467-17477.

Lund-Johansen, F. & Terstappen, L. W. (1993). Differential surface expression of cell adhesion molecules during granulocyte maturation. J. Leukoc. Biol. 54, 47-55.

Luscinskas, F. W., Cybulsky, M. I., Kiely, J. M., Peckins, C. S., Davis, V. M., & Gimbrone Jr., M. A. (1991). Cytokine-activated human endothelial monolayers support enhanced neutrophil transmigration via a mechanism involving both endothelial-leukocyte adhesion molecule-1 and intercellular adhesion molecule-1. J. Immunol. 146, 1617-1625

Luscinskas, F. W., Ding, H., & Lichtman, A. H. (1995). P-selectin and VCAM-1 mediate rolling and arrest, respectively, of $CD4^+$ T-lymphocytes on TNF-α activated vascular endothelium under flow. J. Exp. Med. 181, 1179-1186.

Luscinskas, F. W., Kansas, G.S ., Ding, H., Pizcueta, P., Schleiffenbaum, B., Tedder, T.F., & Gimbrone Jr., M. A. (1994). Monocyte rolling, arrest and spreading on IL-4-activated vascular endothelium under flow is mediated via sequential action of L-selectin, β_1-integrins, and β_2-integrins. J. Cell Biol. 125, 1417-1427.

Majuri, M.-L., Pinola, M., Niemela, R., Tiisala, S., Natunen, J., Renkonen, O., & Renkonen, R. (1994). α2,3-Sialyl and α1,3-fucosyltransferase-dependent synthesis of sialyl Lewis x, an essential oligosaccharide present on L-selectin counterreceptors, in cultured endothelial cells. Eur. J. Immunol. 24, 3205-3210.

Maly, P., Thall, A. D., Petryniak, B., Rogers, C. E., Smith, P. L., Marks, R. M., Kelly, R. J., Gersten, K. M., Cheng, G., Saunders, T. L., Camper, S. A., Camphausen, R. T., Xullivan, F. X., Isogai, Y., Hindsgaul, O., von Andrian, U. H., & Lowe, J. B. (1996). The α(l,3) fucosyltransferase Fuc-TVII controls leukocyte trafficking through an essential role in L-, E-, and P-selectin ligand biosynthesis. Cell 86, 643-653.

Mayadas, T .N., Johnson, R. C., Rayburn, H., Hynes, R. O., & Wagner, D. D. (1993). Leukocyte rolling and extravasation are severely compromised in P selectin-deficient mice. Cell 74, 541-554.

McEver, R. P., Beckstead, J. H., Moore, K. L., Marshal-Carlson, L., & Bainton, D. F. (1989). GMP-140, a platelet alpha granule membrane protein, is also synthesized by vascular endothelial cells and is localized in Weibel-Palade bodies. J. Clin. Invest. 84, 92-99.

Meier, W., Leone, D. R., Miatkowski, K., Lobb, R., & Goelz, S. E. (1993). The ELAM ligand fucosyltransferase, ELFT, directs E-selectin binding to a secreted scaffold protein: a method to produce and purify large quantities of specific carbohydrate structures. Biochem. J. 294, 25-30.

Michie, S. A., Streeter, P. R., Bolt, P. A., Butcher, E. C., & Picker, L. J. (1993). The human peripheral lymph node vascular addressin. An inducible endothelial antigen involved in lymphocyte homing. Am. J. Pathol. 143, 1688-1698.

Migaki, G. I., Kahn, J., & Kishimoto, T. K. (1995). Mutational analysis of the membrane-proximal cleavage site of L-selectin: Relaxed sequence specificity surrounding the clevage site. J. Exp. Med. 182, 549-557.

Mihelcic, D., Schleiffenbaum, B., Tedder, T. F., Harlan, J. M., & Winn, R. K. (1994). Inhibition of leukocyte L-selectin function with a monoclonal antibody attenuates reperfusion injury to the rabbit ear. Blood 84, 2322-2328.

Milstone, D. S., Fukumura, D., Padgett, R. C., O'Donnell, P. E., Davis, V. M., Benavidez, O. J., Monsky, W. L., Melder, R. J., Jain, R. K., & Gimbrone Jr., M. A. (1998). Mice lacking E-selectin show normal numbers of rolling leukocytes but reduced leukocyte stable arrest on cytokine-activated microvascular endothelium. Microcirculation 5, 153-171.

Mitsuoka, C., Kawakami-Kimura, N., Kasugai-Sawada, M., Hiraiwa, N., Toda, K., Ishida, H., Kiso, M., Hasegawa, A., & Kannagi, R. (1997). Sulfated Sialyl Lewis X, the putative L-selectin ligand, detected on endothelial cells of high endothelial venules by a distinct set of anti-sialyl Lewis X antibodies. Bioche. Biophys. Res. Comm. 230, 546-551.

Montgomery, K. F., Osborn, L., Hession, C., Tizard, R., Goff, D., Vassallo, C., Tarr, P. I., Bomsztyk, K., Lobb, R., Harlan, J. M., & Pohlman, T. H. (1991). Activation of ELAM-1 gene transcription. Proc. Natl. Acad. Sci. USA 88, 6523-6527.

Moore, K. L., Eaton, S. F., Lyons, D. E., Lichenstein. H. S.. Cummings, R. D., & McEver, R. P. (1994). The P-selectin glycoprotein ligand from human neutrophils displays sialylated, fucosylated, *O*-linked poly-*N*-acetyllactosamine. J. Biol. Chem. 269, 22318-22327.

Moore, K. L., Patel, K. D., Breuhl, R. E., Li, F., Johnson, D. A., Lichenstein, H. S., Cummings, R. D., Bainton, D. F., & McEver, R. P. (1995). P-selectin glycoprotein ligand-1 mediates rolling of human neutrophils on P-selectin. J. Cell Biol. 128, 661-671.

Moore, K. L., Stults, N. L., Diaz, S., Smith, D. F., Cummings, R. D., Varki, A., & McEver, R. P. (1992). Identification of a specific glycoprotein ligand for P-selectin (CD62) on myeloid cells. J. Cell. Biol. 118, 445-456.

Moore, K. L. & Thompson, L. F. (1992). P-selectin (CD62) binds to subpopulations of human memory T lymphocytes and natural killer cells. Biochem. Biophys. Res. Commun. 186, 173-181.

Mulligan, M. S., Polley, M. J., Bayer, R. J., Nunn, M. F., Paulson, J. C., & Ward, P. A. (1992). Neutrophil dependent acute lung injury. Requirement for P-selectin. J. Clin. Invest. 90, 1600-1607.

Mulligan, M. S., Varani, J., Dame, M. K., Lane, C. L.. Smith, C. W., Anderson, D. C., & Ward, P. A. (1991). Role of endothelial-leukocyte adhesion molecule (ELAM-1) in neutrophil-mediated lung injury in rats. J. Clin. Invest. 88, 1396-1406.

Mulligan, M. S., Watson, S. R., Fennie, C., & Ward, P. A. (1993). Protective effects of selectin chimeras in neutrophil-mediated lung injury. J. Immunol. 151, 6410-6417.

Needham, L. K. & Schnaar, R. L. (1993). The HNK-1 reactive sulfoglucuronyl glycolipids are ligands for L-selectin and P-selectin but not E-selectin. Proc. Natl. Acad. Sci. USA 90, 1359-1363.

Nelson, R. M., Ceccioni, O., Roberts, W. G., Aruffo, A., Linhardt, R. J., & Bevilacqua, M. P (1993). Heparin oligosaccharides bind L- and P-selectin and inhibit acute inflammation. Blood 82, 3253-3258.

Nolte, D., Schmid, P., Jager, U., Botzlar. A., Roesken, F., Hecht, R., Uhl, E., Messmer, K., & Vestweber, D. (1994). Leukocyte rolling in venules of striated muscle and skin is mediated by P-selectin, not by L-selectin. Am. J. Physiol. 267, H1637-H1642.

Norgard, K. E., Moore, K. L., Diaz, S., Stults, N. L., Ushiyama, S., McEver, R. P., Cummings, R. D., & Varki, A. (1993). Characterization of a specific ligand for P-selectin on myeloid cells. A minor glycoprotein with sialylated O-linked oligosaccharides. J. Biol. Chem. 268, 12764-12774.

Norgard-Sumnicht, K. E., Varki, N. M., & Varki, A. (1993). Calcium-dependent heparin-like ligands for L-selectin in nonlymphoid endothelial cells. Science 261, 480-483.

Norman, K. E., Moore, K. L., McEver, R. P., & Ley, K. (1996). Leukocyte rolling in vivo is mediated by P-selectin glycoprotein ligand-1. Blood 86, 4417-4421.

Ord, D. C., Ernst, T. J., Zhou, L. J., Rambaldi, A., Spertini, O., Griffin, J. D., & Tedder, T. F. (1990). Structure of the gene encoding the human leukocyte adhesion molecule-1 (TQ1, Leu-8) of lymphocytes and neutrophils. J. Biol. Chem. 265, 7760-7767.

Oxley, S. M. & Sackstein, R. (1994). Detection of an L-selectin ligand on a hematopoietic progenitor cell line. Blood 84, 3299-3306.

Paavonen. T. & Renkonen, R. (1992). Selective expression of sialyl-Lewis X and Lewis A epitopes, putative ligands for L-selectin, on peripheral lymph-node high endothelial venules. Am. J. Pathol. 141, 1259-1264

Palabrica, T., Lobb, R., Furie, B. C., Aronovitz, M., Benjamin, C., Hsu, Y.-M., Sajer, S. A., & Furie, B. (1992). Leukocyte accumulation promoting fibrin deposition is mediated in vivo by P-selectin on adherent platelets. Nature 359, 848-851.

Palecanda, A., Walcheck, B., Bishop, D. K., & Jutila, M. A. (1992). Rapid activation-independent shedding of leukocyte L-selectin induced by cross-linking of the surface antigen. Eur. J. Immunol. 22, 1279-1286.

Pan, J. & McEver, R. P. (1994). Characterization of the promoter of the human P-selectin gene. J. Biol. Chem. 268, 22600-22608.

Patel, K. D., Moore, K. L., Nollert, M. U., & McEver, R. P. (1995a). Neutrophils use both shared and distinct mechanisms to adhere to selectins under static and flow conditions. J. Clin. Invest. 96, 1887-1896.

Patel, K. D., Nollert, M. U., & McEver, R. P. (1995b). P-selectin must extend a sufficient length from the plasma membrane to mediate rolling of neutrophils. J. Cell Biol. 131, 1893-1902

Pavalko, F. M., Walker, D. M., Graham, L., Goheen, M., Doerschuk, C. M., & Kansas, G. S. (1995). The cytoplasmic domain of L-selectin interacts with cytoskeletal proteins via alpha-actinin: receptor positioning in microvilli does not require interaction with alpha-actinin. J. Cell Biol. 129, 1155-1164.

Petzelbauer, P., Bender, J. R., Wilson, J & Pober, J. S. (1993). Heterogeneity of dermal microvascular endothelial cell antigen expression and cytokine responsiveness in situ and in cell culture. J. Immunol. 151, 5062-5072.

Phillips, M. L., Nudelman, E., Gaeta, F. C. A., Perez, M., Singhal, A. K., Hakomori, S.-I., & Paulson, J. C. (1990). ELAM-1 mediates cell adhesion by recognition of a carbohydrate ligand, Sialyl-Le^X. Science 250, 1130-1132.

Picker. L. J., Michie, S. A., Rott, L. S., & Butcher, E. C. (1990a). A unique phenotype of skin-associated lymphocytes in humans: preferential expression of the HECA-452 epitope by benign and malignant T cells at cutaneous sites. Am. J. Pathol. 136, 1053-1068.

Picker, L. J., Terstappen, L. W. M. M., Rott, L. S., Streeter, P. R., Stein, H., & Butcher, E. C. (1990b). Differential expression of homing-associated adhesion molecules by T cell subsets in man. J. Immunol. 145, 3247-3255.

Picker, L. J., Treer, J. R., Ferguson, D. B., Collins, P. A., Bergstresser, P. R., & Terstappen, L. W. N. N. (1993). Control of lymphocyte recirculation in man. Differential regulation of the cutaneous lymphocyte-associated antigen, a tissue-selective homing receptor for skin-homing T cells. J. Immunol. 150, 1122-1136.

Picker, L. J., Warnock, R. A., Burns, A. R., Doerschuk, C. M., Berg, E. L., & Butcher, E. C. (1991). The neutrophil selectin LECAM-1 presents carbohydrate ligands to the vascular selectins ELAM-1 and GMP-140. Cell 66, 921-933.

Pizcueta, P. & Luscinskas, F. W. (1994). Monoclonal antibody blockade of L-selectin inhibits mononuclear leukocyte recruitment to inflammatory sites in vivo. Am. J. Path. 145, 461-469.

Polley, M. J., Phillips, M. L., Wayner, E., Nudelman, E., Singhal, A. K., Hakomori, S.-I., & Paulson, J. C. (1991). CD62 and endothelial cell-leukocyte adhesion molecule 1 (ELAM-1) recognize the same carbohydrate ligand, sialyl-Lewis X. Proc. Natl. Acad. Sci. USA 88, 6224-6228.

Postigo, A. A., Marazuela, M., Sanchez-Madrid, F., & de Landazuri, M. O. (1994). B lymphocyte binding to E- and P-selectin is mediated through the de novo expression of carbohydrates on in vitro and in vivo activated human B cells. J. Clin. Invest. 94, 1585-1596.

Ponyani, T. & Seed, B. (1995). PSGL-1 recognition of P-selectin is controlled by a tyrosine sulfation consensus at the PSGL-1 amino terminus. Cell 83, 333-343.

Preece, G., Murphy, G., & Ager, A. (1996). Metalloproteinase-mediated regulation of L-selectin levels on leucocytes. J. Biol. Chem. 271, 11634-11640.

Prieto, J., Eklund, A., & Patarroyo, M. (1994). Regulated expression of integrins and other adhesion molecules during differentiation of monocytes into macrophages. Cell. Immunol. 156, 191-211.

Puri, K. D., Finger, E. B., Gaudernack, G., & Springer, T. A. (1995). Sialomucin CD34 is the major L-selectin ligand in human tonsil high endothelial venules. J. Cell Biol. 6, 261-270.

Rabb, H. A., Agosti, S. J., Bittle, P. A., Ramirez, G., & Tedder, T. F. (1995). Soluble L-selectin increases during hemodialysis and is accompanied by a fall in neutrophil surface L-selectin. J. Am. Soc. Nephrol. 6, 1445-1450.

Read, M. A., Neish, A. S., Gerritsen, M. E., & Collins, T. (1996). Postinduction transcriptional repression of E-selectin and vascular cell adhesion molecule-1. J. Immunol. 157, 3472-3479.

Rosen, S. D. & Bertozzi, C. R. (1994). The selectins and their ligands. Curr. Opin. Cell Biol. 6, 663-673.

Rosen, S. D., Chi, S. I., True, D. D., Singer, M.S., & Yednock, T.A. (1989). Intravenously injected sialidase inactivates attachment sites for lymphocytes on high endothelial venules. J. Immunol. 142, 1895-1902.

Rosen, S. D., Singer, M. S., Yednock, Y. A., & Stoolman, L. M. (1985). Involvement of sialic acid on endothelial cells in organ-specific lymphocyte recirculation. Science 228, 1005-1007.

Rosenman, S. J., Ganji, A. A., Tedder, T. F., & Gallatin, W. M. (1993). *Syn*-capping of human T lymphocyte adhesion/activation molecules and their redistribution during interaction with endothelial cells. J. Leukoc. Biol. 53, 1-10.

Rossiter, H., van Reijsen, F.. Mudde, G. C.. Kalthoff, F., Bruijnzeel-Koomen, C. A., Picker, L. J., & Kupper, T. S. (1994). Skin disease-related T cells bind to endothelial selectins: expression of cutaneous lymphocyte antigen (CLA) predicts E-selectin but not P-selectin binding. Eur J Immunol. 24, 205-210.

Saeland, S., Duvert, V., Caux, C., Pandrau, D., Favre, C., Valle, A., Durand, I., Charbord, P., de Vries, J., & Banchereau, J. (1992). Distribution of surface-membrane molecules on bone marrow and cord blood $CD34^+$ hematopoietic cells. Exp. Hematol. 20, 24-33.

Sako, D., Chang, X.-J., Barone, K. M., Vachino, G., White, H. M., Shaw, G., Veldman, K. M., Bean, K. M., Ahern, T. J., Furie, B., Cumming, D. A., & Larsen, G. R. (1993). Expression cloning of a functional glycoprotein ligand for P-selectin. Cell 75, 1179- 1186.

Sako, D., Comess, K. M., Barone, K. M., Camphausen, R. T., Cumming, D. A., & Shaw, G. D. (1995). A sulfated peptide segment at the amino terminus of PSGL-1 is critical for P-selectin binding. Cell 83, 323-331.

Sanders, W. E., Wilson, R. W., Ballantyne, C. M., & Beaudet, A. L. (1992). Molecular cloning and analysis of in vivo expression of murine P-selectin. Blood 80, 795-800.

Sasaki, K., Kurata, K., Funayama, K., Nagata, M., Watanabe, E., Ohta, S., Hanai, N., & Nishi, T. (1994). Expression cloning of a novel alpha 1,3-fucosyltransferase that is involved in biosynthesis of the sialyl Lewis x carbohydrate determinants in leukocytes. J. Biol. Chem. 269, 14730-14737.

Sawada, M., Takada, A., Ohwaki, I., Takahashi, N., Tateno, H., Sakamòto, J., & Kannagi, R. (1993). Specific expression of a complex sialyl Lewis X antigen on high endothelial vunules of human lymph nodes: possible candidate for L-selectin ligand. Biochem. Biophys. Res. Commun. 193, 337-347.

Schindler, U. & Baichwal, V. R. (1994). Three NF-κB binding sites in the human E-selectin gene required for maximal TNFα-induced expression. Mol. Cell. Biol. 14, 5820-5831.

Schleiffenbaum, B. E., Spertini, O., & Tedder, T. F. (1992). Soluble L-selectin is present in human plasma at high levels and retains functional activity. J. Cell Biol. 119, 229-238.

Shimizu, Y., Newman, W., Graber, N., Horgan, K. J., Beall, L. D., Gopal, T. V., van Seventer, G. A., & Shaw, S. (1991). Four molecular pathways of T cell adhesion to endothelial cells: Roles of LFA-1, VCAM-1, and ELAM-1 and changes in pathway hierarchy under different activation conditions. J. Cell Biol. 113, 1203-1212.

Shimizu, Y. & Shaw, S. (1993). Cell adhesion: mucins in the mainstream. Nature 366, 630-631.

Siegelman, M. H.. Cheng, I. C., Weissman, I. L., & Wakeland, E. K. (1990). The mouse lymph node homing receptor is identical with the lymphocyte cell surface marker Ly-22: Role of the EGF domain in endothelial binding. Cell 61, 611-622.

Siegelman, M. H., van de Rijn, M., & Weissman, I. L. (1989). Mouse lymph node homing receptor cDNA clone encodes a glycoprotein revealing tandem interaction domains. Science 243, 1165-1172.

Silber, A., Newman, W., Reimann, K. A., Hendricks, E., Walsh, D., & Ringler, D. J. (1994a). Kinetic expression of endothelial adhesion molecules and relationship to leukocyte recruitment in two cutaneous models of inflammation. Lab. Invest. 70, 163-175.

Silber, A., Newman, W., Sasseveille, V. G., Pauley, D., Beall, D., Walsh, D. G., & Ringler, D. J. (1994b). Recruitment of lymphocytes during cutaneous delayed hypersensitivity in nonhuman primates is dependent on E-selectin and vascular cell adhesion molecule 1. J. Clin. Invest. 93, 1554-1563.

Simmons, D. L., Satterthwaite, A. B., Tenen, D. G., & Seed, B. (1992). Molecular cloning of a cDNA encoding CD34, a sialomucin of human hematopoietic stem cells. J. Immunol. 148, 267-271.

Simon, S. I., Chambers, J. D., Butcher, E., & Sklar, L. A. (1992). Neutrophil aggregation is β2-integrin- and L-selectin-dependent in blood and isolated cells. J. Immunol. 149, 2765-2771.

Smeets, E. F., de Vries, T., Leeuwenberg, J. F. M., van den Eijnden, D. H., Buurman, W. A., & Neefjes, J. J. (1993). Phosphorylation of surface E-selectin and the effect of soluble ligand (Sialyl $Lewis^X$) on the half-life of E-selectin. Eur. J. Immunol 23, 147-151.

Smith, C. W., Kishimoto, T. K., Abbass, O., Hughes, B., Rothlein, R., McIntire, L. V., Butcher, E., & Anderson, D. C. (1991). Chemotactic factors regulate lectin adhesion molecule 1 (LECAM-1)-dependent neutrophil adhesion to cytokine-stimulated endothelial cells in vitro. J. Clin. Invest. 87, 609-618.

Smith, P. L., Gersten, K. M., Petryniak, B., Kelly, R. J., Rogers, C., Natsuka, Y., Alford III, J. A., Scheidegger, E. P., Natsuka, E. P., & Lowe, J. B. (1996). Expression of the α(l,3)fucosyltransferase Fuc-TVII in lymphoid aggregate high endothelial venules correlates with expression of L-selectin ligands. J. Biol. Chem. 271, 8250-8259.

Spertini, O., Callegari, P., Cordey, A.-S., Hauert, J., Joggi, J., von Fliedner, V., & Schapira, M. (1994). High levels of the shed form of L-selectin (sL-selectin) are present in patients with acute leukemia and inhibit blast cell adhesion to activated endothelium. Blood 84, 1249-1256.

Spertini, O., Cordoy, A.-S., Monai, N., Giuffre, L., & Schapira' M. (1996). P-selectin glycoprotein ligand 1 is a ligand for L-selectin on neutrophils, monocytes, and $CD34^+$ hematopoietic progenitor cells. J. Cell Biol. 135, 523-531.

Spertini, O., Freedman, A. S., Belvin, M. P., Penta, A. C., Griffin, J. D., & Tedder, T. F. (1991a). Regulation of leukocyte adhesion molecule-1 (TQ1, Leu-8) expression and shedding by normal and malignant cells. Leukemia 5, 300-308.

Spertini, O., Kansas, G. S., Munro, J. M., Griffin, J. D., & Tedder, T. F. (1991b). Regulation of leukocyte migration by activation of the leukocyte adhesion molecule-1 (LAM-1) selectin. Nature 349, 691-694.

Spertini, O., Kansas, G. S., Reimann, K. A., Mackay, C. R., & Tedder, T. F. (1991c). Functional and evolutionary conservation of distinct epitopes on the leukocyte adhesion molecule-1 (LAM-1) that regulate leukocyte migration. J. Immunol. 147, 942-949.

Spertini, O., Luscinskas, F. W., Gimbrone Jr., M. A., & Tedder, T. F. (1992a). Monocyte attachment to activated human vascular endothelium in vitro is mediated by Leukocyte Adhesion Molecule-1 (L-selectin) under non-static conditions. J. Exp. Med. 175, 1789-1792.

Spertini, O., Luscinskas, F. W., Kansas, G. S., Munro, J. M., Griffin, J. D., Gimbrone Jr., M. A., & Tedder, T. F. (1991d). Leukocyte adhesion molecule-1 (LAM-1, L-selectin) interacts with an inducible endothelial cell ligand to support leukocyte adhesion. J. Immunol. 147, 2565-2573.

Spertini, O., Schleiffenbaum, B., White-Owen, C., Ruiz Jr.. P., & Tedder, T. F. (1992b). ELISA for quantitation of L-selectin shed from leukocytes in vivo. J. Immunol. Methods 156, 115-123.

Springer, T. A. (1995). Traffic signals on endothelium for lymphocyte recirculation and leukocyte emigration. Annu. Rev. Physiol. 57, 827-872.

Sriramarao, P., von Andrian, U. H., Butcher, E. C., Burdon, M. A., & Broide, D. H (1994). L-selectin and very late antigen-4 integrin promote eosinophil rolling at physiological shear rates in vivo. J. Immunol. 153, 4238-4246.

Staite, N. D., Justen, J. M., Sly, L. M., Beaudet, A. L., & Bullard, D. C. (1996). Inhibition of delayed-type contact hypersensitivity in mice deficient in both E-selectin and P-selectin. Blood 88, 2973-2979.

Steeber, D. A., Engel, P., Miller, A. S., Sheetz, M. P., & Tedder, T. F. (1997). Ligation of L-selectin through conserved regions within the lectin domain activates signal transduction pathways and integrin function in human, mouse and rat leukocytes. J. Immunol. 159, 952-963.

Steeber, D. A., Green, N. E., Sato, S., & Tedder, T. F. (1996a). Humoral immune responses in L-selectin deficient mice. J. Immunol. 157, 4899-4907.

Steeber, D. A., Green, N. E., Sato, S., & Tedder, T. F. (1996b). Lymphocyte migration in L-selectin-deficient mice: altered subset migration and aging of the immune system. J. Immunol. 157, 1096-1106.

Steeber, D. A., Campbell, M. A., Basit, A., Ley, K., & Tedder, T. F. (1998). Optimal selectin-mediated rolling of Leukocytes during inflammation in vivo requires intercellular adhesion molecule-1expression. Proc. Natl. Acad. Sci. USA. 95, 7562-7567.

Steegmaler, M., Levinovitz, A., Isenmann, S., Borges, E., Lenter, M., Kocher, H.P., Kleuser, B., & Vestweber, D. (1995). The E-selectin-ligand ESL-1 is a variant of a receptor for fibroblast growth factor. Nature 373, 615-620.

Steinberg, P. E., McEver, R. P., Shuman, M. A., Jacques, Y. V., & Bainton, D. F. (1985). A platelet alpha-granule membrane protein (GMP-140) is expressed on the plasma membrane after activation. J. Cell Biol. 101, 880-886.

Stoolman, L. M., Tenforde, T. S., & Rosen, S. D. (1984). Phosphomannosyl receptors may participate in the adhesive interactions between lymphocytes and high endothelial venules. J. Cell Biol. 99, 1535- 1540.

Stoolman, L. M., Yednock, T. A., & Rosen, S. D. (1987). Homing receptors on human and rodent lymphocytes- evidence for a conserved carbohydrate-binding specificity. Blood 70, 1842-1850.

Streeter, P. R., Rouse, B. T. N., & Butcher, E. C. (1988). Immunologic and functional characterization of a vascular addressin involved in lymphocyte homing into peripheral lymph nodes. J. Cell Biol. 107, 1853-1862.

Subramaniam, M., Koedam, J. A., & Wagner, D. D. (1993). Divergent fates of P- and E-selectins after their expression on the plasma membrane. Mol. Biol. Cell 4, 791-801.

Subramaniam, M., Saffaripour, S., Watson, S. R., Mayadas, T. N., Hynes, R. O., & Wagner, D. D. (1995). Reduced recruitment of inflammatory cells in a contact hypersensitivity response in P-selectin-deficient mice. J. Exp. Med. 181, 2277-2282.

Sueyoshi, S., Tsuboi, S.. Sawada-Harai, R.. Dang, U. N.. Lowe, J. B., & Fukuda, M. (1994). Expression of distinct fucosylated oligosaccharides and carbohydrate-mediated adhesion efficiency directed by two different alpha-1,3-fucosyltransferases. Comparison of E- and L-selectin-mediated adhesion. J. Biol. Chem. 269, 32342-32350.

Tang, M. L. K., Hale, L. P., Steeber, D. A., & Tedder, T. F. (1997). L-selectin is involved in lymphocyte migration to sites of inflammation in the skin: delayed rejection of allografts in L-selectin-deficient mice. J. Immunol. 158, 5191-5199.

Tang, M. L. K., Steeber, D. A., Zhang, X. Q., & Tedder, T. F. (1998). Intrinsic differences in L-selectin expression levels affect T and B lymphocyte subset-specific recirculation pathways. J. Immunol. 160, 5113-5121.

Tedder, T. F. (1991). Cell surface receptor shedding: a means of regulating function. Am. J. Respir. Cell Mol. Biol. 5, 305-306.

Tedder, T. F., Cooper, M. D., & Clement, L. T. (1985). Human lymphocyte differentiation antigens HB-10 and HB-11. II. Differential production of B cell growth and differentiation factors by distinct helper T cell subpopulations. J. Immunol. 134, 2989-2994.

Tedder, T. F., Engel, P., Wagner, N., & Freedman, A. S. (1993a). Adhesion receptors of B lymphocytes: expression and function on normal and malignant cells. In Lymphocyte Adhesion Molecules, Y. Schmizu, ed. (Austin, TX: R. G. Landes Co.), pp. 280-312.

Tedder, T. F., Ernst, T. J., Demetri, G. D., Isaac s, C. M., Adler, D. A., & Disteche, C. M. (1989). Isolation and chromosomal localization of cDNAs encoding a novel human lymphocyte cell-surface molecule, LAM1: Homology with the mouse lymphocyte homing receptor and other human adhesion proteins. J. Exp. Med. 170, 123-133.

Tedder, T. F., Green, N. E., Miller, A., & Steeber, D. A. (1995a). L-selectin function and inflammatory disease. In Leukocyte Recruitment in Inflammatory Disease, G. Peltz, ed (Austin: R.G. Landes), pp. 165-210.

Tedder, T. F., Luscinskas, W., & Kansas, G. S. (1993b). Regulation of leukocyte migration by L-selectin: mechanisms, domains and ligands. Behring Inst. Mitt. 92, 165-177.

Tedder, T. F., Matsuyama, T., Rothstein, D. M., Schlossman, S. F., & Morimoto, C. (1990a). Human antigen-specific memory T cells express the homing receptor necessary for lymphocyte recirculation. Eur. J. Immunol. 20, 1351-1355.

Tedder, T. F., Penta, A. C., Levine, H. B., & Freedman, A. S. (1990b). Expression of the human leukocyte adhesion molecule, LAM1. Identity with the TQ1 and Leu-8 differentiation antigens. J. Immunol. 144, 532-540.

Tedder, T. F., Steeber, D. A., Chen, A., & Engel, P. (1995b). The selectins: Vascular adhesion molecules. FASEB J. 9, 866-873.

Tedder, T. F., Steeber, D. A., & Pizcueta, P. (1995c). L-selectin deficient mice have impaired leukocyte recruitment into inflammatory sites. J. Exp. Med. 181, 2259-2264.

Thorlacius, H., Raud, J., Rosengren-Beezley, S., Forrest, M. J., Hedqvist, P., & Lindbom, L. (1994). Mast cell activation induces P-selectin-dependent leukocyte rolling and adhesion in postcapillary venules in vivo. Biochem. Biophys. Res. Commun. 203, 1043-1049.

Tiemeyer, M., Swiedler, S. J., Ishihara, M., Moreland, M., Schweingruber, H., Hirtzer, P., & Bradley, B. K. (1991). Carbohydrate ligands for endothelial-leukocyte adhesioin molecule 1. Proc. Natl. Acad. Sci. USA 88, 1138-1142.

Tozeren, A. & Ley, K. (1992). How do selectins mediate leukocyte rolling in venules? Biophys. J. 63, 700-709.

Tu, L., Chen, A., Delahunty, M. D., Moore, K. L., Watson, S., McEver, R. P., & Tedder, T. F. (1996). L-selectin binds to P-selectin glycoprotein ligand-1 on leukocytes. Interactions between the lectin, EGF and consensus repeat domains of the selectins determine ligand binding specificity. J. Immunol. 156, 3995-4004.

Tu, L., Delahunty, M.D., Ding, H., Luscinskas, F.W., & Tedder, T.F. (1999). The cutaneous lymphocyte antigen is an essential component of the L-selectin ligand induced on human vascular endothelial cells. J. Exp. Med. 189, (in press).

Turunen, J. P., Paavonen, T., Majuri, M.-L., Tiisala, S., Mattila, P., Mennander, A., Gahmberg, C. G., Hayry, P., Tamatani, T., Miyasaka, M., & Renkonen, R. (1994). Sialyl Lewis x and L-selectin-dependent site specific lymphocyte extravasation into renal transplants during acute rejection. Eur. J. Immunol. 24, 1130-1136.

Tyrrell, D., James, P., Rao, N., Foxall, C., Abbas, S., Dasgupta, F., Nashed, M., Hasegawa, A., Kiso, M., Asa, D., Kidd, J., & Brandley, B. K. (1991). Structural requirements for the carbohydrate ligand of E-selectin. Proc. Natl. Acad. Sci. USA 88, 10372-10376.

Ushiyama, S., Laue, T. M., Moore, K. L., Erickson, H. P., & McEver, R. P. (1993). Structural and functional characterization of monomeric soluble P-selectin and comparison with membrane P-selectin. J. Biol. Chem. 268, 15229-15237.

Vachino, G., Chang, X.-J., Vledman, G. M., Kuman, R., Sako, D., Fouser, L. A., Berndt, M. C., & Cumming, D. A. (1995). P-selectin glycoprotein ligand-1 is the major counter-receptor for P-selectin on stimulated T cells and is widely distributed in non-functional form on many lymphocytic cells. J. Biol. Chem. 270, 21966-21974.

van Huijsduijnen, R. H., Whelan, J., Pescini, R., Becker-Andre, M., Schenk, A. M., & DeLamarter, J. F. (1992). A T-cell enhancer cooperates with NF-kB to yield cytokine induction of E-selectin gene transctiption in endothelial cells. J. Biol. Chem. 267, 22385-22391.

Varki, A. (1994). Selectin ligands. Proc. Natl. Acad. Sci. USA 91, 7390-7397.

Veldman, G. M., Bean, K. M., Cumming, D. A., Eddy, R. L., Sait, S. N. J., & Shows, T. B. (1995). Genomic organization and chromosomal localization of the gene encoding human P-selectin glycoprotein ligand. J. Biol. Chem. 270, 16470-16475.

von Andrian, U. H., Chambers, J. D., Berg, E. L., Michie, S. A., Brown, D. A., Karolak, D., Ramezani, L., Berger, E. M., Arfors, K. E., & Butcher, E. C. (1993). L-selectin mediates neutrophil rolling in inflamed venules through sialyl Lewis^X-dependent and -independent recongition pathways. Blood 82, 182-191.

von Andrian, U. H., Chambers, J. D., McEvoy, L. M., Bargatze, R. F., Arfors, K.-E., & Butcher, E. C. (1991). Two-step model of leukocyte-endothelial cell interaction in inflammation: distinct roles for LECAM-1 and the leukocyte β2 integrins in vivo. Proc. Natl. Acad. Sci. USA 88, 7538-7542.

von Andrian, U. H., Hasslen, S. R., Nelson, S. L., Erlandsen, S. L., & Butcher, E. C. (1995). A central role for microvillus receptor presentation in leukocyte adhesion under flow. Cell 82, 989-999.

von Asmuth, E. J. U., Smeets, E. F., Ginsel, L. A., Onderwater, J. J. M., Lecuwenberg, J. F. M., & Buurman, W. A. (1992). Evidence for endocytosis of E-selectin in human endothelial cells. Eur. J. Immunol. 22, 2519-2526.

Wagers, A. J., Stoolman, L. M., Kannagi, R., Craig, R., & Kansas, G. S. (1997). Expression of leukocyte fucosyltransferases regulates binding to E-selectin. Relationship to previously implicated carbohydrate epitopes. J. Immunol. 159, 1917-1929.

Walcheck, B., Kahn, J., Fisher, J. M., Wang, B. B., Fisk, R. S.. Payan. D. G., Feehan, C., Betageri, R., Darlak, K., Spatola, A. F., & Kishimoto, T. K. (1996a). Neutrophil rolling altered by inhibition of L-selectin shedding in vitro. Nature 380, 720-723.

Walcheck, B., Moore, K. L., McEver, R. P., & Kishimoto, T. K. (1996b). Neutrophil-neutrophil interactions under hydrodynamic shear stress involve L-selectin and PSGL-1. A mechanism that amplifies initial leukocyte accumulation on P-selectin *in vitro*. J. Clin. Invest. 98, 1081-1087.

Walcheck, B., Watts, G., & Jutila, M. A. (1993). Bovine γ/δ T cells bind E-selectin via a novel glycoprotein receptor: first characterization of a lymphocyte/E-selectin interaction in an animal model. J. Exp. Med. 178, 853-863.

Wallace, D. L. & Beverley, P. C. (1993). Characterization of a novel subset of T cells from human spleen that lacks L-selectin. Immunology 78, 623-628.

Walz, G., Aruffo, A., Kolanus, W., Bevilacqua, M., & Seed, B. (1990) Recognition by ELAM-1 of the Sialyl-Le^X determinant on myeloid and tumor cells. Science. 250, 1132-1134.

Watson, M. L., Kingsmore, S. F., Johnston, G. I., Siegelman, M. H., Le Beau, M. M., Lemons, R. S., Bora, N. S., Howard, T. A., Weissman, I. L., McEver, R. P., & Seldin, M. F. (1990). Genomic organization of the selectin family of leukocyte adhesion molecules on human and mouse chromosome 1. J. Exp. Med. 172, 263-272.

Watson, S. R., Fennie, C., & Lasky, L. A. (1991a). Neutrophil influx into an inflammatory site inhibited by a soluble homing receptor-IgG chimera. Nature 349, 164-167.

Watson, S. R., Imai, Y., Fennie, C., Geoffrey, J., Singer, M., Rosen, S. D., & Lasky, L. A. (1991b). The complement binding-like domains of the murine homing receptor facilitate lectin activity. J. Cell Biol. 115, 235-243.

Westermann, J., Nagahori, Y., Walter, S., Heerwagen, C., Miyasaka, M., & Pabst, R. (1994). B and T lymphocytes subsets enter peripheral lymph nodes and Peyer's patches without preference in vivo: no correlation occurs between their localization in different types of high endothelial venules and the expression of CD44, VLA-4, LFA-1, ICAM-1, CD2 or L-selectin. Eur. J. Immunol. 24, 2312-2316.

Whelan, J., Ghersa, P., van Huijsduijnen, H. R., Gray, J., Chandra, G., Talabot, F., & DeLamarter, J.F. (1991). An NF-κB-like factor is essential but not sufficient for cytokine induction of ELAM-1 gene transcription. Nucleic Acids Res. 19, 2645-2653.

Whitley, M. Z., Thanos, D., Read, M. A., Maniatis, T., & Collins, T. (1994). A striking similarity in the organization of the E-selectin and β-IFN gene promoters. Mol. Cell. Biol. 14, 6464-6475.

Wilkins, P. P., Moore, K. L., McEver, R. P., & Cummings, R. D. (1995). Tyrosine sulfation of P-selectin glycoprotein ligand-1 is required for high affinity binding of P-selectin. J. Biol. Chem. 270, 22677-22680.

Winn, R.K., Liggitt, D., Vedder, N.B., Paulson, J.C., & Harlan, J.M. (1993). Anti-P-selectin monoclonal antibody attenuates reperfusion injury to the rabbit ear. J. Clin. Invest. 92, 2042-2047.

Witherden, D. A., Abernethy, N. J., Kimpton, W. G., & Cahill, R. N. (1994). Changes in thymic export of L-selectin^{+} γδ and αβ T cells during fetal and postnatal development. Eur. J. Immunol. 24, 1234-1239.

Wong, C. S. Gamble, J.R., Skinner, M.P., Lucas, C.M., Berndt, M.C., & Vadas, M.A. (1991). Adhesion protein GMP-140 inhibits superoxide anion release by human neutrophils. Proc. Natl. Acad. Sci. USA 88, 2397-2401.

Xu, J., Grewal, I.S., Geba, G.P., & Flavell, R.A. (1996). Impaired primary T cell responses in L-selectin-deficient mice. J. Exp. Med. 183, 589-598.

Yednock, T.A., Butcher, E.C., Stoolman, L.M., & Rosen, S.D. (1987a). Receptors involved in lymphocyte homing: relationship between a carbohydrate-binding receptor and the MEL-14 antigen. J. Cell Biol. 104, 725-731.

Yednock, T.A., Stoolman, L.M., & Rosen, S.D. (1987b). Phosphomannosyl derivatized beads detect a receptor involved in lymphocyte homing. J. Cell Biol. 104, 713-723.

Yoshida, M., Szente, B.E., Kiely, J.-M., Rosenzweig, A., & Gimbrone Jr., M.A. (1998). Phosphorylation of the cytoplasmic domain of E-selectin is regulated during leukocyte-endothelial adhesion. J. Immunol. 161, 933-941.

Yoshida, M., Westlin, W. F., Wang, N., Ingber, D. E., Rosenzweig, A., Resnick, N., & Gimbrone Jr., M. A. (1996). Leukocyte adhesion to vascular endothelium induces E-selectin linkage to the actin cytoskeleton. J. Cell Biol. 133, 445-455.

Yuen, C.-T., Bezouska, K., O'Brien, J., Stoll, M., Lemoine, R., Lubineau, A., Kiso, M., Hasegawa, A., Bockovich, N.J., Nicolaou, K.C., & Feizi, T. (1994). Sulfated blood group Lewis^{a}: a superior oligosaccharide ligand for human E-selectin. J. Biol. Chem. 269, 1595-1598.

Yuen, C.-T., Lawson, A. M., Chai, W., Larkin, M., Stoll, M. S., Stuart, A. C., Sullivan, F. X., Ahern, T. J., & Feizi, T. (1992). Novel sulfated ligands for the cell adhesion molecule E-selectin revealed by the neoglycolipid technology among O-linked oligosaccharides on an ovarian cystadenoma glycoprotein. Biochemistry 31, 9126-9131.

Zetterberg, E. & Richter, J. (1993). Correlation between serum level of soluble L-selectin and leukocyte count in chronic myeloid and lymphocytic leukemia and during bone marrow transplantation. Eur. J. Haematol. 51, 113-119.

Zhou, Q., Moore, K. L., Smith, D. F., Varki, A., McEver, R. P., & Cummings, R. D. (1991). The selectin GMP-140 binds to sialylated, fucosylated lactosaminoglycans on both myeloid and nonmyeloid cells. J. Cell Biol. 115, 557-564.

THE IMMUNOGLOBULIN SUPERFAMILY

David L. Simmons

Advances in Molecular and Cell Biology
Volume 28, pages 113-132.

ISBN: 0-7623-0495-2

I. BASIC PRINCIPLES

The ideas and conventions defining the domain organization of this family were established by Alan Williams and Neil Barclay in a series of papers in the late 1980s (Williams and Barclay, 1988; Williams et al., 1989). The basic Ig domain is built from a tightly packed barrel of β-strands, either 7 strands in C1 and C2 domains (arranged in 2 layers of 3 and 4 strands designated ABE and CC'FG respectively) or 9 strands in V domains (arranged in two layers of 4 and 5 designated ABDE and CC'C"FG respectively). Several Ig superfamily domain structures have been solved over the last 5 years including CD2 (solution nuclear magnetic resonance, Driscoll et al., 1991; Withka et al., 1993; and crystallography at 2.8 Å resolution, Jones et al., 1993); CD4 domains 1-2 (crystallography at 2.6 Å resolution, Ryu et al., 1990; Wang et al., 1990; Garret et al., 1993) CD4 domains 3-4 (crystallography at 2.8 A resolution, Brady et al., 1993) and most recently the complete structure of domains 1-4 (Wu et al., 1997); the single V domain of CD8 (crystallography at 2.6 Å resolution, Leahy et al., 1992); class II (crystallography at 2.8 Å resolution Brown et al., 1993); VCAM domains 1 and 2 (crystallography at 1.8 Å resolution Jones et al., 1995); ICAM-2 (Casanovas et al., 1997).

Most surprising recently was the structure of E cadherin derived from crystallography in 1995 (Overduin et al., 1995), which revealed a domain fold pattern strikingly similar to the fundamental β strand arrangements of Ig like domains even though no one had discerned any potential tertiary similarity from primary sequence alignments.

Thus, the strict definition of the Ig superfamily is now considerably broadened and can even include other previously separate adhesion molecule families such as the cadherins. Although, historically the origins of the Ig superfamily were obviously derived from an analysis of the domain characteristics of the Ig cluster of B immunoglobulins and T cell receptor, more recent work may lead to a reassessment of the basic features of the IgSF. It is now possible to discern an Ig-like fold from sequences that bear little significant primary sequence homologies. In fact, we may be returning to the original way in which the IgSF was defined by Williams and Barclay in the mid 1980s, based on secondary and tertiary structure predictions. They deduced the essential secondary structural elements of the Ig fold of alternating β strands and searched for these elements in other leukocyte cell surface molecules being cloned in the mid 1980s. This early pioneering approach led to and was supplanted by primary sequence alignment scores as arbiters of membership of the family. This level of analysis has held sway for the last 10 years; new members of the family have been allowed entry on the basis of statistically significant matches to the prototypic cannonical templates of V, C1, and C2 type domains. Sequences were aligned to a mini database of card carrying IgSF members using randomized Monte Carlo algorithms to establish statistical significance to the scores from the search results. Established cut off points defined whether a new sequence was or was not a member of the IgSF or contained an IgSF like domain among other modules.

However, in the light of the cadherin structures, it is now clear that the essential Ig fold can be produced by convergence of disparate primary sequences as well as divergence from primordial ancestral sequences. Thus, we will need to include secondary and tertiary structural searching algorithms as well as the primary sequence search tools to fully identify future IgSF members in the flood of new genes identified in the high-throughput cDNA sequence projects currently nearing their fruition.

As an adjunct to nmr and crytallographic structural studies, there have been extensive site directed mutagenesis screens to define key functional residues in individual Ig domains involved in adhesion over the last six years. These now include CD2 which binds CD58 (LFA-3) in humans and the related molecule CD48 in mice (Somoza et al., 1993; Arulanandam et al., 1993), CD4 which binds class II (Moebius et al., 1993), ICAM-1 and ICAM-3 which bind LFA-1 (Staunton et al., 1990; Berendt et al., 1992; Fisher et al., 1997; Holness et al., 1995; Sadhu et al., 1994; Klickstein et al., 1996), VCAM-1 which binds VLA-4 (Vonderhide et al., 1994; Osborn et al., 1994), and members of the sialoadhesin family which bind glycan based ligands (CD22, van der Merwe et al., 1996; Vinson et al., 1996).

The Ig domain most commonly found in surface molecules involved in recognition and or adhesion conforms to the short C2 type 7 β strand plan (e.g. ICAM-1, ICAM-2, ICAM-3, VCAM, CD31/PECAM-1), although there are examples of the longer 9 β strand domain being used (e.g. CD2, CD4. CD28, CTLA4). Also, in general, compared with antibody V domains, adhesion domains seem to have longer β strands and shorter loop regions. The antigen combining site of antibodies consists primarily of the CDR loops (complementarily determining regions) 1, 2 and 3 which are largely made from connecting loops. In contrast, adhesion binding surfaces seem to be made from the β strands themselves arranged in flat faces. This has been arrived at by a combination of NMR and crystallography studies and site directed mutagenesis screens. Thus for the ICAM family (ICAMs 1, 2, and 3, Staunton et al., 1991; Berendt et al., 1992; Sadhu et al., 1995; Holness et al., 1995), VCAM (Vonderheide et al., 1994), CD22 (van der Merwe et al., 1996), and sialoadhesin (Vinson et al., 1996), key functionally important residues involved in ligand/receptor interactions are located on flat faces made up of β strands comprising the CC'DFG surface.

This is, of course, a simplification and exceptions exist for both classical antibodies and adhesion molecules. For antibodies, the precise shape of the framework region of β strands is key to the organization of the CDR loop regions. The CDRs are not completely autonomous units and are highly dependent on the underlying β strand framework regions. As witness to this fundamental architecture, it has taken some considerable time to perfect the mutagenesis schemes needed to successfully humanize murine template monoclonal antibodies (so called CDR grafting) and to generate high affinity in vitro antibody repertoires using degenerate PCR based strategies. Likewise, for recognition/adhesion molecules, there are exceptions to the dominance of β strand face interactions as some key residues involved in adhesion binding sites have been found in loops. In fact, the loops may contain key, specificity determining residues akin to the specificity of the CDR loops of classical cannonical antibodies.

The packing and orientation of domains in adhesion molecules seems to be remarkably uniform. The contiguous Ig domains are arranged as a nearly fully extended rod, with only small angles of pitch between adjacent domains. For example, the packing arrangement of domains 1 and 2 of CD2 and CD4 is very similar compared with that seen in constant domains of antibodies. So far, no antibody-like pattern, with heavy chain V domain bent at a sharp 90° angle relative to the contiguous CH1 domain, has been found. The recently determined crystal structure of the entire extracellular domain of CD4 again showed that the individual domains are arranged in a near to linear extended order (Wu et al., 1997). In addition, electron microscopy studies (negative staining or more powerfully, platinum rotary shadowing) has revealed that multi-domain IgSF members exist with contiguous Ig domains arranged like beads on a string. For example, ICAM-1 (Staunton et al., 1990) and ICAM-3 (Sadhu et al., 1995) (5 C2 type domains each), VCAM (Vonderheide et al., 1994) (7 C2 type domains), NCAM (5 C2 type domains and 2 type III fibronectin domains) and CD3 1 (6 C2 type domains) have classical globular Ig domains of 3.5 nm stacked linearly one on the other giving overall rod like extensions of between 20 and 30 nm depending on the number of domains.

However, in spite of these common structural features, there is a surprising diversity in the detailed modes of interaction among members of the IgSF i.e. which domains are used for receptor engagement. Fortunately, however, they can be grouped into two basic types.

1. Homophilic binding, where the IgSF member binds to itself. This is thought to be the ancestral and prototypic mode of adhesion interactions among members of the IgSF. Surprisingly however, this mode manifests itself in a large number of different structural binding arrangements. Homophilic binding can be mediated either by N-terminal domains (e.g. domain 1 of CD31/CD31, domain 1 of CD66/BGP1) or by "internal" domains requiring anti-parallel alignment of interdigitated molecules (e.g. NCAM/NCAM which uses domain 3 for interactions between apposed molecules or CEA/CEA which uses a large number of domains for the anti-parallel alignments).
2. Heterophilic binding, where the IgSF member binds a different ligand or counter-receptor. This mode of binding seems to be largely mediated by the N-terminal Ig domain or domains. Significant examples are ICAM-1, ICAM 1-2 and ICAM-3 interacting with the integrin LFA-1 (CDT 1 a/CD 18), VCAM interacting with VLA-4 (CD49d/CD29), and sialoadhesin binding to its sialic acid based glycan ligands.

Although the focus of this chapter will be IgSF members involved in cell adhesion and/or recognition, the Ig fold has also evolved and been employed to bind small ligands such as growth factors (e.g., platelet derived growth factor, PDGF), stem cell factor (SCF), and vascular endothelial cell growth factor (VEGF). Some of the recent work mapping their binding sites, which tend to be discrete binding pockets in internal domains, will be discussed for completeness.

II. HOMOPHILIC ADHESION INVOLVING N TERMINAL AND INTERNAL DOMAINS

The structural themes emerging from homophilic CAM interactions are surprisingly diverse when compared with heterotypic modes (see below). The surprise here is that the homophilic mode is thought to be the ancestral prototypic pattern that diversified with domain mutation and duplication to yield heterophilic patterns. Thus, it would be assumed that the homophilic mode should be relatively simple, employing the same domain in the same interaction pattern. Whereas, the heterotypic mode would be more diverse, using different domains for each specific example. However, the converse is true. The homophilic binding sites seem to be located in widely different domains specific for each homophilic CAM, sometimes in N-terminal domains and sometimes employing "internal" domains. This latter type requires an extensive interdigitating alignment of anti-parallel molecules for example, NCAM/NCAM, CEA/CEA (see Figure 1).

A. Neural Cell Adhesion Molecule (NCAM)

Definition of the binding site in NCAM, for so long the prototypic CAM, has lagged behind research on many other CAMs. The extracellular domain of NCAM consists of 5 Ig domains and 2 type III fibronectin repeat modules. Recently, a region in a C' strand (KYSFNYDGSE) in a CDR 2 like region in domain 3 has been shown to mediate homophilic binding (Rao et al., 1992). It is not known how the two NCAM molecules need to be aligned to enable this homophilic site to bind, that is whether they need to be aligned in a parallel or anti-parallel mode.

B. *Dictyostelium Discoideum*. gp80

An early model for how homophilic binding may occur came from the work on gp80, the major adhesin expressed in the late multicellular phase of the slime mould *Dictyostelium discoideum*. gp80 is not thought to be a member of the IgSF, but a short motif (YKLNVNDS) was defined as mediating homophilic adhesion. By aligning two gp80 molecules in an anti-parallel manner, an ionic salt bridge bond could form between the two oppositely charged ends of this motif YKLNVNDS to SDNVNLKY; K $\rightarrow$ D and vice versa D $\rightarrow$ K. This is an attractive model, though yet to be formally proven.

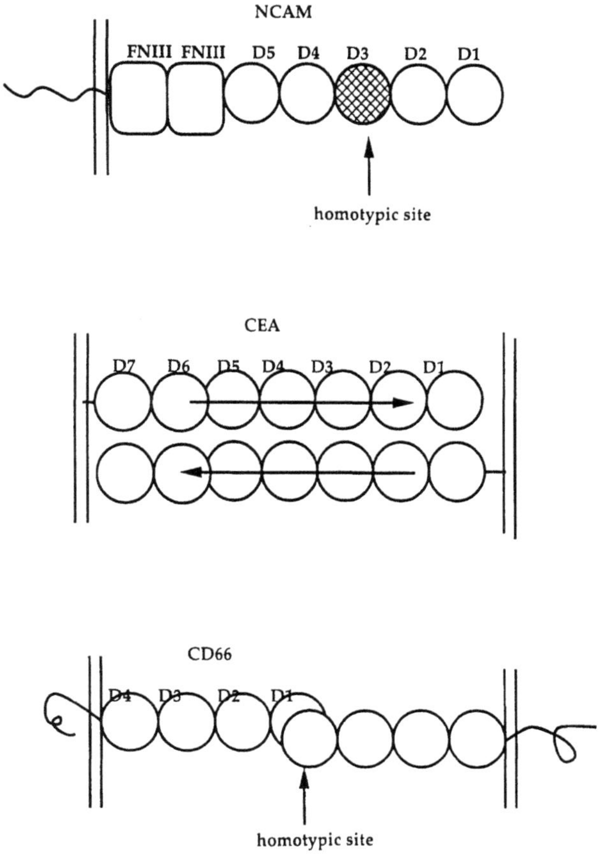

Figure 1. Binding sites in homophilic CAMs. NCAM/NCAM, CEA-CEA anti parallel interdigitation, and CD66-CD66 head-to-head, dominance of the N-terminal domain 1.

C. CD66/ Carcinoembryonic Antigen (CEA) Family

Certain members of the carcinoembryonic antigen family (CEA, CD66a,b,c,d,e,f,g), which amounts to over 30 closely related IgSF molecules, for example, CEA-1 (Carcinoembryonic antigen 1); BGP-1 (Biliary Glycoprotein 1); and PSG-1 (Pregnancy specific glycoprotein) seem to bind in an anti-parallel mode requiring a reciprocal interdigitating alignment of two molecules expressed on apposing cells (Zhou et al., 1993). However, the precise domains mediating this binding are not known, though internal domains must be involved. However, an analysis of "shorter" members of the CEA family, containing 4 Ig domains, pointed to a dominant role for the N-terminal domain, with merely an accessory role played by the other domains related to appropriate presentation and accessibility of the binding site to the apposed CAM (Cheung et al., 1993). This agrees with work on CD66a/biliary glycoprotein 1 (BGP1) which contains 4 Ig domains. The N-terminal V like domain contains the dominant binding site and the remaining 3 C2 type domains merely act as a stalk to present this site away from the cell surface glycocalyx (Watt et al., 1995; Teixera et al., 1995).

Thus even within a tightly clustered and highly homologous sub family of the IgSF, there are apparently very different modes of homophilic adhesion depending on the number of Ig domains in the molecule; ranging from dominant N terminal interactions to the involvement of N-terminal and internal, membrane proximal domains.

D. Platelet Endothelial Cell Adhesion Molecule (PECAM/CD31)

A different pattern is provided by platelet endothelial cell adhesion molecule (PECAM CD31), which is an abundantly expressed endothelial homophilic CAM, whose extracellular region consists of 6 Ig domains (Fawcett et al., 1995; Newton et al., 1997). Recent work supports a dominant role for the N-terminal domain 1 interacting in an anti-parallel mode with domain 1 of an apposed CD31 molecule. There is an additional role for the membrane proximal domains in dimerising CD31 on the surface of one cell and the dimerisation is essential for full manifestation of homophilic adhesion mediated by the N terminal domains. The case of CD31/CD31 will be discussed further to provide a detailed example of how these interactions can be dissected.

E. Case Study of Homophilic Mode of CD31/CD31 Adhesion

Over the last five years, the CD31/CD31 homophilic adhesion mechanism has been dissected using a range of in vitro assays employing recombinant IgG chimeric constructs and heterologous COS cell adhesion. The major findings may be summarized as follows. Firstly, the extracellular domain of CD31 is capable of supporting homophilic adhesion, independent of any contribution from either the transmembrane or the cytoplasmic tail sequences. Secondly, that CD31 domain 1 is necessary to mediate homophilic adhesion, but it is not sufficient to support adhesion when expressed together with domains 2 and 3 out of the context of the whole extracellular domain. More recent evidence suggests that the membrane proximal domains 4-6 allow dimerisation of CD31 molecules in cis i.e. in the plane of the same membrane. Thus domain 1 is necessary, but not sufficient, for homophilic binding. Thirdly, 5 individual residues in domain 1 which are essential to mediate homophilic adhesion have been defined.

The model of CD31/CD31 interaction has evolved in the light of continuing experimental dissection. Initially it was envisaged that the N-terminal, membrane distal domains 1 and 2 of CD31 engaged the membrane proximal domains 5 and 6 on an apposing cell surface and vice versa. To further explore the role of the N-terminal domains, a series of chimeric ICAM-3/CD31 constructs were made, in which a nested series of replacements of CD31 domains by the equivalent domains from I(AM-3 were allowed to adhere to full length CD31, demonstrating that domain 1, in addition to domains 2 and 3 is involved in homophilic adhesion. An important prediction of a fully interdigitating anti-parallel model for homophilic binding is that removal of one of the two binding sites on each molecule would be expected to reduce, rather than eliminate binding. Replacement of the N-terminal domain 1 alone completely ablates binding. In principle, this could arise because removing one binding site in a two site interaction reduces the affinity of the interaction below the detectable threshold of the assay employed. The obvious alternative interpretation is that homophilic binding results from a direct interaction between N-terminal domains, excluding a role for the membrane proximal portion of the molecule. This latter interpretation is consistent with the results obtained by Sun and colleagues (1996), who used an elegant series of loss-of-function and gain-of-function mutants to demonstrate that homophilic adhesion required the presence of domains 1 and 2, and that it was possible to confer the ability to bind human CD31 on mouse CD31 by replacing its first two domains with the equivalent domains from human CD31. In order to determine whether the N-terminal domains alone are homophilically competent in a non-CD31 backbone, constructs were produced in which the N-terminal domains of CD31 were expressed on a stalk consisting of ICAM-3 Ig domains. These chimeras preserved the overall length of the extracellular domain as compared to wild type CD31, thus removing any steric constraints arising from the use of truncated CD31-Fc chimeric proteins.

Importantly, these reagents revealed that while the N-terminal domains are necessary, they are not sufficient to support homophilic adhesion outside of the context of an accessory function provided by the membrane proximal CD31 domains. A model was proposed in which homophilic contact between the N-terminal domains of CD31 requires that these binding domains are held in the correct special orientation—a function provided by the membrane proximal domains. This would explain the positive gain-of-function results obtained by Sun et al. (1996) described above. In this construct, the binding functions would be provided by the human N-terminal domains, while the murine domains perform an accessory role in maintaining the human domains in the correct orientation. Interestingly a number of studies have suggested a positive role for domain 6 in CD31 mediated interactions. For example, Fab fragments of the antibody 4G6, which recognizes the epitope CAVNEG in domain 6 of CD31 (Yan et al., 1995) has been shown to enhance CD31 homophilic adhesion (Sun et al., 1996).

The mechanism by which the membrane proximal domains modulate the binding domains is currently under investigation. One attractive possibility is that these domains mediate cis-interactions between CD31 on the same cell to dimerise or cluster multiple low affinity binding domains, resulting in an increase in overall avidity, as has been shown for ICAM-1 (Miller et al., 1995) and N-cadherin (Shapiro et al., 1995). In addition, it has also been demonstrated that immunoglobulin domains distinct from the ligand binding domains are capable of mediating homodimerisation, for example the fourth immunoglobulin domain of the stem cell factor receptor c-kit (Blechman et al., 1995).

Therefore, it is possible to envisage a model in which CD31 homophilic engagement involves both cis- and trans-interactions. In our model, low affinity trans-interactions mediated by N-terminal domains are clustered to achieve higher avidity by cis-interactions mediated by the membrane proximal domains, forming a zipper-like interaction (see Figure 2). This model provides interesting regulatory possibilities. A puzzling aspect of CD31 biology is that although CD31 is widely expressed on mature, circulating hemopoietic cells, CD31+ cells do not undergo spontaneous CD31-mediated homophilic aggregation. This could be explained in terms of regulation of the cis-interactions of CD31, for example by cytoskeletal association and cell surface distribution. The recent finding that cytokine treatment of endothelial monolayers induces a re-distribution of CD31 from sites of cell–cell contact to the apical surface are particularly suggestive and could also provide a mechanism for switching between homophilic and heterophilic interactions. Alternatively, conformational changes in the membrane proximal domains may be

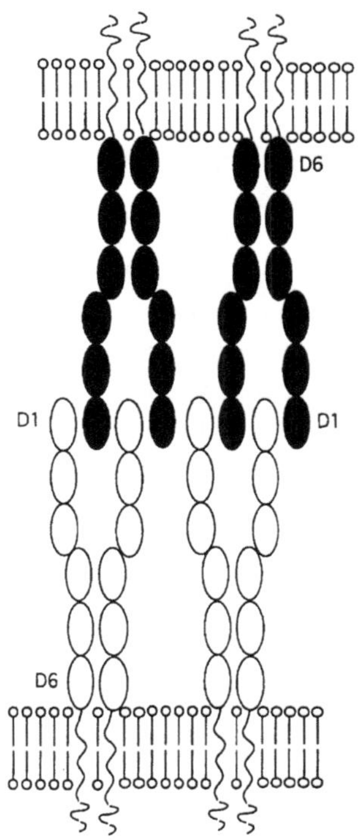

Figure 2. CD31–CD31 A zipper type model; dominance of domain 1 and role for dimerization via domains 4–6.

directly translated into conformational shifts in the binding domains and consequent changes in affinity in a fashion analogous to integrins.

In antigen–antibody complexes, antigen recognition and binding is mediated by loops at the N-terminal end of the fold. A recurring theme among integrin-IgSF interactions is the utilization of acidic residues by the loop joining the CC' β-strands on the CC'FG face of the Ig fold. For example, the LETS motif identified in ICAM-3 (Holness et al., 1995), and the QIDS motif in VCAM (Voderheide et al., 1995). In the case of CD2/CD48, a pseudo-homophilic interaction, mutagenesis data implicate an adhesive surface comprised of the equivalent GFCC'C" faces of the immunoglobulin fold in both CD2 and CD48 (van der Merwe et al., 1995). To date, no systematic site-directed mutagenesis of any fully homophilic IgSF interaction has been undertaken. There are some limited data on the interaction of NCAM based on the ability of a decapeptide derived from its third immunoglobulin domain to block homophilic adhesion (Rao et al., 1993). Likewise spontaneous mutations in L1 give rise to neurological disease such as X-linked hydrocephalus (Jouet et al., 1994). However, many of these mutations result in frameshifts and premature truncation, and only 26 result in missense mutations. Of these, 13 are predicted to affect key structural residues or domain boundaries and destabilize the protein, while others introduce cysteines and may promote intermolecular disulphides. Hence, spontaneous mutations have so far yielded few insights into the nature of homophilic interactions. The analysis of homophilic CD31/CD31 represents the most extensive directed analysis of the structures mediating IgSF homophilic adhesion.

Significantly, site-directed mutagenesis of CD31 has identified residues on both faces of the immunoglobulin fold as being involved in the homophilic contact sites. These are a cluster of two acidic residues, D11 and D33, which lie on the predicted A and B strands, K89 which lies at the top of the FG loop, and two residues K50 and D5 1 on the C-C' loop. This finding represents a novel mode of interaction between immunoglobulin folds and are consistent with the zipper model proposed above, where each CD31 molecule interacts with two others on an apposing cell surface, thus requiring two distinct binding faces.

CD31-mediated adhesion is complex both in terms of the number of potential ligands, both homophilic and heterophilic, and in terms of the domains involved in each interaction. Molecular studies have identified a role for both N-terminal and membrane proximal domains of CD31 in homophilic adhesion and the zipper model incorporates the current available data. This model is consistent with mutagenesis data, which indicate that both faces of immunoglobulin domain 1 are involved in the homophilic interaction, and provide a framework for future biophysical analysis of homophilic adhesion.

F. Summary of Homophilic Mode

In general, the interaction of homophilic CAMs may be a two-part process. The first is in an initial alignment of apposed CAMs, most probably in an anti-parallel orientation with initial engagement of dominant interaction or recognition motifs in key domains. Secondly, this is followed by docking of additional domains or dimerisation of molecules in cis by membrane proximal domains that lead to stabilization of the homophilic interaction

III. HETEROPHILIC BINDING; THE DOMINANCE OF N TERMINAL DOMAINS

Some of the most detailed molecular pictures of IgSF/receptor interactions have come from heterophilic CAMs largely involving leukocyte cell surface molecules, specifically CD2/CD58 CD4/class II, VCAM/VLA-4 and ICAM-1, -2, -3/LFA-1. Combinations of mutagenesis screens with modelling and/or structural analysis have localized binding sites to particular domains, faces and key residues. The results indicate that in both these receptor/ligand pairs, the IgSF binding surface is localized to one domain, namely the N terminal domain 1, one Ig fold face, namely the CFG face and also a specific set of key residues are located within that face and at its edges (see Figure 3).

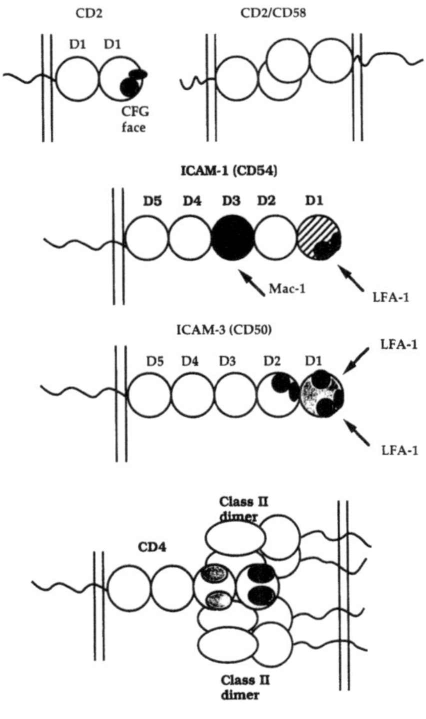

Figure 3. Model of CD31 domain 1 with key residues involved in homophilic binding.

A. CD2/LFA-3

CD2 has 2 Ig domains and interacts with the closely related LFA-3 which also has 2 Ig domains (Driscoll et al., 1991; Springer, 1991). In rodents, there does not appear to be an excat homologue of LFA-3 but an alternative, OX48, is used. For human CD2, residues clustered in the β strands C (K34 and E36) and C' (R48 and K49) and also charged residues in the connecting loop between the F and G strands (K91 and N92) have been shown to be critical in this interaction (Somoza et al., 1993; Arulanandam et al., 1993).

B. ICAM/LFA-1

Three ICAMs have been characterised in detail. All bind to the leukocyte integrin LFA-1. The extracellular regions of ICAM-1 (Simmons et al., 1988; Staunton et al., 1988), ICAM-2 (Staunton et al., 1989) and ICAM-3 (Fawcett et al., 1992; Vazeux et al., 1993; de Fougerolles, 1993) have 5, 2 and 5 C2 type Ig domains respectively. Similarly located charged residues have been identified as key components of the binding site on ICAMs 1 and 3 for the leukocyte integrin LFA-1. For human ICAM-1 and ICAM-3, residues E37 (E34, ICAM-1) in the CC' strand and Q75 (Q72, ICAM-1) within domain 1 of ICAMs -1 and 3 are essential components of the LFA-1 binding site (Staunton et al., 1990; Holness et al., 1995; Klickstein et al., 1997; Fisher et al., 1997). In addition, Li et al. (1993) have defined a peptide from domain 1 of ICAM-2 (covering residues 21–42, Staunton et al. 1989) which inhibits binding of ICAM-2 to LFA-1. This peptide encompasses β strands B, C and C' and thus confirms the importance of this region for all the ICAMs as they interact with LFA-1. The key residues, E34 and Q73, are conserved across species and are present in chimpanzee ICAM-1 (Hammond and McClelland, 1993), rat ICAM-1 (Kite et al., 1992) mouse ICAM-1 (Horley et al., 1989) and mouse ICAM-2 (Xu et al., 1992).

Indeed, this motif may be generally important for the interaction of integrins with members of the immunoglobulin superfamily (IgSF). To date, there are three examples of IgSF/integrin interactions: ICAMs with LFA-1 and/or Mac-1 (Simmons et al., 1988; Staunton et al., 1989; Diamond et al., 1990); VCAM with VLA-4 (α4β1) and α4β7; and MadCAM-1 with the α4β7 integrin (Berlin et al., 1993) (see below).

ICAM-1 can also bind to the major group of human rhinoviruses and to an as yet unidentified receptor expressed on the surface of *Plasmodium falciparum* infected erythrocytes. In both cases the interaction surfaces are much larger than the LFA- 1 binding face. ICAM-1 is thought to fit into a deep canyon on the surface of rhinovirus and the binding site encompasses a large area of the ICAM. Similarly, the *P. falciparum* receptor makes multiple contacts on many faces of domain 1, some overlapping with the LFA- 1 contacts and some additional malaria specific ones.

C. VCAM/VLA-4

VCAM, with seven C2 type domains, engages the integrin VLA-4 ($\alpha4\beta1$) using either domain 1 or domain 4, but employs the same pentapeptide motif, QIDSP, in the exposed CD loop region. The key residue is the charged aspartic acid, which occurs at equivalent locations on the domains D1 and D4, i.e. within the CD loop (Jones et al., 1995; Vonderheide et al., 1994).

Thus, a structural motif consisting of two elements, an acidic residue (either D or E) located in a CD loop and additional residues in the FG loop region, is emerging as a theme in IgSF interactions.

D. Extended Faces

Recent work has revealed that, in addition to the major interaction face located in the N-terminal domain, the contiguous domain, D2, also contributes to the binding site. For CD2, it has long been known that combinations of mAbs that map to domains 1 and 2 need to bind in order to activate T cells. Mutations in the CC' loop of domain 2 (so-called region 3) affect mAb and ligand binding (Peterson and Seed, 1987). Now for ICAM-3, mutations in domain 2 (R127 in the C-C' loop and D166 within the F-G loop) reduce binding to LFA-1 (Holness et al., 1995). Interestingly, these are in equivalent regions to the key residues identified in domain 1.

However, it is clear that the N-terminal two domains in both these cases do not contribute equally to the binding site. Domain 1 is dominant with respect to binding, with domain 2 playing a less critical role. Mutation of the key residues in domain 1 prevents interaction with ligand, while a reduced level of binding is still retained after mutation of the important sites in domain 2. From this work, we speculate that binding may be (at least) a two step process, an initial weak/low affinity binding or docking with domain 2, followed by a high affinity engagement of key residues in domain 1.

The interaction of CD4 with MHC class II, and CD8 with MHC class I, offers a more complex mode of interaction where binding occurs over very extended surfaces that involve multiple domains. In addition, there may be multiple interactions with dimeric forms of the molecules, especially single CD4 molecules binding to dimers of class II molecules and homodimers of CD8 binding to single class I molecules. The crystal structures of class II revealed a dimeric interaction between class II molecules, i.e. a dimer of dimers (Brown et al., 1993). It was proposed that this dimer could interact with 2 CD4 molecules simultaneously, and this agrees with the accumulated data from mutagenesis studies that were previously contradictory but which can now be reconciled. Mutagenesis screens had indicated that all lateral faces (the ABE face and the CFG face) of CD4 seem to bind to class II. If two class II molecules can bind together as a dimer, a single CD4 can engage them via these two opposite faces (see Figure 4) (Moebius et al., 1993). This contrasts with the

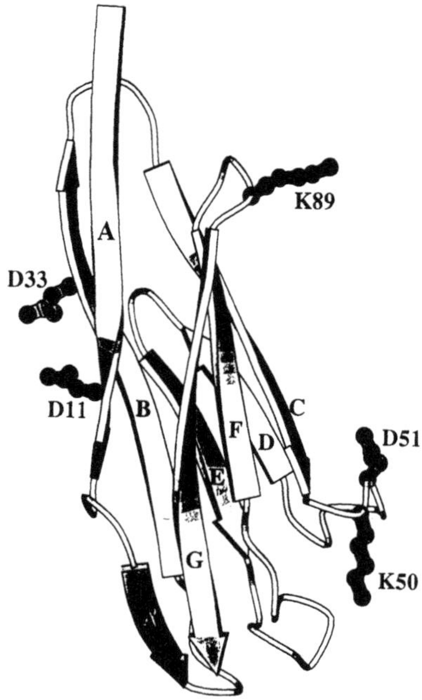

Figure 4. Binding sites in heterophilic CAMs. CD2, and the interaction of CD2 with CD58, ICAMs 1 and 3 interacting with the integrin LFA-1 via domain 1, and CD4 interacting with a dimer of MHC class II molecules.

restricted binding site for HIV gp 120 which involves only the CDR 3 like loop that is on one face of CD4 (Peterson and Seed, 1988).

Recently, one group has proposed that CD4 can self-dimerise. There is some evidence for the role of a CDR3 loop in domain 1 acting as a low affinity dimerisation indicia; a peptide derived from CD4 domain 1 in a CDR3 like loop (D I residues 87-98; EDQKEEVQLLVF) binds to CD4-Fc in an *in vitro* solid phase adhesion system (Langedijk et al., 1993). However, there is little evidence for the existence of CD4 dimers at the cell surface, and, although dimers were evident in CD4 crystals (Ryu et al., 1990; Wang et al., 1990; Brady et al., 1993), the area of contact in these dimers was deemed to be insignificant. This contrasts with CD8, where there is clear evidence for the existence of CD8 homodimer at the cell surface, and the crystal homodimers had extensive surface area in contact (Leahy et al., 1992).

Thus, the overall pattern of homophilic IgSF interactions that has emerged from recent studies points to a more complex mode than for heterophilic adhesion where dominant domains(s) are located in membrane distal N-terminal positions and interact head to head. Homophilic IgSF CAMs display a range of binding modes from the simple head-to-head interactions to more complex interdigitating interactions requiring correct alignments of the CAMs along their lengths.

IV. OLIGOSACCHARIDES

Before moving on from this type of interaction, some mention should be made of the potential influence that oligosaccharide chains have on binding. Many IgSF members are very heavily glycosylated and are especially rich in N-linked glycans (e.g., half of the mass of ICAM-3 is due to glycan decoration and there are seven putative N-linked glycan sites in the N-terminal domains that contain the LFA-1 binding site). There is little experimental evidence for the role of these oligosaccharide chains in binding, but as there are so many of them, they must influence docking interactions. The one clear example of the influence of glycans is in the binding of ICAM-1 to its β2 integrin receptor Mac-1 (CD11b/CD18) (Diamond et al., 1992). N-linked glycans decorating domain 3 of ICAM-1 negatively affect Mac-1 binding; removing them by either enzymatic cleavage or use of inhibitors of glycosylation increased ICAM-1/Mac-1 adhesion. However, in other cases the glycans may not play such a significant role since they may not impinge on the direct interaction face. This appears to be true for CD2 where the three dimensional structure of the fully glycosylated form of domain 1 of human CD2, determined by nuclear magnetic resonance spectroscopy, places the oligosaccharide at the top of the connecting loops between the β-strands, at the perimeter of the CD58 ligand binding site (Withka et al., 1993). Clearly, more experimental studies on the influence of glycans on binding and structural models of their positions relative to binding surfaces are needed.

V. GROWTH FACTOR RECEPTOR BINDING POCKETS

The main focus of this chapter has been IgSF members involved in cell adhesion or cell–cell recognition. However, the extracellular domains of many growth factor receptors (MCSF 1 -R, PDGF-R, FGF-R) contain IgSF domains. The ligand binding sites of some of these receptors have been analyzed recently and reveal a common yet distinct binding pattern. The M-CSF-1 receptor, whose extracellular domain consists of five Ig domains, binds its ligand in the third Ig domain (Wang et al., 1993). Similarly, the PDGF receptor (extracellular domain also consisting of five Ig domains, Heidaran et al., 1990) and some of the FGF receptor family bind their ligands via D3 (Yayon et al., 1992). All these ligands are small and the binding pocket is therefore likely to be confined to a single domain. The ligand/Ig domain contact points have not yet been defined. Certain members of the FGF receptor family use a novel mechanism for altering ligand specificity based on differential splicing of a 50 amino acid module into the C-terminal half of D3 (Yayon et al., 1992). So far this remains unique to these FGF receptors and a similar mechanism for altering ligand specificity has not been found in any of the adhesion molecule IgSF members.

For many growth factors, a key event induced by ligand binding is receptor dimerisation. The dimerisation of the extracellular domains is thought to lead to pairing of the cytoplasmic tails which then activates intrinsic tyrosine kinases or allows association of intracellular kinases. Apart from the proposed model for class II dimers binding to CD4 (dimers ?), this mechanism for adhesion molecule signal transduction has not yet been studied. As recognition/adhesion events between adjacent or apposed cells lead to mass pairings of CAMs, it is possible that the cytoplasmic tails of these multimerised CAMs undergo conformational changes allowing association with signalling and cytoskeletal machinery.

VI. DYNAMICS OF IgSF INTERACTIONS

This chapter has focussed on the static structural aspects of CAM interactions. However, adhesion is a dynamic process and a recent quantitative study of the affinities and kinetics of the interactions of rat CD2 with its ligand CD48 showed that the K_d for this binding was very low (60–90 mM) (van der Merwe et al., 1993). This was due to an extremely rapid off- rate constant ($K_{off} \geq 6\ s^{-1}$), whereas the on-rate constant was as unremarkable ($K_{on} \geq 10^5\ M^{-1}s^{-1}$). Thus, the binding between individual CAMs may be very weak, and only yields significant adhesive strength through multi-valent presentation as occurs at cell surfaces in contact. Another element of dynamics not touched on here is the conformational changes that can occur within CAMs as a result of ligand binding. This has been clearly established for antibodies as they bind antigen; major rearrangements can be induced in Ig domains even when small peptide ligands are bound (Stanfield et al., 1993). There is no structural data yet available for shape changes in the Ig domains of CAMs as they interact with their ligands, but developments in this area are eagerly awaited.

VII. CONCLUSIONS

Inspite of the very conserved patterns of domain folding and domain organization seen in members of the IgSF, there is a surprising diversity in their modes of adhesive interactions. Binding sites can be localized to single domains or involve extended faces encompassing several domains. The sites can be located in N-terminal membrane distal domains or internal and even membrane proximal domains. Molecules can interact via N-terminal membrane distal domains in a head to head manner, or require full inter-digitation of many domains of the CAMs in an anti-parallel orientation on apposed cell surfaces.

ACKNOWLEDGEMENTS

D L. Simmons is a Wellcome Trust Senior Research Fellow in Basic Biomedical Science. The Cell Adhesion Group is funded by the Wellcome Trust, the Imperial Cancer Research

Fund, the Medical Research Council, the Arthritis and Rheumatism Research Council, the Yamanouchi Research Institute, and the British Heart Foundation. I would like to thank all members of the lab (in particular John Fawcett, Claire Holness, Chris Buckley, Justin Newton and Elaine Bell) for contributing data and discussions on which this chapter is based

REFERENCES

Arulanandam, A. R. N., Withka, J. M., Wyss, D. F., Wagner, G., Kister, A., Pallai, P., Recny, M. A., & Reinherz, E. L. (1993). The CD58 (LFA-3) binding site is a localized and highly charged surface area on the AGFCC'C" face of the human CD2 adhesion domain. Proc. Natl. Acad. Sci. USA. 90, 11613-11617.

Berendt, A. R., McDowall, A., Craig, A. G., Bates, P., Sternberg, M. J. E., Marsh, K., Newbold, C. I., & Hogg, N. (1992). The binding site on ICAM-1 for Plasmodium falciparum-infected erythrocytes overlaps, but is distinct from, the LFA-1 binding site. Cell 68, 71-81.

Berlin, C., Berg, E. L., Briskin, M. J., Andrew, D. P., Kilshaw, P. J., Holzmann, B., Weissman, I. L., Hamann, A., & Butcher, E. C. (1993). $\alpha 4\beta 7$ integrin mediates lymphocyte binding to the mucosal vascular addressin MAdCAM-1. Cell 74, 185-195.

Blechman J. M., Lev, S., Barg, J., Eisenstein, J., Vaks, B., Vogel, Z., Givol, D., & Yarden, Y. M. (1995). The fourth immunoglobulin domain of the stem cell factor receptor couples ligand binding to signal transduction. Cell 80, 103-113.

Brown, J. H., Jardetsky, T. S., Gorga, J. C., Stern, L. J., Urban, R. G., Strominger, J. L., & Wiley, D. C. (1993). Three dimensional structure of the human class II histocompatibility antigen HLA-DR1. Nature 364, 33-39.

Brady, R. L., Dodson, E. J., Dodson G. G., Lange, G., Davis, S. J., Williams, A. F., & Barclay, A. N. (1993) Crystal structure of domains 3 and 4 of rat CD4: relation to the NH_2-terminal domains. Science 260, 979-983.

Casanovas, J. M., Springer, T. A., Liu, J., Harrison, S. C., & Wang J. (1997). Crystal structure of ICAM-2 reveals a distinctive integrin recognition surface. Nature 387, 321-314.

Cheung, P. H., Luo, W., Qiu, Y., Zhang, X., Earley, K., Millirons, P., & Lin, S.-H. (1993). Structure and function of C-CAM-1. The first immunoglobulin domain is required for intercellular adhesion. J. Biol. Chem. 268, 24303-24319.

Diamond, M. S., Staunton D. E., Marlin S. D., & Springer T. A. (1991). Binding of the integrin Mac-1 (CD11b/CD18) to the third Ig-like domain of ICAM-1 (CD54) and its regulation by glycosylation. Cell. 65 961-971.

Driscoll, P. C., Cyster, J. G., Campbell, I. D., & Williams, A. F. (1991). Structure of domain 1 of rat T lymphocyte CD2 antigen. Nature, 353, 762-765.

de Fougerolles, A. R., & Springer, T. A. (1993). Intercellular adhesion molecule 3 a third counter-receptor for leukocyte function associated antigen 1 on resting lymphocytes. J. Exp. Med. 175, 185-190.

Fawcett, J., Holness, C. L. L., Needham, L. A., Turley, H., Gatter, K. C., Mason, D. Y., & Simmons, D. L. (1992). Molecular cloning of ICAM-3, a third ligand for LFA- 1, constitutively expressed on resting leukocytes. Nature 360, 481-484.

Fawcett, J., Buckley, C. D., Holness, C. L., Bird, I. N., Spragg, J. H., & Simmons, D. L. (1995). Mapping the homotypic binding sites in CD31 and the role of CD31 adhesion in the formation of inter-endothlelial cell contacts. J. Cell Biol. 128, 1229-1241.

Fisher, K. L., Lu, J., Kim., K. J., Presta, L. G., & Bodary, S. C. (1997). Identification of the binding site in ICAM-1 for its receptor, LFA-1. Mol. Biol. of the Cell 8, 501-510.

Garrett, T. P. J., Wang, J., Yan, Y., Liu, J., & Harrison, S. C. (1993). Refinement and analysis of the structure of the first two domains of human CD4. J. Mol. Biol. 234, 763-778.

Heidaran, M. A., Pierce, J. H., Jensen, R. A., Matsui, T., & Aaronson, S. A. (1990). Chimaeric alpha- and beta-platelet derived growth factor (PDGF) receptors define three immunoglobulin-like domains of the alpha-PDGF receptor that determine PDGF-AA binding specificity. J. Biol. Chem. 265, 18741-18744.

Hammond, L. & McClelland, A. (1993) Genbank Release 80.0. National Center for Biotechnology Information. National Library of Medicine, NIH, Bethesda, MD.

Holness, C. L., Bates, P. A., Littler, A. J., Buckley, C. D., McDowall, A., Bossy, D., Hogg, N., & Simmons D. L. (1995). Analysis of the binding site on ICAM-3 for th eleukocyte integrin LFA-1. J. Biol. Chem. 270, 877-884.

Horley, K. J., Carpenito, C. Baker, B., & Takei, F. (1989). Molecular cloning of murine intercellular adhesion molecule (ICAM-1). EMBO J. 8, 2889-2896.

Jones, E. Y., Harlos, K., Bottomley, M. J., Robinson, R. C., Driscoll, P. C., Edwards, R. M., Clements, J. M., Dudgeon, T. J., & Stuart, D. I. (1995). Crystal structure of an integrin binding fragment of VCAM-1 at 1.8 A resolution. Nature 373, 539-544.

Jones, E. Y., Davis, S. J., Williams, A. F., Harlos, K., & Stuart, D. I. (1992) Crystal structure at 2.8 A resolution of a soluble form of the cell adhesion molecule CD2. Nature 360, 232-239.

Jouet, M., Rosenthal, A., Armstrong, G., MacFarlane, J., Stevenson, R., Paterson, J., Metzenberg, A., Ionasescu, V., Temple, K., & Kenwick, S. (1994) Nature Genetics 7, 402-407.

Klickstein, L. B., York, M. R., de Fougerolles, A. R., & Springer T. A. (1996). Localisation of the binding site in ICAM-3 for LFA-1. J. Biol. Chem. 271, 23920-23928.

Kita, Y., Takashi, T., Iigo, Y., Tamatari, T., Miyasaka, M., & Horinchi, T. (1992). Sequence and expression of rat ICAM-1. Biochim. Biophys. Acta. 1131, 108-110.

Langedijk, J. P. M., Puijk, W. C., van Hoorn, W. P., & Meloen, R. H. (1993). Location of CD4 dimerization site explains critical role of CDR3-like region in HIV-1 infection and T-cell activation and implies a model for complex of co-receptor-MHC. J. Biol. Chem. 268, 16875-16878.

Leahy, D. J., Axel, R., & Hendrickson, W. A. (1992). Crystal structure of a soluble form of the human T cell coreceptor CD8 at 2.6 A resolution. Cell 68, 1145-1162.

Li, R., Nortamo, P., Valmu, L., Tolvanen, M., Huuskonen, J., Kantor, C., & Gahmberg, C. G. (1993). A peptide from ICAM-2 binds to the leukocyte integrin CD11a/CD18 and inhibits endothelial cell adhesion. J. Biol. Chem. 268, 17513-17518.

van der Merwe, P. A., Crocker, P. R., Vinson, M., Barclay, A. N., Schauer, R., & Kelm, S. (1996). Localisation of the putative sialic acid binding site on the immunoglobulin superfamliy cell surface molecule CD22. J. Biol. Chem. 271, 9273-9279.

van der Merwe, P. A., Brown M. H., Davis, S. J., & Barclay, A. N. (1993). Affinity and kinetic analysis of the interaction of the cell adhesion molecules rat CD2 and CD48. EMBO J., 12, 4945-4954.

Miller, J. Knorr, R., Ferrone, M., Houdei, R., Carron, C. P., & Dustin, M. L. (1995). ICAM-1 dimerization and its consequences for adhesion mediated by LFA-1. J. Exp. Med. 182, 1231-1241.

Moebius, U., Pallai, P., Harrison, S. C., & Reinherz, E. L. (1993) Delineation of an extended surface contact area on human CD4 involved in class II major histocompatibility complex binding. Proc. Natl. Acad. Sci. U.S.A. 90 8259-8263.

Newton J. P., Buckley, C. D., Jones, E. Y., & Simmons, D. L. (1990). Residues on both faces of the first immunoglobulin fold contribute to homophilic binding sites of PECAM-1/CD31. J. Biol. Chem. 272, 20555-20563.

Osborn, L., Vassalo, C., & Benjamin, C. D. (1992). Activated endothelium binds lymphocytes through a novel binding site in the alternatively spliced domain of VCAM-1. J. Exp. Med. 176, 99-107.

Overduin, M., Harvey, T. S., Bagby, S., Tong, K. I., Takeichi, M., & Ikura, M. (1995). Solution structure of the epithelial cadherin domain responsible for selective cell adheison. Science 267, 386-390.

Peterson, A. & Seed, B. (1987). Monoclonal antibody and ligand binding sites of the T cell erythrocyte receptor (CD2). Nature, 329, 842-846.

Peterson, A. & Seed, B. (1988). Genetic analysis of monoclonal antibody and HIV binding sites on the human lymphocyte antigen CD4. Cell 54, 65-72.

Rao, Y., Wu, X.-F., Gariepy, J., Rutishauser, U., & Siu, C. H. (1992). Identification of a peptide sequence involved in homophilic binding in the neural cell adhesion molecule NCAM. J. Cell Biol. 118, 937-949.

Ryu, S.-E., Kwong, P. D., Truneh, A., Porter, T. G., Arthos, J., Rosenberg, M., Dai, X., Xuong, N.-h., Axel, R., Sweet, R. W., & Hendrickson, W. A. (1990). Crystal structure of an HIV-binding recombinant fragment of human CD4. Nature. 348, 419-426.

Sadhu, C., Lipsky, B., Erickson, H. P., Hayflick, J., Dick, K. O., Gallatin, M. G., & Staunton, D. E. (1994). LFA-1 binding site in ICAM-3 contains a conserved motif and non-contiguous amino acids. Cell Adhesion & Communication. 2, 429-440.

Shapiro, L., Fannon, A. M., Kwong, P. D., Thompson, A., Lehmann, M. S., Grubel, G., Legrand, J. F., Alsnielsen, J., Colman, D. R., & Hendrickson, W. A. (1995). Structural basis of cell-cell adhesion by cadherins. Nature 374, 327-337.

Simmons, D. L., Makgoba, M. W., & Seed, B. (1988) ICAM-1, an adhesion ligand of LFA-1, is homologous to the neural cell adhesion molecule NCAM. Nature 331, 624-627.

Somoza, C., Driscoll, P. C., Cyster, J. G., & Williams, A. F. (1993) Mutational analysis of the CD2/CD58 interaction: the binding site for CD58 lies on one face of the first domain of human CD2. J. Exp. Med. 178, 549-558.

Springer, T. A. (1991) A birth certificate for CD2. Nature 353, 704-705.

Springer, T. A. (1990) Adhesion receptors of the immune system. Nature 346, 425-434.

Stanfield, R. L., Takimoto-Kamimura, M., Rini, J. M., Profy, A. T., & Wilson, I. A. (1993). Major antigen-induced domain rearrangements in an antibody. Structure. 1, 83-93.

Staunton, D. E., Marlin, S. D., Stratowa, C., Dustin, M. L., & Springer, T. A. (1988) Primary structure of intercellular adhesion molecule 1 (ICAM- 1) demonstrates interaction between members of the immunoglobulin and integrin supergene families. Cell 52, 925-933.

Staunton, D. E., Dustin, M. L., Erickson, H. P., & Springer, T. A. (1990) The arrangement of the immunoglobulin-like domains of ICAM- 1 and the binding sites for LFA- 1 and rhinovirus. Cell 61, 243-254.

Staunton, D. E., Merluzzi, V. J., Rothlein, R., Barton, R., Marlin, S. D., & Springer, T. A. (1989) A cell adhesion molecule ICAM-1, is the major surface receptor for rhinovirus. Cell, 56 849-853.

Staunton, D. E., Dustin, M. L., & Springer, T. A. (1989) Functional cloning of ICAM-2, a cell adhesion ligand for LFA-1 homologous to ICAM-1. Nature 339, 361-364.

Sun, J., Williams, J., Yan, H.-C., Amin, K. M., Albelda, S. M., & DeLisser, H. M. (1996). PECAM-1 homophilic adhesion is mediated by immunoglobulin-like domains 1 and 2 and depends on the cytoplasmic domain and level of surface expression. J. Biol. Chem. 271, 11090-11098.

Teixeira, A. M., Fawcett, J., Simmons, D. L., Watt, S. M., (1994). The N-domain of the biliary glycoprotein (BGP) adhesion molecule mediates homotypic binding: domain interactions and epitope analysis of BGPc. Blood 84, 211 -219.

Vazeux, R., Hoffman, P. A., Tomita, J. K., Dickinson, E. S., Jasman, R. L., St. John, T., & Gallatin, W. M. (1992). Cloning and characterisation of a new intercellular adhesion molecule ICAM-R. Nature 360. 485-488.

Vinson, M., van der Merwe, P. A., Kelm, S., May, A., Jones, E. Y., & Crocker, P. R. (1996). Characterisation of binding site the sialic acid binding site in sialoadhesin by site directed mutagenesis. J. Biol. Chem. 271, 9267-9272

Vonderheide., R. H., Tedder, T. F., Springer, T. A., Staunton, D. E. (1994) residues within a conserved amino acid motif of domains 1 and 4 of VCAM-1 are required for binding to VLA-4. J. Cell Biol. 125, 215-222.

Wang. J., Yan, Y., Garrett, T. P. J., Liu, J., Rodgers, D. W., Garlick, R. L., Tarr, G. E., Husain, Y., Reinherz, E. L., & Harrison, S. C. (1990). Atomic structure of a fragment of human CD4 containing two immunoglobulin-like domains. Nature 348, 411-418.

Wang, Z., Myles, G. M., Brandt, C. S., Lioubin, M. N., & Rohrschneider, L. (1993). Identification of the ligand-binding regions in the macrophage colony-stimulating factor receptor extracellular domain. Mol. Cell. Biol. 13, 5348-5359.

Watt, S. M., Fawcett, J., Murdoch, S. J., Teixera, A. M., Gschmeissner, S. E., & Simmons, D. L. (1994). CD66 identifies the biliary glycoprotein (BGP) adhesion molecule; cloning. expression and adhesion functions of the BGPc splice variant. Blood 84, 200-210.

Williams, A. F., Davis, S. J., He, Q., & Barclay, A. N. (1989). Structural diversity in domains of the immunoglobulin superfamily. Cold Spring Harbor Symp. on Quantitative Biol. 54, 637647.

Williams, A. F. & Barclay, A. N. (1988). The immunoglobulin superfamily-domains for cell surface recognition Annu. Rev. Immunol. 6, 381-412

Withka, J. M., Wyss, D. F., Wagner, G., Arulanandam, A. R. N., Reinherz, E. L., & Recny, M. (1993) Structure of the glycosylated adhesion domain of human T lymphocyte glycoprotein CD2. Structure 1, 69-81.

Wu, H., Kwong, P. D., & Hendrickson, W. A. (1997). Dimeric association and segemental variability in the structure of human CD4. Nature 387, 527-530.

Xu, H., Tong, I. L., de Fougerolles, A. R., & Springer, T. A. (1992). Sequence and gene organization of murine ICAM-2 J. Immunol. 149, 2650-2655.

Yan, H. C., Pilewski, J. M., Zhang, Q., Delisser, H. M., Romer, L., & Albelda, S. M. (1995). Localisation of multiple functional domains on human PECAM-1 (CD31) by monoclonal antibody epitope mapping. Cell Adhesion & Communication 3, 45-66.

Yayon, A., Zimmer, Y., Guo-Hong, S., Avivi, A., Yarden, Y., & Givol, D. (1992), A confined variable region confers ligand specificity on fibroblast growth factor receptors: implications for the origin of the immunoglobulin fold. EMBO J. 11, 1885-1890.

Zhou, H., Fuks, A., Alcaraz, G., Bolling, T. J., & Stanners, C. P. (1993). Homophilic adhesion between Ig superfamily carcinoembryonic antigen molecules involves double reciprocal bonds. J. Cell Biol. 122, 951-960.

PART II

ORGANIZATION OF ADHESION COMPLEXES

FOCAL ADHESIONS AND ADHERENS JUNCTIONS:
THEIR ROLE IN TUMORIGENESIS

Avri Ben-Ze'ev

Advances in Molecular and Cell Biology
Volume 28, pages 135-163.

ISBN: 0-7623-0495-2

I. INTRODUCTION

In multicellular organisms, special mechanisms have evolved that allow adhesion of their constituent cells to either the extracellular matrix (ECM) or to their neighbors. The adhesive sites of cells with the ECM and other cells (also known as junctions) allow the assembly of individual cells into ordered multicellular tissues and organs. Cells have developed a wide variety of adhesive interactions, at the molecular and structural levels, which determine cellular and tissue morphogenesis, and also influence major cellular processes including motility, growth, differentiation, and survival. The adhesions of cells with their environment comprise three major domains: (i) adhesion receptors, consisting of transmembrane molecules that are linked to (ii) the internal cytoskeleton by (iii) a submembranous plaque consisting of a multimolecular complex that contains both structural and signaling molecules (Geiger et al., 1995; Ben-Ze'ev and Bershadsky 1997). This molecular connection between the outside and inside of the cell provides the potential for transmitting mechanical forces and is also involved in the bidirectional signaling from the outside to the inside of the cell—and vice versa. Cell adhesion is now viewed as fulfilling not only a structural role, but also having a role in the interpretation of the information in the DNA into the three-dimensional pattern of cells and tissues. Several excellent reviews on this role of cell adhesion in signaling have recently been published (Juliano and Haskill, 1993; Clark and Brugge, 1995; Geiger et al., 1995; Schwartz et al., 1995; Jockusch et al., 1995; Gumbiner, 1996; Burridge and Chrzanowska-Wodnicka, 1996; Craig and Johnson, 1996; Yamada and Geiger, 1997; Bryant, 1997).

In this chapter, I describe studies on adhesive junctions that are linked to the microfilament system, both at cell–cell contact areas (adherens junctions) and cell–ECM adhesion sites (focal adhesions) (Geiger et al., 1995). Other types of cell–cell and cell–ECM junctions that associate with intermediate filaments are beyond the scope of this chapter (for desmosomes, see chapter by Garrod et al. in this volume).

The adherens type junctional complex (both cell–cell and cell–ECM; see Figure 1) will be addressed as a functional unit in this report. The biochemical and molecular composition of these junctions and the regulation of their assembly will briefly be described. Their potential role in regulating the tumorigenic ability of cells, in conjunction with mechanisms of adhesion-mediated signaling, will be discussed.

II. THE STRUCTURE OF FOCAL ADHESIONS

Typical focal adhesions are developed by many cells in culture (Abercrombie et al., 1971; Heath and Dunn, 1978) and are known to contain vinculin which was discovered as the first junctional plaque protein (Geiger, 1979; Burridge and Feramisco, 1980; see Figure 1c).

Vinculin is localized in the cytoplasm at the tips of microfilament bundles (see Figure 1b) at areas of tight adhesion with the underlying ECM (see Figure la). In addition to

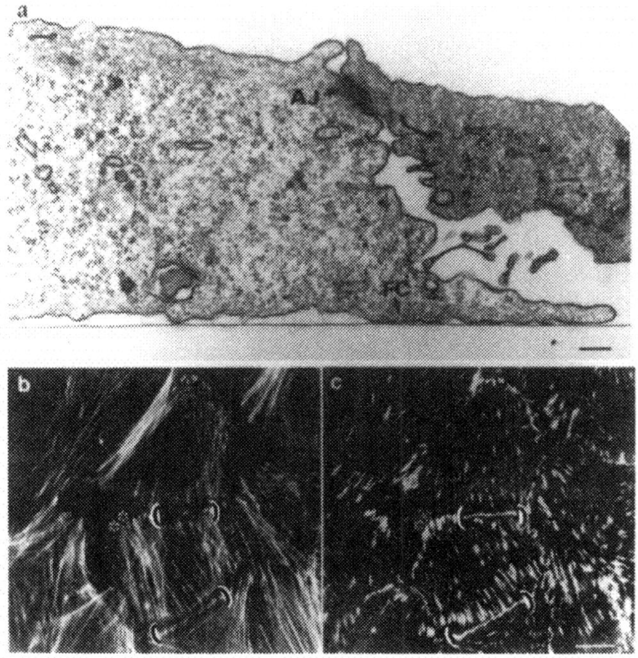

Figure 1. Organization of focal adhesions and adherens type cell–cell junctions in lens epithelial cells. (**a**) Electron micrograph showing the dense submembranal plaque of focal adhesion (FC) and adherens junctions (AJ) representing the abundance of microfilament bundles at these sites of adhesion. Double immunofluorescence for actin (**b**) and vinculin (**c**) demonstrates that stress fibers terminate at vinculin-rich focal adhesions (**), and colocalize at the adherens cell-cell junction belt (marked by parenthesis). Bars in (**a**) 0.2 μm; in (**b**) and (**c**) 10 μm. (From Geiger et al., 1995).

vinculin, a large number of junctional proteins have been purified and characterized, and these include both structural and signaling molecules (Luna and Hitt, 1992; Geiger et al., 1995; Ben-Ze'ev and Bershadsky, 1997). Focal adhesions are mainly characteristic of nonmotile cells in culture (Burridge, 1981) and are only rarely detected in organisms, except in cells involved in wound closure and under shear stress (Byers and Fujiwara 1982; Gabbiani et al., 1983). They are believed to be analogous to the electron-dense areas of dense plaques in smooth muscle and myotendinous junctions that contain similar junctional plaque components (Burridge et al., 1988).

A. The Transmembrane Integrin Receptors

The outermost components of focal adhesions are the transmembrane receptors of the integrin family, a large family of heterodimeric glycoproteins consisting of α and β subunits (Hynes, 1992; see also the chapter in this volume by Green and

Humphries). Most β chains can pair with different α chains and there are over 20 combinations of integrins, with β_1 and β_3 integrins being the main types involved in focal adhesion formation. Integrin gene disruption has strong effects on development (Yang et al., 1993).

The type of α integrin in the heterodimer can determine the ECM molecule that these receptors recognize. Multiple integrins are expressed on the surface of the same cell and sometimes different integrins for the same ECM component are expressed on a single cell, allowing a very wide choice of adhesive interactions for the cell with the ECM. In many ECM molecules, a major motif recognized by integrins is the Arg-Gly-Asp (RGD) sequence that is present on fibronectin, laminin, vitronectin, and other ECM molecules (Ruoslahti, 1996). The development of focal adhesions depends on the density of integrin ligands at the site of adhesion, suggesting that they need to be clustered to form focal adhesions (Yamada and Miyamoto, 1995). Most cells express the type of integrin that can interact with the matrix component that they encounter. Thus, on fibronectin, cells usually express and assemble $\alpha5\beta1$ integrin, the fibronectin receptor, while on vitronectin they express $\alpha v\beta3$ integrin in focal adhesions (Fath et al., 1989).

The cytoplasmic domains of integrins are essential for focal adhesion formation. Deletions or partial mutagenesis of these domains has shown that for both β_1 and β_3 integrins the cytoplasmic region is necessary for their targeting to focal adhesions (Solowska et al., 1989; Hayashi et al., 1990). The α integrin cytoplasmic domain plays an inhibitory role in focal adhesion targeting of the integrin heterodimer that is relieved upon integrin binding to its extracellular ligand (LaFlamme et al., 1992). This binding is believed to involve a conformational change that is transmitted to the cytoplasmic domain of integrin and allows its association with cytoplasmic proteins (Schwartz et al., 1995).

B. The Junctional Plaque Area

The association of the cytoplasmic domain of β integrin with the cytoskeletal plaque proteins is strongly stimulated by the occupancy of integrin by its ECM ligands (Miyamoto et al., 1995a b). Studies by these authors suggest that there is a hierarchy of interactions following integrin clustering that includes first the recruitment of focal adhesion kinase (FAK), tensin and several signaling molecules. Only after integrin clustering and other structural plaque proteins such as talin, vinculin and α-actinin are recruited, with paxillin and actin joining last, tyrosine phosphorylation takes place (Yamada and Miyamoto, 1995a,b). Such stepwise assembly of molecules at focal adhesions may enable the cell to build junctional complexes that are suitable for different functions, for example motility, versus stable adhesion.

The number of molecules identified at focal adhesions is constantly increasing. An attempt to describe some of the complex interactions among both structural and signaling molecules in focal adhesions is shown in Figure 2. For a more detailed description of these interactions the reader is referred to recent reviews (Jockusch et

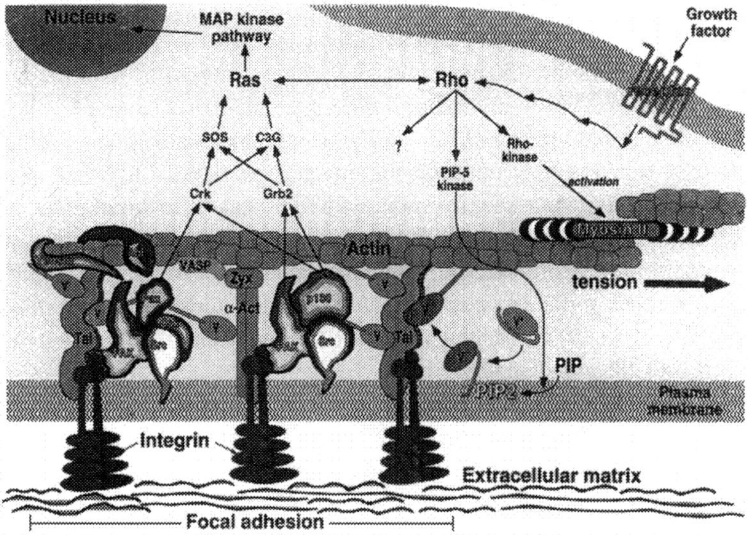

Figure 2. A scheme depicting the organization of focal adhesions and interactions between signaling by focal adhesions and growth factor receptors. Actin filaments are linked to the extracellular matrix via transmembrane integrin receptors by talin (Tal) and α-actinin (α-Act). Vinculin (V) has talin- and α-actinin-binding sites and can stabilize the link between these structural components of focal adhesions. Vinculin can also fold into a conformation where its binding sites are masked (V*). Zyxin (Zyx) is an α-actinin-binding protein that can bind to VASP which can bind to actin filaments.

The assembly of focal adhesions depends on the activation of Rho that is activated by a variety of growth factors. Rho has multiple targets in the cell, some directly related to assembly of focal adhesions and actin bundles. It binds and activates Rho-kinase that phosphorylates a phosphatase of myosin II light chain. This increases contraction (tension) that is central for the formation of actin bundles and focal adhesions. Rho also activates PIP5-kinase to generate PIP2 from PIP. PIP2 can interact with a number of actin-binding proteins to promote actin polymerization and can stimulate the transition of vinculin from a non-active conformation (V*) to one capable of talin- and actin-binding. Protein tyrosine kinases are also localized at focal adhesions. Among them are: focal adhesion kinase (FAK) that binds the cytoplasmic domain of β integrin, Src and Src-family kinases, Csk (a negative regulator of Src kinases), and Abl (binds actin). Other components of focal adhesions, including paxillin (Pax), tensin, and pl30^{cas} (pl30), are common substrates for tyrosine phosphorylation. The adhesion of cells to the extracellular matrix induces tyrosine phosphorylation of FAK, increases its tyrosine kinase activity, and induces tyrosine phosphorylation of paxillin, tensin and p130^{cas}. Cell adhesion can activate the Ras-MAP kinase signaling pathway: Tyrosine phosphorylated FAK can bind the SH2/SH3 adaptor protein Grb2, tyrosine phosphorylated paxillin, that can bind another SH2/SH3 adapter protein- Crk. p130^{cas} can bind Crk and Grb2. The SH3 domains of both Crk and Grb2 bind to the guanine nucleotide exchange factors SOS and C3G that can activate Ras and thus stimulate the MAP kinase pathway. (From Ben-Ze'ev, 1997).

al., 1995; Geiger et al., 1995; Burridge and Chrzanowska-Wodnicka, 1996; Craig and Johnson, 1996; Ben-Ze'ev and Bershadsky, 1997).

Actin and vinculin are the most abundant components of focal adhesions (see Figures 1b and c), and they can also bind to a large number of other molecules residing in the submembrane plaque. More minor components such as tensin, talin, paxillin, and α-actinin also have multiple binding partners. These different molecules can either fulfill redundant roles in linking actin filaments to focal adhesions or may have different roles in various types of focal adhesions that cells assemble. In cells where vinculin expression was eliminated by vinculin gene disruption, focal adhesions were formed in the absence of vinculin and contained α-actinin, talin, and paxillin (Coll et al., 1995; Volberg et al., 1995), suggesting that alternative pathways for focal adhesion assembly exist.

In vitro studies have indicated that integrin can be linked to actin by α-actinin that has both actin binding and integrin binding domains (Otey et al., 1990), or by talin that also has such sites (Horwitz et al., 1986; Muguruma et al., 1990: Hemmings et al., 1996; McCann and Craig, 1997).

Vinculin can interact with actin, α-actinin, talin, paxillin, and with itself (Jockusch et al., 1995; Craig and Johnson, 1996). Recent studies have shed new light on the controversy regarding the binding of vinculin to actin, when it became apparent that this binding site(s) on vinculin is masked by an intramolecular head to tail association (Menkel et al., 1994; Johnson and Craig 1995a,b; Jockusch et al., 1995; Yamada and Geiger 1997). This interaction masks both the talin and the actin binding sites on vinculin and the regulation of the unmasking of these binding sites is apparently central to focal adhesion assembly.

Vinculin was also shown to bind acidic phospholipids (Niggli et al., 1990; Niggli and Gimona 1993; Johnson and Craig, 1995a), especially PIP2 (Fukami et al., 1994). The PIP2 binding site on vinculin is between the head and tail domains of the molecule (Johnson and Craig, 1995b). PIP2 is thought to produce a conformational change in vinculin that allows its unfolding and the exposure of the actin and talin binding sites on the molecule, and of a phosphorylation site for PKC on the tail domain, thus allowing the involvement of vinculin in focal adhesion assembly (Weekes et al., 1996; Gilmore and Burridge, 1996; see Figure 2). Recent studies have identified two separate actin binding sites on the tail domain of vinculin (Hüttelmaier et al., 1997) that coincide with the PIP2 binding sites, and PIP2 binding was shown to induce vinculin oligomerization in vitro (Hüttelmaier et al., 1998).

Zyxin is a component found at low abundance in focal adhesions (Beckerle, 1986). It can bind to α-actinin and to another protein, CRP, that like zyxin, is rich in LIM domains (Sadler et al., 1992) and is also localized in focal adhesions (Crawford et al., 1992). LIM domains are zinc fingers rich in histidine and cysteine, and are believed to be involved in protein–protein interactions (Schmeichel and Beckerle, 1994). These domains have been identified in transcription factors that play a role in muscle development, but zyxin has no obvious nuclear localization sequences. Interestingly, truncation of its N-terminus induces nuclear translocation of zyxin (Nix and Beckerle, 1995). Zyxin can interact with the SH3 (src homology 3) domain of the proto oncogene vav (Hobert et al., 1996), raising the interesting possibility that zyxin may shuttle between focal adhesions and the nucleus to convey information from adhesive interactions into the nucleus.

Another protein that interacts with zyxin is VASP (vasodilator stimulated phosphoprotein) (Reinhard et al., 1995a). This is a substrate for various protein kinases and is localized at both focal adhesions and on stress fibers (Reinhard et al., 1992; Haffner et al., 1995). VASP was recently identified in dense plaques and dense bodies of smooth muscle cells (Markert et al., 1996). In addition, VASP can interact with the proline-rich domain of vinculin (Brindle et al., 1996; Reinhard et al., 1996) and with profilin (Reinhard et al., 1995a). Further in vitro studies have indicated that VASP binding to vinculin is stronger when vinculin is in an oligomeric form induced by vinculin-PIP2 binding (Hüttelmaier et al., 1997b), supporting a role for PIP2 in the formation of VASP-vinculin oligomers that may provide precursors for focal adhesion and stress fiber assembly.

This notion is supported by studies demonstrating that the intracellular movement of bacteria such as *Listeria* and *Shigella* involve actin polymerization and form an "actin tail" at one end of these bacteria (Pollard, 1995). Profilin and VASP are localized at this site and influence bacterial movement by interacting with the bacterial protein ActA in *Listeria* and IcsA in *Shigella* (Theriot et al., 1994; Smith et al., 1996; Chakraborty et al., 1995; Pistor et al., 1995). In addition, vinculin and α-actinin were also recently detected at these sites (Zeile et al., 1996; Suzuki et al., 1996) and vinculin was shown to affect *Shigella* entry into epithelial cells (Tran Van Nhieu et al., 1997). The role of vinculin in bacterial motility inside the cell is, however, still under debate (Goldberg, 1997).

The current view is that VASP, zyxin, and profilin are important for regulating the nucleation of actin assembly at both the motile bacterial tail and focal adhesions and that more mature focal adhesions contain abundant vinculin but contain less VASP (Bubeck et al., 1997) and profilin (Buss et al., 1992; Mayboroda et al., 1997).

C. Signaling Molecules at Focal Adhesion Sites

Focal adhesions are enriched in various components participating in signal transduction (reviewed in Geiger et al., 1995; Clark and Brugge, 1995; Craig and Johnson, 1996; Ben-Ze'ev and Bershadsky, 1997). Among these are protein kinases and phosphatases, their target adaptor proteins, and other enzymes involved in different signaling cascades (see Figure 2). The assembly of focal adhesions requires the recruitment of a multimolecular complex containing both structural and signaling molecules. Focal adhesions are the major site for the localization of phosphotyrosinated proteins in the cell, and tyrosine phosphorylation is, therefore, believed to play a crucial role in focal adhesion assembly and in the propagation of signals triggered by adhesion via the integrin receptors (Geiger et al., 1995;

Schwartz et al., 1995; Burridge and Chrzanowska-Wodnicka, 1996; Craig and Johnson, 1996; Yamada and Geiger, 1997).

A very important component participating in signal transduction at focal adhesions is focal adhesion kinase $pp125^{FAK}$ (FAK) (Schaller and Parsons, 1994). The activation of this kinase is an early event in adhesion-mediated signaling involving the integrin receptors (Burridge et al., 1992; Richardson and Parsons, 1995; Parsons, 1996). FAK can bind to the cytoplasmic domain of β integrin and other components of focal adhesions including paxillin (Turner and Miller, 1994; Rankin and Rozengurt, 1994), talin (Chen et al., 1995) and $p130^{Cas}$ (Nojima et al., 1995). Cells also express a truncated FAK form $pp41/43^{FRNK}$ lacking the protein kinase domain. This protein was suggested to act as an endogenous inhibitor of FAK activation since FRNK overexpression results in delayed assembly of focal adhesions in cells attaching to fibronectin (Richardson and Parsons, 1996). Another member of this family is PYK2 (proline-rich tyrosine kinase 2), or cell adhesion kinase (CAK) (Lev et al., 1995; Sasaki et al., 1995; Avraham et al., 1995). Both FAK and PYK2 signaling can be activated by an integrin- triggered process leading to phosphorylation of Tyr397 on these molecules. This creates a binding site for another tyrosine kinase Src that binds via its SH2 domain to form a bipartite kinase complex consisting of FAK and Src, or FAK and Fyn (Cobb et al., 1994; Schaller and Parsons, 1994; Xing et al., 1994; Parsons and Parsons, 1997). The formation of such a bipartite kinase complex is believed to be crucial for focal adhesion assembly and integrin mediated signaling and includes the recruitment of downstream components of the Grb2/SOS pathway (Schlaepfer et al., 1994). This view is supported by the finding that FAK overexpression increases tyrosine phosphorylation of paxillin and tensin (Schaller and Parsons, 1994), but a kinase-deficient FAK that cannot bind Src is unable to induce these events. In addition, when FRNK, which lacks the catalytic domain of FAK, is coexpressed with the catalytic domain of Src, it can restore tyrosine phosphorylation of paxillin and tensin (Parsons and Parsons, 1997). Fibroblasts from FAK-deficient mouse embryos, however, are able to form focal adhesions that are tyrosine phosphorylated, and which are even more abundant and stable than in control cells (Ilic et al., 1995). In addition, microinjection of cells with a FAK construct lacking the kinase domain does not inhibit the formation of focal adhesions, but inhibits motility and DNA synthesis (Gilmore and Romer, 1996). These results suggest that the major function of FAK may not be in focal adhesion assembly, but in the transduction of signals from these structures downstream, inside the cell. For example, activated forms of FAK can rescue certain epithelial cells from apoptosis, and transform MDCK cells when overexpressed in these cells (Frisch et al., 1996a).

D. Focal Adhesions and Rho

An important regulator of focal adhesions and stress fiber assembly is Rho, a member of the family of small GTPases (Hall, 1994; Takai et al., 1995). Microinjection of the activated form of Rho into quiescent mouse 3T3 fibroblasts that lack stress fibers and focal adhesions induces a rapid assembly of these structures in the injected cells (Ridley and Hall, 1992). This effect is similar to that of serum factors and neuropeptides that, in addition to inducing the assembly of these structures, also stimulate tyrosine phosphorylation of paxillin, FAK, and $p130^{Cas}$ while coupling their effect with Rho activation (Seufferlein and Rozengurt, 1994; see Figure 2). The pathways leading to Rho activation are not completely understood, but involvement of tyrosine phosphorylation by a tyrosine kinase upstream of Rho was inferred (Nobes et al., 1995). Downstream of Rho, Rho-kinase that is activated by Rho, can mimic the effects of Rho on stress fiber and focal adhesion formation in quiescent fibroblasts (Amano et al., 1997; see Figure 2). Several additional potential targets of Rho have been identified, an important one being the regulation of PIP2. The levels of this phosphoinositide are also regulated by cell adhesion to the substrate, and cells placed in suspension culture show decreased amounts of PIP2. The presence of growth factors in the suspension culture medium does not stimulate growth of suspended cells, as these are unable to respond to PIP2 hydrolysis under these conditions (McNamee et al., 1993). PIP2 synthesis is stimulated by activated Rho (Chong et al., 1994) by an interaction between Rho and PIP5-kinase isoforms (Kimura et al., 1996; Ren et al., 1996). Increases in PIP2 levels stimulate the dissociation of profilin and gelsolin from actin and thus enhance actin polymerization (Lassing and Lindberg, 1985; Janmey and Stossel, 1987). In addition, the conformation of vinculin changes from an inactive to an active form by dissociating the head to tail intramolecular folding (Jokusch and Ruddiger, 1996; Johnson and Craig, 1995a,b; see Figure 2), thus exposing its actin and talin binding sites (Gilmore and Burridge, 1996). Both α-actinin (another focal adhesion protein that can link actin to integrin) and vinculin were reported to show enhanced actin binding upon PIP2 binding (Fukami et al., 1994), thus leading to stress fiber and focal adhesion formation. Since PIP2 levels were shown to increase when cells adhere to the ECM via integrin binding, and this increase depends on Rho activation (Chong et al., 1994), it was suggested that integrin activation by adhesion can stimulate Rho, resulting in stress fiber and focal adhesion assembly. An alternative possibility was inferred from studies showing that quiescent fibroblasts (with low levels of active Rho) plated on fibronectin did not assemble focal adhesions or stress fibers in the first hour, (Ben-Ze'ev et al., 1990; Hotchin and Hall, 1995; Barry et al., 1997), unless Rho was activated (Hotchin and Hall, 1995). This may suggest that a cooperation between serum growth factors and integrin-mediated adhesion is necessary for the stimulation of focal adhesion and stress fiber assembly (see Figure 2).

E. Focal Adhesions, Anchorage Dependence, and Tumorigenesis

Cell adhesion to a substrate has long been recognized as an important factor in the regulation of cell proliferation. MacPherson and Montagnier (1964) first reported that non-transformed fibroblasts cannot proliferate in semi-solid medium, while virus-transformed cells multiply and form colonies in soft agar. The term "anchorage dependence" was coined by Stoker et al. (1968) to define the requirement in normal cells to attach and spread on a solid substrate to allow their proliferation. This phenomenon was shown to have an adhesion aspect that is measurable since the addition of small glass fibers that adhere to cells in suspension can induce their proliferation (Stoker et al., 1968; Maroudas, 1973). Later studies have established that anchorage dependent growth is characteristic of normal cells, while its loss is a reliable feature that distinguishes neoplastically transformed cells from their normal counterparts (Shin et al., 1975: for a review, see Ben-Ze'ev, 1985, 1986, 1991, 1997; Vasiliev and Gelfand, 1981).

Studies on the mechanism of anchorage dependence employing substrates with variable adhesiveness have shown that DNA synthesis increases exponentially following linear increases in cell spreading on various ECM components, irrespective of the ECM molecules that were used to induce the change in cell shape (Folkman and Moscona, 1978; Ingber, 1993). Moreover, DNA and RNA synthesis were shown to be more sensitive to the extent of cell spreading on the ECM than protein synthesis (Ben-Ze'ev et al., 1980). While initial contacts with the substrate are sufficient to allow protein synthesis, more stable adhesion that is associated with extensive cell spreading on the substrate is required for DNA and RNA synthesis (Benecke et al., 1978; 1980; Farmer et al., 1978; Ben-Ze'ev et al., 1980; Ben-Ze'ev and Raz, 1981).

Adhesive islands of variable size were used to examine the relationship between the degree of cell spreading, focal adhesion formation, and the probability of such cells to enter the cell cycle (O'Neil et al., 1990). In addition to the area of the adhesive islands, their configuration is also an important parameter affecting the proliferation of cells. Cells plated on elongated adhesive islands form more focal adhesions per unit of cell spreading area than cells on squares or circles and these cells are also more efficient in DNA synthesis than cells on circular or square islands having the same area. The total area of "focal adhesions" is larger on elongated adhesive islands and apparently correlates with the degree of DNA synthesis (O'Neil et al., 1990). Using a similar approach, Heckman et al. (1993) concluded that, when the adhesive area was increased beyond the maximum required for cell spreading, there was no saturation in the proliferative ability of cells, and an even higher level of DNA synthesis occurs when the substrate area is further increased, implying that the ability of cells to locomote on the substrate is also important for growth. Similar conclusions were reached for keratinocyte proliferation that also requires cell motility for optimal DNA synthesis (Barrandon and Green, 1987). The relationship between growth and focal adhesion assembly is also demonstrated with fibroblasts and primary hepatocytes cultured on collagen gels that proliferate only when the gel is anchored and the cells can spread out and form stress fibers, but not when the gels are free-floating and the cells are unable to spread and generate isometric tension (Ben-Ze'ev et al., 1988; Grinnell, 1994).

Adhesion-mediated signals in normal cells regulate a wide range of signaling pathways, including those induced by growth factors and cytokines. The increase in pH and Ca^{2+} that were correlated with growth are mediated by integrins and the Na^+/H^+ antiporter, to regulate pH elevation (Schwartz et al., 1991), and by an integrin-associated protein that controls Ca^{2+} elevation (Schwartz et al., 1993). These pathways are also believed to involve the hydrolysis of PIP2 that occurs in response to growth factor stimulation, while the reduced PIP2 levels found in nonadherent cells is correlated with their failure to respond to growth factors, even though the growth factors can bind to their receptor tyrosine kinases (McNamee et al., 1993; Burridge and Chrzanowska-Wodnicka, 1996).

Recent studies have shown that epithelial and endothelial cells require attachment to the extracellular matrix (ECM) not only for proliferation, but also for survival, as deprivation of "proper" contact with the substrate rapidly induces a sequence of apoptotic events leading to cell death (Meredith et al., 1993; Frish and Francis, 1994; Re et al., 1994). Moreover, the apoptotic process can be prevented in certain cell types by addition of ECM-coated beads to cells (Boudreau et al., 1995). The involvement of integrin in this response was inferred from studies showing that the survival of cells in the absence of serum factors is increased when they express a transfected integrin subunit ($\alpha5$) and thus show enhanced binding to fibronectin (Zhang et al., 1995). In muscle cells $\alpha5$ integrin cooperates with bFGF in preventing apoptosis (Sastry et al., 1996), and the activation of integrin signaling in epithelial cells induces resistance to apoptosis (Frisch et al., 1996b). On the other hand, disruption of the integrin receptor for vitronectin was shown to enhance apoptosis in several cell types (Bates et al., 1994; Brooks et al., 1994; Montgomery et al., 1994; Frisch and Francis, 1994).

The relationships between integrin-mediated signaling, cell transformation, and apoptosis were studied by examining the possible involvement of FAK in these processes. Earlier studies have shown that, in anchorage independent tumor cells, FAK is constitutively activated (Guan and Shalloway, 1992), suggesting an uncoupling between adhesion mediated signaling and growth control. More recent studies have shown elevated FAK expression in certain invasive tumors (Owens et al., 1995). In several tumor cell lines, but not in normal cells, the targeted decrease in FAK levels by an antisense approach resulted in disruption of cell substrate adhesion and increased apoptosis (Xu et al., 1996). Other investigators however found that disruption of FAK organization or expression in normal chicken cells induce apoptosis (Hungerford et al., 1996). MDCK epithelial cells that undergo apoptosis in suspension by activating the Jun kinase (JNK) pathway (Frisch et al., 1996a) and MEKK-1 (Cardone et al., 1997), could be rescued by transfection of activated forms of FAK (Frisch et al., 1996b). The current view suggests that FAK plays a major role in conveying intracellularly the adhesive signal to the cell and this signaling is disrupted in tumor cells so that they can proliferate in the absence of adhesion.

F. Focal Adhesion Proteins and Tumor Suppression

The effects of integrins and junctional plaque proteins on tumor cell behavior are complex. Initial findings described a decrease in the expression of the fibronectin receptor $\alpha_5\beta_1$ in transformed cells (Plantefaber and Hynes, 1989) and demonstrated suppression of tumorigenicity upon restoration of $\alpha_5\beta_1$ expression (Giancotti and Ruoslahti, 1990). Growth activation is however promoted by α5 integrin on certain ECM components (Varner et al., 1995), suggesting that the state of integrin (ligation or lack of ligation) can confer opposing growth signals. In addition, different integrins generate different responses in both normal and transformed cells (reviewed in Varner and Cheresh, 1996; Ben-Ze'ev, 1997).

Earlier studies have also identified changes in the assembly of the cytoskeleton and junctional proteins of cancer cells (Ben-Ze'ev, 1985, 1986) and a frequent reduction in the expression of several actin-associated proteins including various tropomyosin (TM) isoforms, gelsolin, α-actinin and vinculin (Matsumura and Yamashiro-Matsumura, 1986; Vandekerckhove et al., 1986; Glück et al., 1993; Rodríguez Fernández et al., 1992a). Recent studies have shown that restoration of vinculin and α-actinin expression in tumor cells results in the suppression of the tumorigenic and metastatic ability of such cells (Rodríguez Fernández et al., 1992a; Glück et al., 1993). Since vinculin overexpression causes the assembly of larger focal adhesions (Geiger et al., 1992) and inhibits cell motility (Rodríguez Fernández et al., 1992b), while elimination of vinculin results in enhanced motility (Rodríguez Fernández et al., 1993; Coll et al., 1995) and reduced adhesion and spreading, these changes in cell morphology, adhesion, and motility can be viewed as the cause, rather than the effect, of malignant transformation. This notion is supported by studies demonstrating that targeted reduction in the level of vinculin confers anchorage independent growth of 3T3 cells (Rodríguez Fernández et al., 1993), and the specific inhibition of α-actinin expression causes tumorigenicity in 3T3 cells (Glück et al., 1994).

Overexpression of several types of high molecular weight TM isoforms (TM-2 and -3) in oncogene-transformed cells that lack these isoforms, restored stress fiber formation and growth properties characteristic of normal fibroblasts, such as contact inhibition and requirement for serum to grow (Gimona et al., 1996; Janssen and Mier, 1997). Inhibition of TM-1 expression resulted in loss of stress fibers and anchorage independence (Boyd et al., 1995), while TM-1 overexpression in v-ras transformed cells suppressed the tumorigenic phenotype of these cells (Prasad et al., 1993). Since the expression of the oncogenes and the pathways activated by them are apparently not altered when the tumorigenicity of cells is suppressed after transfecting these cytoskeletal proteins (Janssen and Mier, 1997), it is possible that this tumor suppressive activity operates by alternative pathways.

Since vinculin and α-actinin are major PIP2-binding proteins (Fukami et al., 1994), and PIP2 levels were reported to be reduced in suspended cells that are growth arrested (McNamee et al., 1993), it was proposed that the availability of PIP2 for response to growth factor stimulation will increase in cells with reduced vinculin and α-actinin, (Burridge and Chrzanowska-Wodnicka, 1996). If the integration of these junctional proteins into focal adhesions will result in the release of their bound PIP2, increasing its availability for signaling by growth factors, will further support the link between adhesion and growth factor signaling.

III. THE STRUCTURE OF CELL–CELL ADHERENS JUNCTIONS

The most direct effect of cell adhesion between neighboring cells is on morphogenesis for example, the assembly of individual cells into highly ordered tissues and organs through cell–cell junctions (Takeichi, 1991; Gumbiner, 1996; Larue et al., 1996). The maintenance of these adhesions is essential for tissue organization and physiological function. The specific adhesive interactions between cells involve transmembrane cell adhesion receptors of the cadherin family (Takeichi, 1991, 1993, 1995; Kemler, 1992; Geiger and Ayalon, 1992), but effective adhesion and junction formation requires an association of the cadherin receptors with the cytoskeleton which is mediated by junctional plaque proteins (Kemler, 1993; Geiger et al., 1995; Knudsen et al., 1995; Rimm et al., 1995). Cadherin-mediated cell–cell junctions are linked to either actin filaments in adherens junctions (via catenins, α-, β-, and γ-catenin, or plakoglobin (Takeichi, 1991; Kemler, 1992; Knudsen and Wheelock, 1992), or to intermediate filaments in desmosomes (via plakoglobin, desmoplakins, plakophilins, and other molecules; Schmidt et al., 1994). Plakoglobin is a plaque component common to both types of cell–cell junction (Cowin et al., 1986; Franke et al., 1989; Klymkowsky and Parr, 1995; Wahl et al., 1996) that is apparently required for the sorting out of desmosomes and adherens junctions (at least in the embryonic heart). Its elimination by gene disruption in mice results in the collapse of this segregation in the heart, and the development of extended adherens junctions that contain desmosomal proteins. This leads to the "broken heart" phenotype and embryonal death (Ruiz et al., 1996; Bierkamp et al., 1996).

Adherens junctions are found in almost all cell types and consist of an electron-dense submembranous plaque (see Figure la) that is connected to the actin-cytoskeleton by plaque proteins such as vinculin (see Figures 1b and c) and α-actinin (see Figure 3). In some cells, these junctions are found in small patches, while in epithelial cells adherens junctions form a complete belt known as the *zonula adherens* around the apical surface.

A. The Cadherin Receptors

Focal adhesions and adherens type cell–cell junctions share several structural components (i.e. a vinculin-rich submembrane plaque that also contains α-actinin and is associated with microfilaments). There are, however, remarkable molecular differences between these structures, the most obvious being the transmembrane adhesion receptors. In adherens junctions, these are the cadherin family that mediates the Ca^{2+}-dependent homophilic cell–cell interactions while those in focal adhesions are the integrin family (Geiger et al., 1995).

Constitutive expression and function of cadherin receptors are essential for the development and maintenance of epitheial cell–cell interaction. Targeted genetic inactivation of E-cadherin in mice produces an embryonic lethal effect, with the embryonal cells dissociating and failing to form the trophectoderm (Larue et al., 1994; Riethmacher et al., 1995). In *Drosophila* development, loss of E-cadherin inhibits morphogenetic movements more dramatically than epithelial stability, suggesting that the formation of new cell–cell junctions during morphogenetic movements requires zygotic cadherin expression (Tepass et al., 1996; Bryant, 1997).

Recent high resolution determination of the structure of the N-terminus of E-cadherin shows that it forms a dimer with the adhesive binding surfaces interacting with dimers on neighboring cells to form a linear "zipper" of interacting molecules at the intercellular adhesion area (Overduin et al., 1995; Shapiro et al., 1995; Gumbiner, 1996).

E-cadherin is believed to play an important tumor suppresive role in the invasion and metastasis of carcinoma (Birchmeier and Behrens, 1994). E-cadherin is often reduced in many carcinomas, including those of the head and neck, esophagus, skin, thyroid, lung, breast, stomach, liver, kidney, pancreas, colon, bladder, prostate, and female genital track (Takeichi, 1993; Birchmeier and Behrens, 1994). This decrease in E-cadherin usually correlates with malignancy and lower survival. In vitro studies have demonstrated that, in cultured cells, restoration of E-cadherin levels by transfection leads to suppression of invasiveness and the tumorigenic ability of cells (Vleminckx et al., 1991; Frixen et al., 1991), and can even decrease protease secretion by the tumor cells (Miyake et al., 1995). Mutations in the E-cadherin gene also correlate with malignancy in certain tumors and 50% of diffuse type gastric carcinomas were shown to contain a mutation that affects the Ca^{2+}-binding site of E-cadherin (Becker et al., 1994). Deletions in the extracellular domain of E-cadherin were reported in infiltrative lobular breast carcinoma (Berx et al., 1995). The transcription of E-cadherin was also inhibited in cultured mammary carcinoma cells transfected with the erb-B2 receptor, whose expression correlates with the metastatic ability of these human tumors (D'Souza et al., 1994).

B. The Catenins

Functional adhesion requires the association of cadherins with cytoplasmic plaque proteins of the catenin family that are involved in linking E-cadherin to the cytoskeleton (Kemler, 1993; Geiger and Ayalon, 1992; see Figure 3). α-Catenin has structural similarity to vinculin and actin binding properties, suggesting that it links the catenin complex to the actin-cytoskeleton (Rimm et al., 1995). Restoration of α-catenin levels in lung carcinoma cells restored cell–cell adhesion and the assembly of various intercellular junctions (Watabe et al., 1994). In prostate cancer cells

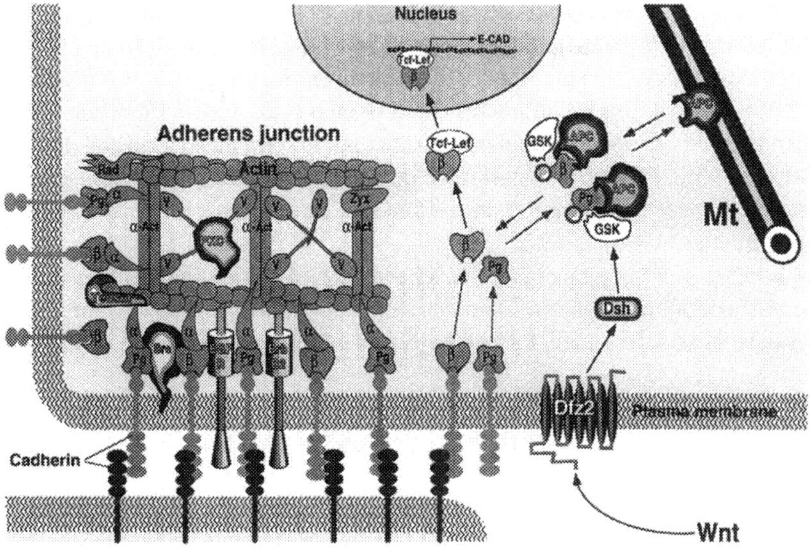

Figure 3. A model for the structure and signaling by cell–cell adherens junctions that interacts with signaling by soluble factors, in which β-catenin/plakoglobin serve as a "signaling center." Signals generated by cell-cell adhesion via cadherins and their assembly with catenins can modulate the level of free β-catenin (β), or plakoglobin (Pg) that is available for interaction with the APC tumor suppressor molecule, or with transcription factors such as LEF-I (Tcf-Lef). The complex between LEF-1 and β-catenin, or plakoglobin, can translocate into the nucleus and directly bind to the 5' end of E-cadherin (E-CAD) and other genes to regulate their expression. Signaling that involves β-catenin and plakoglobin can also be elicited by the Wnt signaling pathway, that includes the Wnt receptor (Dfz2 in *Drosophila*), Dsh, and glycogen synthase kinase (GSK). The association between APC and β-catenin, or plakoglobin and GSK, regulates the stability of β-catenin (by phosphorylation, (P)), and thus controls its translocation into the nucleus. The binding of APC to microtubules (Mt) may, in addition, transduce a negative effect on cell growth. An association between receptor tyrosine kinases such as the EGF receptor (EGFR) and erb-B2-receptor with β-catenin and plakoglobin may also affect signaling. Note the presence of protein kinases such as Src and PKC in the junctional plaque. α, α-catenin; Rad, radixin; Zyx, zyxin: α-Act. α-actinin: V, vinculin; PKC, protein kinase C. (From Ben-Ze'ev, 1997).

and in a human ovarian carcinoma cell line (Bullions et al., 1997), wild-type α-catenin expression resulted in the induction of E-cadherin function, cell–cell adhesion and suppression of tumorigenesis in nude mice (Ewing et al., 1995). The β-catenin binding sites on α-catenin have recently been mapped (Nieset et al., 1997; Huber et al., 1997). The role of β-catenin, and the closely related molecule plakoglobin, in this linkage between the cytoskeleton and E-cadherin is suggested to be regulatory, since α-catenin fused to the E-cadherin tail can support cell–cell adhesion independent of other catenins (Nagafuchi et al., 1994). In addition, the interaction of both β-catenin and plakoglobin with adherens junctions can be regulated by phosphorylation in response to treatment with growth factors (Hoschutzky et al., 1994) and during cell transformation (Kinch et al., 1995). Embryos of β-catenin null mice can undergo implantation, but fail to form mesoderm and die (Haegel et al., 1995). Plakoglobin null mice develop to mid gestation before dying of heart defect cause by absence of desmosomes in cardiac muscle (Ruiz et al., 1996; Bierkamp et al., 1996).

A number of additional components have been identified in recent years in the adherens junctional plaque area and their interactions with various plaque components have been determined. Part of these associations are shown schematically in Figure 3.

C. Signaling Involving the Catenins and its Role In Tumorigenesis

In addition to their function in cell adhesion, β-catenin and plakoglobin are highly homologous to *Drosophila* armadillo that is also found in adherens junctions of flies (Peifer and Weischaus, 1990), and they belong to the armadillo family (Peifer et al., 1994a). Armadillo in *Drosophila* and β-catenin/plakoglobin in *Xenopus* have been shown to play an additional role to adhesion: They transduce signals initiated by the extracellular glycoprotein wg/Wnt that regulates cell growth, differentiation and fate (Gumbiner, 1995; 1996; Peifer et al., 1994b; Huber et al., 1996a; Miller and Moon, 1996; Rubenstein et al., 1997). Activation of this pathway results in elevation of cytolasmic β-catenin levels and its nuclear localization in a complex with the TCF/LEF family of transcription factors (Behrens et al., 1996, Molenaar et al., 1996; Huber et al., 1996b; see Figure 3), suggesting that β-catenin may have a role in regulating gene expression by transactivating target genes (van de Wetering et al., 1997; Riese et al., 1997; see Klymkowsky, 1997; Merriam et al., 1997 for an alternative view). In the absence of wg/Wnt signaling, excess β-catenin is degraded in mammalian cells by a process involving the adenomatous polyposis cold (APC) tumor suppressor protein (Powell et al., 1992; Polakis, 1995) and the ubiquitin-proteasome degrading pathway (Aberle et al., 1997). Mutations in the APC gene that constitute the primary genetic defect in inherited colon cancer and certain melanomas result in the accumulation of β-catenin (Munemitsu et al., 1995, 1996; Rubinfeld et al., 1996; Papkoff et al., 1996; Yost et al., 1996; Peifer, 1996), and most probably cause inappropriate activation of target genes by the β-catenin-LEF/TCF complex (Korinek et al., 1997; Morin et al., 1997; Rubinfeld et al., 1997). In contrast, the involvement of plakoglobin in suppressing tumorigenesis was inferred from studies showing loss of heterozygosity of the plakoglobin gene in certain types of tumors (Aberle et al., 1995), its reduction in several tumor cell types (Sommers et al., 1994, Navarro et al., 1993; Simcha et al., 1996), and by demonstrating that plakoglobin overexpression can suppress the tumorigenicity of mouse and human cells, while localized in the nuclei of such cells

(Simcha et al., 1996). The regulation of β-catenin and plakoglobin level may, therefore, be a key element in their nuclear localization and signal transduction.

Plakoglobin and β-catenin bind in a mutually exclusive manner to cadherins, APC, and transcription factors (Hülsken et al., 1994; Butz and Kemler, 1994; Nathke et al., 1994; Rubinfeld et al., 1995; Hüber et al., 1996b). Interestingly, overexpression of plakoglobin leads to a decrease in β-carenin level, and plakoglobin competes with β-catenin for N-cadherin binding, thus directing the displaced β-catenin molecules for degradation by the ubiquitin-proteasome system (Salomon et al., 1997). Inhibition of the proteasome degradation system in these cells leads to the accumulation of both catenins in the nucleus (Salomon et al., 1997).

Wnt-induced signaling during development includes the accumulation of β-catenin in the cell, but artificially elevated cadherin expression in *Xenopus* can antagonize the propagation of the Wnt signal, by sequestering "free pools" of β-catenin into a complex with cadherin, and thus probably limiting its function in extra-junctional signaling (Heasman et al., 1994; Fagotto et al., 1996; Yost et al., 1996). Therefore, plakoglobin may serve as an additional regulator of β-catenin level acting upstream of the APC-GSK-3β step, by competing on the cadherin binding site, and thus releasing β-catenin and exposing it to the degradation fate.

The accumulation of β-catenin and its nuclear translocation in complex with transcription factors, its aberrant effect on the transcription of genes during development of colon cancer and melanoma (Korinek et al., 1997; Morin et al., 1997; Rubinfeld et al., 1997), as well as the ability of plakoglobin to influence the tumorigenicity of cells when overexpressed and localized in the nucleus (Simcha et al., 1996), highlight the importance of mechanisms that regulate the level of β-catenin in the cell. Interestingly, in tumor cells where plakoglobin overexpression resulted in suppression of the tumorigenic ability (Simcha et al., 1996), the level of β-catenin was reduced (Salomon et al., 1997). This may indicate that plakoglobin confers a tumor suppressive phenotype in these cells by decreasing the level of β-catenin, whose abnormally increased level can be oncogenic (Peifer 1997; Korinek et al., 1997; Morin et al., 1997; Rubinfeld et al., 1997).

The challenges for future studies are to determine whether elevated β-catenin can confer tumorigenicity on nontransformed cells, to elucidate the physiological conditions associated with the regulated expression of β-catenin and plakoglobin and their translocation into the nuclei of mammalian cells, and to identify the target genes whose expression is modulated by transactivation involving complexes that contain these junctional plaque proteins.

ACKNOWLEDGMENTS

I thank Drs. B. Geiger and A. Bershadsky for advice and for supplying Figure 1, and Figures 2 and 3, respectively. I also thank Drs. B. Jockusch and S. Craig for suggestions and for communicating results before publication. The studies from the author's laboratory were supported in part by grants from the USA–Israel Binational Foundation (BSF), the Forchheimer Center for Molecular Genetics, the Pasteur-Weizmann Research Program, and the German-Israeli Foundation for Scientific Research and Development (GIF).

REFERENCES

Aberceombie, M. J., Heaysman, J. E. M., & Pegrum, S. M. (1971). The locomotion of fibroblasts in culture. IV Electron microscopy of the leading lamella. Exp. Cell Res. 67, 359-367.

Aberle, H., Bauer, A., Stappert, J., Kispert, A., & Kemler, R. (1997). β-catenin is a target for the ubiquitin-proteasome pathway. EMBO J. 16, 3797-3804.

Aberle, H., Bierkamp, C., Torchard, D., Serova, O., Wagener, T., Natt, E., Wirsching, J., Heidkämper, C., Montagna, M., Lynch, H. T., Lenoir, G. M., Scherer, G., Fuenteun, J., & Kemler, R. (1995). The human plakoglobin gene localizes on chromosome 17q21 and is subject to loss of heterozygosity in breast and ovarian cancer. Proc. Natl. Acad. Sci. USA. 92, 6384-6388.

Amano, M., Chihara, K., Kimura, K., Fukata, Y., Nakamura, N., Matsuura, Y., & Kaibuchi, K. (1997). Formation of actin stress fibers and focal adhesions enhanced by rho-kinase. Nature 275, 1308-1311.

Avraham, S. London, R., Fu, Y., Ota, S., Hiregowdara, D., Li, J., Jiang, S., Pasztor, L. M., White, R. A., Groopman, J. E., & Avraham, H. (1995). Identification and characterization of a novel related adhesion focal tyrosine kinase (RAFT) from megakaryocytes and brain. J. Biol. Chem. 270, 27742-27751.

Barrandon, Y., & Green, H. (1987). Cell migration is essential for sustained growth of keratinocyte colonies. Cell 50, 1131-1137.

Barry, S. T., Critchley, D. R., Flinn, H. M., Humphries, M. J., and Ridley, A. J. (1997). Requirement for Rho in integrin signaling. Cell Adhesion Commun. 6, 387-398.

Bates, R. C., Buret, A., van Helden, D. F., Horton, M. A., & Burns, G. F. (1994). Apoptosis induced by inhibition of intercellullar contact. J. Cell Biol. 125, 403-415.

Becker, K. F., Atkinson, M. J., Reich, I., Nekarda, H., Siewert, J. R., & Hofter, H. (1994). E-cadherin gene mutations provide clues to diffuse type gastric carcinomas. Cancer Res. 54, 3845-3852.

Beckerle, M. C. (1986). Identification of a new protein localized at sites of cell-substrate adhesion. J. Cell Biol. 103, 1679-1687.

Behrens, J., von Kries, J. P., Kuhl, M., Bruhn, L., Wedlich, D., Grosschedl, R., & Birchmeier, W. (1996). Functional interaction of β-catenin with the transcription factor LEF-1. Nature 382, 638-642.

Ben-Ze'ev, A. (1985). The cytoskeleton in cancer cells. Biochim. Biophys. Acta 780. 197-212.

Ben-Ze'ev, A. (1986). The relationship between cytoplasmic organization, gene expression and morphogenesis. Trends Biochem. Sci. 11, 478-481.

Ben-Ze'ev, A. (1991). Animal cell shape changes and gene expression. BioEssays 13, 207-212.

Ben-Ze'ev, A. (1997). Cytoskeletal and adhesion proteins as tumor suppressors. Curr. Opin. Cell Biol. 9, 99-108.

Ben-Ze'ev, A., & Bershadsky, A. D. (1997). The role of the cytoskeleton in adhesion-mediated signaling and gene expression. Adv. Mol. Cell Biol. 24, 125-163.

Ben-Ze'ev, A., & Raz, A. (1981). Multinucleation and inhibition of cytokinesis in suspended cells: reversal upon reattachment to a substrate. Cell 26, 107-115.

Ben-Ze'ev, A., Farmer, S. R., & Penman, S. (1980). Protein synthesis requires cell-surface contact while nuclear events respond to cell shape in anchorage-dependent fibroblasts. Cell 21, 365-372.

Ben-Ze'ev, A., Reiss, R., Bendori, R., & Gorodecki, B. (1990). Transient induction of vinculin gene expression in 3T3 fibroblasts stimulated by serum growth factors. Cell Regul. 1, 621-636.

Ben-Ze'ev, A., Robinson, G. S., Bucher, N. L. R., & Farmer, S. R. (1988). Cell-cell and cell-matrix interactions differentially regulate the expression of hepatic and cytoskeletal genes in primary cultures of rat hepatocytes. Proc. Natl. Acad. Sci. USA. 85, 2161-2165.

Benecke, B.-J., Ben-Ze'ev, A., & Penman, S. (1978). The control of mRNA production, translation and turnover in suspended and reattached anchorage-dependent fibroblasts. Cell 14, 931-939.

Benecke, B.-J., Ben-Ze'ev, A., & Penman, S. (1980). The regulation of RNA metabolism in suspended and reattached anchorage-dependent 3T6 fibroblasts. J. Cell. Physiol. 103, 247-254.

Berx, G., Cleton-Jansen A. M., Nollet, F., de Leeuw W. J., van de Vijver M., Cornelisse, C., & van Roy, F. (1995). E-cadherin is a tumor/invasion suppressor gene mutated in human lobular breast cancers. EMBO J. 14, 6107-6115.

Bierkamp, C., Mclaughlin, K. J., Schwartz, H., Huber, O., & Kemler, R. (1996). Embryonic heart and skin defects in mice lacking plakoglobin. Dev. Biol. 180, 780-785.

Birehmeier, W. & Behrens, J. (1994). Cadherin expression in carcinomas: role in the formation of cell junctions and the prevention of invasiveness. Biochim. Biophys. Acta 198, 11-26.

Boudreau, N., Sympson, C. J., Werb, Z., & Bissell, M. (1995). Suppression of ICE and apoptosis in mammary epithelial cells by extracellular matrix. Science 267, 891-893.

Boyd, J., Risinger, J. I., Wiseman, R. W., Merrick, A. B., Selkirk, J. K., & Barrett, J. C. (1995). Regulation of microfilament organization and anchorage-independent growth by tropomyosin 1. Proc. Natl. Acad. Sci. USA. 92, 11534-11538.

Brindle, N. P., Holt, M. R., Davies, J. E., Price, C. J., & Critchley, D. R. (1996). The focal-adhesion vasodilator-stimulated phosphoprotein (VASP) binds to the proline-rich domain in vinculin. Biochem. J. 318, 753-757.

Brooks, P. C., Montgomery, A. M., Rosenfeld, M, Reisfeld R. A., Hu, T., Klier, G., & Cheresh, D. A. (1994). Integrin $\alpha v\beta 3$ antagonists promote tumor regression by inducing apoptosis of angiogenic blood vessels. Cell 79, 1157-1164.

Bryant, P. J. (1997). Junction genetics. Dev. Gen. 20, 75-90.

Bubeck, P., Pistor, S., Hehland, J., & Jockusch, B. M. (1997). Ligand recruitment by vinculin domains in transfected cells. J. Cell Sci. 110, 1361-1371.

Bullions, L. C., Notterman, D. A., Chung, L. S., & Levine, A.J. (1997). Expression of wild type α-catenin protein in cells with mutant α-catenin gene restores both growth regulation and tumor suppressor activities. Mol. Cell. Biol. 17, 4501-4508.

Burridge, K. (1981). Are stress fibers contractile? Nature 294, 691-692.

Burridge, K. & Chrzanowska-Wodnicka, M. (1996). Annu. Rev. Cell Dev. Biol. 12, 463-519.

Burridge, K. & Feramisco, J. R. (1980). Microinjection and localization of a 130K protein in living fibroblasts: a relationship to actin and fibronectin. Cell 19, 587-595.

Burridge, K., Fath, K., Kelly, T., Nuckolls, G., & Turner, C. (1988). Focal adhesions: transmembrane junctions between the extracellular matrix and the cytoskeleton. Annu. Rev. Cell. Biol. 4, 487-525.

Burridge, K., Turner, C. E., & Romer, L. H. (1992). Tyrosine phosphorylation of paxillin and $pp125^{FAK}$ accompanies cell adhesion to extracellular matrix: a role in cytoskeletal assembly. J. Cell Biol. 119, 893-903.

Buss, F., Temm-Grove, C., Henning, S., & Jockusch, B. M. (1992). Distribution of profilin in fibroblasts correlates with the presence of highly dynamic actin filaments. Cell Motil. Cytoskel. 22, 51-61.

Butz, S. & Kemler, R. (1994). Distinct cadherin-catenin complexes in Ca^{2+}-dependent cell-cell adhesion. FEBS Lett. 355, 195-200.

Byers, H. & Fujiwara, K. (1982). Stress fibers in cells in situ: immunofluorescence visualization with antiactin, antimyosin, and anti-alpha actinin. J. Cell Biol. 115, 465-483.

Cardone, M. H., Salvesen, G. S., Widmann, C., Johnson, G., & Frisch, S. M. (1997). The regulation of anoikis: MEKK-1 activation requires cleavage by caspases. Cell 90, 315-323.

Chakraborty, T., Ebel, F., Domann, E., Neibuhr, K., Gerstel, B., Pistor, S., Temm-Grove, C. J., Jockusch, B. M., Reinhard, M., Walter, U., & Wehland, J. (1995). A focal adhesion factor directly linking intracellularly motile *Listeria monocytogenes* and *Listeria ivanovii* to the actin-based cytoskeleton of mammalian cells. EMBO J. 14, 1314-1321.

Chen, H. C., Appeddu, P. A., Parsons, J. T., Hildebrand, J. D., Schaller, M. D., & Guan, J. L. (1995). Interaction of focal adhesion kinase with the cytoskeletal protein talin. J. Biol. Chem. 270, 16995-16999.

Chong, L. D., Traynor-Kaplan, A., Bokoch, G. M., & Schwartz, M. A. (1994). The small GTP-binding protein Rho regulates a phosphatidylinositol 4-phosphate 5-kinase in mammalian cells. Cell 79, 507-513

Clark, E. A. & Brugge, J. S. (1995). Integrins and signal transduction pathways: the road taken. Science 268, 233-239.

Cobb, B. S., Schaller, M. D., Leu, T.-H., & Parsons, J. T. (1994). Stable association of $pp60^{src}$ and $pp59^{fyn}$ with the focal adhesion associated protein tyrosine kinase $ppl25^{Fak}$. Mol. Cell. Biol. 14, 147-155.

Coll, J-L., Ben-Ze'ev, A., Ezzell, R. M., Rodríguez Fernández, J. L., Baribault, H., Oshima, R. G., & Adamson, E. D. (1995). Targeted disruption of vinculin genes in F9 and ES cells changes cell morphology, adhesion and locomotion. Proc. Natl. Acad. Sci. USA 92, 9161-9165.

Cowin, P., Kapprell, H. P., Franke, W. W., Tamkun, J., & Hynes, R. O. (1986). Plakoglobin: a protein common to different kinds of intercellular adhering junctions. Cell 46, 1063-1073.

Craig, S. W. & Johnson, R. P. (1996). Assembly of focal adhesions: Progress, paradigms. and portents. Curr. Opin. Cell Biol. 8, 74-85.

Crawford, A. W., Michelsen, J. W., & Beckerle, M. C. (1992). An interaction between zyxin and α-actinin. J. Cell Biol. 116, 1381-1393.

D'Souza, B. & Taylor-Papadimitriou, J. (1994). Overexpression of erb-B2 in human mammary epithelial cells signals inhibition of transcription of the E-cadherin gene. Proc. Natl. Acad. Sci. USA 91, 7202-7206.

Ewing, C. M., Ru, N., Morton, R. A., Robinson, J. C., Wheelock, M. J., Johnson, K. R., Barrett, J. C., & Isaacs, W. B. (1995). Chromosome 5 suppresses tumorigenicity of PC3 prostate cancer cells: correlation with re-expression of α-catenin and restoration of E-cadherin. Cancer Res. 55, 4813-4817.

Fagotto, F., Funayama, N., Glück, U., & Gumbiner, B. (1996). Binding of cadherins antagonizes the signaling activity of β-catenin during axis formation in *Xenopus*. J. Cell Biol. 132, 1105-1114.

Farmer, S. R., Ben-Ze'ev, A., Benecke, B. J., & Penman, S. (1978). Altered translatability of messenger RNA from suspended anchorage-dependent fibroblasts: reversal upon cell attachment to a surface. Cell 15, 931-939.

Fath, K. R., Edgell, C. J., & Burridge, K. (1989). The distribution of distinct integrins in focal contacts is determined by substratum composition. J. Cell Sci. 92, 67-75.

Folkman, J. & Moscona, A. (1978). Role of cell shape in growth control. Nature 273, 345-349

Franke, W. W., Goldschmidt, M. D., Zimbelmann, R., Mueller, H. M., Schiller, D. L., & Cowin, P. (1989). Molecular cloning and amino acid sequence of human plakoglobin, the common junctional plaque protein. Proc. Natl. Acad. Sci. USA. 86, 4027-4031.

Frisch, S. M. & Francis, H. (1994). Disruption of epithelial cell-matrix interactions induces apoptosis. J. Cell Biol. 124, 619-626.

Frisch, S. M., Vuori, K., Ruoslahti, E., & Chan-Hui, P-Y. (1996a). Control of adhesion-dependent cell survival by focal adhesion kinase. J. Cell Biol. 123, 1-7.

Frisch, S., Vuori, K., Kelaita, D., & Sicks, S. (1996b). A role for N-terminal kinase in anoikis; suppression by bcl-2 and crmA. J. Cell Biol. 135, 1377-1382.

Frixen, U. H., Behrens, J., Sachs, M., Eberle, G., Voss, B., Warda, A., Löchner, D., & Birchmeir, W. (1991). E-cadherin-mediated cell-cell adhesion prevents invasiveness of human carcinoma cells. J. Cell Biol. 113. 173-185.

Fukami, K., Endo, T., Imamura, M., & Takenawa, T. (1994). Alpha-actinin and vinculin are PIP2-binding proteins involved in signaling by tyrosine kinase. (1994). J. Biol. Chem. 269, 1518-1522.

Gabbiani, G., Gabbiani, F., Lombard, D., & Schwartz S. M. (1983). Organization of actin cytoskeleton in normal and regenerating arterial endothelial cells. Proc. Natl. Acad. Sci. USA. 80, 2361-2364.

Geiger, B. (1979). A 130K protein from chicken gizzard: its localization at the termini of microfilament bundles in cultured chicken cells. Cell 18, 193-205.

Geiger, B. & Ayalon, O. (1992). Cadherins. Annul Rev. Cell Biol. 8, 307-332.

Geiger, B., Ayalon, O., Ginsberg, D., Volberg, T., Rodríguez Fernández, J. L., Yarden. Y., & Ben-Ze'ev, A. (1992). Cytoplasmic control of cell adhesion. Cold Spring Harb. Symp. Quant. Biol. 57, 631-642.

Geiger, B., Yehuda-Levenberg, S., & Bershadsky, A. D. (1995). Molecular interactions in the submembrane plaque of cell-cell and cell-matrix adhesions. Acta Anat. 154, 46-62.

Giancotti, F. G. & Ruoslahti, E. (1990). Elevated levels of the $\alpha_5\beta_1$ fibronectin receptor suppress the transformed phenotype of chinese hamster ovary cells. Cell 69, 849-859.

Gimona, M., Kazzaz, J. A., &Helfman, D. M. (1996). Forced expression of tropomyosin 2 or 3 in v-Ki-ras-transformed fibroblasts results in distinct phenotypic effects. Proc. Natl. Acad. Sci. USA 93, 9618-9623.

Gilmore, A. P. & Burridge, K. (1996). Regulation of vinculin binding to talin and actin by phosphatidyl-inositol-4-5-bisphosphate. Nature 381, 531-535.

Gilmore, A. P. & Romer, L. H. (1996). Inhibition of focal adhesion kinase (FAK) signaling in focal adhesions decreases cell motility and proliferation. Mol. Biol. Cell 7, 1209-1224.

Glück, U. & Ben-Ze'ev, A. (1994). Modulation of α-actinin levels affects cell motility and confers tumorigenicity on 3T3 cells. J. Cell Sci. 107, 1773-1782.

Glück, U., Kwiatkowski, D. J., & Ben-Ze'ev, A. (1993). Suppression of tumorigenicity in Simian virus 40-transformed cells transfected with α-actinin cDNA. Proc. Natl. Acad. Sci. USA. 90, 383-387.

Goldberg, M. B. (1997). *Shigella* actin-based motility in the absence of vinculin. Cell Motil. Cytoskel. 37, 44-53.

Grinnell, F (1994). Fibroblasts, myofibroblasts and wound contraction. J. Cell Biol. 124, 401-404.

Guan, J. L. & Shalloway, D. (1992). Regulation of focal adhesion-associated protein tyrosine kinase by both cellular adhesion and oncogenic transformation. Nature 358, 690-692.

Gumbiner, B. M. (1995). Signal transduction by β-catenin. Curr. Opin. Cell Biol. 7, 634-640.

Gumbiner, B. M. (1996). Cell adhesion: the molecular basis of tissue architecture and morphogenesis. Cell 84, 345-357.

Haegel, H. Larue, L., Oshugi, M., Federov, L., Herrenknecht, K., & Kemler, R. (1995). Lack of beta-catenin affects mouse development at gastrulation. Development 121, 3529-3537.

Haffner, C., Jurcau, T., Reinhard, M., Hoppe, J., Lohmann, S. M., & Walter, U. (1995). Molecular cloning, structural analysis and functional expression of the proline-rich focal adhesion and microfilament-associated protein VASP. EMBO J. 14, 19-27.

Hall, A. (1994). Small GTP-binding proteins and the regulation of the actin-cytoskeleton. Annu. Rev. Cell Dev. Biol. 10, 31-54.

Hayashi, Y., Haimovich, B., Reszka, A., Boettiger, D., & Horwitz, A. (1990). Expression and function of chicken β1 subunit and its cytoplasmic domain mutants in mouse NIH 3T3 cells. J. Cell Biol. 110, 175-184.

Heasman, J. Crawford, A., Goldstone, K., Garner-Hamrick, P., Gumbiner, B. M., McCrea, P., Kintner, C., Noro, C. Y., & Wylie, C. (1994). Overexpression of cadherins and underexpression of β-catenin inhibit dorsal mesoderm induction in *Xenopus* embryos. Cell 79, 791-803.

Heath, J. P. & Dunn, G. A. (1978). Cell-to-substratum contact of chick fibroblasts and their relation to the microfilament system: A correlated interference reflexion and high voltage election microscopy study. J. Cell Sci. 19, 197-212

Heckman, C. A., Oravecz, K. I., Schwab, D., & Ponten, J. (1993). Ruffling and locomotion: role in cell resistance to growth factor-induced proliferation. J. Cell. Physiol. 154, 554-565.

Hemmings, L., Rees, D. J. G., Ohanian, V., Bolton, S. J., Gilmore, A. P., Patel, B., Priddle, H., Trevithick, J. E., Hynes, R. O., & Critchley, D. R. (1996). Talin contains three actin-binding sites each of which is adjacent to a vinculin-binding site. J. Cell Sci. 109, 1 2715-2726.

Hobert, O., Schilling J. W., Beckerle, M. C., Ullrich, A., & Jallal, B. (1996). SH3 domain-, dependent interaction of the proto-oncogene product Vav with the focal contact protein zyxin. Oncogene 12, 1577-1581

Horwitz, A., Duggan, K., Buck, C., Beckerele, M. C., & Burridge, K. (1986). Interactions of plasma membrane fibronectin receptor with talin - a transmembrane linkage. Nature 320, 531-533.

Hoschuetzky, H., Aberle, H., & Kemler, R. (1994). β-catenin mediates the interaction of the cadherin-catenin complex with epidermal growth factor receptor. J. Cell Biol. 124, 729-741.

Hotchin, N. A. & Hall, A. (1995). The assembly of integrin adhesion complexes requires both extracellular matrix and intracellular RHO/RAC GTPases. J. Cell Biol. 131, 1857-1865.

Huber, O., Bierkamp, C., & Kemler, R. (1996a). Cadherins and catenins in development. Curr. Opin. Cell Biol. 8, 685-691.

Huber, O., Korn, R., McLaughlin, J., Oshugi, M. Herrmann, B. G., & Kemler, R. (1996b). Nuclear localization of β-catenin by interaction with the transcription factor LEF-1. Mech. Dev. 59, 3-10.

Huber, O., Krohn, M., & Kemler, R. (1997). A specific domain in α-catenin mediates binding to β-catenin and plakoglobin. J. Cell Sci. 110, 1759-1765.

Hungerford, J. E., Compton, M., Matter, M., Hoffstrom, B. G., & Otey, C. (1996). Inhibition of $ppl25^{FAK}$ in cultured fibroblasts results in apoptosis. J. Cell Biol. 135, 1383-1390.

Hülsken, J., Birchmeier, W., & Behrens, J. (1994). E-cadherin and APC compete for the interaction with β-catenin and the cytoskeleton. J. Cell Biol. 127, 2061-2069.

Hüttelmaier, S., Bubeck, P., Rudiger, M., & Jockusch, B. (1997). Characterization of two F-actin-binding and oligomerization sites in the cell-contact protein vinculin. Eur. J. Biochem. 247, 1136-1142.

Hüttelmaier, S., hardback, B., Jarchau, T., Jockusch, B. M., & Rüdiger, M. (1998). The interaction of the microfilament protein VASP and vinculin is regulated by the signaling molecule PIP2. Curr. Biol. 9, 479-488.

Hynes, R. O. (1992). Integrins: versatility, modulation and signaling in cell adhesion. Cell 69, 11-25.

Ilic, D., Furuta, Y., Kanazawa, S., Takeda, N., Sobue, K., Nakatsuji, N., Nomura, S., Fujimoto, J., Okada, M., Yamamoto, T., & Aizawa, S. (1995). Reduced cell motility and enhanced focal adhesion contact formation in cells from FAK-deficient mice. Nature 377, 539-543.

Ingber, D. E. (1993). Cellular tensegrity: defining new rules of biological design that govern the cytoskeleton. J. Cell Sci. 104, 613-627.

Janmey, P. A. & Stossel, T. P. (1987). Modulation of gelsolin function by phosphatidylinositol 4,5-bisphosphate. Nature 325, 362-364.

Janssen, R. A. J. & Mier, J. W. (1997). Tropomyosin-2 cDNA lacking the 3' untranslated region riboregulator induces growth inhibition of v-Ki-ras-transformed fibroblasts. Mol. Biol. Cell 8, 897-908.

Jockusch, B. M., Bubeck, P., Giehl, K., Kroemker, M., Moschner, J., Rothkegel, M., Rüdiger, M., Schlüter, K., Stanke, G., & Winkler, J. (1995). The molecular architecture of focal adhesions. Annu. Rev. Cell Dev. Biol. 11, 379-416.

Jockusch, B. M. & Rudiger, M. (1996). Crosstalk between cell adhesion molecules: vinculin as a paradigm for regulation by conformation. Trends Cell Biol. 6, 311-315.

Johnson R. P. & Craig S. W. (1995a). The carboxy-terminal tail domain of vinculin contains a cryptic binding site for acidic phospholipids. Biochem. Biophys. Res. Commun. 210, 159-164.

Johnson, R. P. & Craig, S. W. (1995b). F-actin binding site masked by the intramolecular association of vinculin head and tail domains. Nature 373, 261-264.

Juliano, R. L. & Haskill, S. (1993). Signal transduction from the extracellular matrix. J. Cell Biol. 120, 577-585.

Kemler, R. (1992). Classical cadherins. Semin. Cell Biol. 3, 149-155.

Kemler, R. (1993). From cadherins to catenins: cytoplasmic protein interactions and regulation of cell adhesion. Trends Genet. 9, 317-321.

Kimura, K., Ito, M., Amano, M., Chihara, K., Fukata, Y. Nakafuku, M., Yamamori, B., Feng, J., Nakano, T., Okawa, K., Iwamatsu, A., & Kaibuchi, K. (1996). Regulation of myosin phosphatase by Rho and Rho-associated kinase (Rho-kinase). Science 273, 245-248.

Kinch, M. S., Clark, G. J., Der, C. J., & Burridge, K. (1995). Tyrosine phosphorylation regulates the adhesion of Ras-transformed breast epithelia. J. Cell Biol. 130, 461-471.

Klymokowsky, M. W., & Parr, B. (1995). A glimpse into the body language of cells: the intimate connection between cell adhesion and gene expression. Cell 83, 5-8

Klymkowsky, M. W. (1997). Minireviews, minidogmas and mythinformation. BioEssays 19, 537-539.

Knudsen, K. A., Soler, A. P., Johnson, K. R., & Wheelock, M. J. (1995). Interaction of α-actinin with the cadherin-catenin cell-cell adhesion complex via α-catenin. J. Cell Biol. 130, 67-77.

Knudsen, K. A. & Wheelock, M. J. (1992) . Plakoglobin, or an 83-kD homologue distinct from β-catenin, interacts with E-cadherin and N-cadherin. J. Cell Biol. 118, 671-679.

Korinek, V., Backer, N., Morin, P. J., van Wichen, D., de Weger, R., Kinzler, K. W., Vogelstein, B., & Clevers, H. (1997). Constitutive transcriptional activation by a β-catenin-Tcf complex in $APC^{-/-}$ colon carcinoma. Science 275, 1784-1787.

La Flamme, S. E., Akiyama, S. K., & Yamada, K. M. (1992). Regulation of fibronectin receptor distribution. J. Cell Biol. 117, 437-447.

Larue, L., Antos, C., Butz, S., Huber, O., Delmas, V., Dominis, M., & Kemler, R. (1996). A role for cadherins in tissue formation. Development 122, 3185-3194.

Larue, L., Oshugi, M., Hichenhain & Kemler, R. (1994). E-cadherin null mutant embryos fail to form a trophectoderm epithelium. Proc. Natl. Acad. Sci. USA. 91, 8263-8267.

Lassing, I. & Lindberg, U. (1985). Specific interaction between phosphatidylinositol 4,5-bisphophate and profilactin. Nature 314, 472-474

Lev, S., Moreno, H., Martinez, R., Canoll, P., Peles E., Musacchio, J. M., Plowmann, G. D., Rudy, B., & Schlessinger, J. (1995). Protein tyrosine kinase PYK2 involved in Ca^{2+}-induced regulation of ion channel and MAP kinase functions. Nature 376, 737-745.

Luna, E. J. & Hitt, A. L. (1 992). Cytoskeletal-plasma membrane interactions. Science 258, 955-964.

MacPherson, 1., & Montagnier, L. (1964). Agar suspension culture for selective assay of cells transformed by polyoma virus. Virology 23, 291-294.

Markert, T., Krenn, V., Leebmann, J., & Walter, U. (1996). High expression of the focal-adhesion and microfilament-associated protein VASP in vascular smooth muscle and endothelial cells of the intact human vessel wall. Basic Res. Cardiol. 91, 337-343.

Maroudas, N. G. (1973). Growth of fibroblasts on linear and planar anchorages of limiting dimensions. Exp. Cell Res. 81, 104-110.

Matsumura, F. & Yamashiro-Matsumura, S. (1986). Tropomyosin in cell transformation. Cancer Rev. 6, 21-39.

Mayboroda, O., Schluter, K., & Jockusch, B. M. (1997). Differential colocalization of profilin with microfilaments in PtK2 cells. Cell Motil. Cytoskel. 37, 166-177.

McCann, R. O. & Craig, S. W. (1997). The I/LWEQ module: a conserved sequence that signifies F-actin binding in functionally diverse proteins from yeast to mammals. Proc. Natl. Acad. Sci. USA 94, 5679-5684.

McNamee, H. P., Ingber, D. E., & Schwartz, M. A. (1993). Adhesion to fibronectin stimulates inositol lipid synthesis and enhances PDGF-induced inositol lipid breakdown. J. Cell Biol. 121, 1231-1240.

Menkel, A. R., Kroemker, M., Bubeck, P., Ronsiek, M., Nikolai, G., & Jockusch, B. M. (1994). Characterization of an F-actin-binding domain in the cytoskeletal protein vinculin. J. Cell Biol. 126. 1231-1240.

Meredith, J. E., Fazeli, B., & Schwarz, M. A. (1993). The extracellular matrix as a cell survival factor. Mol. Biol. Cell 4, 953-961.

Merriam, J. M., Rubenstein, A. B., & Klymkowsky, M. W. (1997). Cytoplasmically anchored plakoglobin induces a WNT-like phenotype in *Xenopus*. Dev. Biol. 185, 67-81.

Miller, J. R. & Moon, R. T. (1996). Signal transduction through β-catenin and specification of cell fate during embryogenesis. Genes Dev. 10, 2527-2539.

Miyake, M., Tanaka, K., Kikuchi-Yanoshita, R., Muraoka, M., Konishi, M., & Takeichi, M. (1995). Increased cell-substartum adhesion, and decreased gelatinase secretion and cell growth, induced by E-cadherin transfection of human colon carcinoma. Oncogene 11, 2547-2552.

Miyamoto, S., Akiyama, K., & Yamada, K. M. (1995b). Synergistic roles for receptor occupancy and aggregation in integrin transmembrane function. Science 267, 883-885.

Miyamoto, S., Teramoto, H., Coso, O. A., Gutkind, J. S., Burbelo, P. D., Akiyama, S. K., & Yamada, K. M. (1995a). Integrin functions: Molecular hierarchies of cytoskeletal and signaling molecules. J. Cell Biol. 131, 791-805.

Molenaar, M., van de Wetering, M., Oosterwegel, M., Peterson-Maduro, J., Godsave, S., Korinek, V., Roose, J., Destree, O., & Clevers, H. (1996). XTcf-3 transcription factor mediates β-catenin-induced axis formation in *Xenopus* embryos. Cell 86, 391-399.

Montgomery, A. M., Reisfeld, R. A., & Cheresh, D. A. (1994). Integrin $\alpha_v\beta_3$ rescues melanoma cells from apoptosis in three-dimensional dermal collagen. Proc Natl. Acad. Sci. USA. 91, 8856-8860.

Morin, P. J., Sparks, A. B., Korinek, V., Barker, N., Clevers, H., Vogelstein, B., & Kinzler, K. W. (1997). Activation of β-catenin Tcf-signaling in colon cancer by mutations in β-catenin or APC. Science 275, 1787-1790.

Muguruma, M., Matsumura, S., & Fukazawa, T. (1990). Direct interactions between talin and actin. Biochem. Biophys. Res. Commun. 171, 1217-1223.

Munemitsu, S., Albert, I., Souza, B., Rubinfeld, B., & Polakis, P. (1995). Regulation of intracellular beta-catenin levels by the adenomatous polyposis cold (APC) tumor suppressor protein. Proc. Natl. Acad. Sci. USA 92, 3046-3050.

Munemitsu, S., Albert, I., Rubinfeld, B., & Polakis, P. (1996). Deletion of an amino-terminal sequence stabilizes β-catenin in vivo and promotes hyperphosphorylation of the adenomatous polyposis cold tumor suppressor protein. Mol. Cell. Biol. 16, 4088-4094.

Nagafuchi, A., Ishihara, S., & Tsukita, S. (1994). The roles of catenins in the cadherin-mediated cell adhesion: functional analysis of E-cadherin-α-catenin fusion molecules. J. Cell Biol. 127, 235-245

Nathke, S. I., Hinck, L., Swedlow, J. R., Papkoff, J., & Nelson, W. J. (1994). Defining interactions and distributions of cadherin and catenin complexes in polarized cells. J. Cell Biol. 125, 1341-1352.

Navarro, P., Lozano, E., & Cano, A. (1993). Expression of E- or P-cadherin is not sufficient to modify the morphology and the tumorigenic behavior of murine spindle carcinoma cells: possible involvement of plakoglobin. J. Cell Sci. 105, 923-934.

Nieset, J. E., Redfield, A. R., Jun, F., Knudsen, K. A., Johnson, K. R., & Wheelock, M. J. (1997). Characterization of the interaction of α-catenin with α-actinin and β-catenin/plakoglobin. J. Cell Sci. 110, 1013-1022.

Niggli, V. & Gimona, M. (1993). Evidence for a ternary interaction between α-actinin, (meta) vinculin and acidic phospholipid bilayers. Eur. J. Biochem. 213, 1009-1015.

Niggli, V., Sommer, L., Brunner, J., & Burger, M. M. (1990). Interaction in situ of the cytoskeletal protein vinculin with bilayers studied by introducing photoactivatable fatty acid into living chicken embryo fibroblasts. Eur. J. Biochem. 187, 117-117.

Nix, D. A. & Beckerle, M. C. (1995). Inducible nuclear localization of the focal contact protein, zyxin. Mol. Biol. Cell. 6, A63.

Nobes, C. D. & Hall, A. (1995). Rho, Rac and Cdc42 GTPases regulate the assembly of multimolecular focal complexes associated with stress fibers, lamellipodia, and filopodia. Cell 81, 53-62.

Nojima, Y., Morino, N., Mimura, T., Hamasaki, K., Furuya, H., Sakai, R., Sato, T., Tachibana, K., Morimoto, C., Yazaki, Y., & Hirai, H. (1995). Integrin-mediated cell adhesion promotes tyrosine phosphorylation of $pl30^{Cas}$, a Src homology-3 containing molecule having multiple Src homology 2-binding motifs. J. Biol. Chem. 270, 15398-15402.

O'Neil, C., Jordan, P., Riddle, P., & Ireland, G. (1990). Narrow linear strips of adhesive substratum are powerful inducers of both growth and total focal contact area. J. Cell Sci. 95, 577-586.
Otey, C. A., Pavalko, F. M., & Burridge, K. (1990). An interaction between α-actinin and the beta 1 integrin subunit in vitro. J. Cell Biol. 111, 721-729.
Overduin, M., Harvey, T. S., Bagby, S., Tong, K. I., Yau, P., Takeichi, M., & Ikura, M. (1995). Solution structure of the epithelial cadherin domain responsible for selective cell adhesion. Science 267, 386-389.
Owens, L, Xu, I., Craven, R., Dent, G., Weiner, T., Kornberg, L., Liu, E., & Cance, W. (1995). Overexpression of the focal adhesion kinase (pl25FAK) in invasive human tumors. Cancer Res. 55, 2752-2755.
Parsons, J. T. (1996). Integrin-mediated signaling: regulation by protein tyrosine kinases and small GTP-binding proteins. Curr. Opin. Cell Biol. 8, 146-152.
Parsons, J. T. & Parsons, S. J. (1997). Src family protein tyrosine kinases: cooperating with growth factor and adhesion signaling pathways. Curr. Opin. Cell Biol. 9, 187-192.
Papkoff, J., Rubinfeld, B., Schryver, B., & Polakis, P. (1996) Wnt-1 regulates free pools of catenins and stabilizes APC-catenin complexes. Mol. Cell. Biol. 16, 2128-2134.
Peifer, M. (1996). Regulating cell proliferation: as easy as APC. Science 272, 974-975.
Peifer, M. (1997). β-Catenin as oncogene: the smoking gun. Science 275, 1752-1753.
Peifer, M., Berg, S., & Reynolds, A. B. (1994a). A repeating amino acid motif shared by proteins with diverse cellular roles. Cell 76, 789-791.
Peifer, M., Sweeton, D., Casey, M., & Wieschaus, E. (1994b). Wingless signal and zeste white 3 kinase trigger opposing changes in the intracellular distribution of armadillo. Development 120, 369-380.
Peifer, M. & Wieschaus, E. (1990). The segment polarity gene armadillo encodes a functionally modular protein that is the *Drosophila* homolog of human plakoglobin. Cell 63, 1167-1176.
Pistor, S., Chakraborty, T., Walter, U., & Wehland, J. (1995). The bacterial actin nucleator protein ActA of Listeria monocytogenes contains multiple binding sites for host microfilament proteins. Curr. Biol. 5, 517-552.
Plantefaber, L. C. & Hynes, R. O. (1 989). Changes in integrin receptors of oncogenically transformed cells. Cell 56, 281-290.
Polakis, P. (1995). Mutations in the APC gene and their implications for protein structure and function. Curr. Opin. Gen. Dev. 5, 66-71.
Pollard, T. D. (1995). Actin cytoskeleton. Missing link for intracellular bacterial motility? Curr. Opin. Cell Biol. 5, 837-840.
Powell, S. M., Zilz, N., Beizer-Barclay, Y., Bryan, T., Hamilton, S., Thibodean, S., Vogelstein, B., & Kinzler, K. (1992). APC mutations occur during colorectal tumorigenesis. Nature 359, 235-237.
Prasad, G. L., Fulder, R. A., & Cooper, H. L, (1993). Expression of transduced tropomyosin 1 cDNA suppresses neoplastic growth of cells transformed by the ras oncogene. Proc. Natl. Acad. Sci. USA 90, 7039-7043.
Rankin, S. & Rozengurt, E. (1994). Platelet-derived growth factor modulation of focal adhesion kinase ($pl25^{FAK}$) and paxillin tyrosine phosphorylation in Swiss 3T3 cells. J. Biol. Chem. 269, 704-710.
Re, F., Zanetti, A., Sironi, M., Polentarutti, N., Lanfrancone, L., Dejana, E., & Colotta, F. (1994). Inhibition of anchorage-dependent cell spreading triggers apoptosis in cultured human endothelial cells. J. Cell Biol. 127, 537-546.
Reinhard, M., Giehl, K., Abel, K., Haffner, C., Jarchau, T., Hoppe, V., Jockusch, B. M., & Walter, U. (1995a). The proline-rich focal adhesion and microfilament protein VASP is a ligand for profilins. EMBO J. 14, 1583-1589.
Reinhard, M., Halbrügge, M., Scheer, U., Wiegand, C., Jokusch, B. M., & Walter, U. (1992). The 46/50 kDa phosphoprotein VASP purified from human platelets is a novel protein associated with actin filaments and focal contacts. EMBO J. 11, 2063-2070.

Reinhard, M., Jouvenal, K., Tripier, D., & Walter, U. (1995b). Identification, purification, and characterization of a zyxin-related protein that binds the focal adhesion and microfilament protein VASP (vasodilator-stimulated phosphoprotein). Proc. Natl. Acad. Sci. USA 92, 7956-7960.

Reinhard, M., Rudiger, M., Jockusch, B. M., & Walter, U. (1996). VASP interaction with vinculin: a recurring theme of interactions with proline-rich motifs. FEBS Lett. 399, 103-107.

Ren, X. D. Bokoch, G. M., Traynor-Kaplan, A., Jenkins, G. H., Anderson, R. A., & Schwartz, M. A. (1996). Physical association of the small GTPase rho with a 68 -kDa phosphatidylinositol 4-phosphate 5-kinase in Swiss 3T3 cells. Mol. Biol. Cell 7, 435-442.

Richardson, A. & Parsons, J. T. (1995). Signal transduction through integrins: a central role for focal adhesion kinase? BioEssays 17, 229-236.

Richardson, A. & Parsons, J. T. (1996). A mechanism for regulation the adhesion-associated protein tyrosine kinase $ppl25^{FAK}$. Nature 380, 538-540.

Ridley, A. J. & Hall, A. (1992). The small GTP-binding protein rho regulates the assembly of focal adhesions and actin stress fibers in response to growth factors. Cell 70, 389-399.

Riethmacher, D., Brinkmann, V., & Birchmeier, C. (1995). A targeted mutation in the mouse E-cadherin gene results in defective preimplantation development. Proc. Natl. Acad. Sci. USA. 92, 855-859.

Riese, J., Yu, X., Munnerlyn, A., Eresh, S., Hsu, S-C., Grosschedl, R., & Bienz, M. (1997). LEF-1, a nuclear factor coordinating signaling inputs from wingless and decapentaplegic. Cell 88, 777-787.

Rimm, D. L., Koslov, E. R., Kebriaei, P., Cianci, C. D., & Morrow, J. S. (1995). α_1(E)-catenin is a novel actin binding and bundling protein mediating the attachment of F-actin to the membrane adhesive complex. Proc. Natl. Acad. Sci. USA 92, 8813-8817.

Rodríguez Fernández, J. L., Geiger, B., Salomon, D., Sabanay, I., Zöller, M., & Ben-Ze'ev, A. (1992a). Suppression of tumorigenicity in transformed cells after transfection with vinculin cDNA. J. Cell Biol. 119, 427-438.

Rodríguez Fernández J. L., Geiger B., Salomon D., & Ben-Ze'ev A. (1992b). Overexpression of vinculin suppresses cell motility in Balb/c 3T3 cells. Cell Motil. Cytoskel. 22, 127-134.

Rodríguez Fernández, J. L., Geiger, B., Salomon, D., & Ben-Ze'ev, A. (1993). Suppression of vinculin expression by antisense transfection confers changes in cell morphology, motility and anchorage-dependent growth of 3T3 cells. J. Cell Biol. 122, 1285-1294.

Rubenstein, A., Merriam, J., & Klymkowsky, M. (1997). Localizing the adhesive and signaling functions of plakoglobin. Dev. Genetics 20, 91-102.

Rubinfeld, B., Albert, I., Porfiri, E., Fiol, C., Munemitsu, S., & Polakis, P. (1996). Binding of GSK3β to the APC-β-catenin complex and regulation of complex assembly. Science 272, 1023-1026.

Rubinfeld, B., Robbins, P., El-Gamil, M., Albert, 1., Porfiri, E., & Polakis, P. (1997). Stabilization of β-catenin by genetic defects in melanoma cell lines. Science 275, 1790-1792

Rubinfeld, B., Souza, B., Albert, 1., Munemitsu, S., & Polakis, P. (1995). The APC protein and E-cadherin form similar but independent complexes with α-catenin, β-catenin, and plakoglobin. J. Biol. Chem. 270, 5549-5555.

Ruiz, P., Brinkmann, V., Ledermann, B., Behrend, M., Grund, C., Thalhammer, C., Vogel, F., Birchmeier, C., Günthert, U., Franke, W. W., & Birchmeier, W. (1996). Targeted mutation of plakoglobin in mice reveals essential functions of desmosomes in the embryonic heart. J. Cell Biol. 135, 215-225.

Ruoslahti, E. (1996). RGD and other recognition sequences for integrins. Annu. Rev. Cell Dev. Biol. 12, 697-715

Sadler, I., Crawford, A. W., Michelsen, J. W., & Beckerle, M. C. (1992). Zyxin and cCRP: two interactive LIM domain proteins associated with the cytoskeleton. J. Cell Biol. 119, 1573-1587.

Salomon, D., Sacco, P. A., guha Roy, S., Simcha, 1., Johnson, K. R., Wheelock, M. J., & Ben-Ze'ev, A. (1997). Regulation of β-catenin levels and localization by overexpression of plakoglobin and inhibition of the ubiquitin-proteasome system. J. Cell Biol. 139, 1325-1335.

Sasaki, H., Nagura, K., Ishino, M., Tobioka, H., Kotani, K., & Sassaki, T. (1995). Cloning and characterization of cell adhesion kinase, a novel protein-tyrosine kinase of the focal adhesion kinase subfamily. J. Biol. Chem. 270, 21206-21219.

Sastry, S. K., Lakonishok, M., Thomas, D. A., & Horwitz, A. F. (1996). Integrin alpha subunit ratios, cytoplasmic domains and growth factor synergy regulate muscle proliferation and differentiation. J. Cell Biol. 133, 169-184.

Schaller, M. D. & Parsons, T. J. (1994). Focal adhesion kinase and associated proteins. Curr. Opin. Cell Biol. 6. 705-710.

Schlaepfer, D., Hanks, S. K., Hunter, T., & van der Geer, P. (1994). Integrin-mediated signal transduction linked to Ras pathway by GRB2 binding to focal adhesion kinase. Nature 372, 786-791.

Schmeichel K. L. & Beckerle, M .C. (1994). The LIM domain is a modular protein-binding interface. Cell 79, 211-219.

Schmidt, A., Heid, H. W., Schäfer, S., Nuber, U. A., Zimbelman, R., & Franke, W. W. (1994). Desmosomes and cytoskeletal architecture in epithelial differentiation: cell type specific plaque components and intermediate filament anchorage. Eur. J. Cell Biol. 65, 229-245.

Schwartz, M. A., Brown, E. J., & Fazeli, B. (1993). A 50 kDa integrin-associated protein is required for integrin-regulated calcium entry in endothelial cells. J. Biol. Chem. 268, 19931-19934.

Schwartz, M. A., Lechene, C., & Ingber, D. E. (1991). Insoluble fibronectin activates the Na/H antiporter by clustering and immobilizing integrin $\alpha_5\beta_1$, independent of cell shape. Proc. Natl. Acad. Sci. USA 88, 7849-7853.

Schwartz, M. A., Schaller, M. D., & Ginsberg, M. H. (1995). Integrins: emerging paradigms of signal transduction. Annu. Rev. Cell Biol. 11, 549-599.

Seufferlein, T. & Rozengurt, E. (1994). Lysophosphatidic acid stimulates tyrosine phosphorylation of focal adhesion kinase, paxillin, and pl30. Signaling pathways and cross-talk with platelet-derived growth factor. J. Biol. Chem. 269, 9345-9351.

Shapiro, L., Fannon, A. M., Kwong, P. D., Thompson, A., Lehmann, M. S., Grubel, G., Legrand, J.-F., Als-Neilsen, J., Colman, D. R., & Hendrickson, W. A. (1995). Structural basis of cell-cell adhesion by cadherins. nature 374, 327-337.

Shin, S.-I., Freedman, V. H., Risser, R., & Pollack, R. (1975). Tumorigenicity of virus-transformed nude mice is correlated specifically with anchorage independent growth in vitro. Proc. Natl. Acad. Sci. USA 72, 4435-4439.

Simcha, I., Geiger, B., Yehuda-Levenberg, S. Salomon, D., & Ben-Ze'ev, A. (1996). Suppression of tumorigenicity by plakoglobin: an augmenting effect of N-cadherin. J. Cell Biol. 133, 199-209.

Smith, G. A., Theriot, J. A., & Portnoy, D. A. (1996). The tandem repeat domain in the *Listeria monocytogenes* ActA protein controls the rate of actin-based motility, the percentage of moving bacteria, and the localization of vasodilator-stimulated phosphoprotein and profilin. J. Cell Biol. 135, 647-660.

Solowska, J. Guan, J. J., Marcantonio, E. E., Trevithick, J. E., Buck, C. A., & Hynes, R. O. (1989). Expression of normal and mutant avian integrin subunits in rodent cells. J. Cell Biol. 109, 853-861.

Sommers, C. L., Gelmann, E. P., Kemler, R., Cowin, P., & Byers, S. W. (1994). Alterations in β-catenin phosphorylation and plakoglobin expression in human breast cancer cells. Cancer Res. 54, 3544-3552.

Stoker, M., O'Neil, C., Berryman, S., & Waxman, V. (1968). Anchorage and growth regulation in normal and virus-transformed cells. Int. J. Cancer 3, 683-693.

Suzuki, T., Saga, S., & Sasakawa, C. (1996). Functional analysis of *Shigella* VirG domains essential for interaction with vinculin and actin-based motility. J. Biol. Chem. 271, 21878-21885.

Takai, Y., Sasaki, T., Tanaka, K., & Nakanishi, H. (1995). Rho as a regulator of the cytoskeleton. Trends Biochem. Sci. 20, 227-231.

Takeichi, M. (1991). Cadherin cell adhesion receptors as a morphogenetic regulator. Science 251, 1451-1455.

Takeichi, M. (1993). Cadherins in cancer: implications for invasion and metastasis. Curr. Opin. Cell Biol. 5, 806-811.

Takeichi, M. (1995). Morphogenetic roles of classic cadherins. Curr. Opin. Cell Biol. 7, 619-627.

Tepass, U., Gruszynski-DeFeo, E., Haag, T. A., Omatyar, L., Torok, T., & Hartenstein, V. (1996). Shotgun encodes *Drosophila* E-cadherin and is preferentially required during cell rearrangement in the neuroectoderm and other morphogenetically active epithelia. Genes Dev. 10, 672-685.

Theriot, J. A., Rosenblatt, J., Portnoy, D. A., Goldschmidt-Clermont, P. J., & Mitchison, T. J. (1994). Involvement of profilin in the actin-based motility of *L. monocytogenes* in cells and in cell-free extracts. Cell 76, 505-517.

Tran Van Nhieu, G., Ben-Ze'ev, A., & Sansonetti, P. J. (1997). Modulation of bacterial entry into epithelial cells by association between vinculin and the *Shigella* IpaA invasin. EMBO J. 16, 2717-2729.

Turner, C. E. & Miller, J. T. (1994). Primary sequence of paxillin contains putative SH2 and SH3 domain binding motifs and multiple LIM domains: identification of a vinculin and $ppl25^{FAK}$-binding region. J. Cell Sci. 107, 1583-1591.

Vandekerckhove, J., Bauw, G. K., Vancompernolle, G., Honoré, B., & Celis J. (1990). it Comparative two-dimensional gel analysis and microsequencing identifies gelsolin as one of the most prominent downregulated markers of transformed human fibroblast and epithelial, cells. J. Cell Biol. 111, 95-102.

van de Wetering, M., Cavallo, R., Dooijes, D., van Beest, M., van Es, J. Loureiro, J., Ypma, A., Hursh, D., Jones, T., Bejsovec, A., Peifer, M., Mortin, M., & Clevers, H. (1997). Armadillo coactivates transcription driven by the product of the *Drosophila* segment polarity gene *dTCF*. Cell 88, 789-799.

Varner, J. A. & Cheresh, D. A. (1996). Integrins and cancer. Curr. Opin. Cell Biol. 8, 724-730.

Varner, J. A., Emerson, D. A., & Juliano, R.L. (1995). Integrin $\alpha_5\beta_1$ expression negatively regulates cell growth-reversal by attachment to fibronectin. Mol. Biol. Cell, 6,725-740.

Vasiliev, J. & Gelfand. I. (1981). Neoplastic and normal cells in culture. Cambridge Univ. Press, Cambridge.

Vleminckx, K., Vakaet, L. J., Mareel, M., Fiers, W., & Van Roy, F. (1991). Genetic manipulation of E-cadherin expression by epithelial tumor cells reveals an invasion suppressor role. Cell 66, 107-119.

Volberg, T., Geiger, B., Kam, Z., Pankov, R., Simcha, I., Sabanay, H., Coll, J-L., Adamson, E., & Ben-Ze'ev, A. (1995). Focal adhesion formation by F9 embryonal carcinoma cells after vinculin gene disruption. J. Cell Sci. 108, 2253-2260.

Weekes, J., Barry, S. T., & Critchley, D. R. (1996). Acidic phospholipids inhibit the intramolecular association between the N- and C-terminal regions of vinculin, exposing actin-binding and protein kinase C phosphorylation sites. Biochem. J. 314, 827-832.

Wahl, J. K., Sacco, P. A., McGranahan-Sadler, T. M., Sauppe, L. M., Wheelock, M. J., & Johnson, K. R. (1996). Plakoglobin domains that define its association with the desmosomal cadherins and the classical cadherins: identification of unique and shared domains. J. Cell Sci. 109, 1143-54

Watabe, M., Nagafuchi, A., Tsukita, S., & Takeichi, M. (1994). Induction of polarized cell-cell association and retardation of growth by activation of the E-cadherin-catenin adhesion system in a disperced carcinoma cell line. J. Cell Biol. 127, 247-256.

Xing, Z., Chen, H. C., Nowlen, J. K., Taylor, S. J., Shalloway, D., & Guan, J. L. (1994). Direct interaction of v-Src with the focal adhesion kinase mediated by the Src SH2 domain. Mol. Biol. Cell. 5, 413-421.

Xu, L., Owens, L. V., Sturge, G. C., Yang, X., Liu, E. T., Craven, R. J., & Cance, W. G. (1996). Attenuation of the expression of the focal adhesion kinase induces apoptosis in tumor cells. Cell Growth Diff. 7, 413-418.

Yamada, K. M. & Geiger, B. (1997). Molecular interactions in cell adhesion complexes. Curr. Opin. Cell Biol. 9, 76-85.

Yamada, K. M. & Miyamoto, S. (1995). Integrin transmembrane signaling and cytoskeletal control. Curr. Opin. Cell Biol. 7, 681-689.

Yang, J. T., Rayburn, H., & Hynes, R. O. (1993). Embryonic mesodermal defects in α_5 integrin-deficient mice. Development 119, 1093-1105.

Yost, C., Torres, M., Miller, J. R., Huang, E., Kimelman, D., & Moon, R. T. (1996). The axis inducing activity, stability, and subcellular distribution of β-catenin is regulated in *Xenopus* embryos by glycogen synthase kinase 3. Genes Dev. 10. 1443-1454.

Zeile, W. L., Purich, D. L., & Southwick, F. S. (1996). Recognition of two classes of oligoproline sequences in profilin-mediated acceleration of actin-based *Shigella* motility. J. Cell Biol. 133, 49-59.

Zhang, J., Flack, J. R., Reddy, K. R., Abrams, C. S., Zhao, W., & Rittenhouse, S. E. (1995). Phosphatidylinositol-3,4,5-triphosphate stimulates phosphorylation of pleckstrin in human platelets. J. Biol. Chem. 270, 22807-22810.

DESMOSOMAL ADHESION

David R. Garrod, Chris Tselepis, Sarah K. Runswick,
Alison J. North, Sarah R. Wallis, and
Martyn A. J. Chidgey

Advances in Molecular and Cell Biology
Volume 28, pages 165-202.

ISBN: 0-7623-0495-2

I. INTRODUCTION

The intercellular junctions known as desmosomes or *maculae adherentes* are consistent features of the epithelial phenotype (Davies and Garrod, 1997) and are also present in cardiac muscle and follicular dendritic cells of the lymphoid system. They link the intermediate filament (IF) cytoskeletons of adjacent cells and mediate intercellular adhesion, thus providing a continuous filamentous scaffold throughout tissues. The principal molecular components of desmosomes have been well characterized enabling much recent investigation into the mechanisms of intermediate filament attachment and adhesion.

The functional importance of desmosomes in tissue structure is evident in several human diseases. Autoantibody-induced loss of desmosomal adhesion gives rise to the epidermal blistering disease pemphigus, while the first human desmosomal mutation to be discovered, in the plakophilin 1 gene, gives rise to epidermal fragility, absence of hair and abnormality of nails. Evidence is also accumulating for a possible involvement of desmosomes in suppression of malignant invasion and metastasis.

A surprising feature of desmosomal adhesion is that it is calcium independent in its mature form, even though the desmosomal adhesion molecules are cadherin family members. Its calcium dependence can be regulated by "inside-out" signals involving protein kinase C, a feature that may be important in the modulation of desmosomal adhesion to permit junction turnover and cell motility. The mechanisms of desmosome assembly and disassembly are little understood but would seem to be of crucial importance in epithelial dynamics.

Of further surprise is the finding that the desmosomal glycoproteins, desmocollin and desmoglein, both occur as three different isoforms which, in epidermis, show overlapping, reciprocally-graded, differentiation-related distributions. Where they overlap, the different isoforms occur in the same desmosomes. This extraordinary distribution, coupled with the knowledge that the glycoprotein genes are closely linked on the same chromosome and that the cytoplasmic desmosomal proteins plakoglobin and plakophilin are members of the armadillo family of junctional/signalling proteins, raises questions about how desmosomes are involved in epithelial differentiation.

II. DESMOSOMAL ULTRASTRUCTURE

Desmosomes are disc-shaped, symmetrical junctions, normally up to 0.5 ,μm in diameter, formed between the plasma membrane of adjacent cells (see Figure 1). They consist of two principal domains: (I) a 20–30 nm wide extracellular core domain, the desmoglea, between apposing plasma membranes; (II) dense cytoplasmic plaques, 15–20 nm thick, adjacent to the plasma membranes and slightly separated from them by more electron lucent zones (Kelly, 1966). These ultrastructurally distinct domains

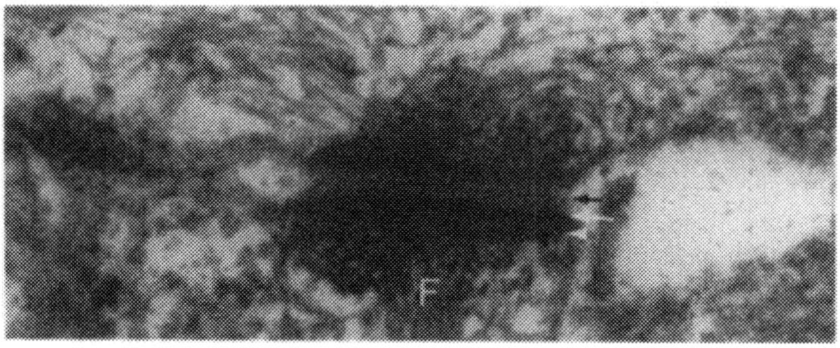

Figure 1. Electron micrograph of desmosome from mouse tongue epithelium showing the mid-line (black arrow), the outer plaque (white arrow), the inner plaque (white arrowhead), and intermediate filaments (IF).

mediate the functions of adhesion and intermediate filament (IF) attachment respectively.

The extracellular domain is bisected by an electron-dense mid-line. Lanthanum infiltration or ruthenium red staining reveal that the mid-line is connected to the plasma membranes by quadratic arrays of side-arms, the two arrays being staggered with respect to each other (Kelly, 1966; Rayns et al., 1969). Desmosomes stained by these methods thus display a "zipper"-like appearance. This core region is composed of the extracellular domains of the desmosomal cadherins, the desmocollins and the desmogleins. It is therefore tempting to speculate that the arrangement of the desmosomal cadherins may be similar to that reported for the classical type N-cadherin (Shapiro et al., 1995), but it is unclear how the desmosomal cadherins combine to produce such an array.

The cytoplasmic dense plaque consists of two regions, an outer, more electron-dense plaque, and an inner plaque, located further from the plasma membrane, into which the keratin IFs insert. IFs appear to converge toward the plaque, then loop away from it at a distance of 20–40 nm from the cell membrane (Fawcett, 1961; Kelly, 1966). Freeze-fracture reveals a second population of finer filaments, known as "traversing" filaments (McNutt and Weinstein, 1973; Leloup et al., 1979; Kelly and Kuda, 1981) which intervene between the IF loops and the desmosomal membranes. The composition of these 4–5 nm wide traversing filaments, their mode of attachment to IFs, and the mechanism by which tensile forces might be transmitted between the two filament populations remain to be clarified.

III. THE MOLECULAR COMPOSITION OF DESMOSOMES

Desmosomes are multimolecular complexes the principal components of which are two types of transmembrane glycoproteins, the desmosomal cadherins, desmocollin (Dsc) and desmoglein (Dsg), and the cytoplasmic plaque components plakoglobin (PG), plakophilin (PP) and desmoplakin (DP) (see Figure 2). There are also a number of minor components (see below). The structure and function of these components will now be discussed.

A. The Desmosomal Glycoproteins

The desmosomal glycoproteins, Dsc and Dsg, are members of the cadherin family of calcium-dependent adhesion molecules. Each occurs as three distinct isoforms, the products of different genes. Sequence data are available for the human, bovine and murine genes (for a review see Garrod et al., 1996). All six human Dsg and Dsc genes are located at chromosome 18q12, the mouse desmoglein genes are closely linked in the proximal region of chromosome 18 and the bovine desmocollin genes cluster on chromosome 24q21/q22 (Arnemann et al., 1992; Buxton et al.,

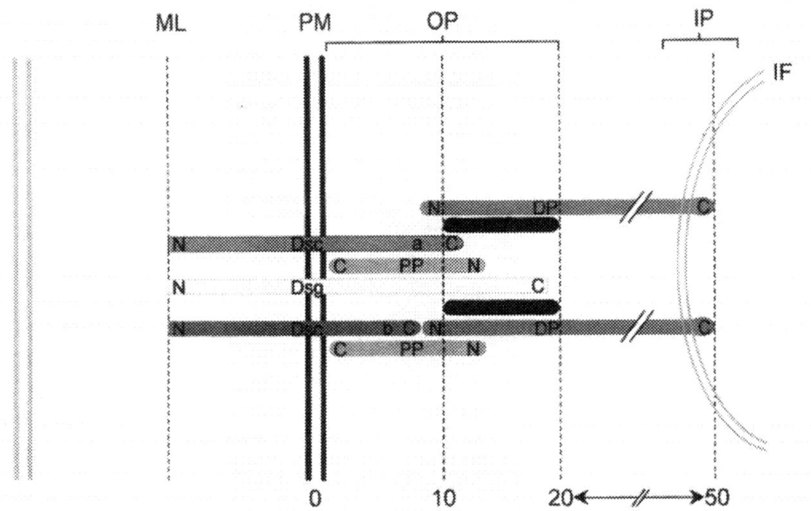

Figure 2. Cartoon showing major components of desmosomes. ML, mid-line; PM, plasma membranes; OP, outer plaque; IP, inner plaque; IF, intermediate filaments. The numbers at the bottom of the picture indicate distances in nanometres from the plasma membranes of the various plaque components as determined by quantitative measurements from immunogold labelling with domain-specific antibodies. Dsc, desmocollin; a, "a" form, b, "b" form; Dsg, desmoglein; DP, desmoplakin; PG, plakoglobin; PP, plakophilin; N, N-terminus; C, C-terminus. Only one half of the desmosome is shown: it is symmetrical about ML. The intercellular space is not drawn to scale as it would be almost 50% wider if shown at the same magnification as the plaque region.

1994; Wang et al., 1994; King et al., 1995; Simrak et al., 1995; Solinos-Toldo et al., 1995; Amagai et al., 1995a). This clustering may be required for regulation of gene expression.

Homology between the desmosomal cadherins and the classical cadherins is most extensive in the extracellular domain. This region contains four cadherin-like repeats which include N-glycosylation sites and several acidic putative calcium-binding sites, and an extracellular anchor. Intracellularly, the cytoplasmic portion of the desmocollins and desmogleins differ from each other and from the classic cadherins. The desmogleins have a larger cytoplasmic domain than the desmocollins. This is partly due to the presence of a number of unique 29-amino acid repeats, which have been predicted to adopt an antiparallel β-pleated sheet structure (Koch et al., 1990, 1991; Nilles et al., 1991; Wheeler et al., 1991; Franke et al., 1992; Schäfer et al., 1993a,b). There are five such repeats in Dsg1, six in Dsg2 and two in Dsg3. Each desmocollin isoform is subject to alternative splicing in the cytoplasmic domain, giving rise to the longer "a" and the shorter "b" form. In Dsc1 and 2 the "b" form occurs due to the inclusion of a 46bp exon containing an in-frame stop codon in the mRNA (Collins et al., 1991; Parker et al., 1991; Buxton et al., 1993). In Dsc3 the mini-exon consists of 43bp (Legan et al., 1994; Yue et al., 1995). Alternative splicing in the cytoplasmic domain is unique to the desmocollins within the cadherin superfamily.

The functional significance of splicing in desmocollins is unclear. Transfection of human A-431 carcinoma cells, with constructs encoding chimeric proteins consisting of the transmembrane region of connexin 32 fused with tails of Dsc1a and Dsc1b showed that the 'a' form is able to assemble novel plaque-like structures containing PG and DP, with attached intermediate filaments. Further studies have identified short sequence elements in the Dsc la cytoplasmic domain that support plaque assembly (Troyanovsky et al., 1993, 1994a; Chitaev et al., 1988). In contrast expression of the 'b' form did not result in association of cytoplasmic components and plaque assembly. Hence the function of the Dsc 'b' form remains unknown.

Adhesion

The following evidence suggests that desmosomes mediate strong cell-cell adhesion:

1. Desmosomes are most abundant in tissues such as epidermis that are subject to mechanical stress.
2. Loss of normal cell–cell adhesion in the skin blistering diseases pemphigus foliaceus (PF), pemphigus vulgaris (PV) and IgA pemphigus, is caused by autoantibodies to desmosomal glycoproteins (for reviews see Stanley and Kárpáti, 1994; Chidgey, 1997).
3. Incubation of cultured Madin-Darby bovine kidney cells with anti-desmocollin Fab' fragments inhibit desmosome assembly (Cowin et al., 1984).

4. In several human cancers, reduced expression of desmosomal components appears to correlate with invasion and metastasis (see below and Garrod, 1995 for a review).
5. Calcium-independent desmosomes can mediate cell–cell adhesion in the absence of all other adhesive junctions (Mattey and Garrod, 1986b; unpublished.
6. Mice expressing a null mutation of Dsg3 show perturbed epidermal cell adhesion (Allen et al., 1996).
7. A null mutation of plakoglobin in mice causes death between 12-16 days of embryogenesis due to the absence of desmosomes and disruption of muscle fiber adhesion in the heart (Ruiz et al., 1996; Bierkamp et al., 1996).

Notwithstanding the above evidence, it has proved extremely difficult to furnish a direct experimental demonstration of the adhesive function of desmosomal components by transfection. Most attempts to do this have used the L929 mouse fibroblast system which has previously been used to demonstrate the adhesive property of E-cadherin (Nagafuchi et al., 1987; Chen and Öbrink 1991). These cells do not express cadherins but do produce small amounts of α-, β- and γ-catenin (PG), cytoplasmic associates which are essential for E-cadherin adhesive function (Nagafuchi and Takeichi, 1988; Ozawa et al., 1989). Transfection of Dsc1a and Dsc1b into L929 cells, either alone or in combination, does not confer cell–cell adhesion. Furthermore, a chimeric molecule consisting of the E-cadherin cytoplasmic domain and extracellular domain of Dsc1 is also insufficient for adhesion, though this molecule binds endogenous catenins and presumably interacts with the actin cytoskeleton (Chidgey et al., 1996). Similar experiments have shown that Dsg1 is also insufficient for adhesion (Kowalczyk et al., 1996). In addition, Amagai and coworkers (1994a) have shown that a chimeric molecule made of the extracellular domain of Dsg3 and the cytoplasmic domain of E-cadherin exhibits only weak homophilic adhesion. Surprisingly a chimera of the extracellular domain of E-cadherin and the cytoplasmic domain of Dsg3 produced strong adhesion even though it did not bind catenins, suggesting that the Dsg intracellular domain may support the clustering necessary for cadherin mediated adhesion (Roh and Stanley, 1995). Recently it was shown that even a combination of Dsg1, Dsc2a and plakoglobin transfected into L929 cells does not lead to adhesion. Furthermore, two chimeric cadherins comprising the cytoplasmic domain of E-cadherin and the extracellular domain of either Dsg1 or Dsc2a did not support cell-cell adhesion when expressed either individually or together in L cells (Kowalczyk et al., 1996).

The above results suggest that neither homophilic (Dsc–Dsc, Dsg–Dsg) nor heterophilic (Dsc–Dsg) interaction takes place between desmosomal glycoproteins. However, using human fibrosarcoma HT-1080 cells, Chitaev and Troyanovsky (1997) have recently demonstrated heterophilic interaction between Dsg1 and Dsc2a and have shown that such interaction can generate cell-cell adhesion. In view of this, transfection of non-adhesive L929 cells with multiple desmosomal components should generate cell adhesion. Recently we have shown that transfecting Dsg1, Dsc1a, Dsc1b and plakoglobin into L-cells produces substantial calcium-dependent adhesion (Tselepis et al., 1998). Moreover, this adhesion is specifically inhibited by short peptides (10mers) centred around the putative adhesion sites of glycoproteins. These sites (-tyr-ala-thr- in Dsc1 and -arg-ala-leu- in Dsg1) correspond in location to the -his-ala-val- (HAV) site of the classical cadherins (Blaschuk et al., 1990). Therefore, these experiments provide the first experimental evidence that these sites are of functional significance in desmosomal adhesion.

It is not clear why we have been able to obtain adhesion in these experiments while Kowalczyk et al. (1996) did not. Our transfections included two differences of possible significance from those of Kowalczyk et al. Firstly we used both the "a" and "b" forms of Dsc1. Secondly we used the Dsg1 and Dsc1 isoforms which are expressed together in stratified epithelia whereas they used Dsg1 and Dsc2a which have an overlapping but not completely congruent expression in some tissues, e.g. epidermis. Neither of these factors appears to provide the crucial explanation because Chitaev and Troyanovsky (1997) have demonstrated that Dsg1 and Dsc2a can interact. Furthermore, Marcozzi and colleagues (1998) have succeeded in obtaining adhesion of L929 cells by transfection with the same combination of proteins as Kowlczyk and colleagues. A more probable explanation is that the stoichiometry of the desmosomal components is crucial for adhesion and that the critical amounts of the components have been obtained only fortuitously in some transfected cell lines. Thus, we obtained a few stable cell lines that showed substantial aggregation while others aggregated much more weakly or not at all. Kowalczyk et al. (1996) showed that Dsg binds six times as much PG as Dsc. Thus comparative over-expression of Dsg or under-expression of PG might disturb the stoichiometry sufficiently to inhibit the adhesive function of Dsc, since PG binding to Dsc is deemed to be important for this. Since we were unable to vary the stoichiometry of the different components systematically in our experiments, we have not yet been able to test the stoichiometry hypothesis.

The glycosylphosphoinositol-anchored desmosomal glycoprotein E48 has also been shown by transfection to have an adhesive function (Brakenhoff et al., 1995; Schrijvers et al., 1991). Originally discovered as a squamous cell carcinoma antigen, E48 is only expressed in the suprabasal layers of stratified epithelia. Here it may function to strengthen cell adhesion.

Distribution and Expression of Desmocollins and Desmogleins in Tissues

The three Dsc and three Dsg isoforms show tissue specific expression. Thus Dsg2 and Dsc2 are ubiquitous, common to all desmosome-bearing tissues, while Dsg1, Dsg3, Dsc1, and Dsc3 are primarily restricted to stratified epithelia (Legan et al., 1994; Schäfer et al., 1994, 1996; Nuber et al., 1995). Furthermore, the glycoprotein isoforms have distinct patterns of expression in epidermis and other stratified epithelia (Arnemann et al., 1993; Legan et al., 1994; Yue et al., 1995; North et al., 1996). In bovine epidermis, Dsc1 is associated with the upper, terminally-differentiating layers but absent from the basal layer and the bases of the deep rete ridges. Dsc2 is most strongly expressed in cells immediately above the basal layer that have begun terminal differentiation. Dsc3 is strongly present in the basal layer and gradually fades in the suprabasal layers (Legan et al., 1994; Yue et al., 1995; North et al., 1996). In human epidermis the expression patterns of Dscs are similar (Arnemann et al., 1993; King et al., 1995; 1996). The desmoglein patterns in epidermis are similarly differentiation-related with Dsg3 being most strongly expressed in the basal regions and Dsg1 increasing in expression suprabasally (Arnemann et al., 1993; Shimizu et al., 1995; Amagai et al., 1996).

An intriguing pattern of desmosomal glycoprotein expression is also found in mammary epithelium. Here Dsc1 and Dsg 1 are not expressed. Dsc2 and Dsg2 are present in both lumenal and myoepithelial cells but Dsc3 and Dsg3 are confined to the latter (Runswick et al., 1996), thus again occupying a basal location.

When different isoforms are present in the same cell how are they distributed between desmosomes? Double immunogold labelling of epidermis using isoform-specific antibodies has revealed that, in regions of isoform expression overlap, the individual desmosomes in these cells contain both Dsc1 and Dsc3 in an apparently mixed arrangement. Quantification of isoform labelling within desmosomes at different levels in the epidermis revealed that they are expressed in a reciprocally graded manner with Dsc3 decreasing suprabasally from the basal layer and Dsc1 increasing (North et al., 1996). An inversely graded pattern of Dsg3 and Dsg1 in human epidermis has also been observed (Shimizu et al., 1995). It may be inferred that different Dsg isoforms are also resent in the same junctions since at intermediate levels of epidermis all desmosomes can be labelled using sera specific for either Dsg1 or Dsg3.

These reciprocally-graded expression patterns suggest two conclusions. Firstly, the expression of the different glycoprotein isoforms appears to be linked, the close chromosomal location of their genes possibly being important for regulation of this. Secondly, since their molecular composition changes, the desmosomes in the epidermis are presumably continually turning over as the cells ascend through the layers. At present the functional significance of this is not clear. It may be that changes in cell–cell adhesive properties are important in regulating epidermal morphology and differentiation. This is an important area for future investigation.

B. Plakoglobin and Plakophilins

Plakoglobin and the plakophilins belong to an expanding family of related proteins which includes β-catenin and its *Drosophila* homologue, the segment polarity gene product armadillo (Peifer et al., 1992). These proteins all associate with adhesive intercellular junctions. Interestingly, several functionally-diverse proteins which share sequence homology with the armadillo family proteins have been described. These include the cadherin-associated protein, p120[cas], the APC tumor suppressor protein, a nucleotide exchange factor, GDS, a nuclear pore targeting protein, α-importin, and a suppressor of RNA polymerase I mutations in yeast, SRP1 (Hülsken et al., 1994; Peifer et al., 1994a; Kussel and Frasch, 1995). The roles of β-catenin as a bridging molecule linking the classical cadherins to α-catenin in the adherens junction (Aberle et al., 1996) and in intracellular signalling as a downstream effector of the vertebrate Wnt signalling pathway (for review see Gumbiner, 1995; Huber et al., 1996) have been firmly established. The roles of plakoglobin and the plakophilins in intercellular junctions and intracellular signalling are less well understood. Plakoglobin occurs as a single isoform which is a universal component of the cytoplasmic plaques of desmosomes, whereas the two isoforms of plakophilins have a restricted tissue distribution. Recently, it has emerged that plakoglobin and plakophilins have a key role in modulating the function of desmosomes (see below).

The functions of the armadillo family are achieved through interactions with a range of different proteins. Current evidence shows that the majority of these interactions are mediated by overlapping regions of a centrally located group of so-called armadillo repeats (see Figure 3). The armadillo repeat, or arm motif, consists of 42 amino acids which was first described in armadillo (Peifer, 1995; Klymkowsky and Parr, 1995). Plakoglobin contains 13 consecutive arm repeats flanked by distinct N-terminal and C-terminal domains (Riggleman et al., 1989). It shares 60–70% sequence identity with armadillo and β-catenin. The plakophilins consist of nine complete arm repeats followed by a short C-terminal extension of only 13 (plakophilin 1) or 11 (plakophilin 2) amino acids and no flanking N-terminal domain (Hatzfeld et al., 1994; Heid et al., 1994; Mertens et al., 1996). The plakophilins show amino acid identity of 33% to p120 with which they form a sub-group of armadillo-related proteins, but less than 25% to plakoglobin, β-catenin and armadillo (reviewed by Cowin and Burke, 1996).

Plakoglobin interacts with the cytoplasmic domains of both the desmosomal cadherins, desmocollin and desmoglein, and the classical cadherins (Korman et al., 1989; Knudsen and Wheelock, 1992; Peifer et al., 1992; Piepenhagen and Nelson, 1993; Kowalczyk et al., 1994; Mathur et al., 1994; Troyanovsky et al., 1994a,b; Butz and Kemler, 1994; Jou et al., 1995; Roh and Stanley, 1995; Chitaev et al., 1998; Smith and Fuchs, 1998). Plakoglobin has been shown to bind with high affinity to the cytoplasmic domain of desmoglein, with less affinity to desmocollin "a" forms and only weakly with that of E-cadherin (Chitaev et al., 1996). Plakoglobin also forms cytoplasmic complexes with APC and α-catenin, the latter being a component of the adherens junction (*zonulae adherntes*) (Rubinfeld et al., 1995). Recent deletion analyses shows that the central arm repeats are required for association with classical cadherins, whereas the repeats flanking the central region are essential for desmosomal cadherin binding (Mathur et al, 1994; Troyanovsky et al., 1994a,b; Roh and Stanley, 1995; Chitaev et al., 1996, 1998; Troyanovsky et al, 1996; Wahl et al., 1996; Witcher et al., 1996). Competition for binding domains on

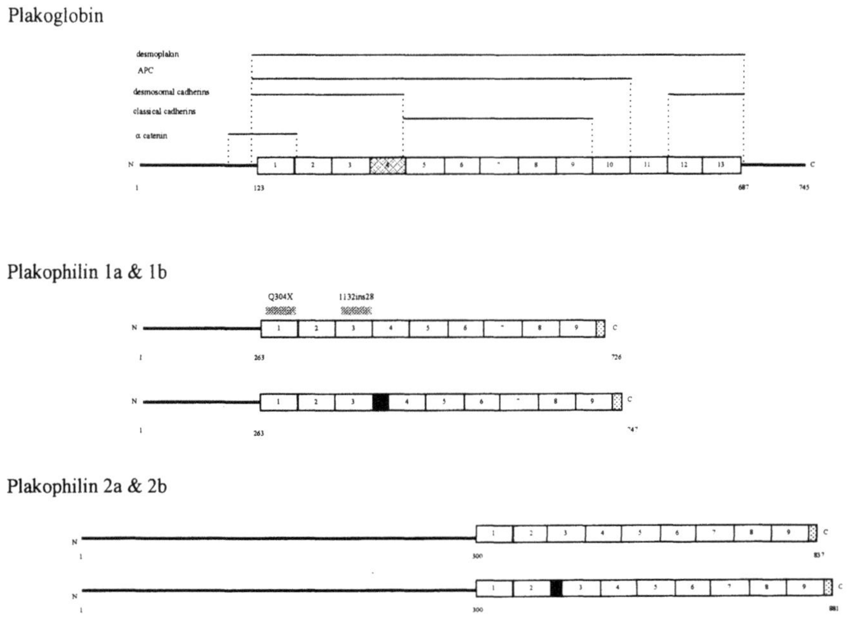

Figure 3. Comparison of the domain structures of plakoglobin (adapted from Wahl et al., 1996) and plakophilins 1a, 1b, 2a, and 2b. The amino acid sequences of human plakoglobin and human plakophilins 1a, 1b, 2a, and 2b are represented. The imperfect armadillo repeats are shown as numbered boxes. The domains of plakoglobin necessary for association with desmosomal cadherins, classical cadherins, α-catenin, APC and desmoplakin as determined by deletion analyses are indicated. The fourth armadillo repeat which is lacking in the alternative form of plakoglobin due to a 120bp deletion is shown (hatched box). The plakophilin 1a mutations which result in ectodermal dysplasia/skin fragility syndrome are shown; mutations Q304X in arm repeat 1 and 1132ins28 in arm repeat 3. The additional 21 and 44 amino acid inserts found in the alternative splice forms, plakophilin 1b and 2b, are indicated (solid box). Incomplete carboxyl-terminal armadillo repeats are shown (patterned boxes). Numbers represent amino acid residues. Diagram is not to scale.

plakoglobin may be an important mechanism for determining which molecular complexes it forms. For example, the desmoglein binding domain in the first three arm repeats overlaps the domain involved in α-catenin binding (Sacco et al., 1995; Chitaev et al., 1996, 1998; Wahl et al., 1996; Witcher et al., 1996; for review see Cowin and Burke, 1996). This overlap may regulate whether plakoglobin associates with desmosomes or with classical cadherins in adherens junctions.

Human, bovine, murine, and rat sequences are available for plakoglobin (Cowin et al., 1986; Franke et al., 1989, Butz et al., 1992; Hiipakka, 1996). The cloning of a variant form of plakoglobin from a human transitional bladder carcinoma-derived cell line has been reported (Ozawa et al., 1995a). The variant plakoglobin contains a

120bp deletion which corresponds to the fourth arm repeat and is thought to result from alternative splicing of the precursor mRNA. The alternative form of plakoglobin binds with less affinity to E-cadherin, desmoglein (Dsg2) and APC than full-length plakoglobin. However, both forms bind α-catenin with the same affinity (Ozawa et al., 1995b). It is not clear where and under what circumstances this alternative form of plakoglobin is expressed in vivo.

Since plakoglobin is the common component of adhesive intercellular junctions, it is a key candidate for a regulatory molecule in junction assembly. One of its roles may be segregating components between adherens junctions and desmosomes. Desmosomes are absent from the intercalated discs of cardiac muscle of plakoglobin null mice which show embryonic lethal structural defects of the developing heart (Ruiz et al., 1996; Bierkamp et al., 1996). In the defective heart muscle, desmoplakin becomes merged into extended junctions with the components of adherens junctions, while desmoglein (Dsg2) becomes scattered over the surface of the cells These mixed junctions are only observed in the heart. It is possible the force exerted by contraction of heart muscle causes intermixing of junctional components. Discrete desmosomes were able to form in epithelial organs but were abnormal. Bierkamp et al. (1996) reported that, in null mice which survived until birth, desmosomes in the epidermis showed abnormal structure resulting in blistering and subcorneal acantholysis. These desmosomes lacked either the inner or the outer plaque, or both, and attachment to intermediate filaments was impaired

The observations from null mutants are consistent with the idea that plakoglobin has an important role in the assembly of plaque components, possibly as a linking molecule between desmosomal cadherins and intermediate filament binding proteins. Recent work utilizing deletion analysis of plakoglobin and yeast two hybrid analysis (Palka and Green, 1997; Kowalczyk et al., 1997) suggests that plakoglobin binds directly to the N-terminus of desmoplakin. The site on plakoglobin which binds desmoplakin has not yet been precisely mapped, but appears to be located in the central arm repeats. Co-expression of a chimera consisting of the extracellular domain of E-cadherin and the cytoplasmic domain of Dsg1, plakoglobin and truncated desmoplakin (containing the N-terminus) in L-cells and COS cells showed that plaque-like junctional structures could be assembled at points of cell–cell contact (Kowalczyk et al., 1997). Plakoglobin, therefore, acts as a link between the cytoplasmic domains of desmoglein and desmoplakin. Whether this is also the case for desmocollin is not yet clear.

Plakoglobin, β-catenin and armadillo share substantial sequence homology in the arm repeat domains but the N- and C-termini are more divergent. This divergence may confer specific functions on the different molecules. To address this question, mutant plakoglobins lacking either or both N- and C- termini were expressed in A-431 cells (Palka and Green, 1997). Deletion of the N-terminus did not affect the assembly or morphology of desmosomes. Conversely, plakoglobin lacking the C-terminus was incorporated into abnormally long desmosomes. This may indicate that the C-terminus is involved in limiting desmosome length. The mechanism by which this operates is not yet clear but the C-terminus may mask upstream sites important in binding other desmosomal components. Consistent with this idea, a sequence of amino acids in arm repeat 13 of plakoglobin interacts and masks cadherin binding sites located further upstream (Troyanovsky et al., 1996). Thus the C-terminus of plakoglobin may mask cryptic binding sites for desmoplakin, so limiting over-recruitment of components and desmosome length (Palka and Green, 1997). It should be remembered, however, that normal desmosomes appear to fuse when brought together by low calcium medium-induced contraction of MDCK cells (Mattey and Garrod, 1986b). Such large desmosomes do not disperse when normal extracellular calcium concentration and normal cell morphology are restored (unpublished observations).

Expression of N-terminally-truncated plakoglobin in both *Xenopus* and cultured epithelial cells increased the cytosolic levels of endogenous and ectopic plakoglobin (Rubenstein et al, 1997; Palka and Green, 1997). Increases in cytosolic levels may reflect enhanced stability of plakoglobin in the non-cadherin associated pool. Stabilization of β-catenin is dependent upon dephosphorylation of a serine/threonine glycogen synthase kinase 3β (GSK3β) consensus site in the N-terminus (Peifer et al., 1994b; Rubenstein et al., 1997). GSK3β is the vertebrate homologue of *Drosophila* Zeste-white-3, a segment polarity gene that is negatively regulated by wingless leading to dephosphorylation, stabilization and increased cytoplasmic levels of armadillo. Plakoglobin also contains a GSK3β consensus site in its N-terminus and deletion of this site results in increased cytosolic levels in *Xenopus* embryos (Rubenstein et al., 1997). Evidence that elevated levels of β-catenin or plakoglobin exert a signalling effect come from overexpression studies. Microinjection of β-catenin or plakoglobin mRNA into *Xenopus* embryos results in duplication of the body axis, replicating similar effects for the overexpression of Wnt-1 (Funayama et al., 1995; Karnovsky and Klymkowsky, 1995). It appears that the balance between cadherin-bound and unbound pools of β-catenin and plakoglobin may be crucial for signalling events during development. Recently, using yeast two hybrid screening, the architectural HMG box transcription factor, LEF-1/XTCF-3 has been shown to interact with β-catenin (Behrens et al., 1996; Huber et al., 1996; Molenaar et al., 1996) and plakoglobin (Huber et al., 1996). In mammalian cells, co-expressed β-catenin/LEF-1 or plakoglobin/LEF-1 form complexes which translocate to the nucleus. The β-catenin/LEF-1 complex can bind to the promoter region of E-cadherin. Thus, it may regulate E-cadherin transcription and, in turn, affect assembly of intercellular junctions.

In addition to its function in intercellular adhesive junctions and role in the Wnt signalling pathway, plakoglobin also forms cytoplasmic complexes with the tumor suppresser protein APC (Rubinfeld et al., 1995). Formation of this complex targets proteins for ubiquitination and degradation, thus down-regulating their expression. APC competes with cadherins for the association with plakoglobin and β-catenin (for review see Polakis, 1995) and down-regulates plakoglobin and β-catenin expression. Therefore, APC may act as a regulator of the levels of cytosolic plakoglobin determining the availability for cadherin binding. Mutant forms of APC, which are unable to down-regulate β-catenin and plakoglobin, are associated with colonic hyperplasia. Hence plakoglobin may be a key regulator in cell adhesion, differentiation, and proliferation.

How the diverse functions of plakoglobin are regulated remains to be answered. The modulation of plakoglobin by tyrosine phosphorylation may be an important mechanism. Growth factor-dependent tyrosine phosphorylation of plakoglobin correlates with a more invasive cell state (Shibamoto et al., 1994) and may be a key factor in regulating adhesive junction assembly. Additionally, it has been reported that the assembly and formation of adherens junctions precedes desmosome assembly (O'Keefe et al., 1987; Wheelock and Jensen, 1992; Lewis et al., 1997). Studies examining the role of plakoglobin in the assembly of the adhesive junctions show that in cells possessing both adherens junctions and desmosomes, the association of plakoglobin with a classical cadherin is necessary before desmosome assembly can take place (Lewis et al., 1997). However, it has also been shown that plakoglobin preferentially assembles into desmosomes rather than adherens junctions (Näthke et al., 1994; Adams et al., 1996). Furthermore, desmosomal adhesion can form in the absence of classical cadherin-mediated cell interactions (Tselepis et al., 1998). The order of events that lead to the formation of intercellular junctions is complex and plakoglobin plays an important modulatory role. It remains to be seen whether steric hindrance between binding partners, molecular conformation, levels of expression, and variation in binding kinetics between cadherin cytoplasmic domains are mechanisms for controlling plakoglobin complex formation.

Plakophilin was originally designated "band 6" in the enriched desmosomal protein preparations from bovine muzzle epithelium (Skerrow and Matoltsy, 1974). It is now known to occur as two isoforms, PP1 and PP2. Sequence data are available for human and bovine plakophilin 1 (Held et al., 1994; Hatzfeld et al., 1994). Alternative splicing of 21 amino acids encoding exon 7 at the beginning of arm repeat 4 of the human plakophilin 1 gene result in two variants, la (726 amino acids) and 1b (747 amino acids)(Schmidt et al., 1997). A second plakophilin isoform occurs as two variants, 2a (837 amino acids) and 2b (881 amino acids), due to alternative splicing of an exon between the second and third arm repeats. Both plakophilin 1 and 2 are ubiquitously expressed proteins that appear to be localized to the nucleus of both epithelial and non-epithelial cells (Mertens et al., 1996; Schmidt et al., 1997). The function of plakophilins in the nucleus is presently unknown.

The complement of plakophilins in the desmosome is cell-type specific. For example, in desmosomes of the suprabasal layers of stratified epithelia, only plakophilin 1 is present, whereas in simple epithelia or myocardial cells only plakophilin 2 can be detected. In other epithelial tissues, both plakophilins 1 and 2 can coexist in desmosomes (Held et al., 1994; Mertens et al. 1996; Schmidt et al., 1997). Only plakophilin la localizes to desmosomes, whereas the 1b variant is restricted to nuclei.

Presumably the additional sequence in the 1b splice variant is responsible for its restricted distribution but this has yet to be demonstrated (Schmidt et al., 1997).

Recent evidence demonstrates that plakophilin 1 has a major role in maintaining desmosome function in epidermis. The first-described human mutations in a desmosomal component result in functional knockout of plakophilin 1 (McGrath et al., 1997). In the affected individual, absence of plakophilin 1 manifests as a condition which features both cutaneous fragility and congenital ectodermal dysplasia, affecting skin, hair, and nails. No abnormalities in other organs have been detected. Immunostaining for plakophilin 1 is completely absent in the skin. Mutational analysis revealed nonsense mutations in each of the plakophilin 1 alleles. One is a paternally inherited point mutation at nucleotide 910 that causes a C-to-T transition, converting a glutamine into a stop codon (Q304X). The other is a maternally inherited 28bp internal duplication at nucleotide position 1132 (1132ins28) resulting in a frame shift and premature stop codon 66bp downstream from the insertion site. It is predicted that these mutations would result in severely truncated forms of plakophilin 1. Desmosomes in the affected skin are small and greatly reduced in number. Moreover, linkage to the keratin IF network is disrupted. Interestingly, immunostaining for desmoplakin showed abnormal diffuse intracellular staining, whereas all other staining for desmosomal components appeared normal. The evidence suggests that plakophilin 1 acts as an important modulator of desmosome function in stratified epithelia possibly linking desmoplakin to the desmosomal cadherins, but also as a mediator of morphogenesis of epidermis and epidermal appendages during embryonic development and beyond.

C. Desmoplakin

DP I and II are major components of the desmosomal plaque, extending across both outer and inner plaque domains (North, Bornslaeger, Green, and Garrod, unpublished results). They are derived by alternative splicing from a single gene, giving products of predicted MW 332 and 260 kDa respectively (see Bornslaeger et al., 1994 for review). Previously thought to be restricted to desmosomes, DP has also been identified as a component of complexus adherentes junctions in epithelial cells (Schmelz et al., 1994) and in endothelial cells (Valiron et al., 1996; Kowalczyk et al., 1998). DPs are constitutive desmosomal components except that DPII is absent from cardiac muscle (Angst et al., 1990). DPI is a homodimer consisting of a central α-helical coiled-coil rod domain flanked by globular end domains (Bornslaeger et al., 1994). By rotary shadowing, the rod domain of DPI is about 130 nm in length, while that of DPII is only about 43 nm long (O'Keefe et al., 1989). Using simultaneous immunogold labelling of both globular end domains in bovine nasal epidermis, we have found the length of DP, calculated perpendicular to the plasma membrane, to be only about 40 nm (North, Bornslaeger, Green and Garrod, unpublished data). It is therefore possible that DPI may be further folded or coiled in tissue. In addition, the rod domain of DPI is characterized by periodic distribution of acidic and basic residues indicating that it may aggregate with itself or with similar proteins to form higher order filamentous structures (Stappenbeck and Green, 1992).

The C-terminal domain of DP consists of three 176 amino acid subdomains, A, B, and C, plus a 68 residue "tail" at the very C-terminus. Each subdomain is composed of a 38 residue repeating motif with the same periodicity of acidic and basic residues as the 1B rod domain of IF proteins, suggesting that DP could interact with the latter by virtue of their common periodicity of charged residues (Bornslaeger et al., 1994). In contrast the N-terminal domain consists of a series of shorter heptad repeats which are predicted to form compact bundles (Virata et al., 1992).

By a series of elegant experiments involving the transfection of specific DP domains into cultured epithelial cells, Green and colleagues have demonstrated that the C-terminus mediates interactions with IF (Stappenbeck and Green, 1992; Stappenbeck et al., 1993; see also below). At least two C-terminal subdomains are involved in this binding, with an additional 48–68 residue region at the C-terminus being critical for association with keratin but not vimentin. This region may thus mediate a tighter association of DP with the keratin network (Stappenbeck et al., 1993). The rod domain is apparently involved in aggregation of desmoplakin molecules and thus contributes to the architecture of the desmosomal plaque (Stappenbeck and Green, 1992; Bornslaeger et al., 1996). Experiments using the yeast two-hybrid system indicate that the carboxyl portion of the rod domain contains a site which is essential for dimerization (Meng et al., 1997). The amino-terminal domain is required to target DP to the cytoplasmic face of the desmosome (Stappenbeck et al., 1993; Bornslaeger et al., 1996; Smith and Fuchs, 1998) and to cluster plakoglobin-desmosomal cadherin complexes into discrete domains at the plasma membrane (Kowalczyk et al., 1997). Consistent with this, immunogold labelling shows the DP N-terminus to lie in the outer plaque close to the plasma membrane and the C-terminus further out in the zone of IF attachment (North, Bornslaeger, Green, and Garrod, unpublished data). A further possible role for the C-terminal domain has emerged from transfection of cells with DP proteins lacking the C-terminus (Bornslaeger et al., 1996). Abnormal junctions are observed in these cells, with merging of desmosomal components (desmoplakin, desmosomal cadherins) and adherens junction components (E-cadherin, α- and β-catenin), suggesting that the C-terminal domain not only contributes to the architecture of the plaque but also to the segregation of desmosomes from other junctions.

D. Desmoplakin Homologues: Envoplakin, Periplakin, Plectin, and BPAG1

Desmoplakin is now known to be a member of a family of related proteins, recently termed the plakin family (Uitto et al., 1996; Ruhrberg and Watt, 1997), which includes plectin, envoplakin, periplakin, and BPAG1 (bullous pemphigoid antigen 1; also referred to as BP230). These proteins all have similar predicted structures consisting of a central rod domain of variable length between two globular end domains (reviewed by Green and Jones, 1996; Ruhrberg and Watt, 1997). The carboxy-terminal domains of these proteins have different numbers of conserved subdomains, originally defined by sequence analysis of desmoplakin (see Ruhrberg and Watt, 1997), which may be involved in binding to different types of IFs (Stappenbeck et al., 1993). The plakins are predicted to form homodimers or heterodimers via the central rod domain which is rich in heptad repeats and is believed to form a parallel α-helical coiled coil with a dimerization partner. The exception is DPII which cannot homodimerize in vitro presumably due to its shorter rod domain (Virata et al., 1992).

Envoplakin (Ruhrberg et al., 1996) is a precursor of the cornified envelope of epidermis and is expressed in stratified squamous epithelia, but not in simple epithelia or in non-epithelial cells. It has a molecular mass of 210 kDa and is derived from a 6.5 kb mRNA transcribed from a single copy gene. Its upregulation during terminal differentiation and its colocalization with desmoplakin at desmosomes and on keratin filaments suggests a role in linking the cornified envelope to the desmosome-IF network.

Periplakin (Ruhrberg et al., 1997) is a 195 kDa protein which associates with the desmosomal plaque and with keratin filaments in the differentiated layers of the epidermis. Like envoplakin it was originally identified as a cornified envelope precursor (Simon and Green, 1984). Periplakin and envoplakin co-immunoprecipitate, and immunolocalization indicates that they form a network radiating out from desmosomes (Ruhrberg et al., 1996; Ruhrberg et al., 1997). It has thus been proposed that together they may provide a scaffolding onto which the cornified envelope is assembled (Ruhrberg et al., 1997).

Plectin is a 300 kDa multifunctional IF-associated protein which is widely distributed among tissues and species, and which forms homotetramers up to 200 nm long and 2–3 nm wide (Foisner and Wiche, 1987, 1991; Svitkina et al., 1996). It is believed to be a versatile cytoplasmic cross-linker which can interact with multiple proteins. Thus it cross-links IFs and connects them to microtubules, actin microfilaments, and membrane adhesion sites including desmosomes and hemidesmosomes (Foisner and Wiche, 1991; Svitkina et al., 1996; Eger et al., 1997). Binding of plectin to DP has recently been demonstrated in vitro (Eger et al., 1997) and a cytokeratin binding site has been identified in the carboxyl terminus (Nikolic et al., 1996). A number of alternatively spliced plectin variants may be involved in these diverse interactions (McLean et al., 1996; Elliott et al., 1997). Defects in the disease muscular dystrophy associated with epidermolysis bullosa simplex (MD-EBS) have been linked to a plectin mutation (Gache et al., 1996; McLean et al., 1996; Pulkinnen et al., 1996; Smith et al., 1996). Electron microscopy of patents' epidermis showed disruption of the inner plaques of hemidesmosomes and detachment of keratin filaments. However, no disruption of desmosomes was observed. Moreover, desmosomes appear unaffected in plectin-deficient mice (Andrä et al., 1997) suggesting that the role of plectin in desmosomes is not indispensable.

BPAG1, a component of hemidesmosomes that is proposed to mediate IF-plaque interactions (reviewed by Green and Jones, 1996), will not be considered further here.

E. IF Attachment and Accessory Plaque Proteins

The principal desmosomal component involved in mediating interactions between IFs and the desmosomal plaque is desmoplakin. Cell transfection experiments using mutant desmoplakin constructs have shown that the C-terminus of desmoplakin can interact with both keratin and vimentin IFs (Stappenbeck and Green, 1992; Stappenbeck et al., 1993) in a phosphorylation-dependent manner (Stappenbeck et al., 1994). Direct interactions between DP and IF proteins have also been demonstrated in vitro (Kouklis et al., 1994; Meng et al., 1997; Smith and Fuchs, 1998). Specific IF types interact with DP via distinct sequences in different IF domains (Meng et al., 1997). Thus type II epidermal keratins bind to the DP C-terminus via sequences in the N-terminal head domain of a single keratin polypeptide chain, while the interaction of DP with simple epithelial keratins requires the presence of both type I and type II partner proteins, indicating the importance of the tertiary structure of the α-helical coiled coil (Meng et al., 1997). The interactions of the DP C-terminus with type III IF proteins (e.g., vimentin) is weaker than with epidermal keratins (Meng et al., 1997), but may be strengthened in vivo by the formation of coiled-coil DP dimers, since strong alignment of DP with vimentin IF in transient transfections has been shown to require the DP rod domain (Stappenbeck and Green, 1992; Stappenbeck et al., 1993). The physiological relevance of DP-IF binding has been confirmed by the expression of DP N-terminal polypeptides in stably transfected cell lines. These polypeptides act in a dominant negative manner to produce disrupted junctional structures which lack associated keratin filaments (Bornslaeger et al., 1996).

Further members of the plakin family are candidates for a role in IF-desmosome attachment, both by their localization to the region where IF insert into the plaque and by their homology to desmoplakin. Plectin has been reported to bind both DP and IF and by immunoelectron microscopy lies further from the desmosomal membrane than DP. Thus, it could function as a linker between them (Eger et al., 1997). Intermediate Filament Associated Protein (IFAP) 300, a vimentin IF-binding protein, localizes to desmosomes and hemidesmosomes at the region where IF attach to the inner plaque and has been demonstrated to bind keratins in vitro (Skalli et al., 1994); the relationship of this protein to plectin remains unclear. The sequence similarity of envoplakin and periplakin to DP, together with their localization to desmosomes and along keratin IF, may also indicate an involvement in anchoring filaments to desmosomes (Ruhrberg et al., 1997). Alternatively these proteins may mediate cross-linking and stabilization of desmosome-associated IF, rather than themselves anchoring the filaments. A further component which may be particularly important in this latter role, namely the organization and/or stabilization of a mature desmosome-IF network is pinin (originally known as the O8L antigen; Ouyang and Sugrue, 1992). This 140 kDa desmosomal accessory protein is located at the periphery of the plaque of only mature desmosomes of epithelia, its temporal appearance here apparently correlating with the establishment of a highly-organized desmosome-IF complex (Ouyang and Sugrue, 1996).

A number of other desmosomal components located nearer to the plasma membrane may also be involved in IF attachment. Visualization of isolated Dsg1 tails by rotary shadowing has revealed a head portion of ~ 4 nm diameter, presumed to reside near the membrane, and a thin tail some 19 nm long (Rutman et al., 1994). It has been proposed that this long tail portion, which contains basic residues, could stretch across the plaque and thus be favorably located to bind IF (Nilles et al., 1991). Such binding has not been demonstrated, but immunogold labelling has recently confirmed that the C-terminus of Dsg3 lies approximately 19 nm from the plasma membrane, coincident with the end of the outer plaque (North and Magee, unpublished data). Plakophilin 1 has been demonstrated to bind cytokeratins in vitro (Kapprell et al., 1988; Hatzfeld et al., 1994; Smith and Fuchs, 1998) yet by immunogold labelling lies within the outer, rather than inner, plaque domain (North, Hatzfeld and Garrod, unpublished results). Desmocalmin (bovine form; the human form is known as keratocalmin, Fairley et al., 1991) is a 240 kDa calmodulin-binding plaque protein which is expressed only in stratified epithelia (Tsukita and Tsukita, 1985). Like plakophilin 1, desmocalmin is localized immediately subjacent to the plasma membrane yet binds to polymerised keratin IF in vitro (Tsukita and Tsukita, 1985).

How can the in vitro binding results be reconciled with apparent physical separation of plakophilin 1 and desmocalmin from the inserting IF? Two possible explanations are that either the in vitro binding to keratin IF of some of the above proteins is not physiologically relevant, or that the keratins may actually approach nearer to the membrane than previously supposed. Leloup and colleagues (1979) suggested that the traversing filaments were unraveled protofilaments from terminating IF. Recent expression studies have shown that DP proteins containing both the rod and C-terminal domains aggregate with IF proteins to form structures that resemble the 4–5 nm traversing filaments of the inner plaque (Stappenbeck and Green, 1992). This meshwork is not seen with DP alone, suggesting that IF may be present in the plaque as an anastomosing network of fine filaments associated with DP (Bornslaeger et al., 1994).

IV. DESMOSOME ASSEMBLY, MATURATION, AND MODULATION

Although desmosomes appear robust in electron micrographs, in reality they must be labile structures that are capable of rapid assembly and modulation so as to facilitate the dynamic behavior of developing and mature epithelia, as well as disease-related cell movements such as wound healing and metastasis. There is almost no evidence relating to desmosomal lability in tissues, although electron microscopy has suggested desmosomal internalization in some tumors (McNutt and Weinstein, 1973; Schenk, 1980; Risinger and Larsen, 1981). Recently it has been shown that the glycoprotein composition of desmosomes changes as cells ascended through the epidermis (North et al., 1996). This provides the first direct evidence suggesting desmosomal turnover in vivo. Studies of desmosomal dynamics have been largely carried out in tissue culture by manipulating extracellular calcium concentration in order to induce desmosome assembly/disassembly (Hennings and Holbrook, 1983). Much of the following discussion is based on evidence derived from such "calcium switch" experiments. It must be borne in mind, however, that this experimental system, though valuable, is entirely artificial and may bear little resemblance to the processes that occur in vivo.

Assembly of functional desmosomes is initiated when neighboring epithelial cells establish contact under near physiological conditions. Hence assembly will not proceed in single cells or in cells cultured at low extracellular calcium concentration (< 0.1 mM). Controversy exists as to whether assembly is initiated along IF or at the cell surface. Punctate aggregates of DP are seen throughout cells at low calcium concentration and several groups have reported a colocalization of these aggregates with keratin IF (Jones and Goldman, 1985; Mattey and Garrod, 1986a; Pasdar and Nelson, 1988a and 1988b; Trevor and Steben, 1992). When desmosome assembly is induced by a calcium switch (see also below) these aggregates apparently relocate to the cell surface in parallel with extension of IF. If keratin IF are necessary for desmosome assembly, cells lacking such filaments should display disrupted desmosomes. However, apparently normal desmosomes were present in epidermis from transgenic mice with a dominant negative K14 mutation (Vassar et al., 1991)and in knockout mice lacking K14 (Lloyd et al., 1995), K10 (Porter et al., 1996; Reichelt et al., 1997) or K8 (Baribault et al., 1993; Baribault et al., 1994), as well as in two rare cases of epidermolysis bullosa simplex lacking functional K14 (Chan et al., 1994; Rugg et al., 1994). Moreover, the observation that C-terminally trunctated DP polypeptides incorporate into desmosomes (Bornslaeger et al., 1996) supports the view that an interaction between DP and IF is not required for assembly. It is possible that IF merely act as temporary cytoplasmic docking targets for assembly intermediates.

Interestingly, DP immunogold labelling has been visualized along keratin filaments in keratinocytes prepared by high pressure freezing followed by freeze substitution, but not by conventional chemical fixation and permeabilisation procedures, suggesting that DP is only weakly associated with IF (Ruhrberg et al., 1996).

An alternative theory is that desmosome assembly is initiated at the cell surface where it then serves to organise the IF cytoskeleton (Bologna et al., 1986). The punctate aggregates of DP, PG and the desmosomal cadherins (Demlehner et al., 1995) in the cytoplasm could be vesicles of internalised desmosomal components (Mattey and Garrod, 1986a; Duden and Franke, 1988; Holm et al., 1993; Burdett, 1993, 1998). Demlehner et al. (1995) propose that desmosomal components are co-assembled into plaque structures ("half-desmosomes") at the plasma membrane of cells in low calcium medium in a continuous cycle of synthesis, assembly, internalization, and degradation. Apparently, normal calcium levels are necessary only for the final step of desmosome assembly, that is the stabilization by adhesion of a symmetrical junction at apposed cell surfaces. Such stabilization would account for the decrease in turnover and increased accumulation of desmosomal components seen on increasing calcium concentration (Penn et al., 1987a, 1987b; Pasdar et al., 1988a, 1988b; Pasdar et al., 1991; Mattey et al., 1990). It seems most likely that the punctate aggregates represent a combination of assembly intermediates and internalized desmosomes. A definitive answer to this question will necessitate the examination of cells transfected with tagged components.

Although fully formed desmosomes may be seen within two hours following calcium-induced desmosome assembly, two subsequent events contribute to maturation of desmosomal adhesion. Firstly, the amount of desmosomal material in the cells continues to increase for at least 24 hours, apparently because of increase in numbers of desmosomes rather than increase in desmosome size (Mattey et al., 1990). Beyond this time, the amount of desmosomal material reaches a plateau suggesting, since synthesis of desmosomal components continues, that cells can regulate desmosome numbers. Secondly, desmosomal adhesion changes from initial calcium dependence to calcium independence. The regulation of calcium dependence of adhesion will now be considered in more detail.

In freshly plated MDCK cells desmosome adhesion is calcium dependent. Thus, when the extracellular calcium concentration is reduced to less than 0.1 mM, desmosomal halves lose mutual adhesion and are internalized (Mattey and Garrod, 1986b). This is not surprising because the desmosomal glycoproteins are members of the calcium dependent cadherin family of adhesion molecules. However, a variety of epithelial cells including MDCK (Mattey and Garrod, 1986b), primary human keratinocytes (Watt et al., 1984), and colorectal carcinoma cell lines (Collins et al., 1990) also exhibit calcium independent desmosomal adhesion. When calcium independent adhesion is present, these cells round up in low calcium medium, but retain long connecting processes. Immunofluorescent labelling and transmission electron microscopy show that the processes are connected by desmosomes. These desmosomes maintain their adhesion even in the presence of chelating agents.

Desmosome adhesion becomes calcium independent when the cells are maintained at confluent density for several days. Confluence is necessary for this maturation step since continuous culture at low cell density does not enable desmosomes to become calcium independent. Disruption of confluence by wounding causes a rapid (< 1 h) change back to calcium dependence of desmosomes between cells at the edge of the wound. This desmosomal change spreads from the edge cells through the monolayer, so that eventually all non-edge cells acquire calcium dependent desmosomes even though they are in all round contact with others. Thus, a signal is generated at the wound edge and is propagated through the epithelial cell layer causing alteration of the adhesive properties of desmosomes.

The precise nature of this wounding signal is unknown, but it involves activation of protein kinase C (PKC). We have shown that the alpha isoform of this family of signalling molecules translocates from the cytoplasm to the membrane in response to wounding, indicating its activation (Wallis, Lloyd, Wise, Ireland, Fleming, and

Garrod: manuscript in preparation). Drugs that affect PKC activity have been shown to alter desmosome adhesion. Thus, activation of PKC with the phorbol ester TPA causes a rapid change from calcium independence to calcium dependence. In addition, PKC inhibitors cause desmosomes to become calcium independent, even in non-confluent cells. Such inhibitors include Gö6976 which is specific for the α and β isoforms of which we find only the former in MDCK cells. Moreover, the serine/threonine phosphatase inhibitor, okadaic acid, also promotes calcium dependent desmosomal adhesion. Thus an increase in serine/threonine phosphorylation, probably initiated by activation of PKCα or inhibition of phosphatases, causes desmosomes to become calcium dependent.

Phosphorylation is an important mechanism in the regulation of cell adhesion. Tyrosine phosphorylation alters the adhesive properties of adherens junctions. E-cadherin and β-catenin become phosphorylated when *v-src* expression is induced in infected cells (Behrens et al., 1993). This leads to disruption of the adherens junctions (and presumably desmosomes), causing cells to undergo epithelial-mesenchymal transition and become more invasive. Furthermore, tyrosine phosphatases are emerging as important regulators of adhesion at adherens junctions, their activity stabilizing junctional adhesion by reducing tyrosine phosphorylation (Brady-Kalnay and Tonks, 1995). PKC has been implicated in the regulation of adhesion at focal contacts and during adhesive interactions of leukocytes (Kolanus and Seed, 1997).

Changes in phosphorylation have also been implicated in desmosome assembly. H7, a general serine/threonine kinase inhibitor caused the formation of cytoplasmic desmosomes on membrane structures inside cultured keratinocytes (Shabana et al., 1996). Pasdar et al. (1995) showed that okadaic acid inhibits calcium-induced desmosome assembly indicating that serine/threonine phosphatase activity is necessary for this process. Sheu et al. (1989) found that kinase inhibitors also inhibited calcium-induced desmosome assembly, suggesting that kinases are also required. Furthermore, they found that TPA treatment of keratinocytes induced desmosome assembly even at low extracellular calcium concentration. TPA also induced desmosome formation in squamous cell carcinoma cells in low calcium medium (Kitajima et al., 1988). Raising calcium to induce desmosome assembly causes a four-fold increase in the cytoplasmic concentration of diacylglycerol, the intracellular activator of PKC (Ziboti et al., 1984). Thus, it may be activation of PKC, rather than increased extracellular calcium concentration per se, that initiates desmosome assembly.

The phosphorylation target(s) of PKCα involved in modulation of desmosomal adhesion is not known, but desmosomal proteins are good candidates. We have shown that all the major desmosomal proteins of MDCK cells are phosphorylated in newly confluent cells, but have not yet identified any component with altered phosphorylation in calcium independent desmosomes. The insolubility of desmosomes presents a problem for the use of immunoprecipitation experiments to determine these targets. Pasdar and colleagues (1995) showed that desmoplakin phosphorylation was increased by treatment with okadaic acid at the time of desmosome assembly, while treatment of hepatocytes with the hepatotoxic serine/threonine phosphatase inhibitor microcystin LR promoted desmoplakin phosphorylation but caused desmosome disassembly (Toivola et al., 1997). Another possibility is that the keratin IF proteins associated with desmosomes are the phosphorylation targets that regulate desmosomal adhesion. Phosphorylation of keratins in response to exogenous EGF and by PKC have been demonstrated (Baribault et al., 1989; Yano et al., 1991; Chou & Omary, 1991). We hypothesize that reduction in phosphorylation of cytoplasmic proteins or the cytoplasmic domains of desmosomal glycoproteins locks the desmosomal glycoproteins into conformational states that do not require extracellular calcium for adhesion. Activation of PKC or inhibition of phosphatases reverses this.

What is the function of the calcium dependence changes in desmosomes? The adhesive changes we have described are revealed by variations in extracellular calcium concentration which do not occur in vivo. However, they appear to reflect changes in the adhesive properties of the junctions signalled by some parameter of the confluence of the cell sheet. We have shown, by exposing tissues to chelating agents and examining their junctions by electron microscopy, that calcium independence is the predominant state of desmosomal adhesion in vivo. We hypothesize that calcium independence represents a more stable adhesive state. The reversion of cells to calcium dependence on wounding may then represent a change to a more readily modulatable state of desmosomes which would facilitate cell migration. Alternatively, the condition of the desmosomes could signal the activity state of the cell. Thus, when increased cellular activity is needed (for example at a wound edge) desmosomes undergo a change, manifested as calcium dependence in culture, so as to communicate their more motile phenotype to neighboring cells. The rapidity of the response of desmosomes to epithelial wounding suggests that this may be involved in "kick-starting" the response of epithelial cells to wounding via the activation of PKC. The ability to modulate desmosomal adhesion is clearly of importance to cells not only in wound healing, but in embryonic development, cancer metastasis and normal tissue dynamics. We believe that our observations provide the first evidence relating to how such modulation may be accomplished.

V. DEVELOPMENT

Desmosomes first form at the 32-cell stage of development in the mouse embryo. They are trophectoderm-specific and initiation of desmosome formation coincides with the onset of cavitation and blastocoel formation suggesting that desmosomes may play a role in stabilizing the trophectoderm as the embryo expands (Fleming et al., 1991). The first desmosomal constituent to appear during murine development is plakoglobin, synthesis of which has been detected at the 8-cell stage (Fleming et al., 1991). Cytokeratin IF assembly also occurs at about this time in embryogenesis

(Chisholm and Houliston, 1987). DPs are first synthesised at the 16-cell stage (Fleming et al., 1991) and Dsc and Dsg proteins have been detected at the 32-cell stage by immunoprecipitation (Fleming et al., 1991). The Dsg isoform present at this time must be either Dsg2 and/or Dsg3; Dsg1 is not expressed until much later in development (King et al., 1996, 1997). The only Dsc isoform present is Dsc2. Embryonic DSC2 gene expression first occurs at the 16-cell stage and is specific to the trophectoderm lineage (Collins et al., 1995). Dsc2 protein appears immediately after detection of its mRNA and coincides with initial desmosome assembly (Fleming et al., 1991; Collins et al., 1995). Desmosome formation between trophectoderm cells may therefore be regulated by transcription of the glycoprotein genes although studies on the DSG genes will be required to confirm this view.

The desmocollin isoforms Dsc1 and Dsc3 are not expressed in the early mouse embryo. Dsc1 is first detected in the epidermis of the external nares at embryonic day of development 13.5 (King et al., 1996, 1997; Chidgey et al., 1997). Dsc3 is initially expressed 12 hours earlier in nasal epidermis, whisker pads, and the most mature vibrissa follicles (Chidgey et al., 1997). Both genes are up-regulated in the general body epidermis at embryonic day 14.5. At this time both isoforms are expressed in the suprabasal layers of the newly stratified epidermis. However by day 18.5 the region of maximum Dsc3 expression becomes basal while that of Dsc1 remains suprabasal (King et al., 1996, 1997; Chidgey et al., 1997). Thus, the adult pattern of Dsc expression is established (see North et al., 1996) and this coincides with the onset of the adult pattern of differentiation including the basal location of cell proliferation and the formation of a stratum corneum. We have suggested that it is the ratio of Dsc1 to Dsc3 expression at different levels in the epidermis which is fundamental to establishing this pattern of differentiation (Chidgey et al., 1997). However it should be noted that the Dsg isoforms may play a role in this process.

VI. DISEASE

Pemphigus is an autoimmune disease of epidermis and mucous membranes. There are two main forms of the disease, pemphigus foliaceus (PF) and pemphigus vulgaris (PV). Patients with either form of pemphigus develop blisters of the skin but only those with PV develop mucous membrane lesions with the oral mucosa being most frequently affected. The only difference in the histology of the blisters is that in PF the loss of adhesion or acantholysis occurs in the granular layer or directly below it whereas in PV this occurs immediately above the basal layer.

The pathogenic role of autoantibodies in PF and PV has been clearly established. Neonates of mothers with PV have transient blistering disease caused by passage of maternal IgG across the placenta (Merlob et al., 1986). Although PF mothers do not have affected babies (Rocha-Alvarez et al., 1992), blistering can be induced by adding either PF or PV IgG to explants of human skin in culture (Schiltz and Michel, 1976; Hashimoto et al., 1983). Similarly, a loss of cellular adhesion results when autoantibodies from PF or PV patients are injected into neonatal mice (Anhalt et al., 1982; Roscoe et al., 1985), even in the absence of complement (Anhalt et al., 1986). The target molecules of the autoantibodies in PF and PV are the desmosomal glycoproteins Dsg1 and Dsg3 respectively. Hence incubation of patients' sera with recombinant forms of these molecules produced in eukaryotic cells abrogates the ability of autoantibodies to cause blistering in the neonatal mouse model (Amagai et al., 1994b; 1995b). The precise molecular mechanism by which binding of autoantibodies to keratinocytes produces blistering remains to be determined. The pathology almost certainly depends at least in part on protease release from affected cells (Hashimoto et al., 1983). However null mutant mice with a targeted disruption in the DSG3 gene have a phenotype resembling that of PV patients (Koch et al., 1997). This is consistent with the idea that autoantibodies in PV directly interfere with adhesive function.

Desmosomal components have been implicated in a number of other blistering disorders. For example, in the subcorneal pustular dermatosis type of IgA pemphigus patients develop blisters in the upper epidermis and Dsc1 is an autoantigen for circulating IgA antibodies (Hashimoto et al., 1997). In paraneoplastic pemphigus, patients exhibit mucosal and cutaneous blistering and have antibodies which immunoprecipitate DPs I and II, the hemidesmosomal protein bullous pemphigoid antigen 1, the cornified envelope protein envoplakin and other as yet uncharacterized antigens (Anhalt et al., 1990; Oursler et al., 1992; Kim et al., 1997). Finally, in erythema multiforme (EM) major patients exhibit clinical and histological similarities to those with paraneoplastic pemphigus and possess autoantibodies to DPs I and II (Foedinger et al., 1995). EM major does not result from defects in cellular adhesion but is caused by death of keratinocytes which evolves into necrosis of the entire epidermis. In IgA pemphigus, paraneoplastic pemphigus and EM major the pathogenic role of the autoantibodies against desmosomal constituents has yet to be established. The appearance of antibodies to cytoplasmic proteins in paraneoplastic pemphigus and EM major may be a consequence rather than a cause of the disease. Other as yet unidentified agents could be responsible for cell lysis and the release of intracellular antigens which elicit an immune reaction.

The first evidence which definitively links a desmosomal component to a human genetic disease has recently been published (McGrath et al., 1997; see above). Mutations in other desmosomal constituents may also occur in genetic disease. The lesion associated with the striated form of palmoplantar keratoderma, a disease with involves thickening of epidermis on palms and soles, has been mapped to chromosome 18ql2, near to the desmosomal cadherin gene cluster (Henries et al., 1995).

Carcinogenesis is a multi-step process involving sequential alterations in oncogenes and tumor suppressor genes. In affected cells, these events can result in cell proliferation, invasion of neighboring tissue and metastasis to distant organs. Some of these steps clearly require modulation of cellular adhesion and desmosomes may therefore play an important role Several immunohistochemical studies of different types of human cancer have shown a correlation between reduced expression of desmosomal constituents and invasive potential. For example a correlation between reduced Dsg expression, poor differentiation, and increased invasiveness has been demonstrated in transitional cell carcinoma of bladder (Conn et al., 1990), oral squamous cell carcinoma (Harada et al., 1992; Imai et al., 1995; Hiraki et al., 1996; Shinohara et al., 1997), and squamous cell carcinoma of skin (Krunic et al., 1996). Similarly a reduction in staining for both DP and Dsc has been shown to correlate with loss of differentiation of the primary tumor, degree of invasion, and lymph node metastases in oral squamous cell carcinomas (Hiraki et al., 1996; Shinohara et al., 1997). It appears that desmosomes may have a tumor invasion and metastasis-suppresser function. In support of this view we have recently demonstrated that adhesion generated by expression of desmosomal components in L929 cells inhibits the invasion of cells into collagen gels and that invasion is restored by specific anti-adhesion peptides (Tselepis et al., 1998). However, it should be noted that in colorectal cancer no correlation was found between the expression levels of various desmosomal constituents, differentiation status, and metastasis (Collins et al., 1990).

Recent data suggest that plakoglobin may represent a tumor suppresser gene. The human plakoglobin gene is located close to the BRCA1 locus on chromosome 17 and is subjected to loss of heterozygosity in breast and ovarian tumors (Aberle et al., 1995). Reduced expression of plakoglobin correlates with malignant transformation in certain cell lines (Navarro et al., 1993; Sommers et al ., 1994). Furthermore, transfection of plakoglobin into cultured cells which either lack or express cadherins and catenins suppresses their tumorigenicity when subsequently injected into nude mice (Simcha et al., 1996).

Abnormalities in desmosome and gap junction organisation have been reported in regions of myofibre disarray in hypertrophic cardiomyopathy (HCM), a myocardial disease characterized by a hypertrophied, non-dilated left ventricle (Sepp et al., 1996). Alterations in desmosomes could affect the elastic properties of the myocardium and so result in diastolic dysfunction, a typical feature of HCM. Severe heart defects are seen in plakoglobin null mutant mice (Ruiz et al., 1996; Bierkamp et al., 1996). However, although mutations in genes encoding components of the cytoskeletal network have been identified in HCM (reviewed by Marian and Roberts, 1995), no abnormalities in desmosomal genes have yet been reported.

VII. CONCLUSION

Many outstanding questions remain about the function, role and regulation of desmosomes. Some of the principal ones appear to be:

1. Why are there different isoforms of Dsc, Dsg and PP, and what are their functions?
2. What signals regulate the assembly and disassembly of desmosomes?

3. How do "inside-out" signals modulate the calcium dependence of desmosomal adhesion and what are the functions of these alternative adhesive states?
4. Do desmosomes transduce "outside-in" signals?
5. Do the armadillo desmosomal proteins PG and PP have a function in the regulation of gene expression as well as in junction structure?
6. How do Dsc and Dsg interact to bring about desmosomal adhesion?
7. Do desmosomes have an invasion and metastasis suppressor function?

ACKNOWLEDGMENTS

Original work report in this manuscript was supported by the Medical Research Council, the Biotechnology and Biological Sciences Research Council, the Wellcome Trust, and the Cancer Research Campaign.

REFERENCES

Aberle, H., Bierkamp, C., Torchard, D., Serova, O., Wagner, T., Natt, E., Wirsching, J., Heidkamper, C., Montagna, M., Lynch, H. T., Lenoir, G. M., Scherer, G., Feunteun, J., & Kemler, R. (1995). The human plakoglobin gene localizes on chromosome 17q21 and is subjected to loss of heterozygosity in breast and ovarian cancers. Proc. Natl. Acad. Sci. USA 92, 6384-6388.

Aberle, H., Schwartz, H., & Kemler, R. (1996). Cadherin-catenin complex protein interactions and their implications for cadherin function. J. Cell Biochem. 61, 514-523.

Adams, C. L., Nelson, W. J., & Smith, S. J. (1996). Quantitative analysis of cadherin-catenin-actin reorganization during development of cell-cell adhesion. J. Cell Biol. 135, 1899-1911.

Allen, E., Yu, Q-C & Fuchs, E. (1996). Mice expressing a mutant desmosomal cadherin exhibit abnormalities in desmosomes, proliferation, and epidermal differentiation. J. Cell Biol. 133, 1367-1382.

Amagai, M., Karpati, S., Klaus-Kovtun, V., Udey, M. C., & Stanley, J. R. (1994a). The extracellular domain of pemphigus vulgaris antigen (desmoglein 3) mediates weak homophilic adhesion. J. Invest. Dermatol. 102, 402-408.

Amagai, M., Hashimoto, T., Shimizu, N., & Nishikawa, T. (1994b). Absorption of pathogenic antibodies by the extracellular domain of pemphigus vulgaris antigen (Dsg3) produced by baculovirus. J. Clin. Invest. 94, 59-67.

Amagai, M., Wang, Y., Minoshima, S., Kawamura, K., Green, K. J., Nishikawa, T., & Shimizu, N. (1995a). Assignment of the human genes for desmocollin 3 (DSC3) and desmocollin 4 (DSC4) to chromosome 18ql2. Genomics 25, 330-332.

Amagai, M., Hashimoto, T., Green, K. J., Shimizu, N., & Nishikawa, T. (1995b). Antigen-specific immunoadsorption of pathogenic autoantibodies in pemphigus foliaceus. J. Invest. Dermatol. 104, 895-901.

Amagai, M., Koch, P. J., Nishikawa, T., & Stanley, J. R. (1996). Pemphigus vulgaris antigen (desmoglein 3) is localized in the lower epidermis, the site of blister formation in patients. J. Invest. Dermatol. 106, 351-355.

Andrä, K., Lassmann, H., Bittner, R., Shorny, S., Fässler, R., Propst, F., & Wiche, G. (1997). Targeted inactivation of plectin reveals essential function in maintaining the integrity of skin, muscle, and heart cytoarchitecture. Genes Dev. 11, 3143-3156.

Angst, B. D., Nilles, L. A., & Green, K. J. (1990). Desmoplakin II expression is not restricted to stratified epithelia. J. Cell Sci. 97, 247-257.

Anhalt, G. J., Labib, R. S., Voorhees, J. J., Beals, T. F., & Diaz, L. A. (1982). Induction of pemphigus in neonatal mice by passive transfer of IgG from patients with the disease. N. Engl. J. Med. 306, 1189-1196.

Anhalt, G. J., Till, G. O., Diaz, L. A., Labib, R. S., Patel, H. P., & Eaglstein, N. F. (1986). Defining the role of complement in experimental pemphigus vulgaris in mice. J. Immunol. 137, 2835-2840.

Anhalt, G. J., Kim, S., Stanley, J. R., Korman, N. J., Jabs, D. A., Kory, M., Izumi, H., Ratrie, H., Mutasim, D., Ariss-Abdo, L., & Labib, R. S. (1990). Paraneoplastic pemphigus. An autoimmune disease associated with neoplasia. N. Engl. J. Med. 323, 1729-1735.

Arnemann, J., Spurr, N. K., Magee, A. I., & Buxton, R. S. (1992). The human gene (DSG2) coding for HDGC, a second member of the desmoglein subfamily of the desmosomal cadherins, is, like DSG1 coding for DG1, assigned to chromosome 18. Genomics 13, 484-486.

Arnemann, J., Sullivan, K. H., Magee, A. I., King, I. A., & Buxton, R. S. (1993). Stratification-related expression of isoforms of the desmosomal cadherins in human epidermis. J. Cell Sci. 104, 741-750.

Baribault, H., Blouin, R., & Marceau, N. (1989). Epidermal growth factor-induced selective phosphorylation of cultured rat hepatocyte 55-kD cytokeratin before filament reorganization and DNA synthesis. J. Cell Biol. 109, 1665-1676.

Baribault, H., Price, J., Miyai, K., & Oshima, R. G. (1993). Mid-gestational lethality in mice lacking keratin 8. Genes Dev. 7, 1191-1202.

Baribault, H., Penner, J., Iozzo, R. V., & Wilson-Heiner, M. (1994). Colorectal hyperplasia and inflammation in keratin 8-deficient FVB/N mice. Genes Dev. 8, 2964-2973.

Behrens, J., Vakaet, L., Friis, R., Winterhager, E., Van-Roy, F., Mareel, M. M., & Birchmeier, W. (1993). Loss of epithelial differentiation and gain of invasiveness correlates with tyrosine phosphorylation of the E-cadherin/β-catenin complex in cells transformed with a temperature-sensitive v-SRC gene. J. Cell Biol. 120, 757-766.

Behrens, J., von Kries, J. P., Kühl, M., Bruhn L., Wedlich, D., Grosschedl, R., & Birchmeier, W. (1996). Functional interaction of β-catenin with the transcription factor LEF-1. Nature 382, 638-642.

Bierkamp, C., McLaughlin, K. J., Schwartz, H., Huber O., & Kemler, R. (1996). Embryonic heart and skin defects in mice lacking plakoglobin. Dev. Biol. 180, 780-785.

Blaschuk, O. W., Sullivan, R., David, S., & Pouliot, Y. (1990). Identification of a cadherin cell adhesion recognition sequence. Dev. Biol. 139, 227-229.

Bologna, M., Allen, R., & Dulbecco, R. (1986). Organization of cytokeratin bundles by desmosomes in rat mammary cells. J. Cell Biol. 102, 560-567.

Bornslaeger, E. A., Stappenbeck, T. S., Kowalczyk, A. P., Palka, H. L., & Green, K. J. (1994). Molecular genetic analysis of desmosomal proteins. In Molecular biology of desmosomes and hemidesmosomes. Collins, J.E., & Garrod, D. R. (eds). RG Landes Company, Austin. Pp. 35-52.

Bornslaeger, E. A., Corcoran, C. M., Stappenbeck, T. S., & Green, K. J. (1996). Breaking the connection: displacement of the desmosomal plaque protein desmoplakin from cell-cell interfaces disrupts anchorage of intermediate filament bundles and alters intercellular junction assembly. J. Cell Biol. 134, 985-1001.

Brady-Kalnay, S. M. and Tonks, N. M. (1995). Protein tyrosine phosphatases as adhesion receptors. Curr. Opin. Cell Biol. 5, 650-657.

Brakenhoff, R. H., Gerretsen, M., Ellen, M. C., Knippels, van Dijk, M., van Essen, H., Weghuis, D. O., Sinke, R. J., Snow, G. B., & van Dongen, G. A. M. S. (1995). The human E48 antigen, highly homologous to the murine Ly-6 antigen ThB, is a GPI-anchored molecule apparently involved in keratinocyte cell-cell adhesion. J. Cell Biol. 129, 1677-1689.

Burdett, I. D. J. (1993). Internalisation of desmosomes and their entry into the endocytic pathway via late endosomes in MDCK cells. J. Cell Sci. 106, 1115-1130.

Burdett, I. D. (1998). Aspects of the structure and assembly of desmosomes. Micron 29, 309-328.

Butz, S., Stappert, J., Weissig, H., & Kemler, R. (1992). Plakoglobin and β-catenin: distinct but closely related. Science. 257, 1142-1144

Butz, S. & Kemler, R. (1994). Distinct cadherin-catenin complexes in Ca^{2+}-dependent cell-cell adhesion. FEBS Lett. 355, 195-200.

Buxton, R. S., Cowin, P., Franke, W. W., Garrod, D. R., Green, K. J., King, I. A., Koch, P. J., Magee, A. I., Rees, D. A., Stanley, J. R., & Steinberg, M. S. (1993). Nomenclature of the desmosomal cadherins. J. Cell Biol. 121, 481-483.

Buxton, R. S., Wheeler, G. N., Pidsley, S. C., Marsden, M. D., Adams M. J., Jenkins N. A., Gilbert, D. J., & Copeland N. G. (1994). Mouse desmocollin (DSC3) and desmoglein (DSG1) genes are closely linked in the proximal region of chromosome 18. Genomics 21, 510-516.

Chan, Y.-M., Anton-Lamprecht, I., Yu, Q.-C., Jäckel, A., Zabel, B., Ernst, J.-P., & Fuchs, E. (1994). A human keratin 14 "knockout": the absence of K14 leads to severe epidermolysis bullosa simplex and a function for an intermediate filament protein. Genes Dev. 8, 2574-2587.

Chen, W. & Öbrink, B. (1991). Cell-cell contacts mediated by E-cadherin (uvomorulin) restrict invasive behaviour of L-cells. J. Cell Biol. 114, 319-327.

Chidgey, M. A. J. (1997). Desmosomes and disease. Histol. Histopath. 12, 1159-1168.

Chidgey, M. A. J., Clarke, J. P., & Garrod, D. R. (1996). Expression of full length desmosomal glycoproteins (desmocollins) is not sufficient to confer strong adhesion in transfected L929 cells. J. Invest. Dermatol. 106, 689-695.

Chidgey, M. A. J., Yue, K. K. M., Gould, S., Byrne, C., & Garrod, D. R. (1997). Changing pattern of desmocollin 3 expression accompanies epidermal organisation during skin development. Dev. Dyn. 210, 315-327.

Chisholm, J. C. & Houliston, E. (1987). Cytokeratin filament assembly in the preimplantation mouse embryo. Development 101, 565-582.

Chitaev, N. A., Leube,. E., Troyanovsky, Eshkind, L. G., & Franke, W. W. (1996). The binding of plakoglobin to desmosomal cadherins: Patterns of binding sites and topogenic potential. J. Cell Biol. 133, 359-369.

Chitaev, N. A. & Troyanovsky, S. M. (1997). Direct Ca^{2+}-dependent heterophilic interaction between desmosomal cadherins, desmoglein and desmocollin, contributes to cell-cell adhesion. J. Cell Biol. 138, 193-201.

Chitaev, N.J., Averbakh, A.Z., Troyanovsky, R.B., & Troyanovski, S.M. (1998). Molecular organization of the desmoglein-plakoglobin complex. J. Cell Sci. 111, 1941-1949.

Chou, C. F. & Omary, M. B. (1991). Phorbol acetate enhances the phosphorylation of cytokeratin 8 and 18 in human colonic epithelial cells. FEBS Lett. 282, 200-204.

Collins, J. E., Legan, P. K., Kenny, T. P., MacGarvie, J., Holton, J. L., & Garrod, D. R. (1991). Cloning and sequence analysis of desmosomal glycoproteins 2 and 3 (desmocollins): cadherin-like desmosomal adhesion molecules with heterogeneous cytoplasmic domains. J. Cell Biol. 113, 381-391.

Collins J. E., Lorimer, J. E., Garrod, D. R., Pidsley, S. C., Buxton, R. S., & Fleming, T. P (1995). Regulation of desmocollin transcription in mouse preimplantation embryos. Development 121, 743-752.

Collins, J. E., Taylor, I., & Garrod, D. R. (1990). A study of desmosomes in colorectal carcinoma. Br. J. Cancer 62, 796-805.

Conn, I. G., Vilela, M. J., Garrod, D. R., Crocker, J., & Wallace, M. A. (1990). Immunohistochemical staining with monoclonal antibody 32-2B to desmosomal glycoprotein 1. Its role in the histological assessment of urothelial carcinomas. Br. J. Urol. 65, 176-180.

Cowin, P., Mattey, D. L., & Garrod, D. R. (1984). Identification of desmosomal surface components (desmocollins) and inhibition of desmosome formation by specific Fab'. J. Cell Sci. 70, 41-60.

Cowin, P., Kapprell, H.-P., Frank, W. W., Tamkun, J., & Hynes, R. O. (1986). Plakoglobin: a protein common to different kinds of intercellular adhering junctions. Cell. 46, 1063- 1073.

Cowin, P. & Burke, B. (1996). Cytoskeleton-membrane interaction. Curr. Opin. Cell Biol. 8, 56-65.

Davies, J. A. & Garrod, D. R. (1997) Molecular aspects of the epithelial phenotype. BioEssays 19, 699-704

Demlehner, M. P., Schäfer, C., Grund, C., & Franke, W. W. (1995). Continual assembly of half-desmosomal structures in the absence of cell contacts and their frustrated endocytosis: a coordinated sisyphus cycle. J. Cell Biol. 131, 745-760.

Duden, R. & Franke, W. W. (1988). Organization of desmosomal plaque proteins in cells growing at low calcium concentrations. J. Cell Biol. 107, 1049-1063.

Eger, A., Stockinger, A., Wiche, G., & Foisner, R. (1997). Polarization dependent association of plectin with desmoplakin and the lateral submembrane skeleton in MDCK cells. J. Cell Sci. 110, 1307-1316.

Elliott, C. E., Becker, B., Oehler, S., Castanon, M. J., Hauptman, R., & Wiche, G. (1997). Plectin transcript diversity: Identification and tissue distribution of variants with distinct first coding exons and rodless isoforms. Genomics 42, 115-125.

Fairley, J. A., Scott, G. A., Jensen, K. D., Goldsmith, L. A., & Diaz, L. A. (1991). Characterization of keratocalmin, a calmodulin-binding protein from human epidermis. J. Clin. Invest. 88, 315-22.

Fawcett, D. W. (1961). Intercellular bridges. Exp. Cell Res. 8, 174-187.

Fleming, T. P., Garrod, D. R., & Elsmore, A. J. (1991). Desmosome biogenesis in the mouse preimplantation embryo. Development 112, 527-539.

Foedinger, D., Anhalt, G. J., Boecskoer, B., Elbe, A., Wolff, K., & Rappersberger, K. (1995). Autoantibodies to desmoplakin I and II in patients with erythema multiforme. J. Exp. Med. 181, 169-179.

Foisner, R. & Wiche, G. (1987). Structure and hydrodynamic properties of plectin molecules. J. Mol. Biol. 198, 515-31.

Foisner, R. & Wiche, G. (1991). Intermediate filament-associated proteins. Curr. Opin. Cell Biol. 3, 75-81.

Franke, W. W., Goldschmidt, M. D., Zimbelmann, R., Mueller, H. M., Schiller, D. L., & Cowin, P. (1989). Molecular cloning and amino acid sequence of human plakoglobin, the common junctional plaque protein. Proc. Nat. Acad. Sci. USA. 86, 4027-4031.

Franke, W. W., Troyanovsky, S. M., Koch, P. J., Troyanovsky, R. B., Fouquet, B., & Leube, R. E. (1992). Desmosomal proteins: Mediators of intercellular coupling and intermediate filament anchorage. Cold Spring Harbor Symp. Quant. Biol. 57, 643-652.

Funayama, N., Fagotto, F., McCrea, P., & Gumbiner, B. M. (1995). Embryonic axis induction by the armadillo repeat domain of β-catenin: evidence for intracellular signalling. J. Cell Biol. 128, 959-968.

Gache, Y., Chavanas, S., Lacour, J. P., Wiche, G., Owaribe, K., Meneguzzi, G., & Ortonne, J. P. (1996). Defective expression of plectin/HD1 in epidermolysis bullosa simplex with muscular dystrophy. J. Clin. Invest. 97, 2289-2298.

Garrod, D. R. Desmosomes and cancer. (1995). In Cancer Surveys Volume 24: Cell adhesion and cancer. Hart I., & Hogg N. (eds). Cold Spring Harbor Laboratory Press, New York. pp 97-111.

Garrod, D., Chidgey, M., & North, A. (1996). Desmosomes; differentiation, development, dynamics and disease. Curr. Opin. Cell Biol. 8, 670-678.

Green, K. J. & Jones, J. C. R. (1996). Desmosomes and hemidesmosomes: structure and function of the molecular components. FASEB J. 10, 871-881.

Gumbiner, B. M. (1995). Signal transduction by β-catenin. Curr. Opin. Cell Biol. 7, 634-640.

Harada, T., Shinohara, M., Nakamura, S., Shimada, M., Oka, M. (1992). Immunohistochemical detection of desmosomes in oral squamous cell carcinomas: correlation with differentiation, mode of invasion, and metastatic potential. Int. J. Oral Maxillofac. Surg. 21, 346-349.

Hashimoto, K., Safran, K. M., Webber, P. S., Lazarus, G. S., & Singer, K. H. (1983). Anti-cell surface pemphigus autoantibody stimulates plasminogen activator activity of human epidermal cells. A mechanism for the loss of epidermal cohesion and blister formation. J. Exp. Med. 157, 259-272.

Hashimoto, T., Kiyokawa, C., Mori, O., Miyasato, M., Chidgey, M. A. J., Garrod, D. R., Kobayashi, Y., Komori, K., Ishii, K., Amagai, M., & Nishikawa, T. (1997). Human desmocollin 1 (Dsc1) is an autoantigen for the subcorneal pustular dermatosis type of IgA pemphigus. J. Invest. Dermatol. 109, 127-131.

Hatzfeld, M., Kristjansson, G. I., Plessmann, U., & Weber, K. (1994). Band 6 protein, a major constituent of desmosomes from stratified epithelia, is a novel member of the armadillo multigene family. J. Cell Sci. 107. 2259-2270.

Heid, H. W., Schmidt, A., Zimbelmann, R., Schäfer, S., Winter-Simanowski, S., Stumpp, S., Keith, M., Figge, U., Schnölzer, M., & Franke, W. W. (1994). Cell-type specific desmosomal plaque proteins of the plakoglobin family: plakophilin (band 6 protein). Differentiation 58, 113-131.

Hennies, H-C., Kuster, W., Mischke, D., & Reis, A. (1995). Localization of a locus for the striated form of palmoplantar keratoderma to chromosome 18g near the desmosomal cadherin gene cluster. Hum. Mol. Genet. 4, 1015-1020.

Hennings, H. & Holbrook, K. A. (1983). Calcium regulation of cell-cell contact and differentiation of epidermal cells in culture. Exp. Cell Res. 143, 127-142.

Hiipakka, R. A. (1996). Molecular cloning and sequencing of the rat plakoglobin cDNA. GenBank Data library (accession no. U58858).

Hiraki, A., Shinohara, M., Ikebe, T., Nakamura, S., Kurahar, S., & Garrod, D. R. (1996). Immunohistochemical staining of desmosomal components in oral squamous cell carcinomas and its association with tumor behaviour. Br. J. Cancer 73, 1491-149.

Holm, P. K., Hansen, S. H., Sandvig, K., & van Deurs, B. (1993). Endocytosis of desmosomal plaques depends on intact actin filaments and leads to a nondegradative compartment. Eur. J. Cell Biol. 62, 362-371.

Huber. O., Korn, R., McLaughlin, J., Ohsugi, M., Herrmann, B. G., Kemler, R. (1996). Nuclear localization of β-catenin by interaction with transcription factor LEF- 1. Mech. Dev. 59, 3-10.

Hülsken, J., Birchmeier W., and Behrens, J. (1994). E-cadherin and APC compete for the interaction with β-catenin and the cytoskeleton. J. Cell Biol. 127, 2061-2069.

Imai, K., Kumagai, S., Nakagawa, K., Yamamoto, E., Nakanishi, I., & Okada, Y. (1995). Immunolocalization of desmoglein and intermediate filaments in human oral squamous cell carcinomas. Head Neck 17, 204-212.

Jones, J. C. R. & Goldman, R. D. (1985). Intermediate filaments and the initiation of desmosome assembly. J. Cell Biol. 101, 506-517.

Jou, T. S., Stewart, D. B., Stappert, J., Nelson, W. J., & Marrs, J. A. (1995). Genetic and biochemical dissection of protein linkages in the cadherin-catenin complex. Proc. Nat. Acad. Sci. USA. 92, 5067-5071.

Karnovsky, A. & Klymkowsky, M. W. (1995). Anterior duplication in Xenopus induced by the over-expression of the cadherin-binding protein plakoglobin. Proc. Natl. Acad. Sci. USA. 92, 4522-4526.

Kapprell, H.-P., Owaribe, K., & Franke, W. W. (1988). Identification of a basic protein of Mr 75,000 as an accessory desmosomal plaque protein in stratified and complex epithelia. J. Cell Biol. 106, 1679-1691.

Kelly, D. E. (1966). Fine structure of desmosomes, hemidesmosomes and an adepidermal globular layer in developing newt epidermis. J. Cell Biol. 28, 51-72.

Kelly, D. E. & Kuda, A. M. (1981). Traversing filaments in desmosomal and hemidesmosomal attachments: Freeze-fracture approaches toward their characterisation. Anat. Rec. 199, 1-14.

Kim, S-C., Kwon, Y. D., Lee, I. J., Chang, S. N., & Lee, T. G. (1997). cDNA cloning of the 210-kDa paraneoplastic pemphigus antigen reveals that envoplakin is a component of the antigen complex. J. Invest. Dermatol. 109, 365-369.

King, I. A., Sullivan, K. H., Bennett, R., & Buxton, R. S. (1995). The desmocollins of human foreskin epidermis: identification and chromosomal assignment of a third gene and expression patterns of the three isoforms. J. Invest. Dermatol. 105, 314-321.

King, I. A., O'Brien, T. J., & Buxton, R. S. (1996). Expression of the "skin-type" desmosomal cadherin DSC1 is closely linked to the keratinization of epithelial tissues during mouse development. J. Invest. Dermatol. 107:531-538.

King, I. A., Angst, B. D., Hunt, D. M., Krugar, M., Arnemann, J., & Buxton, R. S. (1997). Hierarchical expression of desmosomal cadherins during stratified epithelial morphogenesis in the mouse. Differentiation. 62, 83-96.

Kitajima, Y., Inoue, S., Nagao, S., Nagata, K., Yaoita, H., & Nozawa, Y. (1988). Biphasic effects of 12-O-tetradecanoylphorbol-13-acetate on the morphology of low calcium-grown human epidermal carcinoma cells: involvement of translocation and regulation of protein kinase C. Cancer Res. 48, 946-970.

Klymkowsky, M. W. & Parr, B. (1995). The body language of cells: the intimate connection between cell adhesion and behaviour. Cell 83, 5-8.

Knudsen, K. A. & Wheelock, M. J. (1992). Plakoglobin, or an 83kD homologue distinct from β-catenin interacts with E-cadherin and N-cadherin. J. Cell Biol. 118, 671-679.

Koch, P. J., Goldschmidt, M. D., Walsh, M. J., Zimbelmann, R., & Franke W. W. (1991). Complete amino acid sequence of the epidermal desmoglein precursor polypeptide and identification of a second type of desmoglein gene. Eur. J. Cell. Biol. 55. 200-208.

Koch, P. J., Walsh, M. J., Schmelz, M., Goldschmidt, M. D., Zimbelmann, R., & Franke W. W. (1990). Identification of desmoglein, a constitutive desmosomal glycoprotein, as a member of the cadherin family of cell adhesion molecules. Eur. J. Cell. Biol. 53, 1-12.

Koch, P. J., Mahoney, M. G., Ishikawa, H., Pulkkinen, L., Uitto, J., Shultz, L., Murphy, G. F., Whitaker-Menezes, D., & Stanley, J. R. (1997). Targeted disruption of the *Pemphigus vulgaris* antigen (desmoglein 3) gene in mice causes a loss of keratinocyte cell adhesion with a phenotype similar to pemphigus vulgaris. J. Cell Biol. 137, 1091-1102.

Kolanus, W. & Seed, B. (1997). Integrins and inside-out signal transduction; converging signals from PKC and PIP3. Curr. Opin. Cell Biol. 9, 725-731.

Korman, N. J., Eyre, R. W., Klaus-Kovtum, V., & Stanley, J. R. (1989). Demonstration of an adhering-junction molecule (plakoglobin) in the autoantigens of pemphigus foliaceus and pemphigus vulgaris. New Eng. J. Med. 321, 631-635.

Kouklis, P. D., Hutton, E., & Fuchs, E. (1994). Making a connection: direct binding between keratin intermediate filaments and desmosomal proteins. J. Cell Biol. 127, 1049-1060.

Kowalczyk, A. P., Palka, H., Luu, H. N., Nilles, L. A., Anderson, J. E., Wheelock, M. J., & Green, K. J. (1994). Posttranslational regulation of plakoglobin expression: influence of the desmosomal cadherins on plakoglobin metabolic stability. J. Biol. Chem. 269, 31214-31223.

Kowalczyk, A. P., Borgwardt, J. E., & Green, K. J. (1996). Analysis of desmosomal cadherin-adhesive function and stoichiometry of desmosomal cadherin-plakoglobin complexes. J. Invest. Dermatol. 107, 293-300.

Kowalczyk, A. P., Bornslaeger, E. A., Borgwardt, J. E., Palka, H. L., Dhaliwal, A. S., Corcoran, C. M., Denning, M. F., & Green, K. J. (1997). The amino-terminal domain of desmoplakin binds to plakoglobin and clusters desmosomal cadherin-plakoglobin complexes. J. Cell Biol. 139, 773-784.

Kowalczyk, A. P., Navarro, P., Dejana, E., Bornslaeger, E. A., Green, K. J., Kopp, D. S., & Borgwardt, J. E. (1998). VE-cadherin and desmoplakin are assembled into dermal microvascular endothelial intercellular junctions, a pivotal role for plakoglobin in the recruitment of desmoplakin to intercellular junctions. J. Cell Sci. 111, 3045-3057.

Krunic, A. L. J., Garrod, D. R., Smith, N. P., Orchard, G. S., & Cvijetic, O. B. (1996). Differential expression of desmosomal glycoproteins in keratoacanthoma and squamous cell carcinoma of the skin: an immunohistochemical aid to diagnosis. Arch. Derm. Venereol. 76, 394-398.

Kussel, P. & Frasch, M. (1995). Yeast Srp1, a nuclear protein related to Drosophila and mouse pendulin, is required for normal migration, division, and integrity of nuclei during mitosis. Mol. Gen. Genet. 248, 351-363.

Legan, P. K., Yue, K. K. M., Chidgey, M. A. J., Holton, J. L., Wilkinson, R. W., & Garrod, D. R. (1994). The bovine desmocollin family: a new gene and expression patterns reflecting epithelial cell proliferation and differentiation. J. Cell Biol. 126, 507-518.

Leloup, R., Laurent, L., Ronveaux, M. F., Drochmans, P., Wanson, J. C. (1979). Desmosomes and desmogenesis in epidermis of calf muzzle. Biol. Cell 34, 137-152.

Lewis, J. E., Wahl III, J. K., Sass, K. M., Jensen, P. J., Johnson, K. R., & Wheelock, M. J. (1997). Cross-talk between adherens junctions and desmosomes depends on plakoglobin. J. Cell Biol. 136, 919-934.

Lloyd, C., Yu, Q.-C., Cheng, J., Turksen, K., Degenstein, L., Hutton, E., & Fuchs, E. (1995). The basal keratin network of stratified squamous epithelia: defining K15 function in the absence of K14. J. Cell Biol. 129, 1329-1344.

Marcozzi, C., Burdett, I. D., Buxton, R. S., & Magee, A. I. (1998). Co-expression of both types if desmosomal cadherin and plakoglobin confers strong intercellular adhesion. J. Cell Sci. 111, 495-509.

Marian, A. J. & Roberts, R. (1995). Molecular genetics of hypertrophic cardiomyopathy. Annul Rev. Med. 46, 213-222.

Mathur, M., Goodwin, L., & Cowin, P. (1994). Interactions of the cytoplasmic domain of the desmosomal cadherin Dsg1 with plakoglobin. J. Biol. Chem. 269, 14075-14080.

Mattey, D. L. & Garrod, D. R. (1986a). Calcium-induced desmosome formation in cultured kidney epithelial cells. J. Cell Sci. 85, 95-111.

Mattey, D. L. & Garrod, D. R. (1986b). Splitting and internalisation of the desmosomes of cultured kidney epithelial cells by reduction in calcium concentration. J. Cell Sci. 85, 113-124.

Mattey, D. L., Burdge, G., & Garrod, D. R. (1990). Development of desmosomal adhesion between MDCK cells following calcium switching. J. Cell Sci. 97, 689-704.

McLean, W. H., Pulkkinen, L., Smith, F. J., Rugg, E. L., Lane, E. B., Bullrich, F., Burgeson, R. E., Amano, S., Hudson, D. L., Owaribe, K., et al. (1996). Loss of plectin causes epidermolysis bullosa with muscular dystrophy: cDNA cloning and genomic organisation. Genes Dev. 10, 1724-1735.

McGrath, J. A., McMillan, J. R., Shemanko, C. S., Runswick, S. K., Leigh, I. M., Lane, E. B., Garrod, D. R., & Eady, A. J. (1997). Mutations in the plakophilin 1 gene result in ectodermal dysplasia/skin fragility syndrome. Nature Gen. 17, 240-244.

McNutt, N. S. & Weinstein, R. S. (1973). Membrane ultrastructure at mammalian junctions. Prog. Biophys. Mol. Biol. 26, 45-101.

Meng, J.-J., Bornslaeger, E. A., Green, K. J., Steinert, P. M., & Ip, W. (1997). Two-hybrid analysis reveals fundamental differences in direct interactions between desmoplakin and cell type-specific intermediate filaments. J. Biol. Chem. 272, 21495-21503.

Merlob, P., Metzker, A., Hazaz, B., Rogovin, H., & Reisner, S. H. (1986). Neonatal pemphigus vulgaris. Pediatrics 78, 1102-1105.

Mertens, C., Kahn, C., & Frank, W. W. (1996). Plakophilins 2a & 2b: constitutive proteins of dual location in the karyoplasm and the desmosomal plaque. J. Cell Biol. 135, 1009-1025.

Molenaar, M., van de Wetering, M., Oosterwegel, M., Peterson-Maduro, J., Godsave, S., Korinek, V., Roose, J., Destree, O., & Clevers, H. (1996). XTcf-3 transcription factor mediates β-catenin-induced axis formation in xenopus embryos. Cell 86, 391-399.

Nagafuchi, A., Shirayoshi, K., Okazaki, K., Yasuda, K., & Takeichi, M. (1987). Transformation of cell adhesion properties by exogenously introduced E-cadherin cDNA. Nature 329, 341-343.

Nagafuchi, A. & Takeichi, M. (1988). Cell binding function of E-cadherin is regulated by the cytoplasmic domain. EMBO J. 7, 3679-3684.

Näthke, I. S., Hinck, L., Swedlow, J. R., Papkoff, J., & Nelson, J. W. (1994). Defining interactions and distributions of cadherin and catenin complexes in polarized epithelial cells. J. Cell Biol. 125, 1341-1352.

Navarro, P., Lozano, E., & Cano, A. (1993). Expression of E- or P-cadherin is not sufficient to modify the morphology and the tumorigenic behavior of murine spindle carcinoma cells. Possible involvement of plakoglobin. J. Cell Sci. 105, 923-934.

Nikolic, B., MacNulty, E., Mir, B., & Wiche, G. (1996). Basic amino acid residue cluster within nuclear targeting sequence motif is essential for cytoplasmic plectin-vimentin network junctions. J. Cell Biol. 134, 1455-1467.

Nilles, L. A., Parry, D. A. D., Powers, E. E., Angst, B. D., Wagner, R. M., & Green, K. J. (1991). Structural analysis and expression of human desmoglein: a cadherin-like component of the desmosome. J. Cell Sci. 99, 809-821.

North, A. J., Chidgey, M. A. J., Clarke, J. P., Bardsley, W. G., & Garrod, D. R. (1996). Distinct desmocollin isoforms occur in the same desmosomes and show reciprocally graded distributions in bovine nasal epidermis. Proc. Natl. Acad. Sci. USA 93, 7701-7705.

Nuber, U. A., Schäfer, S., Schmidt, A., Koch, P. J., & Franke, W. W. (1995). The widespread human desmocollin Dsc2 and tissue-specific patterns of synthesis of various desmocollin subtypes. Eur. J. Cell Biol. 66:69-74.

O'Keefe, E. J., Briggaman, R. A., & Herman, B. (1987). Calcium-induced assembly of adherens junctions in keratinocytes. J. Cell Biol. 105, 807-817.

O'Keefe, E. J., Erickson, H. P., & Bennett, V. (1989). Desmoplakin I and desmoplakin II. Purification and characterisation. J. Biol. Chem. 264, 8310-8318.

Oursler, J. R., Labib, R. S., Ariss-Abdo, L., Burke, T., O'Keefe, E. J., & Anhalt, G. J. (1992). Human autoantibodies against desmoplakins in paraneoplastic pemphigus. J. Clin. Invest. 89, 1775-1782.

Ouyang, P. & Sugrue, S. P. (1992). Identification of an epithelial protein related to the desmosome and intermediate filament network. J. Cell Biol. 118, 1477-1488

Ouyang, P. & Sugrue, S. P. (1996). Characterisation of pinin, a novel protein associated with the desmosome-intermediate filament complex. J. Cell Biol. 135, 1027-1042.

Ozawa, M., Nuruki, K., Toyoyama, H., & Ohi Y. (1995a). Cloning of an alternative form of plakoglobin (γ-catenin) lacking the fourth armadillo repeat. J. Biochem. 118, 836-840.

Ozawa, M., Terada, H., & Pedraza, C. (1995b). The fourth armadillo repeat of plakoglobin (γ-catenin) is required for its high affinity binding to the cytoplasmic domains of E-cadherin and desmosomal cadherin Dsg2, and the tumor suppressor APC protein. J. Biochem. 118, 1077- 1082.

Ozawa, M., Baribault, H., & Kemler, R. (1989). The cytoplasmic domain of the cell adhesion molecule uvomorulin associates with three independent proteins structurally related in different species. EMBO J. 8, 1711 - 1717.

Palka, H. L. & Green, K. J. (1997). Roles of plakoglobin end domains in desmosome assembly. J. Cell Sci. 110, 2359-2371.

Parker, A. E., Wheeler, G. N., Arnemann, J., Pidsley, S. C., Atoliotis, P., Thomas, C. L., Rees, D. A., Magee, A. I., & Buxton, R. S. (1991). Desmosomal glycoproteins II and III. Cadherin-like junctional molecules generated by alternative splicing. J. Biol. Chem. 266, 10438-10445.

Pasdar, M. & Nelson, W. J. (1988a). Kinetics of desmosome assembly in Madin-Darby canine kidney epithelial cells: temporal and spatial regulation of desmoplakin organization and stabilization upon cell-cell contact. I. Biochemical analysis. J. Cell Biol. 106, 677-685.

Pasdar, M. & Nelson, W. J. (1988b). Kinetics of desmosome assembly in Madin-Darby canine kidney epithelial cells: temporal and spatial regulation of desmoplakin organization and stabilization upon cell-cell contact. II. Morphological analysis. J. Cell Biol. 106, 687-695.

Pasdar, M., Krzeminski, K. A., & Nelson, W. J. (1991). Regulation of desmosome assembly in MDCK epithelial cells: coordination of membrane core and cytoplasmic plaque domain assembly at the plasma membrane. J. Cell Biol. 113, 645-655.

Pasdar, M., Li, Z., & Chan, H. (1995). Desmosome assembly and disassembly are regulated by reversible protein phosphorylation in cultured epithelial cells. Cell Motil. Cytoskel. 30, 108-121.

Peifer, M., McCrea, P. D., Green, K. J., Wieschaus, E., & Gumbiner, B. M. (1992). The vertebrate adhesive junction proteins β-catenin and plakoglobin and the segment polarity gene armadillo form a multigene family with similar properties. J. Cell Biol. 118, 681-691.

Peifer, M., Berg, S., & Reynolds, A. B. (1994a). A repeating amino acid motif defines a new relationship among proteins with roles in signaling, tumor suppression, cell-cell addition and guanine nucleotide exchange. Cell 76, 789-791.

Peifer, M., Pai, L. M., & Casey, M. (1994b). Phosphorylation of the Drosophila adherens junction protein Armadillo: roles for wingless signal and Zeste-white 3 kinase. Dev. Biol. 166, 543-556.

Peifer, M. (1995). Cell adhesion and signal transduction: the Armadillo connection. Trends Cell Biol. 5, 224-229.

Penn, E. J., Hobson, C., Rees, D. A., & Magee, A. I. (1987a). Structure and assembly of desmosome junctions: biosynthesis, processing, and transport of the major protein and glycoprotein in cultured epithelial cells. J. Cell Biol. 105, 57-68.

Penn, E. J., Burdett, I. D. J., Hobson, C., Magee, A. I., & Rees, D. A. (1987b). Structure and assembly of desmosome junctions: biosynthesis and turnover of the major desmosome components of Madin-Darby canine kidney cells in low calcium medium. J. Cell Biol. 105, 2327-2334.

Piepenhagen, P. A., & Nelson, W. J. (1993). Defining E-cadherin-associated protein complexes in epithelial cells: plakoglobin, beta- and gamma-catenin are distinct components J. Cell. Sci. 104, 751-762.

Polakis, P. (1995). Mutations in the APC gene and their implications for protein structure and function. Curr. Opin. Genet. Dev. 5, 66-71.

Porter, R. M., Leitgeb, S., Melton, D. W., Swensson, O., Eady, R. A. J., & Magin, T. M. (1996). Gene targeting at the mouse cytokeratin 10 locus: severe skin fragility and changes of cytokeratin expression in the epidermis. J. Cell Biol. 132, 925-936.

Pulkkinen, L., Smith, F. J. D., Shimizu, H., Murata, S., Yaoita, H., Hachisuka, H., Nishikawa, T., McLean, W. H. I., & Uitto, J. (1996). Hum. Mol. Genet. 5, 1539-1546.

Rayns, D. G., Simpson, F. O., & Ledingham, J. M. (1969). Ultrastructure of desmosomes in mammalian intercalated disc; appearances after lanthanum treatment. J. Cell Biol. 42, 322-326.

Reichelt, J., Bauer, C., Porter, R. M., Lane, E. B., Herzog, V., & Magin, T. M. (1997). Out of balance: consequences of a partial keratin 10 knockout. J. Cell Sci. 110, 2175-2186.

Riggleman, B., Wieschaus, E., & Schedl, P. (1989). Molecular analysis of the armadillo locus: uniformly distributed transcripts and a protein with novel internal repeats are associated with a Drosophila segment polarity gene. Genes Dev. 3, 96-113.

Risinger, M. A. & Larsen, W. J. (1981) Endocytosis of cell-cell junctions and spontaneous cell disaggregation in a cultured human ovarian adenocarcinoma (COLO 316). Tissue Cell 13, 413430.

Rocha-Alvarez, R., Friedman, H., Campbell, I. T., Souza-Aguiar, L., Martins-Castro, R., & Diaz, L. A. (1992). Pregnant women with endemic pemphigus foliaceus (fogo selvagem) give birth to disease-free babies. J. Invest. Dermatol. 99, 78-82.

Roh, J. -Y, & Stanley, J. R. (1995). Plakoglobin binding by human Dsg3 (pemphigus vulgaris antigen) in keratinocytes requires the cadherin-like intracytoplasmic segment. J. Invest. Dermatol. 104, 720-724.

Roscoe, J. T., Diaz, L., Sampaio, S. A. P., Castro, R. M., Labib, R. S., Takahashi, Y., Patel, H., & Anhalt, G. J. (1985). Brazilian pemphigus foliaceus autoantibodies are pathogenic to BALB/c mice by passive transfer. J. Invest. Dermatol. 85, 538-541.

Rubinfeld, B., Souza, B., Albert, I., Munemitsu, S., & Polakis, P. (1995). The APC protein and E-cadherin form similar but independent complexes with α'-catenin, β-catenin and plakoglobin. J. Biol. Chem. 270, 5549-5555.

Rubenstein, A., Merriam, J., & Klymkowsky, M. W. (1997). Localizing the adhesive and signaling functions of plakoglobin. Dev. Genet. 20, 91-102.

Rugg, E. L., McLean, W. H. I., Lane, E. B., Pitera, R., McMillan, J. R., Dopping-Hepenstal, P. J. C., Navsaria, H. A., Leigh, I. M., & Eady, R. A. J. (1994). A functional "knockout" of human keratin 14. Genes Dev. 8, 2563-2573.

Ruhrberg, C., Hajibagheri, M. A. N., Simon, M., Dooley, T. P., & Watt, F. M. (1996). Envoplakin, a novel precursor of the cornified envelope that has homology to desmoplakin. J. Cell Biol. 134, 715-729.

Ruhrberg, C. & Watt, F. M. (1997). The plakin family: versatile organizers of cytoskeletal architecture. Curr. Opin. Genet. Dev. 7, 392-397.

Ruhrberg, C., Hajibagheri, M. A. N., Parry, D. A. D.& Watt, F. M. (1997). Periplakin, a novel component of cornified envelopes and desmosomes that belong to the plakin family and forms complexes with envoplakin. J. Cell Biol. 139, 1835-1849.

Ruiz, P., Brinkmann, V., Ledermann, B., Behrend, M., Grund, C., Thalhammer, C., Vogel, F., Birchmeier, C., Gunthert, U., Franke, W. W., & Birchmeier, W. (1996). Targeted mutation of plakoglobin in mice reveals essential functions of desmosomes in the embryonic heart. J. Cell Biol. 135, 215-225.

Runswick S. K., Garrod, D. R., & Streuli C. H. (1996). The differential expression of desmocollin isoforms in mammary epithelia. Bichem. Soc. Trans. 346S.

Rutman, A. J., Buxton, R. S., & Burdett. I. D. J. (1994). Visualisation by electron microscopy of the unique part of the cytoplasmic domain of a desmoglein, a cadherin-like protein of the desmosome type of cell junction. FEBS Lett. 352, 194-196.

Sacco, P. A., McGranahan, T. M., Wheelock, M. J., & Johnson, K. R. (1995). Identification of plakoglobin domains required for association with N-cadherin and a-catenin. J. Biol. Chem. 270, 20201-20206.

Schäfer, S., Koch, P. J., & Franke, W. W. (1993a). Amino acid sequence of desmoglein Dsg2. EMBL Data library (accession no. Z26317).

Schäfer, S., Troyanovsky, S. M., Heid, H. W., Eshkind, L. G., Koch, P. J., & Franke, W. W. (1993b). Cytoskeletal architecture and epithelial differentiation: molecular determinants of cell interaction and cytoskeletal filamnet anchorage. C.R. Acad. Sci. Paris, Sciences de la vie/Life sciences, 316, 1316-1323.

Schäfer, S., Koch, P. J., & Franke, W. W. (1994). Identification of the ubiquitous human desmoglein, Dsg2, and the expression catalogue of the desmoglein subfamily of desmosomal cadherins. Exp. Cell Res. 211, 391-399.

Schäfer, S., Stumpp, S., & Franke, W. W. (1996). Immunological identification and characterisation of the desmosomal cadherin Dsg2 in coupled and uncoupled epithelial cells and in human tissues. Differentiation 60, 99-108.

Schenk P. (1980) Intracytoplasmic desmosomes in malignant keratinocytes of laryngeal carcinoma. Arch. Otorhinolaryngol. 226, 219-223

Schiltz, J. R., & Michel, B. (1976). Production of epidermal acantholysis in normal human skin in vitro by the IgG fraction from pemphigus serum. J. Invest. Dermatol. 67, 254-260.

Schmelz, M., Moll, R., Kuhn, C., & Franke, W. W. (1994). *Complexus adhaerentes*, new group of desmoplakin-containing junctions in endothelial cells: II. Different types of lymphatic vessels. Differentiation 57, 97- 117.

Schmidt, A., Langbein, L., Rode, M., Prätzel, S., Zimbelmann, R., & Franke, W. W. (1997). Plakophilins 1a and 1b: widespread nuclear proteins recruited in specific epithelial cells as desmosomal plaque components. Cell Tissue Res. 290, 481-499.

Schrijvers, A. H. G. J., Gerretsen, M., Fritz, J., van-Walsum, M., Quak, J. J. Snow, G. B., & van-Dongen, G. A. M. S. (1991). Evidence for a role of the monoclonal antibody E48 defined antigen in cell-cell adhesion in squamous epithelia and head and neck squamous cell carcinomas. Exp. Cell Res. 196, 264-269.

Sepp, R., Severs, N. J., & Gourdie, R. G. (1996). Altered patterns of cardiac intercellular junction distribution in hypertrophic cardiomyopathy. Heart 76, 412-417.

Shabana, A. H. M., Amar, L., Oboeuf, H., Martin, N., Forest, N. (1996). Cytoplasmic desmosome formation by H-7 and EGF treatment in cultured fetal rat kerationocytes. Tissue Cell 28: 537-545

Shapiro, L., Fannon, A. M., Kwong, P. D., Thompson, A., Lehmann, M. S., Grübel, G., Legrand, J.-F., Als-Nielsen, J., Colman, D. R., & Hendrickson, W. A. (1995). Structural basis of cell-cell adhesion by cadherins. Nature 374, 327-337.

Sheu, H. M., Kitajima, Y. And Yoaita, H. (1989). Involvement of protein kinase C in translocation of desmoplakins from cytosol to plasma membrane during desmosome formation in human squamous cell carcinoma cells grown in low to normal calcium concentration. Exp. Cell Res. 185, 176-190

Shibamoto, S., Hayakawa, M., Takeuchi, K., Hori, T., Oka, N., Miyazawa, K., Kitamura, M., Takeichi, M., & Ito, F. (1994). Tyrosine phosphorylation of β-catenin and plakoglobin enhanced by hepatocyte growth factor and epidermal growth factor in human carcinoma cells. Cell Adhesion Commun. 1, 295-305.

Shimizu, H., Masunaga, T., Ishiko., Akira., Kikuchi, A., Hashimoto, T., & Nishikawa, T. (1995). Pemphigus vulgaris and pemphigus foliaceus sera show an inversely graded binding pattern to extracellular regions of desmosomes in different layers of human epidermis. J. Invest. Dermatol. 105, 153-159.

Shinohara, M., Hiraki, A., Kiebe, T., Nakamura, S., Kurahara, S., Shirasuna, K., and Garrod, D. R. (1998). Immunohistochemical study of desmosomes in oral squamous cell carcinoma: correlation with cytokeratin and E-cadherin staining, and with tumor behaviour J. Pathol. 184, 369-381.

Simcha, I., Geiger, B., Yehuda-Levenberg, S., Salomon, D., & Ben-Ze'ev, A. (1996). Suppression of tumorigenicity by plakoglobin: an augmenting effect of N-cadherin. J. Cell Biol. 133, 199-209.

Simon, M. & Green, H. (1984). Participation of membrane-associated proteins in the formation of the cross-linked envelope of the keratinocyte. Cell 36, 827-834.

Simrak, D., Cowley, C. M. E., Buxton, R. S., & Arnemann, J. (1995). Tandem arrangement of the closely linked desmoglein genes on human chromosome 18. Genomics 25, 591-594.

Skalli, O., Jones, J. C. R., Gagescu, R., & Goldman, R. D. (1994). IFAP 300 is common to desmosomes and hemidesmosomes and is a possible linker of intermediate filaments to these junctions. J. Cell Biol. 125, 159-170.

Skerrow, C. J. & Matoltsy, A. G. (1974). Chemical characterization of isolated epidermal desmosomes. J. Cell Biol. 63, 524-531.

Smith, F. J. D., Eady, R. A. J., Leigh, I. M., McMillan, J. R., Rugg, E. L., Kelsell, D. P., Bryant, S. P., Spurr, N. K., Geddes, J. F., Kirtschig, G., Milana, G., de Bono, A. G., Owaribe, K., Wiche, G., Pulkkinen, L., Uitto, J., McLean, W. H. I., & Lane, E. B. (1996). Plectin deficiency results in muscular dystrophy with epidermolysis bullosa. Nature Genet. 13, 450-457.

Smith, E.A. & Fuchs, E. (1998). Defining the interactions between intermediate filaments and desmosomes. J. Cell Biol. 141, 1229-1241.

Solinas-Toldo, S., Troyanovsky, R., Weitz, S., Lichter, P., Franke, W. W., & Freis, R. (1995). Bovine desmocollin genes (DSC1, DSC2, DSC3) cluster on chromosome 24q21/q22. Mamm. Genome. 6, 484-486.

Sommers, C. L., Gelmann, E. P., Kemler, R., Cowin, P., & Byers, S. W. (1994). Alterations in β-catenin phosphorylation and plakoglobin expression in human breast cancer cells. Cancer Res. 54, 3544-3552.

Stanley, J. R. & Karpati, S. (1994). Desmosome, hemidesmosome and disease. In: Molecular biology of desmosomes and hemidesmosomes. Collins, J. E. and Garrod, D. R. (eds) . R. G. Landes Company, Austin. pp 107-125.

Stappenbeck, T. S., & Green, K. J. (1992). The desmoplakin carboxyl terminus coaligns with and specifically disrupts intermediate filaments networks when expressed in cultured cells. J. Cell Biol. 116, 1197-1209.

Stappenbeck, T. S., Bornslaeger, E. A., Corcoran, C. M., Luu, H. H., Virata, M. L. A., & Green, K. J. (1993). Functional analysis of desmoplakin domains: specification of the interaction with keratin versus vimentin intermediate filament networks. J. Cell Biol. 123, 691-705.

Stappenbeck T. S., Lamb, J. A., Corcoran, C. M., & Green, K. J. (1994). Phosphorylation of the desmoplakin COOH terminus negatively regulates its interaction with keratin intermediate filament networks. J. Biol. Chem. 269, 29351-29354.

Svitkina, T. M., Verkhovsky, A. B., & Borisy, G. G. (1996). Plectin sidearms mediate interaction of intermediate filaments with microtubules and other components of the cytoskeleton. J. Cell Biol. 135, 991-1007.

Toivola, D. M., Goldman, R. D., Garrod, D. R. and Eriksson, J. E. (1997) Protein phosphatases maintain the organization and structural interactions of hepatic keratin intermediate filaments. J Cell Sci. 110, 23-33.

Trevor, K. T. & Steben, L. S. (1992). Distribution of desmosomal proteins in F9 embryonal carcinoma cells and epithelial cell derivatives. J. Cell Sci. 103, 69-80.

Troyanovsky, S. M., Eshkind, L. G., Troyanovsky, R. B., Leube, R. E., & Franke, W. W. (1993). Contributions of cytoplasmic domains of desmosomal cadherins to desmosome assembly and intermediate filament anchorage. Cell 72, 561-574.

Troyanovsky, S. M., Troyanovsky, R. B., Eshkind, L. G., Leube, R. E., & Franke, W. W. (1994a). Identification of amino acid sequence motifs in desmocollin, a desmosomal glycoprotein that are required for plakoglobin binding and plaque formation. Proc. Natl. Acad. Sci. USA 91, 10790-10794.

Troyanovsky, S. M., Troyanovsky, R. B., Eshkind, L. G., Leube, R. E., & Franke, W. W. (1994b). Identification of the plakoglobin- binding domain in desmoglein and its role in plaque assembly and intermediate filament storage. J. Cell Biol. 127, 151-160.

Troyanovsky, R. B., Chitaev, N. A., & Troyanovsky, S. M. (1996). Cadherin binding sites of plakoglobin: localization, specificity and role in targeting to adherens junctions. J. Cell Sci. 109, 3069-3078.

Tselepis, C., Chidgey, M., North, A., & Garrod, D. (1998). Desmosomal adhesion inhibits invasive behavior. Proc. Natl. Acad. Sci. USA. 95, 8064-8069.

Tsukita, S., & Tsukita, S. (1985). Desmocalmin: a calmodulin binding high molecular weight protein isolated from desmosomes. J. Cell Biol. 101, 2070-2080.

Uitto, J., Pulkkinen, L., Smith, F. J. H., & McLean, W. H. I. (1996). Plectin and human genetic disorders of the skin and muscle. Exp. Dermatol. 5, 237-246.

Valiron, O., Chevrier, V., Usson, Y., Breviario, F., Job, D., & Dejana, E. (1996). Desmoplakin expression and organization at human umbilical vein endothelial cell-to-cell junctions. J. Cell Sci. 109, 2141-2149.

Vassar, R., Coulombe, P. A., Degenstein, L., Albers, K., & Fuchs, E. (1991). Mutant keratin expression in transgenic mice causes marked abnormalities resembling a human genetic skin disease. Cell 64, 365-380.

Virata, M. L. A., Wagner, R. M., Parry, D. A. D., & Green, K. J. (1992). Molecular structure of the human desmoplakin I and II amino terminus. Proc. Natl. Acad. Sci. USA 89, 544-548.

Wahl, J. K., Sacco, P. A., McGranahan-Sadler, T. M., Sauppe, L. M., Wheelock, M. J., & Johnson K. R. (1996). Plakoglobin domains that define its association with the desmosomal cadherins and the classical cadherins: identification of unique and shared domains. J. Cell Sci. 109, 1143-1154.

Wang, Y. M., Amagai, M., Minoshimia, S., Sakai, K., Green, K. J., Nishikawa, T., & Shimizu, N. (1994). The human genes for desmogleins (DSG1 and DSG3) are located in a small region on chromosome 18ql2. Genomics 20, 492-495.

Watt, F. M., Mattey, D. L., & Garrod, D. R. (1984). Calcium induced reorganisation of desmosomal components in cultured human keratinocytes. J. Cell Sci. 99, 2221-2215.

Wheeler, G. N., Parker, A. E., Thomas, P., Atoliotis, D., Poynter, J., Arnemann, J., Rutman, A. J., Pidsley, S. M., Watt, F. M., Rees, D. A. et al (1991). Desmosomal glycoprotein DG1, a component of intercellular desmosome junctions, is related to the cadherin family of cell adhesion molecules. Proc. Natl. Acad. Sci. USA 88, 4796-4800.

Wheelock, M. J. & Jensen, P. J. (1992). Regulation of keratinocyte intercellular junction organization and epidermal morphogenesis by E-cadherin. J. Cell Biol. 117, 415-425.

Witcher, L. L., Collins, R., Puttagunta, S., Mechanic, S. E., Munson, M., Gumbiner, B., & Cowin, P. (1996). Desmosomal cadherin binding domains of plakoglobin. J. Biol. Chem. 271, 10904-10909.

Yano, T., Tokui, T., Nishi, Y., Nishizawa, K., Shibata, M., Kikuchi, K., Tsuiki, S., Yamauchi, F and Inagaki, M. (1991) Phosphorylation of keratin intermediate filaments by protein kinase C, by calmodulin-dependent kinase and by cAMP-dependent kinase. Eur. J. Biochem. 197, 281-290.

Yue, K. K. M., Holton, J. L., Clarke, J. P., Hyam, J. L. M., Hashimoto, T., Chidgey, M. A. J., & Garrod, D. R. (1995). Characterisation of a desmocollin isoform (bovine DSC3) exclusively expressed in lower layers of stratified epithelia. J. Cell Sci. 108: 2163-2173.

Ziboti, V. A., Isseroff, R. R., Pandey, R. (1984). Phospholipid metabolism in calcium-regulated differentiation in cultured murine keratinocytes. Biochem. Biphys. Res. Comm. 122, 1234-1240.

THE MOLECULAR BASIS FOR THE STRUCTURE, FUNCTION, AND REGULATION OF TIGHT JUNCTIONS

Sandra Citi and Michelangelo Cordenonsi

Advances in Molecular and Cell Biology
Volume 28, pages 203-233.

ISBN: 0-7623-0495-2

I. THE STRUCTURE AND FUNCTION OF TIGHT JUNCTIONS

Tight Junctions (TJ) are the most apical element of the junctional complex of polarized epithelial cells, which comprises also zonulae adhaerentes and desmosomes (Farquhar and Palade, 1963) (see Figure 1, left panel). TJ are also present in capillary endothelial cells (blood-tissue barriers), and in other specialized cell types, such as Sertoli cells (Dym and Fawcett, 1970). The ultrastructure of TJ is characteristic. By transmission electron microscopy, TJ appear as a series of very close appositions of the external leaflets of adjacent membranes (Farquhar and Palade, 1963). By freeze-fracture, these sites are revealed as a complementary network of junctional strands (fibrils) on the P-face of the fractured membrane, and grooves on the E-face (Staehelin et al., 1969; Chalcroft and Bullivant, 1970; Goodenough and Revel, 1970)(for reviews, see (Madara, 1991; Schneeberger and Lynch, 1992; Hirsch and Noske, 1993)).

The main function of TJ is to form a gasket-like, semipermeable "barrier" that prevents the free diffusion of ions, molecules, and cells across the space between epithelial cells (paracellular pathway) (Diamond, 1977). The "barrier" function allows epithelial and endothelial cell sheets to maintain physiological gradients of solutes (ions, nutrients, secretory products, etc.) across different body compartments, a function that is critical for tissues involved in polarized absorption and secretion. In addition, the TJ barrier allows the formation of specialized environments where immune cells, antigens, and antibodies are sequestered. The permeability of TJ to ions and small molecules can be determined in cultured cells by measurement of electrical resistance and tracer fluxes, and in intact tissues and organs by labeled or electron-dense tracer studies (Madara & Pappenheimer, 1987). These studies, together with electron microscopic observations, have revealed two important properties of TJ. First, TJ are highly heterogeneous. TJ permeability and ultrastructure vary from tissue to tissue. Electrical resistance values ranging from < 100 to $>40{,}000\ \Omega\ cm^{-2}$ have been reported (Powell, 1981). Second, TJ permeabilities within each tissue are influenced by specific physiological states. For example, TJ respond to hormones, growth factors, cytokines, neuronal influences, osmotic and mechanical stress. They are also remodeled during a variety of physiological events, such as mitosis, spermatogenesis, secretory activity (of glandular epithelium), and white blood cells migration (reviewed by Polak-Charcon, 1992; Schneeberger, 1994; Anderson and Van Itallie, 1995). Therefore, TJ do not simply provide a static "barrier," but form a selective "gate" that regulates the amount and type of permeating solutes and cells. TJ also have a "fence" function, since they define the

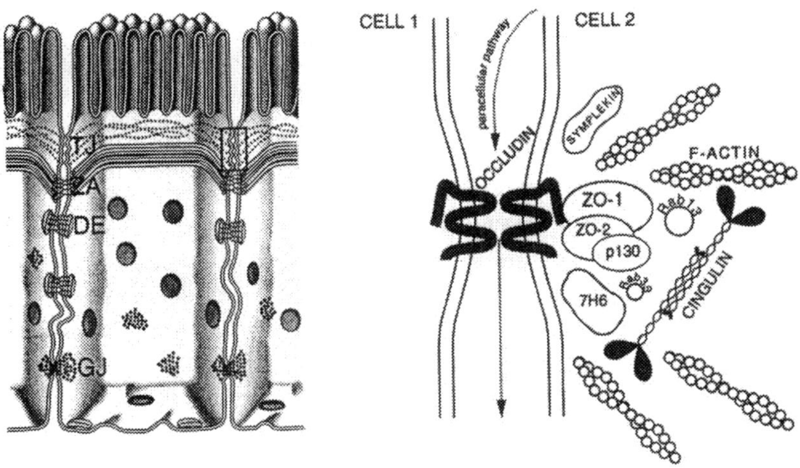

Figure 1. (**Left panel**) Schematic representation of a polarized epithelial cell, with the junctions that connect it to neighbouring cells. TJ, tight junction; ZA, zonula adhaerens; DE, desmosome; GJ, gap junction. The insert on the right encloses the apical TJ area. (**Right panel**) Schematic diagram of the arrangement of TJ proteins with respect to the plasma membranes of adjacent cells. In this scheme, occludin appears as the only transmembrane protein, however claudins (Furuse et al., 1998a and 1998b) and JAM have recently been identified as novel TJ transmembrane proteins. Other TJ proteins are localized in the cytoplasmic plaque domain. F-actin filaments are also schematically shown. The shapes and respective sizes of the proteins are speculative. ZO-1 is depicted as interacting with the C-terminal domain of occludin, and as being complexed with ZO-2 and p130/ZO-3. Cingulin is represented as two interacting dimers, each with two N-terminal and two C-terminal globular regions.

boundary between the apical and basolateral membrane domains of polarized cells, and thus contribute to maintaining an asymmetric distribution of lipids and proteins in these domains (reviewed in Nelson, 1992).

To understand the molecular basis for the heterogeneity and modulation of TJ, it is necessary to determine their protein composition in different tissues and to study how different factors affect their organization.

II. THE MOLECULAR COMPOSITION OF TIGHT JUNCTIONS

The idea that proteins are key constituents of TJ was originally suggested by the observations that protein fixatives, but not detergents, alter the morphology of TJ fibrils (Goodenough and Revel, 1970; van Deurs and Luft, 1979), and that inhibition of protein synthesis blocks TJ assembly (Griepp et al., 1983). The identity of the specific TJ proteins remained unknown until the discovery of ZO-1 (Stevenson et al., 1986). Since then, several TJ-associated proteins have been identified (reviewed by Citi, 1993; Gumbiner, 1993; Anderson and Van Itallie, 1995) (see Table 1). Most TJ proteins have been identified by raising monoclonal antibodies to partially purified junctional or cytoskeletal preparations (e.g., mouse liver junctions for ZO-1, Stevenson et al., 1986), chicken enterocyte actomyosin fraction for cingulin (Citi et al., 1988), rat liver junctions for 7H6 (Zhong et al., 1993), chick liver junctions for occludin (Furuse et al., 1993) and *Xenopus* junctions for 19B1 (Merzdorf and Goodenough, 1997), while others have been identified as polypeptides that co-immunoprecipitated with ZO-1 (e.g. ZO-2 (Gumbiner et al., 1991) and p130/ZO-3 (Balda et al., 1993; Jesaitis and Goodenough, 1994; Haskins et al., 1998). All TJ proteins except a few (rabl3, rab8, and pl30) have been localized to TJ by immunoelectron microscopy (see Table 1).

TJ, like other intercellular junctions, can be schematically represented as comprising three structural domains: membrane-associated, cytoplasmic plaque, and cytoskeletal. Occludin was the only known component of the membrane domain

Table 1. Tight Junction Proteins and Some of their Properties

Name	*M_r (kD)*	*EM-TJ*	*Endoth.*	*Nucleus*	*Non-ep.*	*Phosph.*	*Homol.*
ZO-1	210-225	+	+	+	+	Ser-Tyr	MAGUK
cingulin	140-160	+	+/–	+	–	Ser	(myosin)
BG9.1	192	+	?	?	?	?	?
ZO-2	160	+	?	+	–	Ser-Tyr	MAGUK
p130/ZO-3	130	?	?	?	?	+	MAGUK
occludin	58-66	+	+/–	?	–	Ser-Thr	(connexin)
7H6	155-175	+	+	?	–	?	?
rab8	25	?	?	?	+	?	GTP-BP
rab13	41	?	+	?	+	?	GTP-BP
rab3B	25	+	?	?	+	?	GTP-BP
symplekin	150	+	–	+	+ (nucl.)	?	none
19B1	210	+	?	?	?	?	?
AF-6	180- 195	+/–	?	?	+	?	RB-M-PDZ

Notes: (M_r (kD))=Apparent molecular mass determined by SDS-PAGE in various tissues/species. (EM-TJ)= Localization in epithelial TJ, established by immunoelectron microscopy; (?)=unknown/not determined; (+/-)= EM localization throughout the junctional complex. (Endoth.)= expression in endothelial cell types. (+)= detected in most or all endothelial cell types; (+/-)= detected only in a subset of endothelial cell types; (-)= not detected. (Nucl. loc.)= Nuclear localization; (+)=detected. (Non-epith.)= expression in non-epithelial cell types (e.g. fibroblasts, cardiac muscle cells, neurons, etc). (Phosph.)= phosphorylation; (Ser)=phosphorylated on serine; (Tyr)=phosphorylated on tyrosine; (+)=phosphorylated on unknown amino acid (Seq. hom.)= deduced aminoacid sequence homologies; (MAGUK)= proteins of the MAGUK family (see text); (GTP-BP)= GTP-binding proteins; (RB-M-PDZ)= Ras-binding domain, myosin V-like domain, PDZ domain. Homologies in parentheses (connexin for occludin and myosin for cingulin) indicate weak sequence homologies. See text for additional information and for references.

of TJ (see Table 1) until the recent discovery of claudins (Furuse et al., 1998a and 1998b) and JAM (Martin-Padura et al., 1998). Early transmission electron microscopy indicated that the cytoplasmic surface of TJ is coated by fuzzy and filamentous electron dense material (Farquhar and Palade, 1963; Hull and Staehelin, 1979). It has become clear that in addition to specific TJ proteins (cytoplasmic plaque domain), this material also contains actin (cytoskeletal domain). How the cytoskeletal domain of TJ interacts with and regulates the cytoplasmic plaque domain proteins remains an important question.

A. Molecular Characterization of Tight Junction Proteins: Insights into their Functions

Occludin, Claudins, and JAM

The most compelling evidence that occludin is localized in the membrane domain of TJ is its derived amino acid sequence. The membrane localization is also supported by immunoelectron microscopy, and by the requirement for detergents for the solubilization of occludin (Furuse et al., 1993). The aminoacid sequences of chicken (Furuse et al., 1993), dog, human, rat-kangaroo, and mouse occludin (Ando-Akatsuka et al., 1996) show a protein with four predicted membrane-spanning domains, generating a membrane topology similar to that of gap junctional connexins (Furuse et al., 1993). The putative transmembrane regions of occludin are connected by two extracellular loops that are unusually rich in tyrosine and glycine residues. One of these loops is thought to be involved in the formation of the TJ barrier "pore" (Wong and Gumbiner, 1997)(see below). In chicken occludin, a 57-aminoacid long N-terminus and a 255-amino acid long hydrophylic C-terminus lie on the cytoplasmic face of the membrane. The C-terminal cytoplasmic domain is important for targeting to the TJ and interaction with ZO-1 (see below). The sequences of mammalian occludins are closely related to each other (90% identity), whereas they diverge from those of chicken and rat-kangaroo (50% identity) (Ando-Akatsuka et al., 1996). Occludin homologues have not been detected in septate junctions, the invertebrate homologues of TJ (reviewed in Bryant, 1994). Interestingly, the membrane domain of septate junctions is morphologically quite different from that of vertebrate TJ (Lane, 1992). It will be important to study how expression of occludin and its homologues correlates with structural and functional changes in the evolution of TJ across phyla.

Claudins and JAM have recently been characterized as novel transmembrane proteins of TJ. Claudins show a membrane topology similar to occludin but different in sequence (Furuse et al., 1998a). They are now presumed to be the major molecular constituents of TJ strands, and they interact with occludin (Furuse et al., 1998b). JAM is a single-pass transmembrane protein with sequence homology to the immunolglobulin superfamily of adhesion molecules, and may play a role in modulating the passage of cells through endothelial TJ (Martin-Padura et al., 1998).

MAGUK proteins of Tight Junction: ZO-1, ZO-2 and p130/ZO-3

ZO-1 cDNA sequences from mouse (Itoh et al., 1993) and human (Willott et al., 1993) have been reported. ZO-1 protein has a modular structure (see Figure 2). Its N-terminal portion contains three copies of a ~90-amino acid domain called DHR (for Discs-large Homology Region), now referred to also as PDZ (for p55-DlgA-ZO-1). The PDZ regions are followed by one Src homology 3 domain (SH3). and one domain homologous to yeast guanylate kinase (GuK) (Willott et al., 1993).

The PDZ, SH3 and GuK domains have been identified in a number of membrane and junction-associated proteins (MAGUK = Membrane Associated Guanylate Kinase) (Woods and Bryant, 1993), including the product of the discs large-1 tumor suppressor gene in *Drosophila*, mutations of which lead to loss of epithelial apico-basal polarity and abnormal growth of imaginal discs in fly larvae (Woods and Bryant, 1991), the *C. elegans* vulval induction gene lin-2 (Hoskins et al., 1996), and the product of the tamou gene in *Drosophila*, mutations of which reduce the transcription of a repressor gene and lead to a supernumerary mechanosensory organ phenotype (Takahisa et al., 1996). In MAGUK proteins, the PDZ domain is thought to provide a membrane localization signal and to cluster membrane channels (Kim et al., 1995; Lue et al., 1996). The SH3 domain may mediate association with proteins containing proline-rich motifs, with the actin cytoskeleton or with protein kinases (Balda et al., 1996). The GuK homology domain may affect the activity of small GTP-binding proteins, and control the asymmetric segregation of signaling receptors (Anderson et al., 1995; Kim, 1995). Unlike most other members of the MAGUK family of proteins (but similarly to ZO-2 and the product of the tamou gene), ZO-1 contains a large proline-rich C-terminal domain (see Figure 2). In ZO-1, the proline-rich region contains three alternatively spliced domains (α, $\beta 1$, $\beta 2$), (Anderson et al., 1995)(see Figure 2).

The derived amino acid sequence of ZO-2 (predicted molecular mass 132 kD) is homologous to the N-terminal domain of ZO-1 (containing the PDZ, SH3 and GuK homology regions), whereas its C-terminal region is considerably shorter than ZO-1 (Beatch et al., 1996; Jesaitis and Goodenough, 1994)(see Figure 2). It was recently reported that the derived aminoacid sequence of pl30 (predicted molecular mass 98 kD) is homologous to ZO-1 and ZO-2 and is a novel member of the MAGUK family of proteins (Stevenson et al., 1996). ZO-2 and p130 contain a basic, arginine-rich region linking the PDZ1 and PDZ2 domains (Beatch et al., 1996; Stevenson et al., 1996; Haskins et al., 1998). The function of this region is unknown.

Cingulin

Cingulin from chicken enterocytes migrates by SDS-PAGE as polypeptides with an apparent size of M_r 140 kD and 108 kD. The M_r 108 kD form of cingulin (probably a proteolytic degradation product of the M_r 140 kD form) was purified to homogeneity (Citi et al., 1988; Citi et al., 1989). Purified cingulin was shown to be a heat- and

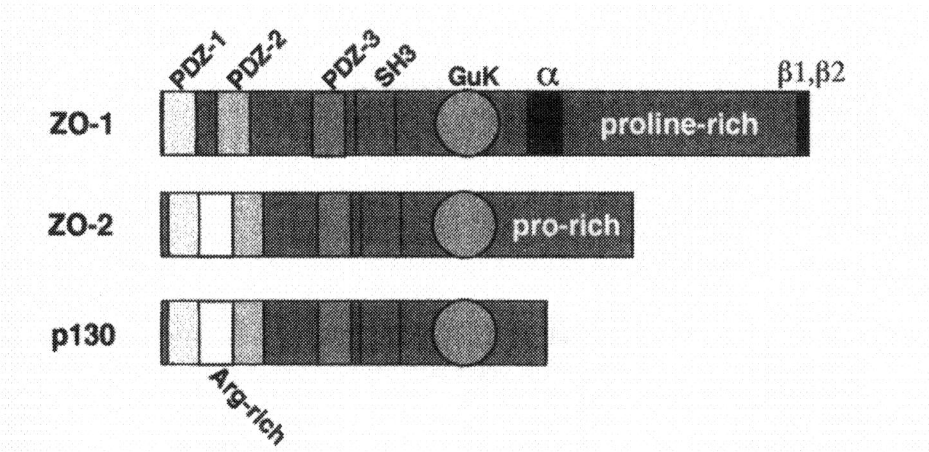

Figure 2. Schematic diagram showing the domain organization of the TJ proteins ZO-1 (Willott et al., 1993), ZO-2 (Beatch et al., 1996) and p130/ZO-3 (Stevenson et al., 1996; Haskins et al., 1998). Regions of homology include the PDZ domains (1, 2, and 3), a Src homology 3 domain (SH3), and a guanylate kinase homology domain (GuK). ZO-1 and ZO-2 contain proline-rich C-terminal regions. In ZO-1, this region contains three alternatively spliced domains, α, β1, and β2. ZO-2 and p130/ZO-3 contain arginine-rich, basic regions linking PDZ1 and PDZ2. The MAGUK family structural motifs are linked to signal transduction events and/or interactions between the plasma membrane and the cortical cytoskeleton (Anderson et al., 1995; Kim, 1995).

ethanol-stable, acidic dimer of two subunits cross-linked by S-S bridges, and by rotary shadowing electron microcopy appeared as a 130 nm long and 2-nm wide flexible rod, similar in shape to the rod regions of conventional myosins (myosins II) (Citi et al., 1988; Citi et al., 1989). This, and the amino acid composition of purified cingulin (Citi et al., 1988) suggested that cingulin, or at least its M_r 108 kD region, has an α-helical, coiled-coil structure. This notion was confirmed by sequencing cDNAs coding for cingulin, showing that cingulin contains a coiled-coil rod domain with weak sequence homologies (about 20% identity) to the coiled-coil regions of myosin heavy chains (Citi et al., 1990)(our unpublished results). In addition to the coiled-coiled domain, cingulin molecules contain a globular, N-terminal domain, whose sequence does not show significant homology to any other protein sequence in current databases, and a small globular domain at the extreme C-terminus (our unpublished results). Thus, cingulin could be depicted schematically as a dimer, consisting of a coiled-coil rod, two globular N-terminal "heads," and two globular C-terminal "tails" (see Figure 1, right panel). The coiled-coil domain may allow cingulin dimers to interact one with the other in a parallel or antiparallel fashion to form higher order polymers (see Figure 1, right panel). We speculate that by forming a filamentous network in the cytoplasmic plaque of TJ, cingulin could stabilize it and provide a link between the membrane/cytoplasmic-plaque domains and the cytoskeleton.

7H6

7H6 was identified as a M_r 175 kD protein localized near TJ in the bile-canaliculus-rich hepatocyte membrane fraction (Zhong et al., 1993). A recent report indicates that 7H6 contains a sequence motif present in yeast chromosomal segregation proteins of the Smc1p family (Mori et al., 1996).

Symplekin

Symplekin cDNA has been cloned from a human adenocarcinoma library (Keon et al., 1996). The derived amino acid sequence (predicted molecular mass 126.5 kD) shows no homologies to other proteins in databases (Keon et al., 1996).

AF-6

The Ras target protein AF-6 was recently identified as a TJ-associated protein (Yamamoto et al., 1997), but was originally described as the fusion partner of the acute lymphoblastic leukemia-1 (ALL-1) protein (Prasad et al, 1993, Kuriyama et al, 1996). AF-6 has a PDZ domain, a myosin-V-like domain, and a Ras-binding domain, and was recently found to localize with and interact with ZO-1 (Yamamoto et al., 1997).

Other Tight Junction proteins

A M_r 192 kD antigen, named BG9. 1, was localized to the cytoplasmic faces of mouse liver TJ membranes (Chapman and Eddy, 1989). Further studies on this protein have not been reported. A number of proteins that may regulate membrane traffic and protein–protein interactions. have been localized to the cytoplasmic surfaces of TJ. These include two small GTP-binding proteins, rabl3 (Zahraoui et al., 1994) and rab3B (Weber et al., 1994), and specific isoforms of G proteins and protein kinase C (Dodane and Kachar, 1996; Izumi et al., 1998). Phosphorylated polypeptides identified in ZO-1 (Stuart and Nigam, 1995) and cingulin (Citi and Denisenko, 1995) immunoprecipitates may represent additional novel structural TJ proteins.

Lipids

No evidence has yet supported the idea that lipids are structural constituents of TJ (Kachar and Reese, 1982; Pinto da Silva and Kachar, 1982). However, lipids play a role in TJ function and in epithelial polarization, as shown by the observations that an antiglycolipid antibody interferes with TJ formation (Zinkl et al., 1996), and that cholesterol and other lipids affect TJ assembly and function (Schneeberger et al., 1988; Ropke et al., 1996; Stankewich et al., 1996)

B. Tissue Expression and Isoforms of Tight Junction Proteins

It has become clear in recent years that certain TJ proteins are expressed in non-epithelial cells and tissues, and/or show species- and tissue-dependent variations in their apparent molecular mass. A detailed analysis of the expression and isotypes of TJ proteins can help to clarify their functions and the molecular basis for TJ heterogeneity and modulation.

ZO-1 is not a TJ-specific protein, although when it was first identified as a hepatocyte antigen, it was named after zonula occludens (ZO) (Stevenson et al., 1986). The association of ZO-1 with TJ and/or with adherens-type junctions appears to depend on cell type. In epithelial cells, ZO-1 has been localized at the level of TJ by immunoelectron microscopy (Stevenson et al., 1986; Anderson et al., 1988; Stevenson et al., 1989). ZO-l is expressed and junctionally localized in a variety of epithelial cells and tissues, including trophectoderm (Fleming et al., 1989), epididymus (Byers et al., 1992), retina (Nabi et al., 1993), cornea (Wang et al., 1993), mammary gland (Zettl et al., 1992; Singer et al., 1994), thyroid (Yap et al., 1994), airway epithelium (Herard et al., 1996), salivary gland (Hieda et al., 1996), corpus luteum (Khan-Dawood et al., 1996), and endometrium (Bowen et al., 1996). In addition, ZO-1 is expressed in a variety of endothelial cells (Li and Poznansky, 1990; Krause et al., 1991; Petrov et al., 1994; Gardner et al., 1996), and in Sertoli cells (Byers et al., 1992; Pelletier et al., 1997). ZO-1 is also expressed in cells that are not known to have TJ, for example kidney podocytes (Schnabel et al., 1990) and astrocytes (Howarth et al., 1992). ZO-1 is identical to a 220 kD protein expressed in zonulae adhaerentes of fibroblasts and intercalated discs of cardiac muscle cells (Itoh et al., 1991; Itoh et al., 1993). ZO-1 is also expressed in several nerve cell types (Miragall et al., 1994; Petrov et al., 1994; Dezawa et al., 1996). In summary, ZO-1 can be considered a TJ-specific protein only in polarized cells that display "typical" TJ. In all other cell types, including epithelial cells from stratified epithelia, cancer cells and so on (see below) ZO-1 expression does not necessarily correlate with the occurrence of TJ.

Unlike ZO-1, other TJ proteins (occludin, cingulin, 7H6, and ZO-2) have not been detected in fibroblasts, nerve cells or cardiac myocytes (see also Citi, 1993), although cingulin has been detected in junctional regions of cultured cells derived from chicken embryo hearts (Eisenberg and Bader, 1995). ZO-1 and cingulin, but not occludin, are detected in multilayered squamous epithelia (Citi et al., 1991; Kimura et al., 1997; Sugrue and Zieske, 1997). Cingulin is not detected in vein endothelial cells (HUVEC) that express ZO-1 (Citi et al., 1991), but it may be expressed in other endothelial cell types (Citi et al., 1989). Occludin expression has been detected in brain endothelium, but not in aortic endothelium (Hirase et al., 1996; Hirase et al., 1997). In the developing brain, occludin expression is lost during the transition from neuroopithelial cells to neurons (Aaku-Saraste et al., 1996), suggesting that occludin is a bona fide epithelial TJ marker. This is confirmed by analysis of the expression of occludin mRNA in epithelial and non-epithelial cells (Saitou et al., 1997). Rab3B has been detected in neurons (Stettler et al., 1995), astrocytes (Madison et al., 1996) and pancreatic islet cells (Regazzi et al., 1996). 7H6 is expressed by cultured endothelial cells (Satoh et al., 1996). Symplekin has been detected in several non-epithelial cell types, including lymphocytes, but not in endothelia (Keon et al., 1996). The presence of certain TJ-associated proteins in non-epithelial cell types suggests that their function is not strictly linked to TJ, and that they may work as versatile adaptors of membrane-cytoskeleton interactions.

The expression of TJ proteins in epithelial and non-epithelial cell types raises the possibility of tissue-specific isoforms. In addition, the presence of TJ protein isoforms could explain the variability in structure and permeability properties of TJ from different epithelial tissues. Indeed, ZO-1 has been detected in two isoforms, distinguishable on the basis of the presence of the alternatively spliced domain α. For example, the α+ ZO-1 isoform (M_r 223 kD) is expressed in typical polarized epithelial cells, and the α- isoform (M_r 214 kD) is expressed in endothelial cells and glomerular podocytes (Kurihara et al., 1992; Balda and Anderson, 1993; Willott et al., 1992). In Sertoli cells, both isoforms are expressed, but are differentially distributed (Pelletier et al., 1997). The function of domain α has not been established, although its presence has been correlated with a higher degree of plasticity of the junction (Balda and Anderson, 1993; Pelletier et al., 1997). In preimplantation mouse embryos, ZO-1 α- is expressed maternally and is directly assembled at sites of cell–cell contact as punctate sites that coalesce into a zonular belt, whereas ZO-1 α+ is first seen co-localized with occludin in perinuclear foci before assembling into TJ (Sheth et al., 1997) . ZO-1 contains other alternatively spliced domains besides domain α (see Figure 2), but nothing is known about the expression of additional ZO-1 isoforms.

It is not clear whether other TJ proteins occur in different isoforms. A single message for occludin was originally reported in epithelial and endothelial tissues (Furuse et al., 1993), but more recent data suggest that several occludin mRNA bands occur, and therefore isoforms may exist (Saitou et al., 1997). For cingulin and 7H6, variations in apparent molecular sizes occur across different species. For example, 7H6 occurs as a M_r 155 polypeptide in the rat, and as a M_r 175 kD polypeptide in MDCK cells (Zhong et al., 1993). Cingulin occurs as a M_r 140 kD polypeptide in chicken, human, and rodent tissues (Citi, 1993), and as a M_r 160 kD polypeptide in *Xenopus* (Cardellini et al., 1996).

Occludin migrates as a series of polypeptides with different electrophoretic mobilities (apparent M_r between 58 and 66 kD) (Furuse et al., 1993). It has been shown that this mobility shift is due, at least in part, to phosphorylation/dephosphorylation (Cordenonsi et al., 1997; Sakakibara et al., 1997)(see below).

C. Subcellular Localization of Tight Junction Proteins

The localization of TJ-associated proteins in polarized epithelia has been established by immunoelectron microscopy or immunofluorescence (see Table 1). However, what is the localization of ZO-1 in cells that do not have TJ? And can TJ proteins be localized at sites other than TJ in epithelial cells? Recent studies have addressed these questions and shown that some TJ proteins are not exclusively detected in TJ but, depending on cell type, culture conditions, and fixation protocols, can be localized at other sites, for example adherens-type junctions and the nucleus.

In non-epithelial cells that form contacts via cadherin-mediated cell-cell adhesion (for example, fibroblasts, NRK cells, cardiac miocytes, etc), ZO-1 is co-localized with cadherin and actin in *zonulae adhaerentes* (Itoh et al., 1991; Howarth et al., 1992; Howarth et al., 1994). In addition, the co-localization of ZO-1 and cadherin occurs in certain types of epithelial or epithelial-derived cells, for example in lung carcinoma cells (Watabe et al., 1994), non-polarized mammary tumor epithelial cells (Yonemura et al., 1995), neural tube neuroepithelial cells (Aaku-Saraste et al., 1996), apoptotic LLC-PK1 kidney epithelial cells (Peralta Soler et al., 1996), and cells of the mid-epithelial level of the cornea (Sugrue and Zieske, 1997). Thus, while ZO-1 appears exclusively complexed with occludin in TJ-containing cells (see below), it also associates with cadherin (or cadherin-associated proteins) in non-epithelial cells, or in epithelial cells with reduced polarity and adhesive properties. For example, ZO-1 is associated with the β-catenin/plakoglobin complex in MDCK cells cultured in low extracellular calcium (Rajasekaran et al., 1996). Whether ZO-1 is associated with TJ or with adherens-type junction may depend on the expression of occludin and the relative affinity of binding of ZO-1 to occludin, or to *zonula adhaerens* proteins.

A second, "atypical" localization of TJ proteins is the nucleus. As shown in Figure 3, cingulin immunoreactivity is detected in both junctions and the nucleus of MDCK cells grown at low culture density. A nuclear localization for ZO-1 has been described in subconfluent cultured epithelial cells and in epithelial tissues in vivo (Gottardi et al., 1996). Nuclear staining for symplekin has been observed in polarized epithelial cells using certain fixation conditions and the nucleus is the only site of symplekin immunoreactivity in non-epithelial cultured cells, non epithelial tissues, and stratified epithelial tissues (Keon et al., 1996). Interestingly, both ZO-1 and symplekin display putative nuclear localization signal sequences. A nuclear localization for other TJ-associated proteins, besides ZO-1, symplekin, and cingulin, has not yet been reported (see Table 1).

What is the role of TJ proteins in the nucleus? It has been suggested that symplekin may have originated as a nuclear protein and becomes recruited to the junctional region in TJ-forming cells (Keon et al., 1996). Another possibility is that TJ proteins are involved in the control of transcription of nuclear genes and thus function as signal transducers from cell–cell junctions to the nucleus. For example, because of its homology to the tamou gene product in *Drosophila*, it has been suggested that ZO-1 may be involved in a signaling pathway that activates the expression of the Id family of genes (Takahisa et al., 1996). There is increasing evidence that components of the cytoplasmic plaque domain of adherens-type junctions translocate to the nucleus and bind to nuclear transcription factors (Behrens et al., 1996). It would therefore not be surprising to find TJ components that play similar roles.

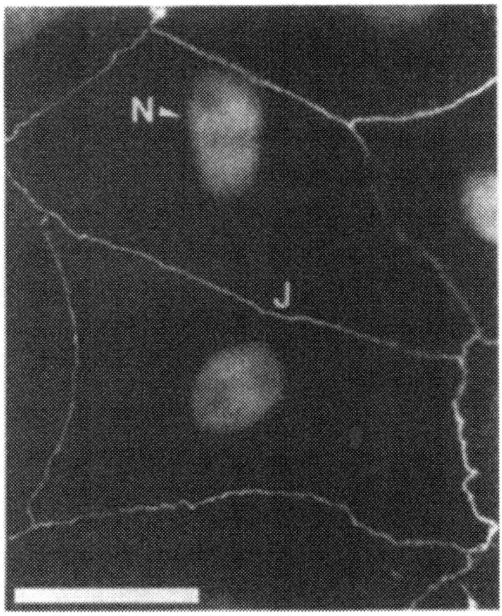

Figure 3. Nuclear localization of cingulin in subconfluent MDCK cells. Cells were permeabilized and fixed with Triton-paraformaldehyde (0.1% Triton X-100, 3% paraformaldehyde) and labelled with rabbit anti cingulin antiserum (R39)(Citi et al., 1988) followed by FITC-conjugated secondary antibody. N, nuclear staining. J, junctional staining. See also (Citi et al., 1994) for cingulin nuclear staining in cytochalasin-treated cells. Bar, 10 μm.

III. THE MOLECULAR BASIS FOR TIGHT JUNCTION FUNCTION AND REGULATION

A. Functions of Tight Junction Proteins: The Membrane Domain

Studies *in vitro* indicate that occludin has some functional role in TJ, since exogeneous expression of occludin affects the barrier and fence functions of TJ in cultured cells (Balda et al., 1996; McCarthy et al., 1996) and in *Xenopus* embryos (Chen et al., 1997). A peptide corresponding to an extracellular region of occludin perturbs the TJ barrier in cultured cells (Wong and Gumbiner, 1997). Expression of chick occludin in transfected MDCK cells leads to an increase in electrical resistance, suggesting that occludin is involved in the formation or regulation of aqueous "pore" channels (Balda et al., 1996; McCarthy et al., 1996). The effects of occludin transfection on resistance are opposite to those observed on the flux of mannitol and other small molecular weight tracers, since these fluxes are actually increased (Balda et al., 1996; McCarthy et al., 1996). Such functional dissociation of paracellular permeability from electrical resistance has been explained by postulating that a tracer has the time to migrate layer by layer through TJ which as a unit are electrically sealed during the process (Balda et al., 1996).

Occludin mediates cell–cell adhesion through its extracellular surface. Direct evidence for this has been provided by experiments in which occludin was transfected into fibroblasts (lacking endogenous occludin) (Van Itallie and Anderson, 1997). The results suggested that the ability of occludin to confer adhesiveness correlates with the ability to co-localize with its cytoplasmic binding protein, ZO-1 (Van Itallie and Anderson, 1997).

Occludin is at least one component of the fibrils observed in freeze-fractured TJ. In the intracellular multilamellar bodies formed in occludin-transfected SF9 cells, opposing membrane leaflets appear fused together in a pattern resembling that of TJ, and freeze-fracture generates short intramembranous fibrils labelled by anti-occludin antibodies (Furuse et al., 1996). Preliminary reports indicate that, when occludin is delivered to the plasma membrane in transfected L-cells, it induces the formation of TJ-like strands (Furuse et al., 1996b). In transfected epithelial cells, it is difficult to distinguish between fibrils generated by endogeneous or transfected occludin. However, expression of transfected occludin causes a slight increase in the number and complexity of TJ fibrils (Balda et al., 1996; McCarthy et al., 1996). In addition, transfected occludin co-localizes with endogeneous occludin, suggesting that it may interact with it (Balda et al., 1996; Furuse et al., 1994). Taken together, these results suggest that occludin may polymerize into intramembranous supramolecular assemblies and is a component of TJ fibrils.

Occludin appears to be also responsible for the "fence" function of TJ. Occludin transfection renders MDCK cells incapable of maintaining a fluorescent lipid in a specifically labeled cell surface domain, suggesting that occludin organization is critical for the structural polarization of the plasma membrane (Balda et al., 1996).

Despite all this evidence, recent analysis of mouse embryonic stem cells lacking occludin has demonstrated that occludin is not necessary for the formation of apparently normal TJ, as determined by morphological and functional analysis (Saitou et al., 1998). The identification and characterization of claudins (Furuse et al., 1998a; Furuse et al., 1998b) and JAM (Martin-Padura et al., 1998) suggests that the membrane domain of TJ is more heterogeneous than previously thought, and that different proteins may contribute to its structure and regulation.

B. Functions of Tight Junction Proteins: The Cytoplasmic Plaque Domain

ZO-1 interacts with occludin, based on *in vitro* binding studies, co-precipitation, and co-localization in transfected epithelial and non-epithelial cells (Balda et al., 1996; Furuse et al., 1996; Furuse et al., 1994). ZO-1-occludin interaction occurs between the C-terminal cytoplasmic domain of occludin and the N-terminal 570 amino acids of ZO-1 (Fanning et al., 1996; Fanning et al., 1998). Immunoprecipitation and co-precipitation experiments have shown that ZO-1 is associated with ZO-2 (Gumbiner et al., 1991) and pl30 (Balda et al., 1993). The PDZ2 motif of

ZO-1 appears to be responsible for the interaction of ZO-1 with ZO-2 (Fanning et al., 1996; Fanning et al., 1998). GST coprecipitation and *in vitro* binding experiments also indicate that ZO-1 interacts with the Ras-binding domain of AF-6, and that this interaction may be inhibited by activated Ras (Yamamoto et al., 1997).

Finally, ZO-1 has recently been reported to interact with connexin-43 (Giepmans and Moolenaar, 1998) and cortactin (Katsube et al., 1998). ZO-3 has been reported to interact with ZO-1 and occludin (Haskins et al., 1998).

Less information is available concerning protein–protein interactions of other TJ proteins. Recently, we observed that a GST fusion protein containing the N-terminal region of cingulin precipitates ZO-1 from MDCKII extracts, indicating that the two proteins interact directly or indirectly (Cordenonsi et al, unpublished results).

In summary, these studies indicate that a complex between occludin, ZO-1, ZO-2, p130/ZO-3, cingulin (and AF-6) may occur at the membrane-cytoplasmic plaque interface in TJ (see Figure 1B).

C. Functions of Tight Junction Proteins: The Cytoskeletal Domain

Actin microfilaments have been detected in the cytoplasmic face of TJ by several electron microscopic techniques: quick-freeze deep etch (Hirokawa and Tilney, 1982), S-1 decoration (Madara, 1987), and immunogold labeling (Drenckhahn and Dermietzel, 1988). The role of the actin cytoskeleton in TJ regulation has long been established, since a number of microfilament-active drugs influence TJ structure and function (Bentzel et al., 1976; Meza et al., 1980; Madara et al., 1986; Stevenson and Begg, 1994).

In epithelial and non-epithelial cells, actin microfilaments are spatially associated with TJ proteins (Citi, 1994; Howarth and Stevenson, 1995) (see Figure 4). This association appears to be crucial in the formation of TJ. For example, when MDCK epithelial cells are trypsinized and plated at low cell density, in single cells cingulin is first detected at free cell edges, where it is restricted to areas that display thick bundles of actin filaments (arrowheads in Figure 4A-A'). These bundles may be similar to the "purse-string" contraction belts seen during epithelial wound repair (Bement et al., 1993). Thus, the focal assembly of actin microfilaments (and perhaps other actin-associated cytoplasmic and membrane proteins) is necessary for the delivery and stabilization of cingulin at cell borders prior to the formation of junctions. When neighboring cells start to form junctions along part of their borders, cingulin is co-localized with a thin bundle of actin in the junction (see arrow in Figure 4B-B'). However, cingulin at free cell edges is restricted to areas that contain thick bundles of actin (see arrowheads in Figure 4B-B').

To understand how the actin cytoskeleton regulates TJ at the molecular level, it is critical to study the interaction of specific TJ proteins with actin and actin-binding proteins. Recent evidence indicates that ZO-1 interacts with the actin cytoskeleton. Basal stress fibers are reduced in MDCK cells transfected with a C-terminal construct of ZO-1 (Fanning et al., 1996; Fanning et al., 1998), and the C-terminal region of ZO-1 binds to actin *in vitro* (Itoh et al., 1997). Since the N-terminal region of

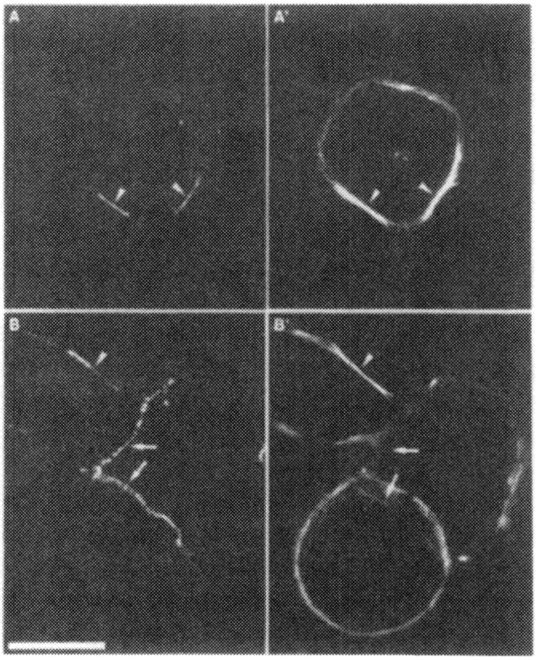

Figure 4. Cingulin (A, B) and actin (A′, B′) co-localization in a single MDCK cell (A-A′) and in developing junctions between three neighbouring cells (B-B′). Arrowheads in A and B (corresponding arrowheads in A′ and B′) pointo to sites of cingulin labeling that correspond to focal assembly of thick actin bundles, in the absence of any junctions. Note that actin filaments are distributed around the entire periphery of the cell (A′), but are thickened only in focal areas. Arrows in B indicate junctional cingulin that co-localizes with relatively weakly stained junctional actin (B′). For this experiment, cells were cultured for 4 hours after trypsinization and plating, and were stained as described for Figure 3. Rhodamine-phallacidin was used to label actin. Bar, 10 μm.

ZO-1 interacts with alpha-catenin *in vitro* (Itoh et al., 1997), it was suggested that ZO-1 is a functional component of cadherin-based adhesion sites and provides a link between the cadherin-catenin complex and actin (Itoh et al., 1997; Rajasekaran et al., 1996). In summary, these studies suggest that the TJ-protein complex (Figure IB) may interact with the actin cytoskeleton and cadherin-based junctions via ZO-l.

A second cytoskeletal protein that has been implicated in the organization of TJ is spectrin (Bennett, 1990). Spectrin/fodrin have been reported to interact with ZO-1, based on coimmunoprecipitation from extracts incubated with anti-ZO-1 antibodies and the formation of complexes visualized by electron microscopy or immunoblotting (Itoh et al., 1991; Tsukamoto and Nigam, 1997). In addition, GST pull-down assays showed that the cytoplasmic domain of occludin is associated with alpha-spectrin, presumably through its interaction with ZO-1 (Furuse et al., 1994).

MAGUK protein family members are known to bind to protein 4.1., which connects integral membrane proteins to the actin/spectrin cytoskeleton (Bennett, 1990). However, it is not known whether ZO-1, ZO-2 or p130 can interact with protein 4.1 or its homologs. Finally, the relevance of spectrin in the organization of TJ and its invertebrate homologue, the septate junction (Bryant, 1994), is questioned by the observation that alpha-spectrin deficient mutants in *Drosophila*, which exhibit cell shape and interaction defects, still show intact septate junctions (Lee et al., 1993).

D. Experimental Modulation of Tight Junction Assembly: The Role of Phosphorylation and the Actin Cytoskeleton

Experimental manipulation of TJ assembly/disassembly in cultured epithelial cells has been useful in studying the mechanisms involved in the delivery of TJ proteins to sites of cell–cell contact and in the establishment and maintenance of a functional TJ barrier. TJ disassembly can be induced by removal of extracellular calcium (Siliciano and Goodenough, 1988) and TJ assembly can be induced by addition of calcium to cells incubated in low-calcium medium for several hours (calcium switch)(Gonzalez-Mariscal et al., 1985).

In the calcium removal-junction disassembly model, protein kinase activity is a critical element in the chain of events connecting breakdown of cadherin-dependent adhesion with changes in cell shape and TJ organization (Citi, 1992). Using the calcium switch (assembly) model, a number of second messengers and signaling pathways have been shown to regulate TJ. These include serine/threonine and tyrosine protein kinases, calcium (both extracellular and intracellular), heterotrimeric G proteins, cAMP, and phospholipase C (Balda et al., 1991; Nigam et al., 1991; Balda, 1992; Denisenko et al., 1994; Stuart and Nigam, 1995; Denker et al., 1996; Stuart et al., 1996).

The identity of the kinases (and substrates) involved in regulation of TJ assembly and disassembly are not known, and cannot be precisely determined only on the basis of the specificity of the kinase inhibitors or activators used. However, recent studies have identified the actin cytoskeleton as a major putative target for these kinases. For example, the effects of protein kinase inhibitors on junction assembly and disassembly occur through modulation of the actin cytoskeleton, because the inhibitors alter the organization and contractility of actin filaments in the absence and in the presence of microfilament-active drugs (Citi et al., 1994; Volberg et al., 1994). In addition, TJ function is perturbed by modulating the activity of the rho family of GTP-binding proteins (Nusrat et al., 1995), and myosin light chain kinase (Hecht et al., 1996: Turner et al., 1997). Phosphorylation of myosin light chains promotes myosin filament assembly and actin-activated Mg-ATPase activity *in vitro*, and increases actomyosin contractility *in vivo* (Citi and Kendrick-Jones, 1987). Taken together, all these results show that TJ function is modulated by the organization and contractility of the actomyosin cytoskeleton, and suggest that many physiological and pathological modulators of TJ may ultimately act through the actin cytoskeleton (Madara, 1989).

Whether TJ are controlled primarily by TJ actin (Hirokawa and Tilney, 1982; Drenckhahn and Dermietzel, 1988; Madara, 1987), or by the perijunctional actin-myosin bundle in the cytoplasmic face of the *zonula adhaerens* (Mooseker, 1985) is not clear. Presumably, contraction of the latter structure would affect TJ organization indirectly, by exerting a mechanical tension, whereas modulation of TJ actin could affect more directly the organization of specific TJ proteins. In this perspective, studying the interactions between TJ proteins and actin microfilaments will be crucial to test possible roles of TJ-associated actin.

The importance of phosphorylation as a possible regulatory mechanism of TJ is underlined by the recent observation that an atypical protein kinase C associates and colocolizes at the epithelial tight junction with a protein that is homologous to the *C. elegans* polarity protein PAR-3 (Izumi et al., 1998).

E. Phosphorylation of Tight Junction Proteins

Since intracellular kinases play an important role in modulating TJ assembly, it is relevant to determine whether regulated phosphorylation of TJ proteins occurs physiologically and if so, what its role is. Several TJ proteins have now been shown to be phosphorylated *in vivo*. For example, ZO-1 and ZO-2 are phosphorylated on Ser residues in MDCK cells (Anderson et al., 1988), and can be phosphorylated on Tyr residues upon various experimental treatments that increase the activity of tyrosine protein kinases or reduce the activity of tyrosine protein phosphatases (Kurihara et al., 1995: Staddon et al., 1995; Takeda and Tsukita, 1995; Van Itallie et al., 1995). Cingulin is phosphorylated in vivo in MDCK and CaCo2 cells on Ser residues (Citi and Denisenko, 1995). Occludin phosphorylation in vivo occurs in *Xenopus* (Cordenonsi et al., 1997) and in cultured epithelial cells (Sakakibara et al., 1997).

The actual in vivo phosphorylation sites of TJ proteins are not known, although in vitro phosphorylation can provide some clues. Recombinant chicken cingulin can be phosphorylated in vitro by protein kinase C, and two Ser phosphorylation sites have been mapped to the C-terminal coiled-coil domain (Rabino et al., 1993). The SH3 domain of ZO-1 binds a serine protein kinase that phosphorylates a region immediately C-terminal to this domain (Balda et al., 1996). The cytoplasmic, C-terminal domain of occludin can be phosphorylated in vitro on Ser residues by cdc2 kinase and by protein kinase CK2 (Cordenonsi et al., 1997). ZO-1, ZO-2, cingulin, occludin, and symplekin sequences display several potential Ser/Thr phosphorylation sites by protein kinase C, cAMP-dependent protein kinase, cdc2 kinase and protein kinase CK2. Occludin sequences from different species contain a conserved motif identical to the sequence of the cytoplasmic domain of insulin receptor kinase that is phosphorylated on Tyr (Cordenonsi et al., 1997). However, Tyr phosphorylation of occludin has not been reported.

What are the effects of TJ protein phosphorylation on TJ function in vivo? A relationship between phosphorylation of ZO-1 or cingulin and permeability functions of TJ has not been clearly established (Balda et al., 1993; Citi and Denisenko, 1995; Stevenson et al., 1989). However, changes in total ZO-1 phosphorylation may be correlated with alterations in its distribution (Howarth et al., 1994). Sakakibara and colleagues (1997) examined occludin phosphorylation in cultured MDCK cells and in chicken tissues and concluded that non- or less phosphorylated occludin is distributed on the basolateral membranes and that highly phosphorylated occludin is selectively concentrated at tight juctions in a NP-40-insoluble form (see also Wong, 1997). We (Cordenonsi et al., 1997) examined occludin phosphorylation in *Xenopus* laevis early embryos, and concluded that occludin dephosphorylation is associated with de novo formation of TJ in vivo. Thus, it appears that the precise role of phosphorylation in promoting (or preventing) TJ assembly may depend on cell type or organism, or developmental state. Since occludin plays a fundamental structural role in TJ, one can speculate that its phosphorylation/dephosphorylation may function as a molecular switch that controls its assembly at sites of cell–cell contact, its polymerization or its interaction with ZO-1.

IV. TIGHT JUNCTION IN DEVELOPMENT AND DISEASE

A. Developmental Regulation of Tight Junction Assembly

Studying the expression and localization of TJ proteins in developing embryos can help to understand TJ regulatory mechanisms and test the functions of TJ proteins. Mouse and *Xenopus laevis* are currently the two major experimental models used to study TJ in development.

In the mouse preimplantation embryo, epithelial differentiation of the trophectoderm is associated with junction formation. Maternal mRNA for ZO-1 (α-isoform) occurs in oocytes and at all preimplantation stages and the protein is localized at punctate sites at the cell–cell contacts of 8-cell embryos (Sheth et al., 1997). Zygotic ZO-1 (α+ isoform) expression and membrane assembly begin at compaction (8-cell stage), coinciding with the onset of cell polarity and the activation of cell–cell adhesion mediated by E-cadherin (Fleming et al., 1989; Fleming and Hay, 1991; Sheth et al., 1997). Maternal cingulin is expressed in eggs and early cleavage stages (up to 16 cells), and shows a cortical localization (Fleming et al., 1993; Javed et al., 1993). Membrane assembly of zygotic cingulin occurs at the 16-cell stage (Fleming et al., 1993). Thus, TJ assembly is regulated at the transcriptional and translational level and involves a sequential assembly of TJ proteins. This notion is also supported by the observation that in developing chick intestinal epithelial cells 7H6 is localized in coarse dots along the basolateral cell membrane, whereas occludin and ZO-1 are localized linearly along the cell border (Kimura et al., 1996).

In early *Xenopus* embryos, the barrier and fence functions of TJ can be detected as early as the 2-cell stage, as determined by electrondense tracer studies (Kalt, 1971) and by the development of cell surface polarity (Roberts et al., 1992). However, a tight electrical seal to the passage of ions using microelectrode probes is not evident until the blastula stage (Slack and Warner, 1973), and electron microscopy studies have failed to reveal typical TJ structures before the 32-cell stage (Muller and Hausen, 1995). Thus, several questions remain unanswered. When are TJ proteins expressed in *Xenopus* development? Where are they localized? Is there a developmentally regulated "switch" that induces TJ assembly?

Using cross-reacting polyclonal antibodies. we have shown that cingulin is expressed as a M_r 160 kD polypeptide in *Xenopus* oocytes and early embryos (Cardellini et al., 1996). Maternal cingulin in oocytes has a cortical localization (similar to early stage mouse embryos) and starting from the first cell division it becomes localized linearly along junctions. Cingulin accumulation into junctions is spatially and temporally related to the deepening of the cleavage furrow and occurs even when cell–cell contacts are reduced by incubating dividing embryos in a low calcium solution (Cardellini et al., 1996). In addition to cingulin, occludin (Cordenonsi et al., 1997) and ZO-1 (Merzdorf and Goodenough, 1996) are expressed maternally in *Xenopus* oocytes. Future studies on TJ proteins in *Xenopus* will certainly provide important insights into the molecular mechanism of TJ assembly.

B. Tight Junctions in Pathology

A variety of drugs (reviewed in (Powell, 1981; Balda et al., 1992; Bentzel et al., 1992; Schneeberger, 1994)), bacteria or bacterial products and toxins (e.g. from.

Clostridium difficile (Moore et al., 1990; Hecht et al., 1992), *Salmonella* (Finely et al., 1988), *Vibrio cholerae* (Balda et al., 1991; Faso et al., 1991; Woo et al., 1996), *Bacteroids fragilis* (Kochi et al., 1996; Obis et al., 1997), *Pseudomonas aeruginosa* (Azghani, 1996), *Escherichia coli* (Philpott et al., 1996), hormones, growth factors and cytokines (Madara and Stafford, 1989; Royall et al., 1989; Mullin and Snock, 1990; Thiery and Boyer, 1992; Kaoutzani et al., 1994; McRoberts and Riley, 1992; Van Itallie et al., 1995; Woo et al., 1996) have been reported to affect the structure and function of TJ, or the organization of TJ proteins and the junctional actin cytoskeleton. The precise action of all these factors on TJ proteins is not known, but at least in some cases it is likely that they affect TJ by interfering with the organization of actin microfilaments. For example, the Rho family of GTP-binding proteins, that is implicated in actin filament organization and several cell signalling events, is the target for covalent modification by many pathogenic bacteria (reviewed in Machesky and Hall, 1996). It is noteworthy that not all bacteria have a TJ-disrupting effect (Pujol et al., 1997).

There is a close correlation between cholestasis and TJ function and an altered localization of TJ proteins has been demonstrated in rat experimental cholestasis (Fallon et al., 1993; Fallon et al., 1995; Anderson, 1996). Abnormal levels of hormones and cytokines, which may occur in disease (Isaacs et al., 1992), can result in altered TJ function in epithelia and endothelia (Elias et al., 1983; Cui et al., 1996; Madsen et al., 1997). This can in turn induce permeation of toxic products and bacteria through barriers and generate local inflammation, infection and other pathological effects.

For example, hepatobiliary complications in inflammatory bowel disease may result from alterations in liver TJ function (Lora et al., 1997). An altered TJ function has been implicated in Crohn's disease (Hollander, 1988) and in pediatric cholestatic liver diseases (De Vos et al., 1975; Weber et al., 1981). Finally, the paracellular barrier may prevent the metastatic spread of cancerous cells (Tobioka et al., 1996).

Although TJ proteins are localized at sites of cell–cell adhesion and recent data indicate that they may be involved in nuclear transcription (Gottardi et al., 1996; Keon et al., 1996) and signal transduction (Anderson and Van Itallie, 1995), few studies have addressed the relationship between TJ protein expression and cancer. When epithelial cell polarity and junction formation is maintained in tumor cells, cingulin, and ZO-1 expression and localization are normal (junctional belt). This has been observed in human intestinal adenomas and well-differentiated adenocarcinomas (Citi et al., 1991), hepatoma cells (Stevenson et al., 1989), and a rat hepatoma-human fibroblast hybrid line (Decaens et al., 1996). However, in poorly differentiated carcinomas, TJ protein expression appears to depend on the tumor type and its adhesive properties. In poorly differentiated human colon adenocarcinomas, ectopic localization of cingulin, at the border between unpolarized cells and surrounding connective tissue, was observed (Citi et al., 1991). The amount of cingulin in these adenocarcinomas was normal or increased, suggesting that cingulin expression could be used to identify undifferentiated epithelial neoplasias (Citi et al., 1991). ZO-1 is expressed in non-adhesive PC9 lung carcinoma cells, where it is associated with cadherin and distributed randomly over the cell surface (Watabe et al., 1994). Activation of adhesion by transfection with alpha-catenin triggers the formation of the junctional complex and the polarized distribution of cell surface proteins and structures (Watabe et al., 1994). In HeLa cells, which are non-adhesive and derive from a poorly differentiated carcinoma of the cervical epithelium, ZO-1 and cingulin are not expressed (Doyle et al., 1995). However, expression and junctional localization of cingulin and ZO-1 can be obtained *in vitro* by functional expression of the Po membrane adhesion protein, that reverses HeLa cells to an epithelial-like phenotype, and induces cadherin-mediated cell-cell adhesion (Doyle et al., 1995). Finally, generation of a polarized cell monolayer, ZO-1 assembly into TJ, and suppression of growth are obtained by administration of dexamethasone to mammary epithelial tumor cells, and these effects are antagonized by TGFα (Buse et al., 1995; Woo et al., 1996; Zettl et al., 1992). Additional studies should pursue the relationship between TJ protein expression, epithelial phenotype and adhesive properties of human tumors.

V. FUTURE PERSPECTIVES

A variety of antibody and cDNA tools are now available to investigate TJ organization at the molecular level. Some of the protein interactions occurring at the cytoplasmic face of TJ have begun to be clarified and further studies on all other TJ proteins will provide a more detailed picture of TJ architecture. Characterization of

TJ protein expression at sites other than TJ and investigating the role of phosphorylation in TJ protein function should provide clues to understanding the mechanisms of modulation and plasticity of TJ. In addition, genetic analysis of TJ proteins by homologous recombination techniques in the mouse will provide critical information about their roles. Progress in the knowledge about the molecular organization of TJ will hopefully lead to the development of rational strategies to modulate TJ function for therapeutical purposes, for example to prevent TJ breakdown induced by pathological factors, or to deliver drugs across blood-tissue barriers.

ACKNOWLEDGMENTS

We gratefully acknowledge the technical assistance of Ms. Natasha Denisenko in immunofluorescence experiments, and the support of Professor David Shore. We also thank EC Biomed Program, Istituto Superiore di Sanita', CNR, MURST and the State of Geneva for financial support.

REFERENCES

Aaku-Saraste, E., Hellwig, A., & Huttner, W. B. (1996). Loss of occludin and functional tight junctions, but not ZO-1, during neural tube closure—remodeling of the neuroepithelium prior to neurogenesis. Dev. Biol. 180, 664-679.

Anderson, J. M. (1996). Leaky junctions and cholestasis: a tight correlation. Gastroenterology 110, 1662-1665.

Anderson, J. M., Fanning, A. S., Lapierre, L., & Van Itallie, C. M. (1995). Zonula occludens (ZO)-1 and ZO-2: membrane -associated guanylate kinase homologues (MAGuKs) of the tight junction. Biochem. Soc. Trans. 23, 470-475.

Anderson, J. M., Stevenson, B. R., Jesaitis, L. A., Goodenough, D. A., & Mooseker, M. S. (1988). Characterization of ZO- 1, a protein component of the tight junction from mouse liver and Madin-Darby Canine Kidney cells. J. Cell Biol. 106, 1141-1149.

Anderson, J. M. & Van Itallie, C. M. (1995). Tight junctions and the molecular basis for regulation of paracellular permeability. Am. J. Physiol. 269, G467-G475.

Ando-Akatsuka, Y., Saitou, M., Hirase, T., Kishi, M., Sakakibara, A., Itoh, M., Yonemura, S., Furuse, M., & Tsukita, S. (1996). Interspecies diversity of the occludin sequence: cDNA cloning of human, mouse, dog and rat-kangaroo homologues. J. Cell Biol. 133, 43-47.

Azghani, A. O. (1996). Pseudomonas aeruginosa and epithelial permeability: role of virulence factors elastase and exotoxin A. Am. J. Respir. Cell Mol. Biol. 15, 132-140.

Balda, M. S. (1992). Intracellular signals in the assembly and sealing of tight junctions. In: Tight Junctions (ea. M. Cereijido), pp. 121-137. CRC Press, Boca Raton, FL.

Balda, M. S. & Anderson, J. M. (1993). Two classes of tight junctions are revealed by ZO-1 isoforms. Am. J. Physiol. 264, C918-C9' 4.

Balda, M. S., Anderson, J. M., & Matter, K. (1996). The SH3 domain of the tight junction protein ZO-1 binds to a serine protein kinase that phosphorylates a region C-terminal to this domain. FEBS Lett 399, 326-332.

Balda, M. S., Fallon, M. B., Van Itallie, C. M., & Anderson, J. M. (1992). Structure, regulation, and patophysiology of tight junctions in the gastrointestinal tract. Yale J. Biol. Med. 65, 725-735.

Balda, M. S., Gonzalez-Mariscal, L., Contreras, R. G., Macias-Silva, M., Torres-Marquez, M. E., Garcia-Sainz, J. A., & Cereijido, M. (1991). Assembly and sealing of tight junctions: possible participation of G-proteins, phospholipase C, protein kinase C and calmodulin. J. Membr. Biol. 122, 193-202.

Balda, M. S., Gonzalez-Mariscal, L., Matter, K., Cereijdo, M., & Anderson, J. M. (1993). Assembly of the tight junction: the role of diacylglycerol. J. Cell Biol. 123, 293302.

Balda, M. S., Whitney, J. A., Flores, C., Gonzalez, S., Cereijido, M., & Matter, K. (1996). Functional dissociation of paracellular permeability and transepithelial electrical resistance and disruption of the apical- basolateral intramembrane diffusion barrier by expression of a mutant tight junction membrane protein. J. Cell Biol. 134, 1031-1049.

Beatch, M., Jesaitis, L. A., Gallin, W. J., Goodenough, D. A., & Stevenson, B. R. (1996). The tight junction protein ZO-2 contains three PDZ (PSD-95/Discs- Large/ZO-I) domains and an alternatively spliced region. J. Biol. Chem. 271, 25723-25726.

Behrens, J., von Kries, J. P., Kuhl, M., Bruhn, L., Wedlich, D., Grosschedl, R., & Birchmeier, W. (1996). Functional interaction of beta-catenin with the transcription factor LEF-1. Nature 382, 638-642.

Bement, W. M., Forscher, P., & Mooseker, M. S. (1993). A novel cytoskeletal structure involved in purse string wound closure and cell polarity maintenance. J. Cell Biol. 121, 565-578.

Bennett, V. (1990). Spectrin: a structural mediator between diverse plasma membrane proteins and the cytoplasm. Curr. Opin. Cell Biol. 2, 51-56.

Bentzel, C. J., Hainau, B., Edelman, A., Anagnostopoulos, T., & Benedetti, E. L. (1976). Effects of plant cytokinins on microfilaments and tight junctional permeability. Nature 264, 666-668.

Bentzel, C. J., Palant, C. E., & Fromm, M. (1992). Physiological and pathological factors affecting the tight junction. In: Tight junctions (ea. M. Cereijido), pp. 151-173. CRC Press, Boca Raton, FL

Bowen, J. A., Newton, G. R., Weise, D. W., Bazer, F. W., & Burghardt, R. C. (1996). Characterization of a polarized porcine uterine epithelial model system. Biol. Reprod. 55, 613-9.

Bryant, P. J. (1994). Molecular structure and function of invertebrate junctions. In: Molecular Mechanisms of Epithelial Cell Junctions: from Development to Disease (ea. S. Citi), pp. R.G. Landes, Austin, TX.

Buse, P., Woo, P. L., Alexander, D. B., Reza, A., & Firestone, G. L. (1995). Glucocorticoid-induced functional polarity of growth factor responsiveness regulates tight junction dynamics in transformed mammary epithelial tumor cells. J. Biol. Chem. 270, 28223-28227.

Byers, S. W., Citi, S., Anderson, J. M., & Hoxter, B. (1992). Polarized functions and permeability properties of rat epididymal epithelial cells in vitro. J. Reprod. Fert. 95, 385-396.

Cardellini, P., Davanzo, G., & Citi, S. (1996). Tight junctions in early amphibian development: detection of junctional cingulin from the 2-cell stage and its localization at the boundary of distinct membrane domains in dividing blastomeres in low calcium. Dev. Dyn. 207, 104- 113.

Chalcroft, J. P. & Bullivant, S. (1970). An interpretation of liver cell membrane and junction structure based on observation of freeze-fracture replicas of both sides of the fracture. J. Cell Biol. 47, 49-60.

Chapman, L. M. & Eddy, E. M. (1989). A protein associated with the mouse and rat hepatocyte junctional complex. Cell Tissue Res. 257, 333-341.

Chen, Y., Merzdorf, C., Paul, D. L., & Goodenough, D. A. (1997). COOH terminus of occludin is required for tight junction barrier function in early xenopus embryos. J. Cell Biol. 138, 891-9.

Citi, S. (1992). Protein kinase inhibitors prevent junction dissociation induced by low extracellular calcium in MDCK epithelial cells. J. Cell Biol. 117, 169-178.

Citi, S. (1993). The molecular organization of tight junctions. J. Cell Biol. 121, 485-489.

Citi, S. (1994). The molecular composition of tight junctions. In: Molecular Mechanisms of Epithelial Cell Junctions: from Development to Disease (ea. S. Citi), pp. 83-106. R.G. Landes, Austin, TX

Citi, S., Amorosi, A., Franconi, F., Giotti, A., & Zampi, G. (1991). Cingulin, a specific protein component of tight junctions, is expressed in normal and neoplastic human epithelial tissues. Am. J. Pathol. 138, 781-789.

Citi, S. & Denisenko, N. (1995). Phosphorylation of the tight junction protein cingulin and the effect of protein kinase inhibitors and activators in MDCK epithelial cells. J. Cell Sci. 108, 2917-2926.

Citi, S. & Kendrick-Jones, J. (1987). Regulation of nonmuscle myosin structure and function. Bioessays 7, 155-159.

Citi, S., Kendrick-Jones, J., & Shore, D. (1990). Cloning of cingulin cDNA: evidence for a tight junction protein with a coiled-coil structure. J. Cell Biol. 111, Pt2, 409a.

Citi, S., Sabanay, H., Jakes, R., Geiger, B., & Kendrick-Jones, J. (1988). Cingulin, a new peripheral component of tight junctions. Nature 333, 272-276.

Citi, S., Sabanay, H., Kendrick-Jones, J., & Geiger, B. (1989). Cingulin: characterization and localization. J. Cell Sci. 93, 107-122.

Citi, S., Volberg, T., Bershadsky, A. D., Denisenko, N., & Geiger, B. (1994). Cytoskeletal involvement in the modulation of cell-cell junctions by the protein kinase inhibitor H-7. J. Cell Sci. 107, 683-692.

Cordenonsi, M., Mazzon, E., De Rigo, L., Baraldo, S., Meggio, F., & Citi, S. (1997). Occludin dephosphorylation in early development of Xenopus laevis. J. Cell Sci. 110, 3131-3139.

Cui, N., Madsen, K. L., Friend, D. R., Stevenson, B. R., & Fedorak, R. N. (1996). Increased permeability occurs in rat ileum following induction of pan- colitis. Dig. Dis. Sci. 41, 405-411.

De Vos, R., DeWolf-Pecters, C., & Desmet, V. (1975). Progressive intrahepatic cholestasis (Byler's disease): case report. Gut 16, 943-950.

Decaens, C., Rodriguez, P., Bouchaud, C., & Cassio, D. (1996). Establishment of hepatic cell polarity in the rat hepatoma-human fibroblast hybrid WIF-B9. A biphasic phenomenon going from a simple epithelial polarized phenotype to an hepatic polarized one. J. Cell. Sci. 109, 1623-1635.

Denisenko, N., Burighel, P., & Citi, S. (1994). Different effects of protein kinase inhibitors on the localization of junctional proteins at cell-cell contact sites. J. Cell Sci. 107, 969-981.

Denker, B. M., Saha, C., Khawaja, S., & Nigam, S. K. (1996). Involvement of a heterotrimeric G protein alpha subunit in tight junction biogenesis. J. Biol. Chem. 271, 25750-25753.

Dezawa, M., Mutoh, T., Dezawa, A., & Ishide, T. (1996). Tight junctions between the axon and Schwann cell during PNS regeneration. Neuroreport 7, 1829-1832.

Diamond, J. M. (1977). The epithelial junction: bridge, gate and fence. Physiologist 20, 10-18.

Dodane, V. & Kachar, B. (1996). Identification of isoforms of G proteins and PKC that colocalize with tight junctions. J. Membr. Biol. 149, 199-209.

Doyle, J. P., Stempak, J. G., Cowin, P., Colman, D. R., & D'Urso, D. (1995). Protein zero, a nervous system adhesion molecule, triggers epithelial reversion in host carcinoma cells. J. Cell Biol. 131, 465-82.

Drenckhahn, D. & Dermietzel, R. (1988). Organization of the actin filament cytoskeleton in the intestinal brush border: a quantitative and qualitative immunoelectron microscope study. J. Cell Biol. 107, 1037-1048.

Dym, M. & Fawcett, D. W. (1970). The blood-testis barrier in the rat and the physiological compartmentation of the seminiferous epithelium. Biol. Reprod. 3, 308-326.

Eisenberg, C. A. & Bader, D. (1995). QCE-6: a clonal cell line with cardiac myogenic and endothelial cell potentials. Dev. Biol. 167, 469-481.

Elias, E., Iqbal, S., Knutton, S., Hickey, A., & Coleman, R. (1983). Increased tight junction permeability: a possible mechanism of oestrogen cholestasis. Eur. J. Clin. Invest. 13, 383-390.

Fallon, M. B., Brecher, A. R., Balda, M. S., Matter, K., & Anderson, J. M. (1995). Altered hepatic localization and expression of occludin after common bile duct ligation. Am. J. Physiol. 269, C1057-C1062.

Fallon, M. B., Mennone, A., & Anderson, J. M. (1993). Altered expression and localization of the tight junction protein ZO-1 after common bile duct ligation. Am J Physiol 264, C1439-C1447.

Fanning, A. S., Jameson, B. T., & Anderson, J. M. (1996). Molecular interactions among the tight junction proteins ZO- I, ZO-2 and occludin. Mol. Biol. Cell 7, 607a (Abstr.).

Fanning, A. S., Jameson, B. J., Jesaitais, L. A., & Anderson, J. M. (1998a). The tight junction protein ZO-1 establishes a link between the transmembrane protein occludin and the actin cytoskeleton. J. Biol. Chem. 273, 29745-29753.

Farquhar, M. G. & Palade, G. E. (1963). Junctional complexes in various epithelia. J. Cell Biol. 17, 375-412.

Fasano, A., Baudry, B., Pumplin, D. W., Wasserman, S. S., Tall, B. D., Ketley, J. M., & Kaper, J. B. (1991). Vibrio cholerae produces a second enterotoxin, which affects intestinal tight junctions. Proc. Natl. Acad. Sci. USA 88, 5242-5246.

Finlay, B. B., Gumbiner, B., & Falkow, S. (1988). Penetration of Salmonella through a polarized Madin-Darby canine kidney epithelial cell monolayer. J. Cell Biol. 107, 221-230.

Fleming, T. P., Hay, M., Javed, Q., & Citi, S. (1993). Localisation of tight junction protein cingulin is temporally and spatially regulated during early mouse development. Development 117, 1135-1144.

Fleming, T. P. & Hay, M. J. (1991). Tissue-specific control of expression of the tight junction polypeptide ZO-1 in the mouse early embryo. Development 113, 295-304

Fleming, T. P., McConnell, J., Johnson, M. H., & Stevenson, B. R. (1989). Development of tight junctions de novo in the mouse early embryo: control of assembly of the tight junction-specific protein, ZO-1. J. Cell Biol. 108, 1407-1418.

Furuse, M., Fujimoto, K., Sato, N., Hirase, T., Tsukita, S., & Tsukita, S. (1996a). Overexpression of occludin, a tight junction-associated integral membrane protein, induces the formation of intracellular multilamellar bodies bearing tight junction-like structures. J. Cell Sci. 109, 429-435.

Furuse, M., Fujimoto, K., Saitou, M., & Tsukita, S. (1996b). Assembly of occludin at cell-cell contact sites of fibroblasts transfected with occludin cDNA. Mol. Biol. Cell 7, 607a (Abstr.).

Furuse, M., Hirase, T., Itoh, M., Nagafuchi, A., Yonemura, S., Tsukita, S., & Tsukita, S. (1993). Occludin: a novel integral membrane protein localizing at tight junctions. J. Cell Biol. 123, 1777-1788.

Furuse, M., Itoh, M., Hirase, T., Nagafuchi, A., Yonemura, S., Tsukita, S., & Tsukita. S. (1994). Direct association of occludin with ZO-1 and its possible involvement in the localization of occludin at tight junctions. J. Cell Biol. 127, 1617-1626.

Furuse, M., Sasaki, H., Fujimoto, K., & Tsukita, S. (1998). A single gene product, claudin-1 or -2, reconstitutes tight junction strands and recruits occludin in fibroblasts. J. Cell Biol. 143, 391-401.

Furuse, M., Fujita, K., Hiiragi, T., Fujimoto, K., & Tsukita, S. (1998b). Claudin-1 and -2: Novel Integral Membrane Proteins Localizing at Tight Junctions with No Sequence Similarity to Occludin. J. Cell Biol. 141, 1539-1550.

Gardner, T. W., Lesher, T., Khin, S., Vu, C., Barber, A. J., & Brennan, W. A., Jr. (1996). Histamine reduces ZO-1 tight-junction protein expression in cultured retinal microvascular endothelial cells. Biochem J 320, 717-721.

Giepmans, B. N., & Moolenaar, W. H. (1998). The gap junction protein connexin43 interacts with the second PDZ domain of zona occludens-1 protein. Curr. Biol. 8, 931-934.

Gonzalez-Mariscal, L., Chavez de Ramirez, B., & Cereijido, M. (1985). Tight junction formation in cultured epithelial cells (MDCK). J. Membr. Biol. 86, 113-125.

Goodenough, D. A. & Revel, J. P. (1970). A fine structural analysis of intercellular junctions in the mouse liver. J. Cell Biol. 45, 272-290.

Gottardi, C. J., Arpin, M., Fanning, A. S., & Louvard, D. (1996). The junction-associated protein, zonula occludens-1, localizes to the nucleus before the maturation and during the remodeling of cell-cell contacts. Proc. Natl. Acad. Sci. USA 93, 10779-10784.

Griepp, E. B., Dolan, W. J., Robbins, E. S., & Sabatini, D. D. (1983). Participation of plasma membrane proteins in the formation of tight junctions by cultured epithelial cells. J. Cell Biol. 96, 693-702.

Gumbiner, B. (1993). Breaking through the tight junction barrier. J. Cell Biol. 123, 1631-1633.

Gumbiner, B., Lowenkopf, T., & Apatira, D. (1991). Identification of a 160 kDa polypeptide that binds to the tight junction protein ZO-1. Proc. Natl. Acad. Sci. USA 88, 3460-3464.

Haskins, J., Gu, L., Wittchen, E. S., Hibbard, J., & Stevenson, B. R. (1998). ZO-3, a novel member of the MAGUK protein family found at the tight junction, interacts with ZO-1 and occludin. J. Cell Biol. 141, 199-208.

Hecht, G., Koutsouris, A., Pothoulakis, C., LaMont, J. T., & Madara, J. L. (1992). Clostridium difficile toxin B disrupts the barrier function of T84 monolayers. Gastroenterol. 102, 416-423.

Hecht, G., Pestic, L., Nikcevic, G., Koutsouris, A., Tripurareni, J., Lorimer, D. D., Nowak, G., Guerriero, V., Jr., Elson, E. L., Lanerolle, P. D. (1996). Expression of the catalytic domain of myosin light chain kinase increases paracellular permeability. Am. J. Physiol. 271, C1678-C1684.

Herard, A. L., Zahm, J. M., Pierrot, D., Hinnrasky, J., Fuchey, C., & Puchelle, E. (1996). Epithelial barrier integrity during in vitro wound repair of the airway epithelium. Am. J. Respir. Cell Mol. Biol. 15, 624-632.

Hieda, Y., Iwai, K., Morita, T., & Nakanishi, Y. (1996). Mouse embryonic submandibular gland epithelium loses its tissue integrity during early branching morphogenesis. Dev. Dyn. 207, 395-403.

Hirase, T., Staddon, J. M., Saitou, M., Ando-Akatsuka, Y., Itoh, M., Furuse, M., Fujimoto, K., Tsukita, S., & Rubin, L. L. (1997). Occludin as a possible determinant of tight junction permeability in endothelial cells. J Cell Sci 110, 1603-13.

Hirase, T., Staddon, J. M., Saitou, M., Furuse, M., Tsukita, S., & Rubin, L. L. (1996). Occludin in endothelial cells. Mol. Biol. Cell 7, 605a (Abstr.).

Hirokawa, N. & Tilney, L. G. (1982). Interactions between actin filaments and between actin filaments and membranes in quick-frozen and deeply etched hair cells of the chicken ear. J. Cell Biol. 95, 249-261.

Hirsch, M. & Noske, W. (1993). The tight junction: structure and function. Micron 3, 325-352.

Hollander, D. (1988). Crohn's disease—a permeability disorder of the tight junction? Gut 29, 1621-1624.

Hoskins, R., Hajnal, A. F., Harp, S. A., & Kim, S. (1996). The C. elegans vulva induction gene lin-2 encodes a member of the MAGUK family of cell junction proteins. Development 122, 97-111.

Howarth, A. G., Hughes, M. R., & Stevenson, B. R. (1992). Detection of the tight junction-associated protein ZO-1 in astrocytes and other nonepithelial cell types. Am. J. Physiol. 262, C461-C469.

Howarth, A. G., Singer, K. L., & Stevenson, B. R. (1994). Analysis of the distribution and phosphorylation state of ZO-1 in MDCK and nonepithelial cells. J. Membr. Biol. 137, 261-270.

Howarth, A. G. & Stevenson, B. R. (1995). Molecular environment of ZO-1 in epithelial and non-epithelial cells. Cell Motil. Cytosk. 31, 323-332.

Hull, B. E. & Staehelin, L. A. (1979). The terminal web. A re-evaluation of its structure and function. J. Cell Biol. 81, 67-82.

Isaacs, K. L., Sartor, R. B., & Haskill, S. (1992). Cytokine messenger RNA profiles in inflammatory bowel disease mucosa detected by polymerase chain reaction amplification. Gastroenterology 103, 1587-1595.

Itoh, M., Nagafuchi, A., Moroi, S., & Tsukita, S. (1997). Involvement of ZO-1 in cadherin-based cell adhesion through its direct binding to alpha catenin and actin filaments. J. Cell Biol. 138, 181-92.

Itoh, M., Nagafuchi, A., Yonemura, S., Yasuda-Kitani, T., Tsukita, S., & Tsukita, S. (1993). The 220kD protein colocalizing with cadherins in non-epithelial cells is identical to ZO-1, a tight junction-associated protein in epithelial cells: cDNA cloning and immunoelectron microscopy. J. Cell Biol. 121, 491-502.

Itoh, M., Yonemura, S., Nagafuchi, A., Tsukita, S., & Tsukita, S. (1991). A 220-kD undercoat-constitutive protein: its specific localization at cadherin-based cell-cell adhesion sites. J. Cell Biol. 115, 1449-1462.

Izumi, Y., Hirose, T., Tamai, Y., Hirai, S., Nagashima, Y., Fujimoto, T., Tabuse, Y., Kemphues, K. J., & Ohno, S. (1998). An atypical PKC directly associates and colocalizes at the epithelial tight junction with ASIP, a mammalian homologue of *Caenorhabditis elegans* polarity protein PAR-3. J. Cell Biol. 143, 95-106.

Javed, Q., Fleming, T. P., Hay, M. J., & Citi, S. (1993). Tight junction protein cingulin is expressed by maternal and embryonic genomes during early mouse development. Development 117, 1145-1151.

Jesaitis, L. A. & Goodenough, D. A. (1994). Molecular characterization and tissue distribution of ZO-2, a tight junction protein homologous to ZO-1 and the *Drosophila* discs-large tumor suppressor protein. J. Cell Biol. 124, 949-961.

Kachar, B. & Reese, T. S. (1982). Evidence for the lipidic nature of tight junction strands. Nature 296, 464-466.

Kalt, M. R. (1971). The relationship between cleavage and blastocoel formation in Xenopus laevis. II. Electron microscopic observations. J. Embryol. Exp. Morphol. 26, 51-66.

Kaoutzani, P., Colgan, S. P., Cepek, K. L., Burkard, P. G., Carlson, S., Delp-Archer, C., Brenner, M. B., & Madara, J. L. (1994). Reconstitution of cultured intestinal epithelial monolayers with a mucosal-derived T lymphocyte cell line. Modulation of epithelial phenotype dependent on lymphocyte-basolateral membrane apposition. J. Clin. Invest. 94, 788-796.

Katsube, T., Takashisa, M., Ueda, R., Hashimoto, N., Kobayashi, M., & Togashi, S. (1998). Cortactin Associates with the Cell–Cell Junction Protein ZO-1 in both *Drosophila* and Mouse. J. Biol. Chem. 273, 29672-29677.

Keon, B. H., Schafer, S., Kuhn, C., Grund, C., & Franke, W. W. (1996). Symplekin, a novel type of tight junction plaque protein. J. Cell Biol. 134, 1003-1018.

Khan-Dawood, F. S., Yang, J., & Dawood, M. Y. (1996). Localization and expression of zonula occludens-1 tight junction- associated protein in baboon (Papio anubis) corpora lutea. Hum. Reprod. 11, 1262-1267.

Kim, E., Niethammer, M., Rothschild, A., Jan, Y. N., & Sheng, M. (1995). Clustering of Shaker-type K+ channels by interaction with a family of membrane-associated guanylate kinases. Nature 378* 85-88.

Kim, S. K. (1995). Tight junctions, membrane-associated guanylate kinases and cell signaling. Curr. Op. Cell Biol. 7, 641-649.

Kimura, M., Sawada, N., Kimura, H., Isomura, H., Hirata, K., & Mori, M. (1996). Comparison between the distribution of 7H6 tight junction-associated antigen and occludin during the development of chick intestine. Cell Struct. Funct. 21, 91-96.

Kimura, Y., Shiobaki, H. Hirao, M., Maeno, Y., Doki, Y., Inoue, M., Monden, T., Ando-Akatsuka, Y., Furuse M., Tsukita, S., Monden, M. (1997). Expression of occludin, tight-junction-associated protein, m human digestive tract. Am J Pathol 151, 45-54.

Koshy, S. S., Montrose, M. H., & Sears, C. L. (1996). Human intestinal epithelial cells swell and demonstrate actin rearrangement in response to the metalloprotease toxin of Bacteroides fragilis. Infect. Immun. 64, 5022-5028.

Krause, D., Mischeck, U., Galla, H. J., & Dermietzel, R. (1991). Correlation of zonula-occludens ZO-1 antigen expression and transendothelial resitance in porcine and rat cultured cerebral endothelial cells. Neurosci. Lett. 128, 301-304.

Kurihara, H., Anderson, J. M., & Farquhar, M. G. (1992). Diversity among tight junctions in rat kidney: glomerular slit diaphragms and endothelial junctions express only one isoform of the tight junction protein ZO- 1. Proc. Natl. Acad. Sci. USA 89, 7075-7079.

Kurihara, H., Anderson, J. M., & Farquhar, M. G. (1995). Increased Tyr phosphorylation of ZO-1 during modification of tight junctions between glomerular foot processes. Am. J. Physiol. 268, F514-F524.

Kuriyama, M., Harada, N., Kuroda, S., Yamamoto, T., Nakafuku, M., Iwamatsu, A., Yamamoto, D., Prasad, R., Croce, C., Canaani, E., Kaibuchi, K. (1996). Identification of AF-6 and canoe as putative targets for Ras. J. Biol. Chem. 271, 607-610.

Lane, N. J. (1992). Anatomy of the tight junction: invertebrates. In: Tight junctions (ea. M. Cereijido), pp. 23-48. CRC Press, Boca Raton, FL.

Lee, J. K., Coyne, R. S., Dubreuil, R. R., Goldstein, L. S. B., & Branton, D. (1993). Cell shape and interaction defects in a-spectrin mutants of Drosophyla melanogaster. J. Cell Biol. 123, 1797-1809.

Li, C. & Poznansky, M. J. (1990). Characterization of the ZO-1 protein in endothelial and other cell lines. J. Cell Sci. 97, 231-237.

Lora, L., Mazzon, E., Martines, D., Fries, W., Muraca, M., Martin, A., d'Odorico, A., Naccarato, R., & Citi, S. (1997). Hepatocyte tight-junctional permeability is increased in rat experimental colitis. Gastroenterology 113, 1347-54.

Lue, R. A., Brandin, E., Chan, E. P., & Branton, D. (1996). Two independent domains of hDlg are sufficient for subcellular targeting: the PDZ1-2 conformational unit and an alternatively spliced domain. J. Cell Biol. 135, 1125-1137.

Machesky, L. M. & Hall, A. (1996). Rho: a connection between membrane receptor signalling and the cytoskeleton. Trends in Cell Biol. 6, 304-310

Madara, J. (1989). Loosening tight junctions. Lessons from the intestine. J. Clin. Invest. 83, 1089-1094.

Madara, J. L. (1987). Intestinal absorptive cell tight junctions are linked to the cytoskeleton. Am. J. Physiol. 253, C171-C1, 5.

Madara, J. L. (1991). Anatomy of the tight junction: vertebrates. In: Tight junctions (ea. M. Cereijido), pp. 15-22. CRC Press, Boca Raton, FL.

Madara, J. L., Barenberg, D., & Carlson, S. (1986). Effects of cytochalasin D on occluding junctions of intestinal absorptive cells: further evidence that the cytoskeleton may influence paracellular permeability and junctional charge selectivity. J. Cell Biol. 102, 2125-2136.

Madara, J. L. & Pappenheimer, J. R. (1987). The structural basis for physiological regulation of paracellular pathways in intestinal epithelia. J. Membr. Biol. 100, 149-158.

Madara, J. L. & Stafford, J. (1989). Interferon-gamma directly affects barrier function of cultured intestinal epithelial monolayers. J. Clin. Invest. 83, 724-727.

Madison, D. L., Kruger, W. H., Kim, T., & Pfeiffer, S. E. (1996). Differential expression of rab3 isoforms in oligodendrocytes and astrocytes. J. Neurosci. Res. 45, 258-268.

Madsen, K. L., Lewis, S. A., Tavernini, M. M., Hibbard, J., & Fedorak, R. N. (1997). Interleukin 10 prevents cytokine-induced disruption of T84 monolayer barrier integrity and limits chloride secretion. Gastroenterology 113, 151-9.

Martin-Padura, I., Lostaglio, S., Schneemann, M., Williams, L., Romano, M., Fruscella, P., Panzeri, C., Stoppacciaro, A., Ruco, L., Villa, A., Simmons, D., & Dejana, E. (1998). Junctional adhesion molecule, a novel member of the immunoglobulin superfamily that distributes at intercellular junctions and modulates monocyte transmigration. J. Cell Biol. 142, 117-127.

McCarthy, K. M., Skare, I. B., Stankewich, M. C., Furuse, M., Tsukita, S., Rogers, R. A., Lynch, R. D., & Schneeberger, E. E. (1996). Occludin is a functional component of the tight junction. J. Cell Sci. 109, 2287-2298.

McRoberts, J. A. & Riley, N. E. (1992). Regulation of T84 cell monolayer permeability by insulin-like growth factors. Am. J. Physiol. 262. C207-C213.

Merzdorf, C. S. & Goodenough, D. A. (1996). Three differentially phosphorylated forms of ZO-1 during maturation and fertilization of Xenopus eggs. Mol. Boll Cell 7, 606a (Abstract).

Merzdorf, C. S. & Goodenough, D. A. (1997). Localization of a novel 210 kDa protein in Xenopus tight junctions. J Cell Sci 110, 1005-12.

Meza, I., Ibarra, G., Sabanero, M., Martinez-Palomo, A., & Cereijido, M. (1980). Occluding junctions and cytoskeletal components in a cultured transporting epithelium. J. Cell Biol. 87, 746-754.

Miragall, F., Krause, D., de Vries, U., & Dermietzel, R. (1994). Expression of the tight junction protein ZO-1 in the olfactory system: presence of ZO-1 on olfactory sensory neurons and glial cells. J. Comp. Neurol. 341, 433-448.

Moore, R., Pothoulakis, C., LaMont, J. T., Carlson, S., & Madara, J. L. (1990). C. difficile toxin A increases intestinal permeability and induces Cl- secretion. Am. J. Physiol. 259, G165-G172.

Mooseker, M. S. (1985). Organisation, chemistry and assembly of the cytoskeletal apparatus of the intestinal brush border. Annul Rev. Cell Biol. 1, 209-241.

Mori, M., Sawada, N. Isomura, H., Ezoe, E., Kokai, Y., & Baba, T. (1996). Signifiance of the 7H6 antigen to the paracellular barrier function of tight junctions-Keystone Symp. Lake Tahoe, CA, March 1996, Abs. 116.

Muller, H. A. & Hausen, P. (1995). Epithelial cell polarity in early Xenopus development. Dev. Dyn. 202, 405-420.

Mullin, J. M. & Snock, K. V. (1990). The effect of tumor necrosis factor on epithelial tight junctions and transepithelial permeability. Cancer Res. 50, 2172-2176.

Nabi, I. R., Mathews, A. P., Cohen-Gould, L., Gundersen, D., & Rodriguez-Boulan, E. (1993). Immortalization of polarized rat retinal pigment epithelium. J Cell Sci 104, 37-49.

Nelson, W. J. (1992). Regulation of cell surface polarity from bacteria to mammals. Science 258, 948-955.

Nigam, S., Denisenko, N., Rodriguez-Boulan, E., & Citi, S. (1991). The role of phosphorylation in development of tight junctions in cultured renal epithelial (MDCK) cells. Biochem. Biophys. Res. Commun. 181, 548-553.

Nusrat, A., Giry, M., Turner, J. R., Colgan, S. P., Parkos, C. A., Carmes, D., Lemichez, E., Boquet, P., & Madara, J. L. (1995). Rho protein regulates tight junctions and perijunctional actin organization in polarized epithelia. Proc. Natl. Acad. Sci. USA 92, 10629-10633.

Obiso, R. J., Jr., Azghani, A. O., & Wilkins, T. D. (1997). The Bacteroides fragilis toxin fragilysin disrupts the paracellular barrier of epithelial cells. Infect Immun 65, 1431-9.

Pelletier, R. M., Okawara, Y., Vitale, M. L., & Anderson, J. M. (1997). Differential distribution of the tight-junction-associated protein ZO-1 isoforms alpha+ and alpha- in guinea pig Sertoli cells: a possible association with F-actin and G-actin. Biol. Reprod. 57, 367-76.

Peralta Soler, A., Mullin, J. M., Knudsen, K. A., & Marano, C. W. (1996). Tissue remodeling during tumor necrosis factor-induced apoptosis in LLC- PK1 renal epithelial cells. Am. J. Physiol. 270, F869-F879.

Petrov, T., Howarth, A. G., Krukoff, T. L., & Stevenson, B. R. (1994). Distribution of the tight junction-associated protein ZO-1 in circumventricular organs of the CNS. Molec. Brain. Res. 21, 235-246.

Philpott, D. J., McKay, D. M., Sherman, P. M., & Perdue, M. H. (1996). Infection of T84 cells with enteropathogenic Escherichia cold alters barrier and transport functions. Am. J. Physiol. 270, G634-G645.

Pinto da Silva. P., & Kachar, B. (1982). On tight-junction structure. Cell 28. 441-450.

Polak-Charcon, S. (1992). Proteases and the tight junction. In: Tight Junctions (ea. M. Cereijido), pp. 257-277. CRC Press, Boca raton, FL.

Powell, D. (1981). Barrier function of epithelia. Am. J. Physiol. 241, G275-G288.

Prasad, R., Gu, Y., Alder, H., Nakamura, T., Canaani, O., Saito, H., Huebner, K., Gale, R. P., Nowell, P.C ., Kuriyama, K. et al. (1993). Cloning of the ALL-1 fusion partner the AF-6 gene, involved in acute myeloid leukemias with the t(6;11) chromosome translocation. Cancer Res. 53, 5624-5628.

Pujol, C., Eugene, E., de Saint Martin, L., & Nassif, X. (1997). Interaction of Neisseria meningitidis with a polarized monolayer of epithelial cells. Infect Immun 65, 4836-42.

Rabino, M., Denisenko, N., & Citi, S. (1993). Phosphorylation of cingulin in vitro by protein kinase C: identification of two phosphorylation sites. Mol. Biol. Cell 4, 440a.

Rajasekaran, A. K., Hojo, M., Huima, T., & Rodriguez-Boulan, E. (1996). Catenins and zonula occludens-1 form a complex during early stages in the assembly of tight junctions. J. Cell Biol. 132, 451 -463.

Regazzi, R., Ravazzola, M., Iezzi, M., Lang, J., Zahraoui, A., Andereggen, E., Morel, P., Takai, Y., & Wollheim, C. B. (1996). Expression, localization and functional role of small GTPases of the Rab3 family in insulin-secreting cells. J Cell Sci 109, 2265-2273.

Roberts, S. J., Leaf, D. S., Moore, H.-P., & Gerhart, J. C. (1992). The establishment of polarized membrane traffic in Xenopus laevis embryos. J. Cell Biol. 118, 1359-1369.

Ropke, M., Hansen, M., Carstens, S., Christensen, P., Danielsen, G., & Frederiksen, O. (1996). Effects of a short-chain phospholipid on ion transport pathways in rabbit nasal airway epithelium. Am. J. Physiol. 271, L646-1 655.

Royall, J. A., Berkow, R. L., Beckman, J. S., Cunningham, M. K., Matalon, S., & Freeman, B. A. (1989). Tumor necrosis factor and interleukin 1-alpha increase vascular endothelial permeability. Am. J. Physiol. 257, L399-L410.

Saitou, M., Ando-Akatsuka, Y., Itoh, M., Furuse, M., Inazawa, J., Fujimoto, K., & Tsukita, S. (1997). Mammalian occludin in epithelial cells: its expression and subcellular distribution. Eur J. Cell Biol. 73, 222-31.

Saitou, M., Fujimoto, K., Doi, Y., Itoh, M., Fujimoto, T., Furuse, M., Takano, H., Noda, T., & Tsukita, S. (1998). Occludin-deficient embryonic stem cells can differentiate into polarized epithelial cells bearing tight junctions. J. Cell Biol., 141, 397-408.

Sakakibara, A., Furuse, M., Saitou, M., Ando-Akatsuka, Y., & Tsukita, S. (1997). Possible involvement of phosphorylation of occludin in tight junction formation. J. Cell Biol. 137, 1393-401.

Satoh, H., Zhong, Y., Isomura, H., Saitoh, M., Enomoto, K., Sawada, N., & Mori, M. (1996). Localization of 7H6 tight junction-associated antigen along the cell border of vascular endothelial cells correlates with paracellular barrier function against ions, large molecules, and cancer cells. Exp. Cell Res. 222, 269-274.

Schnabel, E., Anderson, J. M., & Farquhar, M. G. (1990). The tight junction protein ZO-1 is concentrated along slit diaphragms of the glomerular epithelium. J. Cell Biol. 111, 12551263.

Schneeberger, E. (1994). Tight junctions: their modulation under physiological and pathological states. In: Molecular mechanisms of epithelial cell junctions: from development to disease (ea. S. Citi), pp. 123-140. RG Landes Biomedical Publishers, Austin, TX.

Schneeberger, E. E. & Lynch, R. D. (1992). Structure, function and regulation of cellular tight junctions. Am. J. Physiol. 262, L647-L661.

Schneeberger, E. E., Lynch, R. D., Kelly, C. A., & Rabito, C. A. (1988). Modulation of tight junction formation in clone 4 MDCK cells by fatty acid supplementation. Am. J. Physiol. 254, C432-C440.

Sheth, B., Fesenko, I., Collins, J. E., Moran, B., Wild, A. E., Anderson, J. M., & Fleming, T. P. (1997). Tight junction assembly during mouse blastocyst formation is regulated by late expression of ZO-1 alpha+ isoform. Development 124, 2027-37.

Siliciano, J. D., & Goodenough, D. A. (1988). Localization of the tight junction protein, ZO-1, is modulated by extracellular calcium and cell-cell contact in Madin-Darby canine kidney epithelial cells. JCB 107, 2389-2399.

Singer, K. L., Stevenson, B. R., Woo, P. L., & Firestone, G. L. (1994). Relationship of serine/threonine phosphorylation/dephosphorylation signaling to glucocorticoid regulation of tight junction permeability and ZO-1 distribution in nontransformed mammary epithelial cells. J. Biol. Chem. 269, 16108-16115.

Slack, C. & Warner, A. E. (1973). Intracellular and intercellular potentials in early amphibian embryo. J. Physiol. 232, 313-330.

Staddon, J. M., Herrenknecht, K., Smales, C., & Rubin, L. L. (1995). Evidence that tyrosine phosphorylation may increase tight junction permeability. J. Cell Sci. 108, 609619.

Staehelin, L. A., Mukherjee, T. M., & Williams, A. W. (1969). Freeze-etch appearance of the tight junctions in the epithelium of small and large intestine of mice. Protoplasma 67, 165-184.

Stankewich, M. C., Francis, S. A., Vu, Q. U., Schneeberger, E. E., & Lynch, R. D. (1996). Alterations in cell cholesterol content modulate Ca(2+)-induced tight junction assembly by MDCK cells. Lipids 31, 817-828.

Stettler, O., Nothias, F., Tavitian, B., & Vernier, P. (1995). Double in situ hybridization reveals overlapping neuronal populations expressing the low molecular weight GTPases Rab3a and Rab3b in Rat brain. Eur J Neurosci 7, 702-713.

Stevenson, B. R., Anderson, J. M., Braun, I. D., & Mooseker, M. S. (1989). Phosphorylation of the tight junction protein ZO-1 in two strains of Madin-Darby canine kidney cells which differ in transepithelial resistance. Biochem. J. 263. 597-599.

Stevenson, B. R. & Begg, D. A. (1994). Concentration-dependent effects of cytochalasin D on tight junctions and actin filaments in MDCK epithelial cells. J. Cell. Sci. 107, 367-375.

Stevenson, B. R., Haskins, J., Hibbard, J., Tamber, M., & Weber, D. (1996). P130 is homologous to ZO-1 and ZO-2 and is a novel member of the MAGUK family of proteins. Mol. Biol. Cell 7, 605a (Abstr.).

Stevenson, B. R., Heintzelman, M. B., Anderson, J. M., Citi, S., & Mooseker, M. S. (1989). ZO-1 and cingulin: tight junction proteins with distinct identities and localizations. Am. J. Physiol. 257, C621-C628.

Stevenson, B. R., Siliciano, J. D., Mooseker, M. S., & Goodenough, D. A. (1986). Identification of ZO-1: a high molecular weight polypeptide associated with the tight junction (zonula occludens) in a variety of epithelia. J. Cell Biol. 103, 755-766.

Stuart, R. O., & Nigam, S. K. (1995). Regulated assembly of tight junctions by protein kinase C. Proc. Natl. Acad. Sci. U S A 92, 6072-6076.

Stuart, R. O., Sun, A., Bush, K. T., & Nigam, S. K. (1996). Dependence of epithelial intercellular junction biogenesis on thapsigargin-sensitive intracellular calcium stores. J. Biol. Chem. 271, 13636-13641.

Sugrue, S. P. & Zieske, J. D. (1997). ZO-1 in corneal epithelium: association to the zonula occludens and adherens junctions. Exp. Eye. Res. 64, 11-20.

Takahisa, M., Togashi, S., Suzuki, T., Kobayashi, M., Murayama, A., Kondo, K., Miyake, T., & Ueda, R. (1996). The *Drosophila* tamou gene, a component of the activating pathways of extramacrochaetae expression, encodes a protein homologous to mammalian cell-cell junction-associated protein ZO-1. Genes Dev. 10, 1783-1795.

Takeda, H. & Tsukita, S. (1995). Effects of tyrosine phosphorylation on tight junctions in temperature-sensitive v-src-transfected MDCK cells. Cell Struct. Funct. 20,

Thiery, J. P. & Boyer, B. (1992). The junction between cytokines and cell adhesion. Curr. Op. Cell Biol. 4, 782-792

Tobioka, H., Sawada, N., Zhong, Y., & Mori, M. (1996). Enhanced paracellular barrier function of rat mesothelial cells partially protects against cancer cell penetration. Br. J. Cancer 74, 439-445.

Tsukamoto, T. & Nigam, S. K. (1997). Tight junction proteins form large complexes and associate with the cytoskeleton in an ATP depletion model for reversible junction assembly. J. Biol. Chem. 272, 16133-9.

Turner, J. R., Rill, B. K., Carlson, S. L., Carnes, D., Kerner, R., Mrsny, R. J., & Madara, J. L. (1997). Physiological regulation of epithelial tight junctions is associated with myosin light-chain phosphorylation [In Process Citation]. Am J Physiol 273. C137885.

van Deurs, K. G. & Luft, J. H. (1979). Effects of glutaraldehyde fixation on the structure of tight junctions. A quantitative freeze-fracture analysis. J. Ultrastruct. Res. 68, 160-172.

Van Itallie, C. M. & Anderson, J. M. (1997). Occludin confers adhesiveness when expressed in fibroblasts. J Cell Sci 110, 1113-1121.

Van Itallie, C. M., Balda, M. S., & Anderson, J. M. (1995). Epidermal growth factor induces tyrosine phosphorylation and reorganization of the tight junction protein ZO-1 in A431 cells. J. Cell Sci. 108, 1735-1742.

Volberg, T., Geiger, B., Citi, S., & Bershadsky, A. D. (1994). The effect of protein kinase inhibitor H-7 on the contractility, integrity and membrane anchorage of the microfilament system. Cell Motil. Cytosk. 29, 321-338.

Wang, Y., Chen, M., & Wolosin, J. M. (1993). ZO-1 in corneal epithelium; stratal distribution and synthesis induction by outer cell removal. Exp. Eye. Res. 57, 283-92.

Watabe, M., Nagafuchi, A., Tsukita, S., & Takeichi, M. (1994). Induction of polarized cell-cell association and retardation of growth by activation of the E-cadherin-catenin adhesion system in a dispersed carcinoma line. J. Cell Biol. 127, 247-56.

Weber, A. M., Tuchweber, B., Yousef, I., Brochu, P., Turgeon, C., Gabbiani, G., Morin, C. L., & Roy, C. C. (1981). Severe familial cholestasis in North American Indian children: a clinical model of microfilament dysfunction? Gastroenterology 81, 653-62.

Weber, E., Berta, G., Tousson, A., St. John, P., Green, M. W., Gopalokrishnan, V., Jilling, T., Sorscher, E. S., Elton, T. S., Abrahamson, D. R., & Kirk, K. L. (1994). Expression and polarized targeting of a Rab3 isoform in epithelial cells. J. Cell Biol. 125, 583-594.

Willott, E., Balda, M. S., Fanning, A. S., Jameson, B., Van Itallie, C., & Anderson, J. M. (1993). The tight junction protein ZO-1 is homologous to the *Drosophila* discs-large tumor suppressor protein of septate junctions. Proc. Natl. Acad. Sci. USA 90, 7834-7838.

Willott, E., Balda, M. S., Heintzelman, M., Jameson, B., & Anderson, J. M. (1992). Localization and differential expression of two isoforms of the tight junction protein ZO-1. Am. J. Physiol. 262, C1119-C1124.

Wong, V. & Gumbiner, B. M. (1997). A synthetic peptide corresponding to the extracellular domain of occludin perturbs the tight junction permeability barrier. J. Cell Biol. 136, 399-409.

Wong, V. (1997). Phosphorylation of occludin correlates with occludin localization and function at the tight junction. Am. J. Physiol. 273, C1859-C1867.

Woo, P. L., Cha, H. H., Singer, K. L., & Firestone, G. L. (1996). Antagonistic regulation of tight junction dynamics by glucocorticoids and transforming growth factor-beta in mouse mammary epithelial cells. J. Biol. Chem. 271, 404-412.

Woods, D. F. & Bryant, P. J. (1991). The discs-large tumor suppressor gene of Drosophyla encodes a guanylate kinase homolog localized at septate junctions. Cell 66, 451-464.

Woods, D. F. & Bryant, P. J. (1993). ZO-1, DlgA and PSD95/SAP90: homologous proteins in tight, septate and synaptic cell junctions. Mech. Dev. 44, 85-89.

Wu, Z., Milton, D., Nybom, P., Sjo, A., & Magnusson, K. E. (1996). Vibrio cholerae hemagglutinin/protease (HA/protease) causes morphological changes in cultured epithelial cells and perturbs their paracellular barrier function. Microb. Pathog. 21, 111-23.

Yamamoto, T., Harada, N., Kano, K., Taya, S., Canaani, E., Matsuura, Y., Mizoguchi, A., Ide, C., & Kaibuchi, K. (1997). The Ras Target AF-6 Interacts with ZO-1 and Serves as a Peripheral Component of Tight Junctions in Epithelial Cells. J. Cell Biol. 139, 785-95.

Yap, A. S., Stevenson, B. R., Armstrong, J. W., Keast, J. R., & Manley, S. W. (1994) Thyroid epithelial morphogenesis in vitro: a role for bumetanide- sensitive C1- secretion during follicular lumen development. Exp. Cell Res. 213, 319-26.

Yonemura, S., Itoh, M., Nagafuchi, A., & Tsukita, S. (1995). Cell-to-cell adherens junction formation and actin filament organization: similarities and differences between non-polarized fibroblasts and polarized epithelial cells. J Cell Sci 108, 127-42.

Zahraoui, A., Joberty, G., Arpin, M., Fontaine, J. J., Hellio, R., Tavitian, A., & Louvard, D. (1994). A small rate GTPase is distributed in cytoplasmic vesicles in non polarized cells but colocalizes with the tight junction marker ZO-1 in polarized epithelial cells. J. Cell Biol. 124, 101-115

Zettl, K. S., Sjaastad, M. D., Riskin, P. M., Parry, G., Machen, T. E., & Firestone, G L. (1992). Glucocorticoid-induced formation of tight junctions in mouse mammary epithelial cells in vitro. Proc. Natl. Acad. Sci. USA 89, 9069-9073.

Zhong, Y., Saitoh, T., Minase, T., Sawada, N., Enomoto, K., & Mori, M. (1993). Monoclonal antibody 7H6 reacts with a novel tight junction-associated protein distinct from ZO-1. cingulin and ZO-2. J. Cell Biol. 120. 477-483.

Zinkl, G. M., Zuk, A., van der Bijl, P., van Meer, G., & Matlin, K. S. (1996). An antiglycolipid antibody inhibits Madin-Darby canine kidney cell adhesion to laminin and interferes with basolateral polarization and tight junction formation. J. Cell Biol. 133, 695-708.

PART III

SIGNALING BY ADHESION MOLECULES

ACTIVATION OF INTEGRIN SIGNALING PATHWAYS BY CELL INTERACTIONS WITH EXTRACELLULAR MATRIX

Gwynneth M. Edwards and Charles H. Streuli

Advances in Molecular and Cell Biology
Volume 28, pages 237-268.

ISBN: 0-7623-0495-2

I. INTRODUCTION

Integrins are an important family of cell surface receptors that mediate adhesive interactions between cells and the extracellular matrix (ECM). These transmembrane glycoproteins are composed of non-covalently associated α- and β-subunit heterodimers (Hynes, 1992). In mammals, there are currently 16 different α-chains and 8 β-subunits known which can combine to give more than 22 distinct receptors and integrins, which are presently classified into subgroups depending on the β subunit type (Gianocotti and Mainiero, 1994). Integrins containing either β1, β3, β4, or αv subunits primarily mediate cell adhesion to the extracellular matrix (ECM) and at least 12 separate matrix ligands for these receptors have been identified. The other β subclasses not involving the αv subunit (β2 and β7) are mainly involved in cell–cell contacts via interactions with members of the immunoglobulin and cadherin families (Cepek et al., 1994).

While cellular adhesion to specific components of the ECM is an important function of integrins, signals initiated by ligation of these receptors are critically involved in spreading, migration, proliferation, differentiation, and survival of cells (reviewed in Hynes, 1992; Juliano and Haskell, 1993; Ruoslahti and Reed, 1994; Clarke and Brugge, 1995; Rosales et al., 1995; Richardson and Parsons, 1995; Schwartz et al., 1995; Gumbiner, 1996; Gianocotti, 1997). Furthermore, integrins can regulate the expression of several genes including those for immediate-early genes, collagenase, stromolysin, cytokines, milk proteins, and integrins themselves (Werb et al., 1989; Sporn et al., 1990; Streuli et al., 1991; Yurochko et al., 1992; Miyake et al., 1993; Lin et al., 1994; Lin et al., 1995; Delcommenne and Streuli, 1995; Fan et al., 1995; reviewed in Mondal et al., 1995; Lafrenie and Yamada, 1996).

Since the discovery of integrins over a decade ago, the fundamental importance of these receptors in regulation of cell phenotype has been increasingly appreciated. Much effort has been directed towards understanding how these receptor molecules mediate transmission of signals from the extracellular matrix to influence diverse cell processes. The aim of this chapter is to describe the different signaling pathways that can be activated by ligation of integrins. In particular, the proximal events that lead to alterations in cytoskeletal architecture and to triggering conventional signaling pathways will be discussed. There is increasing evidence that integrin receptors play an integral function in the coordination of signals from the ECM with those initiated by soluble growth and differentiation factors. A model emerges whereby integrin signaling is mediated by a multiprotein signaling complex containing integrins, cytoskeletal proteins, and components of traditional signaling cassettes. Other aspects of integrin signaling, such as the signals that control the affinity of integrin for ECM ligand (inside-out signaling), are covered elsewhere in this volume (see chapter by Du and Ginsberg).

II. DIVERSITY AND REDUNDANCY OF INTEGRIN-ECM INTERACTIONS

Paradoxically, although integrins influence diverse effects on cell processes, these receptors appear to have a high degree of redundancy since an individual integrin can bind several different ECM molecules while several different integrins may recognize the same ECM component. For example, $\alpha v\beta 3$ is one of the most promiscuous of all integrin complexes in terms of its ability to recognize a wide variety of Arg-Gly-Asp- (RDG-) containing extracellular components. On the other hand, the ECM consituent, laminin, is recognized by $\alpha 1\beta 1$, $\alpha 2\beta 1$, $\alpha 3\beta 1$, $\alpha 6\beta 51$, $\alpha 7\beta 1$, and $\alpha 6\beta 4$ integrins. Most mammalian cells constitutively express multiple integrins and many of these may be receptors for the same ligand. Furthermore, cells with divergent functions express the same integrins implying that individual integrins influence distinct cellular responses in different cells. Indeed, gene knockout experiments have demonstrated that integrins with overlapping ligand binding specificities have specific functions in development (Hynes, 1996). Moreover, the composition of the ECM and the type of cell determines whether the cell is induced to differentiate or proliferate (Streuli et al., 1991; Adams and Watt, 1993) or whether the apoptosis program is suppressed (Pullan et al., 1996), indicating that integrins are capable of transmitting different signals.

In adult tissues, epithelial cells are normally stationary and contact each other via adhesive molecules such as cadherins, and contact the ECM through integrins. These interactions generate signals which are crucial for monitoring local changes in cell attachment and detachment, and detecting alterations in the composition of the extracellular matrix. Thus, as in other cell types, epithelial integrins are intimately involved with the processes of proliferation, differentiation and cell survival. Integrin interactions with the ECM are also essential for several specialized epithelial functions including differentiation into secretory cells, establishment of polarity, wound healing, and responses to inflammation. Although specialized epithelia may express a unique pattern of integrins, the basic repertoire of epithelial integrins are typically similar. At least five different integrins are constituitively expressed in both lung airway epithelia and in the skin (reviewed in Sheppard, 1996) including $\alpha 2\beta 1$, $\alpha 3\beta 1$, $\alpha 6\beta 4$, $\alpha 9\beta 1$, and $\alpha v\beta 5$ integrins. In mammary luminal epithelial cells, αv, $\alpha 2$, $\alpha 3$, $\alpha 6$, and $\beta 1$ integrin subunits are expressed, whereas in mammary myoepithelial cells $\alpha 3$, $\alpha 6$, $\beta 1$, and $\beta 4$ integrins are present (Oliver, Klinowska, and Streuli, unpublished). The location of each of these integrins within the cell differs, however. For example, the hemidesmosomal integrin, $\alpha 6\beta 4$, is almost exclusively localized to the basal cell surface consistent with its role of adhesion to laminin in the basement membrane. The laminin/collagen receptor, α3β1 integrin, although concentrated on the basal surface of these cells, is also expressed diffusely on the lateral and apicial surfaces. α2β1, a collagen receptor, is expressed around the cell, which suggests a possible role in cell-adhesion. The integrins expressed by epithelial cells are also present in many other cell types and thus signaling events initiated by individual integrins are likely to be influenced in a cell-specific manner.

III. INTEGRIN EFFECTS ON THE CYTOSKELETON

Adhesive interactions between cells and the ECM results in the co-localization of actin binding proteins with integrins and in the formation of actin microfilaments which anchor the cytoskeleton to the plasma membrane. In tissue culture, ligation of integrin leads to the organization of discreet focal contacts. These structures contain a number of cytoskeletal proteins and signaling molecules at the inner surface of the plasma membrane and serve to link the extracellular matrix to the ends of actin filament stress fibers through integrins (Burridge et al., 1989).

Integrin subunits have large extracellular domains, a single transmembrane domain but the cytoplasmic domains of both α- and β-integrin subunits are short (with the exception of the β4 and subunit) and do not possess intrinsic enzymatic activity. Nevertheless, integrins not only regulate the adhesion and spreading of cells on ECM substrata, but ligation of these receptors also generates a variety of signals conveying information from the exterior of the cell to the nucleus. It is the β integrin cytoplasmic domain that is primarily involved in interactions with the cytoskeleton through actin binding proteins and this domain also functions to couple integrins to signaling proteins such as focal adhesion kinase (FAK), Src-family kineses and members of the Ras signaling pathway. Given the promiscuity of the β1 integrin subunit with many different α subunits, the α subunit serves to provide specificity in integrin signaling. For example, only certain α subunits couple integrin signaling to the control of cell cycle progression via the adapter protein Shc (Wary et al., 1996).

A. Integration of β1 Integrin into Focal Contacts

All the information necessary to localize the β1 integrin receptor to focal contacts is contained within its 47 amino acid cytoplasmic domain (see Figure 1). Three clusters of amino acids within this domain have been identified which contribute to focal contact localization. These are known as the cyto-1, cyto-2, and cyto-3 motifs (Reszka et al., 1992). The cyto-1 motif is an 11 amino acid sequence found in all β subunits except for β4 and β8. The cyto-2 and cyto-3 motifs are comprised of 4 amino acids (NPXY where N, neutral amino acid; P, proline; X, any

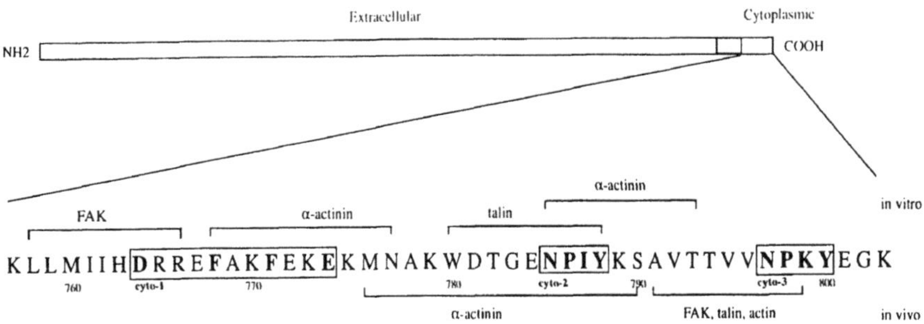

Figure 1. Binding domains within the β1 integrin cytoplasmic tail. The sequence HDRREFAKFEKEK is a highly conserved membrane-proximal region in the cytoplasmic domain of β integrins. The residues outlined in boxes are involved with the localization of β1 integrin to focal contacts (residues 764–774 is cyto-1; 785–788, cyto-2; and 797–800, cyto-3; see text). D764, F768, F771, and E774 in cyto-1 are likely to form a signal occupying one side of an α helix and cannot be changed without reducing localization. NPIY and NPKY form tight turn motifs that when perturbed by adding or removing proline residues or mutating the tyrosine residue reduces integrin localization to focal contacts. In vitro studies indicate that residues 769–777 and 785–794 are important for α-actinin binding and that residues 780–789 bind talin. Transfection studies using mutated β1 integrin cytoplasmic domains have delineated a region between 776–790, which is important for α-actinin binding and region between 791–799, which is responsible for colocalization of FAK, talin, and actin to focal contacts in vivo.

amino acid; and Y, tyrosine). The NPXY motif is highly conserved among different β subunits being present in β1, β2, β3, β5, β6, and β7 cytoplasmic domains whereas cyto-2 is only found in the β1, β2, and β7 subunits. This motif forms a tight turn structure and often acts as an internalization signal in membrane receptors although this does not appear to be its function in integrins (Vignoud et al., 1994). In the β3 chain, the NPXY motif is involved in cell migration and metastasis (Filardo et al., 1995) and controls integrin binding affinity (O'Toole et al., 1995). Focal adhesion assembly strictly depends on talin (Nuckolls et al., 1992; Albiges-Rizo et al., 1995) and the NPXY motifs in the β1 integrin cytoplasmic domain which are critical for localizing integrin to focal contacts also regulate talin binding to integrin although they are distinct from the talin binding sites (Vignoud et al., 1997). In addition, tyrosine phosphorylation within the NPXY motif may regulate β1 integrin assembly into focal contacts (Reszka et al., 1992).

β integrin subunits retain a high degree of conservation between species for each subtype (DeSimone and Hynes, 1988). Furthermore, the β subunits that normally localize to focal adhesions have the conserved focal contact localization motifs in their cytoplasmic domains whereas the β subunits, which do not normally associate with focal contacts (β4, β5, βlb, and β2551c), lack these sequences. The β integrin cytoplasmic domain is also essential for cell spreading. C-terminal deletion mutants of β1 integrin block spreading of cells on ECM (Hayashi et al., 1990) and on antibodies against the extracellular domain of β1 (Guan et al., 1991) whereas α subunit C-terminal deletion mutants have no effect on spreading (Bauer et al., 1993, Ylanne et al., 1993).

B. Interaction between Integrins and the Actin Cytoskeleton

The physical association between cytoskeletal actin filaments and the cytoplasmic domain of β integrins is indirect and involves a number of intermediary proteins (Pavalko and Otey, 1994). The cytoskeletal proteins α- actinin, talin, vinculin, and tensin all colocalize with integrins in focal contacts and although the in vitro binding arrangements between these proteins and the cytoplasmic tail of the integrin β subunit have been investigated, their precise interactions in focal contacts are not fully understood (reviewed in Meredith et al., 1996; see Figure 1). For example, in vitro studies have demonstrated that talin interacts with the β subunit cytoplasmic domain at residues 780–789 (Tapley et al., 1989) whereas in vivo studies point to a requirement for integrin residues 791–799 (Lewis and Schwartz, 1995). Similarly, α-actinin binds integrin cytoplasmic domain residues 768–777 and 785–794 in vitro (Otey et al., 1993) and residues 776–790 in vivo (Lewis and Schwartz, 1995). Despite the colocalization of α-actinin with a mutant β1 integrin, which lacks residues 791–799, this complex did not associate with actin suggesting that the ability of α-actinin to bind actin may be regulated by other focal adhesion components (Lewis and Schwartz, 1995). The interaction between α-actinin and actin is sensitive to the levels of phosphatidylinositol 4,5 bisphosphate (PIP_2) (Lassing and Lindberg, 1985). This lipid is an important precursor for second messenger production whose synthesis is regulated by integrin-mediated adhesion (McNamee et al., 1993; see below) and therefore residues 791-799 may be crucial to regulating integrin signaling.

There are several potential interactions between the proteins α-actinin, talin, vinculin, and tensin that link integrins to the actin microfilaments (see Figure 2). Talin can bind directly to actin but also associates with actin through a vinculin α-actinin link (Belkin and Koteliansky, 1987) as well as through a vinculin-tensin link (Wachsstock et al., 1987). Taken together, these studies highlight the complexities of the protein–protein interactions involving the β integrin subunit cytoplasmic domain and the alternative linkages which may be involved in focal contact formation

C. The Role of Small GTPases in Integrin–Cytoskeletal Interactions

The Rho family of GTPases, Rho, Rac, and Cdc42 play a role in cytoskeleton reorganisation by regulating the polymerisation of actin to form stress fibers, lamellipodia, and filopodia, respectively (Ridley and Hall, 1992; Ridley et al., 1992;

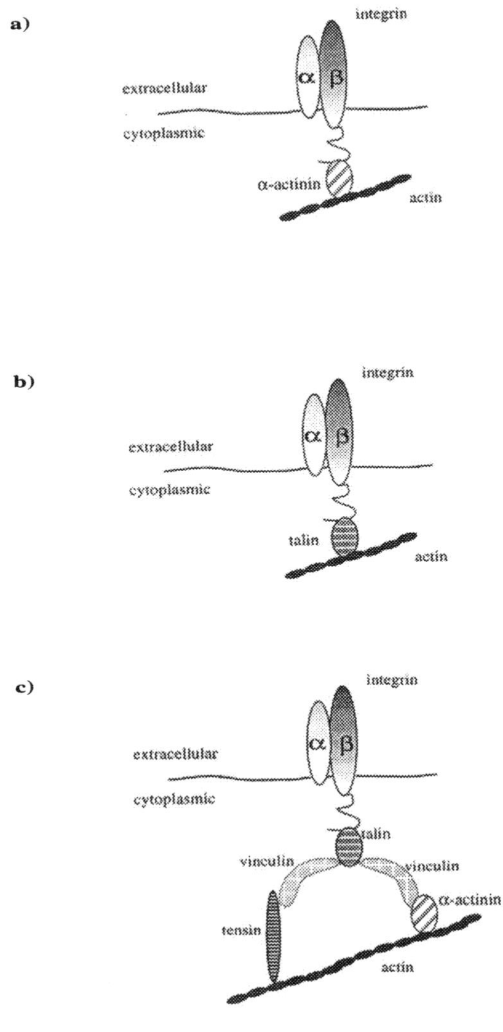

Figure 2. Potential interactions liking integrins to the cytoskeleton. In Vitro studies suggest the potential for several different protein arrangements within the focal contact (**a**) α-actinin binds directly to both the β-subunit of integrin and the F-actin. (**b**) Talin binds directly to both the β-subunit of integrin and the F-actin. (**c**) Talin binds directly to the β-subunit of integrin and to vinculin. Vinculin can bind to tensin or to α-actin, which both in turn bind F-actin.

Kozma et al., 1995; Nobes and Hall, 1995; see Figure 3). The dynamic assembly and disassembly of these GTPase-mediated morphological changes occurs in response to growth factors and requires integrin ligation (McNamee et al., 1993; Chong et al., 1994). The Rho family proteins also activate members of the MAPK

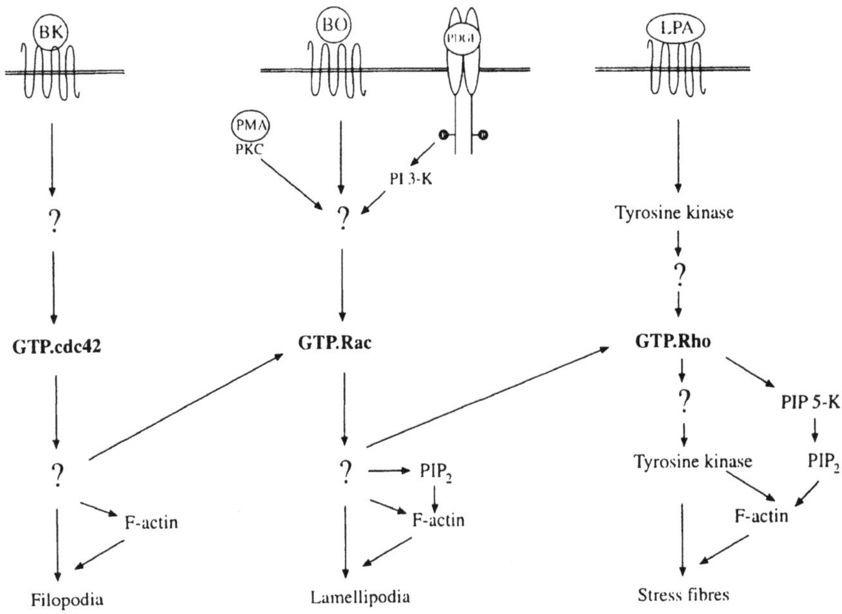

Figure 3. The relationship between the GTPases, cdc42, Rac, and Rho. Rho family GTPases regulate the actin cytoskeleton and influence gene expression by interacting with multiple effectors. In addition to inducing specific changes, the Rho family share some functions and there is cross-talk between family members such that each member can induce focal contact formation and a given agonist can activate more than one member of the family. Thus cdc42 can activate Rac, and Rac can activate Rho, at least in Swiss 3T3 fibroblasts. The extracellular ligands in the diagram are bradykinin (BK), bombesin (BO), platelet-derived growth factor (PDGF), phorbol myristic acetate (PMA), and lysophosphatidic acid (LPA). The receptors for the extracellular ligands are serpentine receptors linked to heterotrimeric G. proteins, tyrosine kinase receptors, or protein kinase C (PKC). The kinases are PIP5-K, PI 3-K (identified by sensitivity to wortmannin), PKC (the target of PMA), and tyrosine kinase (identified by sensitivity or tyrphostin or genistein).

family which are involved with the regulation of immediate early gene expression (Minden et al., 1995; reviewed in Lim et al., 1996). Cdc42 and Rac have been implicated in the activation of the Jnk pathway whereas Rho activates the serum response element (SRE). In addition, Rho has been shown to be involved in integrin-dependent activation of MAPK (Renshaw et al., 1996).

Rho singularly appears to play a key role in integrating the signals induced by integrins and growth factors by regulating the assembly of actin stress fibers and focal contacts. Starving confluent Swiss 3T3 fibroblasts of serum leads to the loss of these structures, although the cells still remain attached to the ECM. Stress fibers and focal contacts rapidly reform (2–5 min) following treatment with serum or lysophosphatidic acid (LPA), the active component of serum (Barry and Critchley, 1994; Chrzanwska-Wodnicka and Burridge, 1994; Ridley and Hall, 1994). Microinjection of activated GTP-Rho also rapidly induces stress fiber and focal contact formation (Ridley and Hall, 1992) whereas injection with C3 exoenzyme, a bacterial toxin which inactivates Rho through ADP-ribosylation, results in the same phenotype as serum-starved cells indicating that LPA exerts its effect on focal contact formation by activating Rho (Nobes and Hall, 1995). Moreover, C3 inhibition of Rho blocks adhesion-dependent activation of FAK (Nobes and Hall, 1995; Chrzanowska-Wodnicka and Burridge, 1994) and scrape loading activated Rho into Swiss 3T3 cells stimulates phosphorylation of FAK, $p130^{cas}$ as well as paxillin (Flinn and Ridley, 1996). These results suggest that Rho is required for FAK activity and that Rho activation is a proximal event at least in those integrin mediated signals which control cytoskeletal organization.

Growth factors such as platelet-derived growth factor (PDGF), epidermal growth factor (EGF), insulin, bombesin, and bradykinin can also stimulate stress fiber and focal contact assembly by activating Rho. This process is slow (15-30 minutes) in comparison with the rapid changes in cytoskeleton induced by LPA and occurs through a separate pathway involving the sequential activation of Cdc42, Rac then Rho (Nobes and Hall, 1995; reviewed in Craig and Johnson, 1996; Zigmond, 1996; see Figure 3). Integrin ligation alone, however, is sufficient to activate some of the pathways regulated by Rho and Rac. As mentioned above, cell adhesion stimulates the production of PIP_2 and this is now known to occur by enhancing phosphatidylinositol 4-phosphate 5-kinase (PIP 5-K) activity through a physical complex between activated GTP-Rho and PIP 5-K (Chong et al., 1994; Ren et al., 1996). Although PIP_2 is involved in protein kinase C (PKC) and protein kinase B (PKB) signaling (see below), it is also involved directly with stress fiber and focal contact formation (see Figure 4). PIP_2 induces the dissociation of actin monomers from gelsolin and profilin complexes (Schafer and Cooper, 1995) making them available for polymerisation into actin filaments and also binds to vinculin stimulating its association with talin and actin (Gilmore and Burridge, 1996). In addition, Rho regulates the activity of myosin light chain (MLC) phosphatase (reviewed in Tapon and Hall, 1997). Activated GTP-bound Rho binds to and activates Rho kinase (Rho-K) (Ishizaki et al., 1996; Matsui et al., 1996) leading to the inactivation of MLC phosphatase through phosphorylation of its myosin-binding subunit (Kimura et al., 1996). The resulting net increase in MLC phosphorylation leads to the formation of actin-myosin bundles into stress fibers. Confirmation of this hypothesis comes from artificial inhibition of MLC-kinase which decreases MLC phosphorylation and results in the loss of stress fibers and focal contacts and in the inhibition of Rho-dependent contraction of fibroblasts (Chrzanowska-Wodnicka and Burridge, 1996). The effect of Rho on stress fiber formation together with its effects on actin, vinculin, and talin implicates this GTPase as a key regulator of focal contact assembly.

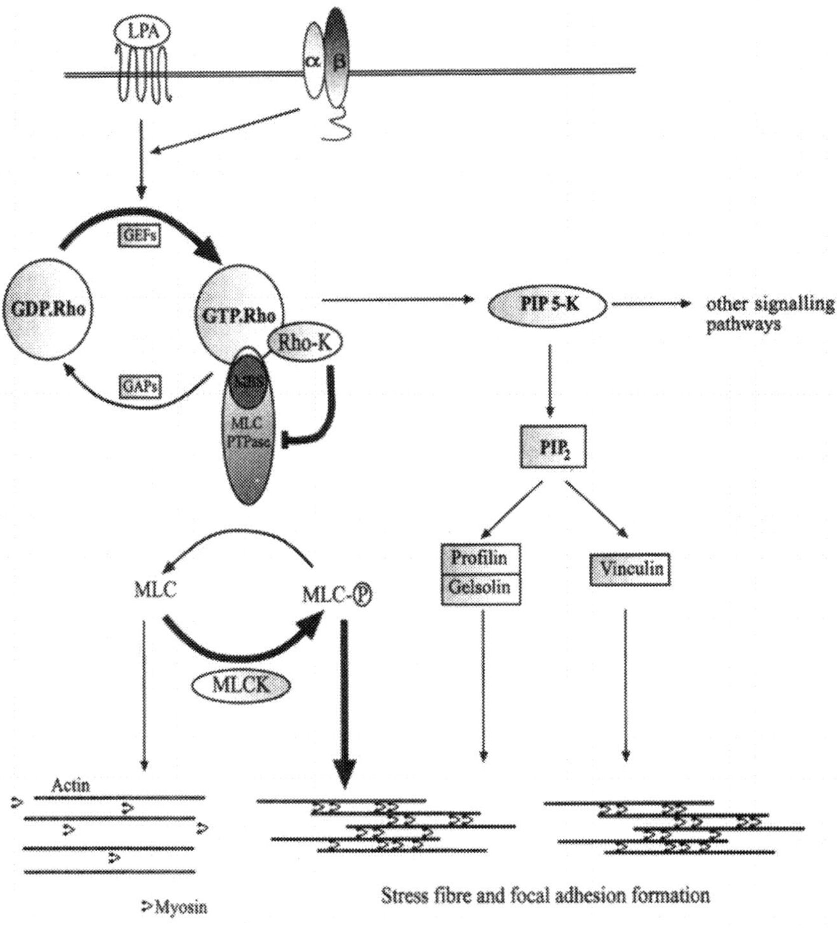

Figure 4. The central role of Rho in formation of stress fibers and focal contact formation. Following stimulation by extracellular ligands (e.g., LPA and ECM), Rho switches to its active GTP-bound state and binds to Rho-kinase (Rho-K) and to the myosin-binding subunit (MBS) of mysosin light-chain phosphatase (MLC PTPase). The activity of Rho-K is stimulated by GTP-Rho leading to the phosphorylation of MBS and inactivation of MLC PTPase. Since myosin light chain (MLC) would no longer be dephosphorylated under these conditions, the action of MLC kinase (MLCK) leads to elevated phosphorylated MLC and its subsequent incorporation with actin filaments into stress fibers. Active GTP-Rho also results in an increase in the levels of PIP_2 through activation of PIP 5-K. PIP_2 can induce the release of actin monomer from profilin and gelsolin and additionally binds to vinculin promoting its interaction with both talin and actin, events that are likely to participate in the assembly of stress fibers into focal contacts.

IV. SIGNALING EVENTS MEDIATED BY INTEGRINS

The clustering of integrins and their associated cytoplasmic proteins serves to link specific components of the ECM to the plasma membrane of the cell. The actin cytoskeleton is required not only for the formation of focal adhesions but also for their maintainance and stability since cytoskeletal-disrupting drugs such as cytochalasin rapidly disassemble these structures (Flinn and Ridley, 1996). The affinity of integrins for ECM ligands in cell adhesion enables a cell to remain permanently attached to the matrix. A striking illustration of this in vivo is the hemidesmosome which functions to resist shear forces in the skin by linking large electron-dense cytoplasmic plaques to the underlying basement membrane through α6β4 integrin. Clearly the adhesive aspect of cell-matrix interactions is an important integrin function. However, diverse intracellular signaling events occur when integrins are ligated (see Figure 5).

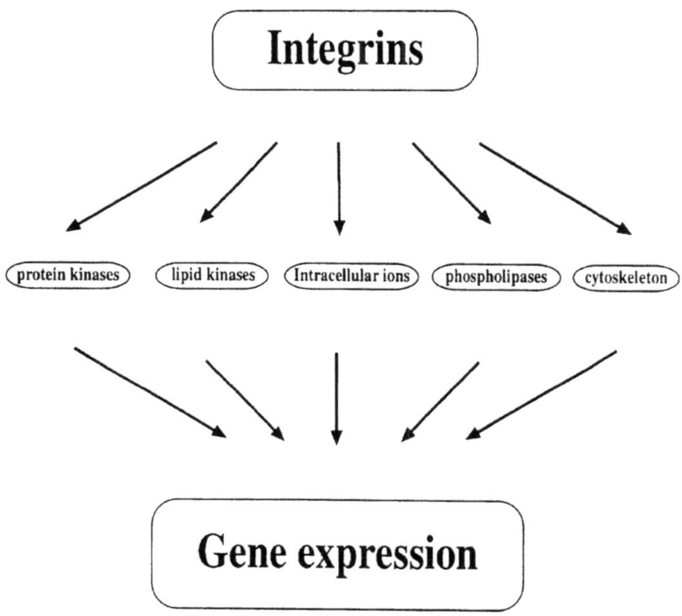

Figure 5. Integrin-mediated signaling. A variety of signaling events that ultimately lead to altered gene expression are mediated by integrins. Integrins are involved in regulating the levels of H^+ and Ca^{2+} within the cell. Integrin action can also lead to the activation of several enzymes including protein kinases (FAK, Src, PKC, PKB, MAPK, JNK, p70 S6 kinase), lipid kinases (PI 3-K, PIP 4-K, PIP 5-K), and phospholipases (PLC-γ, PLA_2). In addition, integrins regulate the reorganization of the cytoskeleton through proteins such as tensin, talin, vinculin, α-actin, and paxillin. Many of these signaling events may be interdependent. Moreover, gene expression in the nucleus may require collaboration with other signals transmitted from the plasma membrane to the nucleus.

A. Regulation of Intracellular Ion Concentration

Integrin-mediated cell adhesion regulates the concentration of certain intracellular ions. The increase in intracellular pH through the activation of the Na^+/H^+ antiporter appears to be a property shared by all integrins. Adhesion of bovine endothelial cells to the ECM components fibrinogen, collagens type III, IV, and V, laminin or vitronectin, results in a significant elevation in intracellular H^+ concentration as does adhesion to specific anti-integrin antibodies (Schwartz et al., 1991). Since other ligands (basic fibroblast growth factor, concavalin A, and thrombin), which bind to cells cultured on plastic do not have this effect, the elevation in intracellular pH is specific to integrins.

Integrin ligation can also trigger an increase in intracellular free calcium ion (Ca^{2+}) concentration (Schwartz, 1993; reviewed in Sjaadstad and Nelson, 1996), an important second messenger regulating protein kinases, phosphatases, and other enzymes (Berridge, 1993) Moreover, the ability of growth factors such as PDGF to induce a transient increase in intracellular Ca^{2+} requires cell attachment (Tucker et al., 1990). Unlike integrin-stimulated elevation of intracellular pH, the increase in intracellular Ca^{2+} triggered by cell adhesion to ECM components is only mediated by certain integrins. In experiments where specific anti-integrin antibodies were bound to glass surfaces and used to induce endothelial cell spreading, only anti-αv—but not anti-$\alpha5\beta1$ antibodies—stimulated a rise in intracellular Ca^{2+} despite the ability of the anti-$\alpha5\beta1$ antibodies to increase intracellular pH (Schwartz and Denninghoff, 1994). Moreover, the αv integrin-mediated effect on Ca^{2+} concentration was specifically regulated by the integrin α subunit itself. Therefore, regulation of intracellular Ca^{2+} levels is integrin-specific and occurs by a mechanism different from that regulating the Na^+/H^+ antiporter, which indicates that different integrin receptors can generate distinct signals.

B. Regulation of Intracellular Lipid Levels

Integrin-mediated cell adhesion is also important in the generation of second messengers through alterations in the metabolism of inositol phospholipids (see Figure 6). Phospholipase C-gamma (PLC-γ), phosphatidylinositol 4-kinase (PI 4-K), phosphatidylinositol-4-phosphate 5-kinase (PIP 5-K), and phosphotidylinositol 3-kinase (PI 3-K) are specifically activated by integrin ligation (Banga et al., 1986; Cybulsky et al., 1993; Kanner et al., 1993; McNamee et al., 1993; Berditchevski et al., 1997; reviewed in Clarke and Brugge, 1995). Currently, it is not clear how integrins activate these enzymes. There is evidence that some integrins activate PI 3-K by recruiting the SH2 domain adaptor protein, Shc, which in turn serves to activate Ras (Mainiero et al., 1997). Other studies indicates that PI 4-K may be indirectly activated by the integrin-associated transmembrane-4 (TM4) proteins (Berditchevski et al., 1997). As noted above, PIP 5-K is reversibly activated by cell adhesion and this occurs through an interaction with GTP-Rho

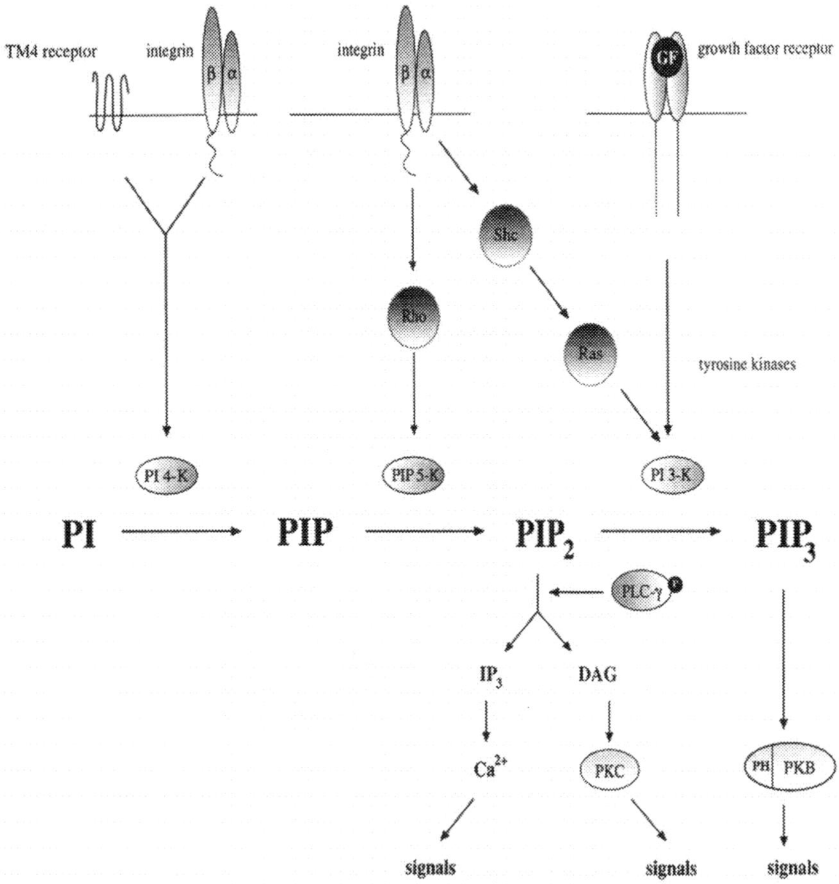

Figure 6. Integrins regulate the synthesis of phospholipids by cooperating with other receptor signaling pathways. Synergism between integrin and growth factor receptor pathways can be achieved by limiting the availability of the important precursor phospholipid, phosphinositol (4,5) bishosphate (PIP_2). Integrin cooperates with TM4 receptors to activate phosphoinositol 4-kinase (PI 4-K) and thus the levels of phospoinositol 4-phosphate (PIP). The synthesis of PIP_2 from PIP is regulated by phosphatidylinositol 4-phosphate 5-kinase(PIP 5-K), which is reversibly activated by cell adhesion. Activated growth factor receptors such as the EGF and PDGF receptors, phosphorylate and activate phospholipase C-γ (PLC-γ), which catalyze the hydrolysis of PIP_2 to inositol triphosphate (IP_3) and diacylglycerol (DAG). The synthesis of PIP_2 catalyzed by PIP 5-K may therefore be the critical rate-limiting step for the IP_3-dependent regulation of intracellular Ca^{2+} and the activation of PKC (protein kinase C) by DAG. Ligation of certain integrins can activate PI 3-K through, for example, Shc and Ras, synergizing with growth factor receptor–mediated activation of PI 3-K leading to the conversion of PIP_2 to PIP_3. Pleckstrin homology (PH) domain-containing proteins such as PKB are recruited by PIP_3 to the plasma membrane.

resulting in integrin-induced PIP_2 production. PIP_2 is the substrate for growth factor-activated PLC-γ which converts it to inositol triphosphate (IP_3) and diacylglycerol (DAG) (Tucker et al., 1990, Schwarz and Lechene, 1992). Therefore, the integrin-mediated synthesis of PIP_2 catalysed by PIP 5-K may be the rate limiting step for the IP_3-dependent regulation of intracellular Ca^{2+} ion concentration and for the activation of PKC by DAG (see Figure 6). PIP_2 also acts as substrate for PI3-K generating the lipid, phosphatidyl inositol-3,4,5- trisphosphate (PIP_3), which recruits cytoplasmic signaling proteins such as PKB to the plasma membrane via Pleckstrin Homology (PH) domains (Harlan et al., 1994).

The metabolism of arachiodonic acid is also effected by integrin-mediated cell adhesion through the activation of phospholipase A_2 (PLA_2). Adhesion of HeLa cells to collagen or to immobilized RGD peptide triggers a cascade of events including activation of PLA_2, release of arachiodonic acid and formation of lipoxygenase metabolites. Adhesion also leads to the activation of PLC-γ and the production of DAG and ultimately results in cell spreading (Chun and Jacobson, 1992, 1993). Thus, integrins may regulate several adhesion-mediated signals by altering the levels of lipids within the cell.

C. Integrin-Triggered Activation of Protein Kinases

Ultimately, the changes described above lead to the activation of several serine-threonine family kineses including protein kinase C (PKC) (Woods and Couchman, 1992; Vuori and Ruoslahti, 1993; Chun and Jacobson, 1993; Auer and Jacobson, 1995; reviewed in Divecha and Irvine, 1995), cyclin A kinase (Symington, 1992), mitogen-activated protein kineses (MAPKs) (Chen Q. et al., 1994; Schlaepfer et al., 1994; Morino et al., 1995; Zhu and Assoian, 1995), Jun-kinases (JNKs) (Miyamoto et al., 1995b) and p70 ribosomal S6 kinase (Malik and Parsons, 1996), and thereby result in changes in gene expression. One of the earliest events detected in response to integrin ligation and clustering, however, is the tyrosine phosphorylation of several proteins associated with focal contact complexes (Hynes, 1992; Juliano and Haskell, 1993; Arroyo et al., 1994).

D. Focal Adhesion Kinase

Integrin engagement with ligand (Guan et al., 1991) or the aggregation of integrins without ligand occupancy (Kornberg et al., 1991) both lead to the tyrosine phosphorylation and enhanced catalytic activity of the 125 kD non-receptor protein tyrosine kinase, focal adhesion kinase (FAK). This highly conserved protein is widely expressed and appears to play a central role in the transduction of signals mediated by many integrin receptors—notably those containing β1 or αv subunits and αIIbβ3 integrin (Lukashev et al., 1994; Akiyama et al., 1994; Schaller and Parsons, 1994).

FAK is a structurally distinct protein tyrosine kinase (Schaller et al., 1992) in that it does not contain the SH2 or SH3 domains characteristic of other cytoplasmic tyrosine kineses which mediate interactions with target or regulating proteins. Nevertheless, FAK interacts with several different proteins through its amino and carboxyl regions which flank a centrally located kinase domain (see Figure 7). The amino terminal domain directly binds cytoplasmic domain peptides of β1 integrin (Schaller et al., 1995a). A region in the carboxyl terminus which is responsible for targeting FAK to focal contacts (Hildebrand et al., 1993) known as the FAT (focal adhesion targeting) sequence, interacts with talin (Chen et al., 1995), and is additionally required for FAK activation (Lewis and Schwartz 1995). The adaptor protein, paxillin, also binds the carboxyl terminus (Hildebrand et al., 1995) through a distinct but overlapping sequence to the talin binding site (Turner and Miller, 1994;

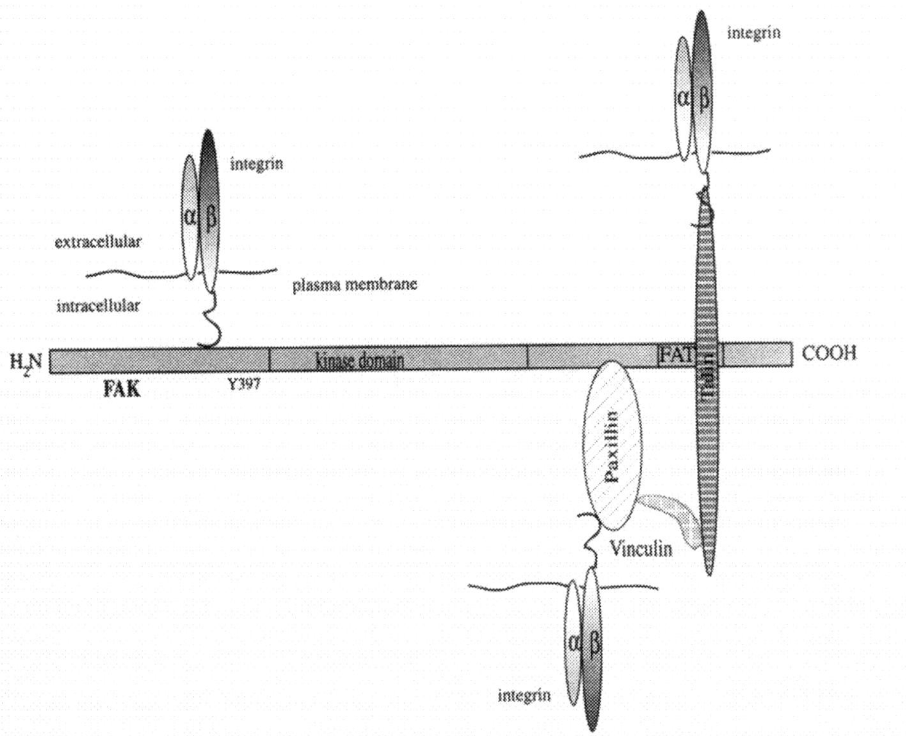

Figure 7. Interactions of FAK with integrins, talin, and paxillin. FAK is a protein tyrosine kinase with a molecular weight of 125 kDa whose catalytic domain is centrally located. The focal adhesion targetting (FAT) sequence in the C-terminus of the enzyme is responsible for directing FAK to focal adhesions and also binds to talin. The N-terminal domain of FAK contains a sequence that binds the cytoplasmic tail of β integrins. Talin, in addition to binding to FAK, also binds vinculin and the cytoplasmic tail of β integrins. Paxillin, which binds to the C-terminus of FAK through a site distinct from the FAT sequence, is also capable of binding vinculin and additionally binds the β1 cytoplasmic domain providing another link of FAK with integrins.

Clark and Brugge, 1995). Paxillin additionally interacts directly with the β1 integrin cytoplasmic tail (Schaller et al., 1995a) and binds vinculin (Turner and Miller, 1994; Wood et al., 1994). Vinculin can interact with talin (Burridge and Mangeat, 1984; Belkin and Koteliansky, 1987, Wachsstock et al., 1987) which, as noted above, can directly bind β1 integrin.

Therefore, several potential direct or indirect interactions between FAK and integrins are possible. Since the N-terminal domain of FAK appears to play a negative autoregulatory role presumably by folding back onto its catalytic domain (Schaepfer and Hunter, 1997), one interpretation to explain the apparent redundancy in interactions between these two molecules is that talin initially recruits inactive FAK to the focal contact. Subsequently, FAK undergoes a conformational change which allows its N-terminal domain to bind directly to the cytoplasmic tail of β integrin which is prerequisite for its activation following integrin clustering (Giannocotti, 1997).

E. FAK and Src are Central in Recruiting Signaling Molecules to Focal Adhesions

Multiple signaling molecules are recruited into the focal contact through SH2 and SH3 domain-mediated interactions (Pawson and Schlessinger 1993) centering around FAK and paxillin which are two of the most heavily phosphorylated proteins in adhering cells (see Figure 8). Integrin clustering induces autophosphorylation of FAK on Y397 creating a high-affinity SH2 domain binding site for Src family kineses (Cobb et al., 1994; Xing et al., 1994). Src then phosphorylates Y576 and Y577 in the catalytic domain of FAK leading to its maximal activation (Calalb et al., 1995). The phosphorylation of Y925 on FAK by Src creates an in vitro SH2 binding site for the guanine nucleotide exchange factor (GEF) Grb2/Sos complex (Schlaepfer et al., 1994) and moreover, the association of c-Src with FAK is required for Grb2 binding in vivo (Schlaepfer and Hunter, 1996) and thus potentially localizes the Ras signaling pathway to focal contacts. Activated FAK or the FAK/Src complex phosphorylates paxillin (Burridge et al., 1992; Turner, 1994) creating a binding site for the SH2 domain of the adapter protein Crk (Birge et al., 1993; Schaller and Parsons 1995b). Crk, in turn, recruits C3G, another GEF, through an SH3 domain interaction (Schaller et al., 1994) providing an additional route for Ras signaling. The tyrosine kinase, Csk, a negative regulator of Src family kinases (reviewed in Superti-Furga and Courtneidge, 1995), also associates with phosphorylated paxillin through an SH2 domain interaction (Clarke and Brugge, 1995) and may regulate Src activity through a Src/paxillin/Csk complex where the SH3 domain of Src binds to a proline-rich domain on paxillin (Weng et al., 1993).

Phosphorylated Y397 on FAK also provides an SH2 binding site for p85, the regulatory subunit of PI3-K, which leads to its phosphorylation and activation (Chen and Guan, 1994; Chen H-C et al., 1996). Other substrates of activated FAK or the FAK/Src complex include tensin (Lo et al., 1994) and p130^{cas} (fetch et al., 1995;

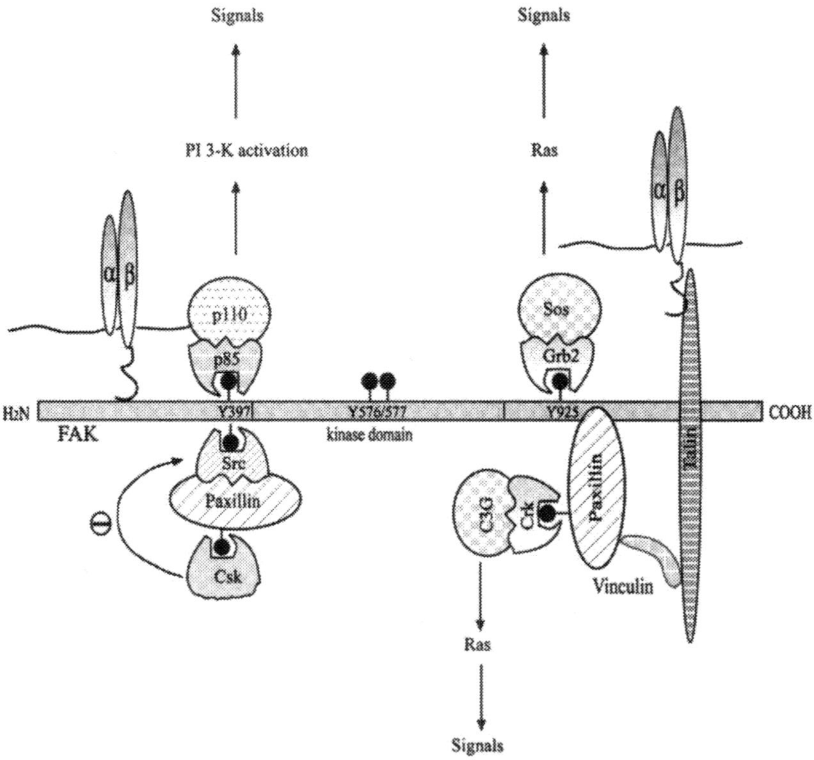

Figure 8. FAK recruits signaling molecules to the focal adhesion. SH2 domain proteins (p85, Src, Grb2, and Csk) bind to specific phosphotyrosine (•) containing sequences either in FAK or paxillin. Some of these proteins also contain SH3 domains (p85, Src, Grb2 and Csk), which in turn interact with proline rich motifs on other signaling molecules (p110, paxillin, C3G, and Sos) to form supramolecular complexes of signaling molecules in the focal contact.

Hamasaki et al., 1996). Integrin-mediated interactions between Csk/FAK, p130^{cas}/FAK and Crk/p130^{cas}/C3G/Sos (Polte and Hanks, 1995; Clarke and Brugge, 1995; Vuori et al., 1996) have also been demonstrated, emphasising the complexity of the multiple protein–protein interactions in focal contacts. Thus, FAK has the potential to integrate extracellular signals for cell growth and gene expression with those for dynamic reorganisation or stabilisation of cytoskeleton–membrane interactions.

Although Src family kineses play a key role in integrin signaling, the initial events which recruit and/or activate Src are not yet clear. The SH3 domain of Src directs it to focal contacts (Kaplan et al., 1994) and may bind potential targeting molecules such as p130^{cas} or paxillin. In normal quiescent cells, however, c-Src appears to be largely absent from focal contacts and is maintained in an inactive state. This is achieved through the intramolecular interaction of phosphorylated Y527 with the amino terminal SH2 domain (reviewed in Superti-Furga and Courtneidge, 1995) and between the SH3 domain with a novel sequence in the linker region between the SH2 and kinase domains (Xu et al., 1997). Disrupting the phosphotyrosine/SH2 interaction, by mutating Y527 or expressing c-Src in cells which lack Csk, causes the redistribution of Src to focal adhesions (Kaplan et al., 1994). This suggests a phosphotyrosine phosphatase stabilizes an active "open" confirmation of Src by keeping Y527 dephosphorylated. Indeed, the phosphatase SHP-1 has been shown to interact with Src via its amino terminal SH2 domain, to be phosphorylated by Src and to preferentially dephosphorylate Src at its inhibitory phosphotyrosine site suggesting that SHP-1 is a positive regulator of Src activation (Somani et al., 1997).

FAK is widely expressed and its amino acid sequence highly conserved among species (Hens and DeSimone, 1995). Although the above discussion suggests that FAK is a central mediator of integrin signaling, gene targetting experiments indicate that there may be a family of proteins with similar function. Thus, FAK knockout mice die at E9.0 but embryonic fibroblasts derived from these mice are still able to form focal contacts (Ilic et al., 1995a, 1995b). However, in cells that are wild type for FAK, over-expression of the C-terminal domain of FAK, pp^{41}/pp^{43FRNK}, which is devoid of tyrosine kinase activity, transiently blocks focal contact formation (Richardson and Parsons, 1996) and inhibits integrin-mediated activation of p70 ribosomal S6 kinase (Malik and Parsons, 1996). Moreover, experiments in which FAK (Cary et al., 1996; Frisch et al., 1996) or a dominant-negative form of FAK (Gilmore and Romer, 1996) were over-expressed indicate that this molecule is important for migration and also for survival and proliferation. It is therefore likely that related molecules such as Pyk2/CAKβ/RAFT or FakB (Sasaki et al., 1995; Avraham et al., 1995; Kanner, 1996) compensate for the loss of FAK activity in the focal contacts of $FAK^{-/-}$ cells.

F. Activation of MAPK by Integrins

The events leading to activation of MAPK in growth factor receptor signaling are well documented (reviewed in Johnson and Vaillincourt, 1994). Early events after ligation of growth factor receptors include autophosphorylation of receptor, accumulation of GTP-Ras, stimulation of phospholipid turnover, and activation of serine/threonine kineses such as PKC, PKB and the MAPK family. MAPKs require phosphorylation on both tyrosine and threonine residues for activation and this is achieved through the action of the tyrosine/threonine kinase MEK. Activation of MEK occurs by the phosphorylation of serine and threonine residues by Raf, following the recruitment of Raf to the plasma membrane where its regulatory domain binds GTP-Ras. GTP-Ras is linked to the activated phosphorylated growth factor receptor through the Grb2/Sos complex. Therefore, Raf links the MAPK signal transduction pathway to Ras activation. Once activated, MAPK subsequently translocates into the nucleus where many of its physiological targets are located including the transcription factors Elk-1, c-Myc, c-Jun, c-Fos and C/EBP (reviewed in Treisman, 1996), and thus MAPKs are key molecules in the transmission of growth factor signals to the nucleus.

The sequence of events leading to the activation of MAPK after cell adhesion are less well understood than for growth factor receptor ligation and may occur by several different routes As with growth factor stimulation, integrin ligation results in the accumulation of GTP-Ras (Kapron-Bras et al., 1993) through proteins recruited to focal contacts. However, in the variety of systems studied it has become apparent that multiple mechanisms are available for integrin-mediated Ras activation. Adhesion of 3T3 cells to fibronectin promotes the association of Grb2/Sos with FAK (Schlaepfer et al., 1994) and FAK-dependent phosphorylation of the adapter protein paxillin creates a high affinity binding site for Crk (Schaller and Parsons, 1995b) which in turn recruits C3G, another GEF for Ras (Tanaka et al., 1994). Similarly, the adapter protein, $p130^{cas}$, associates with Crk and C3G/Sos after integrin-mediated phosphorylation (Vuori et al., 1996). In rat fibroblasts expressing the human insulin receptor, Grb2/Sos complex associates with $\alpha v\beta 3$ integrin in an insulin-dependent manner through the adaptor protein, IRS-1 (Vuori and Ruoslahti, 1994). In A431 cells, ligation of a specific subset of $\beta 1$ and αv integrins activates the MAPK pathway through the adaptor protein, Shc (Wary et al., 1996), which also acts to recruit the Grb2/Sos complex to the plasma membrane (Pawson, 1995). Transient transfection studies in epithelial cells involving the overexpression of FAK, FAK mutants and Ras mutants, show that both Src and Ras activation events are required for maximal signaling to Erk2 following integrin stimulation (Schlaepfer and Hunter, 1997). Moreover, these studies indicate that FAK/Src complexes may be connected to multiple signaling pathways involving $p130^{cas}$, Shc and Grb2.

It is important to note that the Ras-dependent signaling pathway of certain growth factors is not simply mimicked by the integrin-mediated activation of FAK, since the induction of MAPK by cell adhesion can operate separately and independently from the stimulation of FAK (Wary et al., 1996; Lin et al., 1997). For example, expression of a dominant-inhibitory mutant of Ras in serum-starved NIH3T3 cells blocks adhesion-dependent triggering of Erk2 even though FAK is activated in these cells (Clark and Hynes, 1996) indicating that integrin-mediated signals involved in regulating cell morphology appear to diverge from those regulating MAPK activation at a level upstream of Ras activation.

MAPK can also be activated by growth factors through Ras-independent pathways (Burgering et al., 1993; Silvennoinen et al., 1993; Buscher et al., 1995; Coffer et al., 1995) and this also appears to be the case for integrin signaling. For example, in NIH3T3 fibroblasts, MAPK activation following integrin ligation is not accompanied by the accumulation of GTP-loaded Ras that occurs in the EGF signaling pathway (Chen et al., 1996). Moreover, expression of dominant-negative Raf and Ras, fail to interfere with integrin signaling to both MEK and MAPK. In transfected fibroblasts expressing a kinase domain deletion mutant of Src, although Grb2 binding to FAK after fibronectin ligation was blocked, both Erk2 activation and the association of p130^{cas} with FAK and its subsequent phosphorylation were unaffected (Schlaepfer et al., 1997). Furthermore, the SH2 domain-dependent binding of p130^{cas} to another adaptor protein, Nck, was promoted in these cells which may facilitate signaling to Erk2. In transfection studies in Cos-7 cells, the PI 3-K inhibitors wortmannin and LY294002 significantly inhibited integrin-mediated activation of Erk-2, MEK-1 and Raf-1 but did not effect Ras-GTP loading suggesting that PI 3-K may function upstream of Raf-1 but downstream of Ras (King et al., 1997). One possibility is that in some circumstances alternative guanosine triphosphatases (GTPases) may substitute for Ras in integrin-mediated MAPK activation. For example, Rap1, a close relative of Ras, is activated by the GEF, C3G (Gotoh et al., 1995).

V. SPECIFICITY AND DIVERSITY IN INTEGRIN SIGNALING

Integrins participate in a wide range of activities including organization of cell adhesion sites, formation of the actin-containing cytoskeleton, as well as signal transduction. As noted above, many of the different signaling pathways activated by integrins share common downstream targets with growth factor signaling systems such as H+ exchange, Ca^{2+} influx, protein tyrosine, serine, and threonine phosphorylation, changes in phosphoinositide metabolism and changes in gene expression. Moreover, ligation of individual integrins leads to the phosphorylation of a distinct subset of proteins (Jewell et al., 1995). Thus, integrins are now perceived as central regulators of signal transduction pathways (see Figure 5). Given that integrin and growth factor receptors share similar downstream targets, it is possible that a physical cooperation exists between them. Indeed, one function of the integrin-triggered focal contact may be to assemble appropriate signaling molecules into a complex which can respond to growth factor stimulation.

A. Formation of an Integrin Signaling Complex

Integrin signaling usually depends upon clustering of receptor (Kornberg et al., 1991; Yurochko et al., 1992). The use of ligand- or antibody-coated beads in conjunction with inhibitors of tyrosine phosphorylation has identified a number of discreet steps in the generation of downstream signals involving the dual events of occupancy and clustering of integrin receptors (Miyamoto et al., 1995a, 1995b; Plopper et al., 1995; reviewed in Yamada and Miyamoto, 1995; see Figure 9). Specific subsets of cytoskeletal proteins co-localize with integrins depending on whether or not integrin clustering is accompanied by receptor occupancy and also upon whether tyrosine phosphorylation occurs. Integrin receptor occupancy by monovalent ligand causes redistribution to a pre-existing site for focal contact

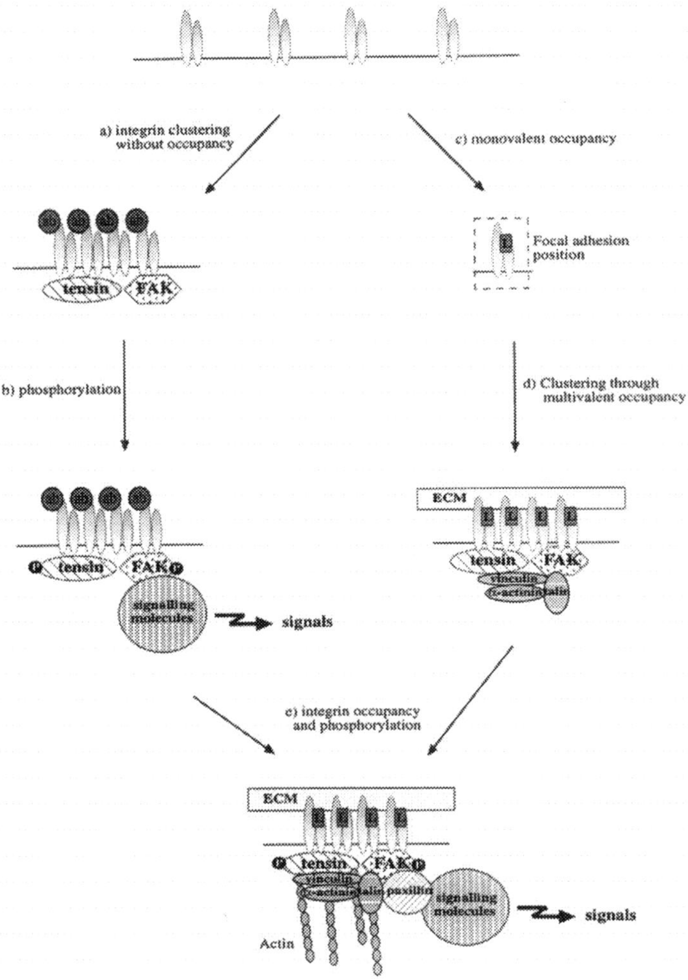

Figure 9. The integrin signaling complex. Assembly of a focal contact can be viewed in distinct stages depending on occupancy of the integrin binding site, clustering of integrin receptor, and tyrosine phosphorylation. (**a**) If Integrin clustering occurs by interaction with antibodies or the ECM through a site distinct from the 'active site', tensin and FAK co-localize with integrin. (**b**) After clustering, FAK becomes phosphorylated and several signaling molecules are recruited to the complex including Src, Grb2, Sos, Crk, PI3-K, Rho, Rac, Ras, MEK, Erk, and Jnk. (**c**) Binding of monovalent ligand in the 'active site' causes integrin to locate to a preexisting focal adhesion site. (**d**) Integrin clustering occurs when multivalent ligand such as ECM binds the 'active site' of integrins. In addition to FAK and tensin, the actin binding proteins vinculin, α-actinin, and talin also co-localize with integrin. (**e**) Clustering through multivalent occupancy leads to phosphorylation of FAK and to the recruitment of the same molecules seen when integrin clusters without occupancy. In addition, paxillin and actin are recruited into the complex resulting in the generation of signals.

formation but is accompanied by only minimal tyrosine phosphorylation. Clustering of unoccupied integrins by anti-integrin antibodies which do not block adhesion leads to co-localization of FAK and tensin with integrin and is accompanied by tyrosine phosphorylation which results in at least 20 different signaling molecules accumulating in the complex including Src, Rho, Rac, Ras, Raf, MEK, ERK, and JNK. The recruitment of the actin binding proteins, talin, α-actinin and vinculin requires both integrin clustering as well as receptor occupancy. In these circumstances, the ensuing tyrosine phosphorylation leads to the co-localization of the signaling molecules which assemble with clustered but unoccupied integrin but also to recruitment of F-actin and phosphorylated paxillin.

Thus, integrin clustering experiments have demonstrated separate and synergistic roles for integrin occupancy and aggregation. The heirarchical accumulation of different proteins within an integrin signaling complex may provide an explanation for how a single transmembrane receptor can mediate such diverse effects in the cell. The multivalency of extracellular matrix molecules provides a direct mechanism for inducing adjacent integrin aggregation and receptor ligation. Therefore, depending on the local environment of the cell and the type of α integrin subunit, integrin occupancy, and clustering allows for selectivity in recruiting cytoskeletal proteins and triggering of different intracellular signaling pathways.

B. Cross-Talk between Integrin Receptors and Receptors for Soluble Factors

Integrins have been shown to cooperate functionally with growth factors in several biological processes (reviewed in Damsky and Werb, 1992; Schwartz et al., 1995; Juliano, 1996). Cell growth and/or differentiation of fibroblasts, mammary epithelial cells, myoblasts, and chondrocytes all require collaborative interactions between integrins and growth factors (Blum et al., 1987; Tucker et al., 1990; Arner and Tortorella, 1995; Zhu and Assoian, 1995; Streuli et al., 1995a; Zhu et al., 1996; Sastry et al., 1996; Streuli and Edwards, 1998). The mechanisms of signal transduction required for such complex biological processes which are brought about by interactions of cells with the ECM and soluble factors are beginning to emerge.

One example of integrin and growth factor receptor cooperation, involves the effect of EGF on hemidesmosomes and migration in A431 cells (Mainiero et al., 1996). In the absence of EGF, ligation of α6β4 integrin with laminin or antibody resulted in the tyrosine phosphorylation of the β4 integrin cytoplasmic domain and promoted its association with the adapter molecules, Shc and Grb2, as well as its association with cytoskeletal elements of the hemidesmosome. However, when integrin was ligated in the presence of EGF, the β4 cytoplasmic domain became tyrosine phosphorylated, but rather than leading to its association with Shc/Grb2 and hemidesmosome assembly, cell migration was stimulated. Thus, EGF-dependent signals suppressed the association of activated α6β4 integrin with both signaling and cytoskeletal molecules, and upregulated α6β4-dependent migration.

As discussed above, MAPK is an important regulator of cell growth, differentiation, and gene activation pathways. This kinase can be controlled separately by integrins and growth factors although its activation by integrins is generally slower and of longer duration than after growth factor stimulation (Zhu and Assoian, 1995). Adhesion of cells to ECM can cooperate with growth factors to activate MAPK over and above the level seen with ligation of either integrin or growth factor receptor alone. Aggregation of growth factor receptors such as those for EGF, PDGF, and basic FGF is transiently induced in the absence of growth factor only when integrin is both clustered and occupied (Miyamoto et al., 1996). Subsequent stimulation by growth factor results in the enhanced tyrosine phosphorylation of these growth factor receptors, presumably because their capacity for transphosphorylation is sterically enhanced. and also leads to a marked but transient increase in MAPK activity.

In confluent endothelial cells, soluble vascular endothelial growth factor (VEGF) stimulates actin stress fiber formation and migration (Abedi and Zachary, 1997). Concomitantly, both FAK and paxillin are tyrosine phosphorylated in response to VEGF and localize to focal contacts although this is critically dependent on actin filament formation since it is abolished in the presence of cytochalasin D. VEGF treatment of these cells also results in a transient increase in MAPK activation, an increase in PLCγ activity but does not effect PI3-K activity but it is unresolved how the VEGF signaling pathway intersects FAK/paxillin signaling.

It now appears that growth factor receptor and integrin cooperation is mediated by specific integrins. For example, fibroblasts stably transfected with the insulin receptor and expressing the vitronectin receptor, $\alpha v\beta 3$ integrin, show enhanced DNA synthesis in response to insulin when plated on vitronectin. Moreover, insulin promotes association of the SH2/SH3 domain adaptor, IRS-1, with $\alpha v\beta 3$ integrin. However, these effects were specific to $\alpha v\beta 3$ since cells expressing the vitronectin receptor, $\alpha v\beta 5$, did not proliferate and IRS-1 did not associate with this integrin in response to insulin (Vuori and Ruoslahti, 1994).

In mammary gland epithelial cells, signals generated through the specific interaction of $\beta 1$ integrin with the basement membrane component laminin 1, but not those generated by interaction with the stromal component collagen I, coordinate with signals provided by prolactin to regulate milk protein expression (Streuli et al., 1991; Streuli, 1995a). The type of ECM encountered by these cells controls milk protein expression by regulating the ability of prolactin to stimulate its receptor and initiate the tyrosine phosphorylation cascade which ultimately leads to the activation of the transcription factor, Stat5, which is required for milk protein gene transcription (Streuli et al., 1995b). Thus, the prolactin receptor cooperates with a laminin but not a collagen integrin receptor to induce differentiation in these cells (Edwards et al., 1997).

In conclusion, the assembly of integrins and growth factor receptors into large signaling and cytoskeletal complexes depends upon both integrin aggregation and occupancy and results in the transient concentration of effectors and substrates providing a mechanism through which growth factors and integrins can synergise to mediate complex biological processes.

ACKNOWLEDGEMENT

Charles Streuli is a Wellcome Senior Fellow in Basic Biomedical Science.

REFERENCES

Abedi, H. & Zachary, I. (1997). Vascular endothelial growth factor stimulates tyrosine phosphorylation and recruitment to new focal adhesions of focal adhesion kinase and paxillin in endothelial cells. J. Biol. Chem. 272, 15442-15451.

Adams, J. C. & Watt, F. M. (1993). Regulation of development and differentiation by extracellular matrix. Dev. 117, 1183-1198.

Akiyama, S. K., Yamada, S. S., Yamada, K. M. & LaFlamme S. E. (1994). Transmembrane signal transduction by integrin cytoplasmic domains expressed in single-subunit chimeras. J. Biol. Chem. 269, 15961-15964.

Albiges-Rizo, C., Frachet, P. & Block, M. R. (1995). Down regulation of talin alters cell adhesion and the processing of the $\alpha5\beta1$ integrin. J. Cell Sci. 108, 3317-3329.

Arner, E. C. & Tortorella, M. D. (1995). Signal transduction through chondrocyte integrin receptors induces matrix metalloproteinase synthesis and synergises with interleukin-l. Arthritis Rheum. 138, 1304-1314.

Arroyo, A. G., Campanero, M. R., Sanchez-Mateos, P., Zapata, J. M., Ursa, M. A., del Pozo, M. A. & Sanchez-Madrid, F. (1994). Induction of tyrosine phosphorylation during ICAM-3 and LFA-1-mediated intercellular adhesion and its regulation by the CD45 tyrosine phosphatase. J. Cell Biol. 126, 1277-1286.

Auer, K. L. & Jacobson, B. S. (1995). $\beta1$ integrins signal lipid second messengers required during cell adhesion. Mol. Biol. Cell. B, 1305-1313.

Avraham, S., London, R., Fu, Y., Ota, S., Hiregowdara, D., Li, J., Jiang, S., Pasztor, L. M., White, R. A. & Groopman, J. E. (1995). Identification and characterisation of a novel related adhesion focal tyrosine kinase (RAFTK) from megakaryocytes and brain. J. Biol. Chem. 270, 27742-27751.

Banga, H. S., Simons, E. R., Brass, L. F. & Rittenhouse, S. E. (1986). Activation of phospholipases A and C in human platelets exposed to epinephrine: role of glycoproteins IIb/IIIa and dual role of epinephrine. Proc. Natl. Acad. Sci. 83, 9197-9201.

Barry, S. T. & Critchley, D. R. (1994). The rhoA-dependent assembly of focal adhesions in Swiss 3T3 cells is associated with increased tyrosine phosphorylation and the recruitment of both $pp125^{FAK}$ and protein kinase C-δ to focal adhesions. J. Cell Sci. 107, 2033-2045.

Bauer, J. S., Varner, J., Schreiner, C., Kornberg, L., Nicholas, R. & Juliano, R. L. (1993). Functional role of the cytoplasmic domain of the integrin $\alpha5$ subunit. J. Cell Biol. 122, 209-221.

Belkin, A. M. & Koteliansky, V. E. (1987). Interaction of iodinated vinculin, metavinculin and α-actinin with cytoskeletal proteins. Febs Lett. 220, 291-294.

Berditchevski, F., Tolias, K. F., Wong, K., Carpenter, C. L. & Hemler, M. E. (1997). A novel link between integrins, transmembrane-4 superfamily proteins (CD63 and CD8 1), and phosphatidylinositol 4-kinase. J. Biol. Chem. 272, 2595-2598.

Berridge, M. J. (1993). Inositiol triphosphate and calcium signaling. Nature. 361, 315-325.

Birge, R. B., Fajardo, J. E., Reichman, C., Shoelson, S. E., Songyang, Z., Cantley, L. C. & Hanafusa, H. (1993). Identification and characterization of a high-affinity interaction between v-Crk and tyrosine-phosphorylated paxillin in CT10-transformed fibroblasts. Mol. Cell. Biol. 13, 4648-4656.

Blum, J. L., Zeigler, M. E. & Witcha, M. S. (1987). Regulation of rat mammary gene expression by extracellular matrix components. Exp. Cell. Res. 173, 322-340.

Burgering, B. M., de Vries-Smits, A., Medema, R. H., van Weeren, P. C., Tertoolen, L. G. & Bos, J. L. (1993). Epidermal growth factor induces phosphorylation of extracellular signal-regulated kinase 2 via multiple pathways. Mol. Cell. Biol. 13, 7248-7256.

Burridge, K. & Mangeat, P. (1984). An interaction between vinculin and talin. Nature. 308, 744-746.

Burridge, K., Fath, K., Kelly, G., Nuckolls, G. & Turner, C. (1989). Focal adhesions Transmembrane junctions between extracellular matrix and the cytoskeleton. Ann. Rev. Cell. Biol. 4, 487-525.

Burridge, K., Turner, C. E. & Romer, L. H. (1992). Tyrosine phosphorylation of paxillin and $pp125^{FAK}$ accompanies cell adhesion to extracellular matrix: a role in cytoskeleton assembly. J Cell. Biol. 119, 893-903.

Buscher, D., Hipskind, R. A., Krautwald, S., Reimann, T. & Baccarini, M. (1995). Ras-dependent and -independent pathways target the mitogen-activated protein kinase network in macrophages. Mol. Cell. Biol. 15, 466-475.

Calalb, M. B., Polte, T. R. & Hanks, S. K. (1995). Tyrosine phosphorylation of focal adhesion kinase at sites in the catalytic domain regulates kinase activity: a role for Src family kineses. Mol. Cell. Biol. 15, 954-963.

Cary, L. A., Chang, J. F. & Guan, J. L. (1996). Stimulation of cell migration by overexpression of focal adhesion kinase and its association with Src and Fyn. J. Cell Sci. 109, 1787-1794.

Cepek, K. L., Shaw, S. K., Parker, C. M., Russell, G. J., Morrow, J. S., Rimm, D. L. & Brenner, M. B. (1994). Adhesion between epithelial cells and T lymphocytes mediated by E-cadherin and the $\alpha^E\beta7$ integrin. Nature. 372, 190-193.

Chen, H-C. & Guan, J-L. (1994). Association of focal adhesion kinase with its potential substrate phosphatidylinositol 3-kinase. Proc. Natl. Acad. Sci. 91, 10148-10152.

Chen, H-C., Appeddu, P. A., Parsons, J. T., Hildebrand, J. T., Schaller, M. D. & Guan, J-L. (1995). Interaction of focal adhesion kinase with cytoskeleton protein talin. J. Biol. Chem. 270, 16995-16999.

Chen, H-C., Appeddu, P. A., Isoda, H. & Guan, J. L. (1996). Phosphorylation of tyrosine 397 in focal adhesion kinase is required for binding phosphatidylinositol 3-kinase. J. Biol. Chem. 271, 26329-26334.

Chen, Q., Kinch, M. S., Lin, T. H., Burridge, K. & Juliano, R. L. (1994). Integrin-mediated cell adhesion activates mitogen-activated protein kineses. J. Biol. Chem. 269, 26602-26605.

Chen, Q., Lin, T. H., Der, C. J. & Juliano, R. L. (1996). Integrin-mediated activation of mitogen-activated protein (MAP) kinase and extracellular signal-related kinase kinase (MEK) is independent of Ras. J. Biol. Chem. 271, 18122-18127.

Chong, L. D., Traynor-Kaplan, A., Bokoch, G. M. & Schwartz, M. A. (1994). The small GTP-binding protein Rho regulates a phosphatidylinoitol 4-phosphate 5-kinase in mammalian cells. Cell. 79, 507-513.

Chrzanowska-Wodnicka, M. & Burridge, K. (1994). Tyrosine phosphorylation is involved in reorganization of the actin cytoskeleton in response to serum or LPA stimulation. J. Cell. Biol. 107, 3643-3654.

Chrzanowska-Wodnicka, M. & Burridge, K. (1996). Rho-stimulated contractility drives the formation of stress fibers and focal adhesions. J. Cell Sci. 133, 1403-1415.

Chun, J. S. & Jacobson, B. S. (1992). Spreading of HeLa cells on a collagen substratum requires a second messenger formed by the lipoxygenase metabolism of arachidonic acid released by collagen receptor clustering. Mol. Biol. Cell 3, 481-492.

Chun, J. S. & Jacobson, B. S. (1993). Requirement for diacylglycerol and protein kinase C in HeLa cell-substratum adhesion and their feedback amplification of arachidonic acid production for optimum cell spreading. Mol. Biol. Cell 4, 271-281.

Clarke, E. A. & Brugge, S. J. (1995). Integrins and signal transduction pathways: The road taken. Science. 26X, 233-239.

Clarke, E. A. & Hynes, O. H. (1996). Ras activation is necessary for integrin-mediated activation of extracellular signal-regulated kinase-2 and cytostolic phosphalipase A_2 but not for cytoskeletal organisation. J. Biol. Chem. 271, 14814-14818.

Cobb, B. S., Schaller, M. D., Leu, T. H. & Parsons, J. T. (1994). Stable association of pp60src and ppS9fyn with the focal adhesion-associated protein tyrosne kinase, ppl25FAK. Mol. Cell. Biol. 14, 147-155.

Coffer, P., Lutticken, C., van Puijenbroek, A., Klop-de Jonge, M., Horn, F. & Kruiger, W. (1995). Transcriptional regulation of the junB promoter: analysis of Stat-mediated signal transduction. Oncogene. 10, 985-994.

Craig, S. W. & Johnson, R. P. (1996). Assembly of focal adhesions: progress, paradigms, and portents. Curr. Opin. Cell Biol. 8, 74-85.

Cybulsky, A. V., Carbonetto, S., Cyr, M. D., McTavish, A. J. & Huang, Q. (1993). Extracellular matrix-stimulated phospholipase activation is mediated by β1 integrin. Am. J. Physiol. 264, C323-C332.

Damsky, C. H. & Werb, Z. (1992). Signal transduction by integrin receptors for extracellular matrix: cooperative processing of extracellular information. Curr. Opin. Cell. Biol. 4, 772-781.

Delcommenne, M. & Streuli, C. H. (1995). Control of integrin expression by extracellular matrix. J. Biol. Chem. 270, 26794-26801.

Divecha, N. & Irvine, R. F. (1995). Phospholipid signaling. Cell. 80, 269-278.

DeSimone, D. W. & Hynes, R. O. (1988). Xenpus laevis integrins. Structural conservation and evolutionary divergence of integrin β subunits. J. Biol. Chem. 263, 5333-5340.

Edwards, G. M., Wilford, F. H., Liu, X, Hennighausen, L., Djiane, J., Streuli, C. H. (1998). Regulation of mammary differentiation by extracellular matrix involves protein tyrosine phosphatases. J. Biol. Chem. 273, 9495-9500.

Fan, S. T., Mackman, N., Cui, M. Z. & Edgington, T. S. (1995). Integrin regulation of an inflammatory effector gene. Direct induction of the tissue factor promoter by engagement of β1 or α4 integrin chains. J. Immunol. 154, 3266-3274.

Filardo, E. J., Brooks, P. C., Deming, S. L., Damsky, C. & Cheresh, D. A. (1995). Requirement of the NPXY motif in the integrin β3 subunit cytoplasmic tail for melanoma cell migration in vitro and in vivo. J. Cell. Biol. 130, 441-450.

Flinn, H. M. & Ridley, A. J. (1996). Rho stimulates tyrosine phosphorylation of focal adhesion kinase, pl30 and paxillin. J. Cell Sci. 109, 1133-1141.

Frisch, S. M., Vuori, K., Ruoslahti, E. & Chan-Hui, P-Y. (1996). Control of adhesion-dependent cell survival by focal adhesion kinase. J. Cell Biol. 134, 793-799.

Gilmore, A. P. & Romer, L. H. (1996). Inhibition of focal adhesion kinase (FAK) signaling in focal adhesions decreases cell motility and proliferation. Mol. Cell. Biol. 7, 1209-1224.

Gilmore, A. P. & Burridge, K. (1996). Regulation of vinculin binding to talin and actin by phosphatidylinositol-4,5-bisphosphate. Nature. 381, 531-535.

Gianocotti, F. G. & Mainiero, F. (1994). Integrin-mediated adhesion and signaling in tumorigenesis. Biochim. Biophys. Acta. 1198, 47-64.

Gianocotti, F. G. (1997). Integrin signaling: specificity and control of cell survival and cell cycle progression. Curr. Opin. Cell Biol. 9, 691-700.

Gotoh, T., Hattori, S., Nakamura, S., Kitayama, H., Noda, M., Takai, Y., Kaibuchi, K., Matsui, H., Hatase, O., Takahashi, H. et al. (1995). Identification of Rapl as a target for the Crk SH3 domain-binding guanine nucleotide-releasing factor C3G. Mol. Cell. Biol. 15, 6746-6753.

Guan, J. L., Trevithick, J. E. & Hynes, R. O. (1991). Fibronectin/integrin interaction induces tyrosine phosphorylation of a 120 kD protein. Cell Regul. 2, 951-964.

Gumbiner, B. M. (1996).Cell adhesion: The molecular basis of tissue architecture and morphogenesis. Cell. 84, 345-357.

Hamasaki, K., Mimura, T., Morino, N., Furuya, H., Nakamoto, T., Aizawa, S. I., Morimoto, C., Yazaki, Y., Hirai, H. & Nojima, Y. (1996). Src kinase plays an essential role in integrin-mediated tyrosine phosphorylation of Crk-associated substrate $p130^{cas}$ Biochem. Biophys. Res. Comm. 222, 338-343.

Harlan, J. E., Hajduk, P. J., Yoon, H. S. & Fesik, S. W. (1994). Pleckstrin homology domains bind to phosphatidylinosiol-4,5-bisphosphate. Nature 371, 168-170.

Hayashi, Y., Haimovich, B., Reszka, A., Boettiger, D. & Horowitz, A. (1990). Expression and function of chicken integrin β1 subunit and its cytoplasmic domain mutants in mouse NIH3T3 cells. J. Cell Biol. 110, 175-184.

Hens, M. D. & DeSimone, D. W. (1995). Molecular analysis and developmental expression of the focal adhesion kinase ppl25FAK in Xenopus laevis. Dev. Biol. 170, 274-288.

Hildebrand, J. D., Schaller, M. D. & Parsons, J. T. (1993). Identification of sequences required for the efficient localization of the focal adhesion kinase, $pp125^{FAK}$, to cellular focal adhesions. J. Cell. Biol. 123, 993-1005.

Hildebrand, J. D., Schaller, M. D. & Parsons, J. T. (1995). Paxillin: a tyrosine phoshorylated focal adhesion-associated protein binds to the carboxyl terminal domain of focal adhesion kinase Mol. Biol. Cell. 6, 637-647.

Hynes, R. O. (1992). Integrins: versatility, modulation, and signaling in cell adhesion. Cell. 69, 11-25.

Hynes, R. O. (1996). Targeted mutations in cell adhesion genes: What have we learnt from them? Dev. Biol. 180, 402-412.

Ilic, D., Furuta, Y., Kanazawa, S., Takeda, N., Sobue, K., Nakatsuji, N., Nomura, S., Fujimto, J., Okada, M., Yamamoto, T. & Aizawa, S. (1995a). Reduced cell motility and enhanced focal adhesion contact formation in cells from FAK-deficient mice. Nature. 377, 539-544.

Ilic, D., Furuta, Y., Suda, T., Atsumi, T., Fujimoto, J., Ikawa, Y., Yamamoto, T. & Aizawa, S. (1995b). Focal adhesion kinase is not essential for in vitro and in vivo differentiation of ES cells. Biochem. Biophys. Res. Comm. 209, 300-309.

Ishizaki, T., Maekawa, M., Fujisawa, K., Okawa, K., Iwamatsu, A., Fujita, A., Watanabe, N., Saito, Y., Kakizuka, A., Morii, N. & Narumiya, S. (1996). The small GTP-binding protein Rho binds to and activates a 160kDa Ser/Thr protein kinase homologous to myotonic dystrophy kinase. EMBO J. 15, 1885-1893.

Jewell, K., Kapron-Bras, C., Jeevaratnam, P. & Dedhar, S. (1995). Stimulation of tyrosine phosphorylation of distinct proteins in response to antibody-mediated ligation and clustering of a3 and a6 integrins. J. Cell Sci. 108, 1165-1174.

Johnson, G. L. & Vaillancourt, R. R. (1994). Sequential protein kinase reactions controlling cell growth and differentiation. Curr. Opin. Cell Biol. 6, 230-238.

Juliano, R. L. & Haskell, S. (1993). Signal transduction from the extracellular matrix. J. Cell Biol. 120, 577-585.

Juliano, R. (1996). Cooperation between soluble factors and integrin-mediated cell anchorage in the control of cell growth and differentiation. Bioessays. 18, 911-917.

Kanner, S. B., Grosmaire, L. S., Ledbetter, J. A. & Damle, N. K. (1993). β2 integrin LFA-1 signaling through phospholipase C-γ1 activation. Proc. Natl. Acad. Sci. 90, 7099-7103.

Kanner, S. B. (1996). Focal adhesion kinase-related FakB is regulated by the integrin LFA-1 and interacts with the SH3' domain of phospholipase C-γ1. Cell Immunol. 171, 164-169.

Kaplan, K. B., Bibbins, K. B., Swedlow, J. R., Arnaud, M., Morgan, D. O. & Varmus, H. E. (1994). Association of the amino terminal half of c-Src with focal adhesions alters their properties and is regulated by phosphorylation of tyrosine 527. EMBO J. 13, 4745-4756.

Kapron-Bras, C., Fitz-Gibbon, L., Jeevaratnam, P., Wilkins, J. & Dedhar, S. (1993). Stimulation of tyrosine phosphorylation and accumulation of GTP-bound p21 ras upon antibody-mediated α2β1 integrin activation in T-lymphoblastic cells. J. Biol. Chem. 268, 20701-20704.

Kimura, K., Ito, M., Amano, M., Chihara, K., Fukata, Y., Nakafuku, M., Yamamori, B., Feng, J., Nakano, T., Okawa, K., Iwamatsu, A. & Kaibuchi, K. (1996). Regulation of myosin phosphatase by Rho and Rho-associated kinase (Rho-Kinase). Science. 273, 245-248.

King, W. G., Mattaliano, M. D., Chan, T. O., Tsichlis, P. N. & Brugge, J. S. (1997) Phosphatidylinositol 3-kinase is required for integrin-stimulated AKT and Raf-l/mitogen-activated protein kinase pathway activaton. Mol. Cell. Biol. 17, 4406-4418.

Kornberg, L. J., Earp, H. S., Turner, C. E., Prockop, C. & Juliano, R. L. (1991). Signal transduction by integrins: increased protein tyrosine phosphorylation caused by clustering of β1 integrins Proc. Natl. Acad. Sci. 88, 8392-8396.

Kozma, R., Ahmed, S., Best, A. & Lim, L. (1995). The Ras-related protein Cdc42Hs and bradykinin promote formation of peripheral actin microspikes and filopodia in Swiss 3T3 fibroblasts. Mol. Cell. Biol. 15, 1942-1952.

Lafrenie, R. M. & Yamada, K. M. (1996). Integrin-dependent signal transduction. J. Cell. Biochem. 61, 543-553.

Lassing, I. & Lindberg, U. (1985). Specific interaction between phosphatidylinositol-4,5-bisphosphate and profilactin. Nature. 314, 472-474.

Lewis, J. M. & Schwartz, M. A. (1995). Mapping in vivo associations of cytoplasmic proteins with integrin β1 cytoplasmic domain mutants. Mol. Biol. Cell. 6, 151-161.

Lim, L., Manser, E., Leung, T. & Hall, C. (1996). Regulation of phosphorylation pathways by p21 GTPases: the p21 Ras-related Rho subfamily and its role in phosphorylation signaling pathways. Eur. J. Biochem. 242, 171-185.

Lin, T. H., Yurochko, A., Kornberg, L., Morris, J., Walker, J. J., Haskill, S. & Juliano, R. L. (1994) The role of protein tyrosine phosphorylation in integrin-mediated gene induction in monocytes J. Cell Biol. 126, 1585-1593.

Lin, T. H., Rosales, C., Mondal, K., Bolen, J. B., Haskill, S. & Juliano, R. L. (1995). Integrin-mediated tyrosine phosphorylation and cytokine message induction in monocyte cells. A possible signaling role for the Syk tyrosine kinase. J. Biol. Chem. 270, 16189-16197.

Lin, T. H., Aplin, A. K., Shen, Y., Chen, Q., Schaller, M., Romer, L., Aukhil, I. & Juliano, R. L. (1997). Integrin-mediated activation of MAP kinase is independent of FAK: Evidence for dual integrin signaling pathways in fibroblasts. J. Cell Biol. 136, 1385-1395.

Lo, S. H., Weisberg, E. & Chen, L. B. (1994). Tensin: a potential link between the cytoskeleton and signal transduction. Bioessays. 16, 817-823.

Lukashev, M. E., Shephard, D. & Pytela, R. (1994). Disruption of integrin function and induction of tyrosine phosphorylation by the autonomously expressed β1 integrin cytoplasmic domain. J. Biol. Chem. 269, 18311-18314.

Mainiero, F., Pepe, A., Yeon, Ren, Y. & Giancotti, F. G. (1996). The intracellular functions of α4β6 integrin are regulated by EGF. J. Cell. Biol. 134, 241-253.

Mainiero, F., Murgia, C., Wary, K. K., Curatola, A. M., Pepe, A., Blumemberg, M., Westwick, J. K., Der, C. J. & Gianocotti, F. G. (1997). The coupling of α6β2554 integrin to Pas-MAP kinase pathways mediated by Shc controls keratinocyte proliferation. EMBO J. 16, 2365-2375.

Malik, R. K. & Parsons, J. T. (1996). Integrin-dependent activation of the p70 ribosomal S6 kinase signaling pathway. J. Biol. Chem. 271, 29785-29791.

Matsui, T., Amano, M., Yamamoto, T., Chihara, K., Nakafuku, M., Ito, M., Nakano, T., Okawa, K., Iwamatsu, A. & Kaibuchi, K. (1996). Rho-associated kinase, a novel serine/threonine kinase, as a putative target for the small GTP-binding protein Rho. EMBO J. 15, 2208-2216.

McNamee, H. P., Ingber, D. E. & Schwarz, M. A. (1993). Adhesion to fibronectin stimulates inositol lipid synthesis and enhances PDGF-induced inositol lipid breakdown. J. Cell. Biol. 121, 673-678.

Meredith, J. E., Winitz, S., McArthur-Lewis, J., Hess, S., Ren, X-D., Renshaw, M. W. & Schwartz, M. A. (1996).The regulation of growth and intracellular signaling by integrins. Endocrine Reviews. 17, 207-220.

Minden, D., Lin, A., Claret, F. X., Abo, A. & Karin, M. (1995). Selective activation of the Jnk signaling cascade and c-Jun transcriptional activity by the small GTPases Rac and cdc42Hs. Cell. 81, 1147-1157.

Miyake, S., Yagita, H., Maruyama, T., Hashimoto, H., Miyasaka, N. & Okumura, K. (1993). β1 integrin-mediated interaction with extracellular matrix proteins regulates cytokine gene expression in synovial fluid cells of rheumatoid arthritis patients. J. Exp. Med. 177, 863-868

Miyamoto, S., Akiyama, S. K. & Yamada, K. M. (1995a). Synergistic roles for receptor occupancy and aggregation in integrin transmembrane function. Science. 267, 883-885.

Miyamoto, S., Teramoto, H., Coso, O. A., Gutkind, J. S., Burbelo, P. D., Akiyama, S. K. & Yamada, K. M. (199Sb). Integrin function: molecular hierachies of cytoskeletal and signaling molecules. J. Cell. Biol. 131, 791-805.

Miyamoto, S., Teramoto, H., Gutkind, J. S. & Yamada, K. M. (1996). Integrins can collaborate with growth factors for phosphorylation of receptor tyrosine kineses and MAP kinase activation: Roles of integrin aggregation and occupancy of receptors. J. Cell. Biol. 135, 1633-1642.

Mondal, K., Lofquist, A. K., Watson, J. M., Morris, J. S., Price, L. K., & Haskill, J. S. (1995). Adhesion and direct integrin engagement differentially regulate gene transcription, transcript stabilization and translocation. Biochem. Soc. Trans. 23, 460-464.

Morino, N., Mimura, T., Hamasaki, K., Tobe, K., Ueki, K., Kikuchi, K., Takehara, K., Kadowaki, T., Yazaki, Y. & Nojima, Y. (1995). Matrix/integrin interaction activates the mitogen-activated protein kinase, p44erk-1 and p42erk-2. J. Biol. Chem. 270, 269-273.

Nobes, C. D., Hawkins, P., Stephens, L. & Hall, A. (1995). Activation of the small GTP-binding proteins rho and rac by growth factor receptors. J. Cell Sci. 108, 225-233.

Nobes, C. D. & Hall, A. (1995). Rho, rac and cdc42 GTPases regulate the assembly of multimolecular focal complexes associated with actin stress fibers, lamellipodia and filopodia. Cell. 81, 53-62.

Nuckolls, G. H., Romer, L. H. & Burridge, K. (1992). Microinjection of antibodies against talin inhibits the spreading and migration of fibroblasts. J. Cell Sci. 102, 753-762.

Otey, C. A., Vasquez, G. B., Burridge, K. & Erikson, B. W. (1993). Mapping of the α-actinin binding site within the β1 integrin cytoplasmic domain. J. Biol. Chem. 268, 21193-21197.

O'Toole, T. E., Ylanne, J. & Culley, B. M. (1995) Regulation of integrin affinity states through an NPXY motif in the β subunit cytoplasmic domain. J. Biol. Chem. 268, 8553-8558.

Pavalko, F. M. & Otey, C. A. (1994). Role of adhesion molecule cytoplasmic domains in mediating interactions with the cytoskeleton. Proc. Soc. Exp. Biol. Med. 205, 282-293.

Pawson, T. & Schlessinger, J. (1993). SH2 and SH3 domains. Curr. Biol. 3, 434-442.

Pawson, T. (1995). Protein modules and signaling networks. Nature. 373, 573-580.

Petch, L. A., Bockholt, S. M., Bouton, A., Parsons, J. T. & Burridge, K. (1995). Adhesion-induced tyrosine phosphorylation of the pl30 Src substrate. J. Cell Sci. 108, 1371-1379.

Plopper, G. E., Mcnamee, H. P., Dike, L. E., Bojanowski, K. & Inger, D. E. (1995). Convergence of integrin and growth factor receptor signaling pathways within the focal adhesion complex. Mol. Biol. Cell. 6, 1394-1365.

Polte, T. R. & Hanks, S. K. (1995). Interaction between focal adhesion kinase and Crk-associated kinase substrate, $p130^{cas}$. Proc. Nat. Acad. Sci. 92, 10678-10682.

Pullan, S., Wilson, J., Metcalfe, A., Edwards, G., Goberdhan, N., Tilly, J., Hickman, J. A., Dive, C. & Streuli, C. H. (1996). Requirement of basement membrane for the suppression of programmed cell death in breast epithelium. J. Cell. Sci. 109, 631-642.

Ren, X-D., Bokoch, G. M., Traynor-Kaplan, A., Jenkins, G. H., Anderson, R. A. & Schwartz, M. A. (1996). Physical association of the small GTPase Rho with a 68 kDa phosphotidylinositol 4-phosphate 5-kinase in Swiss 3T3 cells. Mol. Biol. Cell. 7, 435-442.

Renshaw, M. W., Toksoz, D. & Schwartz, M. A. (1996). Involvement of the small GTPase Rho in integrin-mediated activation of mitogen-activated protein kinase. J. Biol. Chem. 271, 21691-21694.

Reszka, A. A., Hayashi, Y. & Horowitz, A. F. (1992). Identification of amino acid sequences in the integrin β1 cytoplasmic domain implicated in cytoskeletal association. J. Cell Biol. 117, 1321-1330.

Richardson, A. & Parsons, T. (1996). A mechanism for regulation of the adhesion-associated protein tyrosine kinase pp^{125FAK} Nature. 380, 538-540.
Ridley, A. J. & Hall, A. (1992). The small GTP-binding protein rho regulates the assembly of focal adhesions and actin stress fibers in response to growth factors. Cell. 70, 389-399.
Ridley, A. J. & Hall, A. (1994). Signal transduction pathways regulating Rho-mediated stress fiber formation: requirement for a tyrosine kinase. EMBO J. 13, 2600-2610.
Ridley, A. J., Paterson, H. F., Johnston, C. L., Diekmann, D. & Hall, A. (1992). The small GTP-binding protein rac regulates growth factor- induced membrane ruffling. Cell. 70, 401-410
Rosales, C., O'Brien, V., Kornberg, L. & Juliano, R. L. (1995). Signal transduction by cell adhesion receptors. Biochim. Biophys. Acta. 1242, 77-98.
Ruoslati, E. & Reed, J. C. (1994). Anchorage dependence, integrins and apoptosis. Cell. 77, 477-478.
Sasaki, H., Nagura, K., Ishino, M., Tobioka, H., Kotani, K. & Sasaki, T. (1995). Cloning and characterisation of cell adhesion kinase beta, a novel protein tyrosine kinase of the focal adhesion kinase subfamily. J. Biol. Chem. 270, 21206-21219.
Sastry, S. K., Lakonishok, M., Thomas, D. A., Muschler, J. & Horowitz, A. F. (1996). Integrin α subunit ratios, cytoplasmic domains, and growth factor synergy regulate muscle proliferation and differentiation. J. Cell. Biol. 133, 169-184.
Schafer, D. A. & Cooper, J. A. (1995). Control of actin assembly at filament ends. Ann. Rev. Cell Dev. Biol. 11, 497-518.
Schaller, M. D., Borgman, C. A., Cobb, B. S., Vines, R. R., Reynolds, A. B. & Parsons, J. T. (1992). Pp^{125FAK} a structurally distinctive protein tyrosine kinase associated with focal adhesions. Proc. Natl. Acad. Sci. 89, 5192-5196.
Schaller, M. & Parsons, J. T. (1994). Focal adhesion kinase and associated proteins. Curr. Opin. Cell. Biol. 6, 705-710.
Schaller, M. D., Hildebrand, J. D., Shannon, J. D., Fox, J. W., Vines, R. R. & Parsons, J. T. (1994). Autophosphorylation of the focal adhesion kinase pp^{125FAK}, directs SH2-dependent binding of $pp60^{src}$. Mol. Cell. Biol. 14, 1680-1688.
Schaller, M. D., Otey, C. A., Hildebrand, J. D. & Parsons, J. T. (1995a). Focal adhesion kinase and paxillin bind to peptides mimicking β integrin cytoplasmic domains. J. Cell. Biol. 130, 1181-1187.
Schaller, M. D. & Parsons, J. T. (1995b). pp^{125FAK}-dependent tyrosine phosphorylation of paxillin creates a high affinity binding site for Crk. Mol. Cell. Biol. 15, 2635-2645.
Schlaepfer, D. D., Hanks, S. K., Hunter, T. & van der Geer, P. (1994). Integrin-mediated signal transduction linked to Ras pathway by Grb2 binding to focal adhesion kinase. Nature. 372, 786-791.
Schlaepfer, D. D. & Hunter, T. (1996). Evidence for in vivo phosphorylation of the Grb2 SH2-domain binding site on focal adhesion kinase by Src-family protein-tyrosine kineses. Mol. Cell. Biol. 16, 5623-5633.
Schlaepfer, D. D. & Hunter, T. (1997). Focal adhesion kinase overexpression enhances Ras-dependent integrin signaling to Erk2/mitogen-activated protein kinase through interaction with and activation of c-src. J. Biol. Chem. 272,13189-13195.
Schlaepfer, D. D., Broome, M. A. & Hunter, T. (1997). Fibrinogen-stimulated signaling from a focal adhesion kinase-c-Src complex: Involvement of the Grb2, pl30cas, and Nck adaptor proteins. Mol. Cell. Biol. 17, 1702-1713.
Schwartz, M. A., Ingber, D. E., Lawrence, M., Springer, T. A. & Lechene, C. (1991). Multiple integrins share the ability to induce elevation of intracellular pH. Exp. Cell. Res. 195, 533-535.
Schwarz, M. A. & Lechene, C. (1992). Adhesion is required for protein kinase C-dependent activation of the Na-H antiporter by platelet-derived growth factor. Proc. Natl. Acad. Sci. 89, 6138-6141.
Schwartz, M. A. (1993). Spreading of human endothelial cells on fibronectin or vitronectin triggers elevation of intracellular free calcium. J. Cell. Biol. 120, 1003-1010.

Schwarz, M. A. & Denninghoff, K. (1994). αv in integrins mediate the rise in intracellular calcium in endothelial cells on fibronectin even though they play a minor role in adhesion. J. Biol. Chem. 269, 111333-111337.

Schwartz, M. A., Schaller, M. D. & Ginsberg, M. H. (1995). Integrins - emerging paradigms of signal-transduction. Annul Rev. Cell Dev. Biol. 11, 549-599.

Sheppard, D. (1996). Epithelial integrins. Bioessays. 18, 655-660.

Silvennoinen, O., Schindler, C., Schlessinger, J. & Levy, D. E. (1993). Ras-independent growth factor signaling by transcription factor tyrosine phosphorylation. Science. 261, 1736-1739.

Sjaastad, M. D. & Nelson, W. J. (1996). Integrin-mediated calcium signaling and regulation of cell adhesion by intracellular calcium. Bioessays. 19, 47-55.

Somani, A-K., Bignon, J. S., Mills, G. B., Siminovitch, K. A. & Branch, D. R. (1997). Src kinase activity is regulated by the SHP-1 protein tyrosine phosphatase. J. Biol. Chem. 272, 21113-21119.

Sporn, S. A., Eierman, D. F., Johnson, C. E., Morris, J., Martin, G., Ladner, M. & Haskill, S. (1990) Monocyte adherence results in the selective induction of novel genes sharing homology with mediators of inflammation and tissue repair. J. Immunol. 144, 4434-4441.

Streuli, C. H., Bailey, N. & Bissell, M. J. (1991). Control of mammary epithelial differentiation: basement membrane induces tissue-specific gene expression in the absence of cell-cell interaction and morphological polarity. J. Cell Biol., 115, 1383-1395.

Streuli, C. H., Schmidhauser, C., Bailey, N., Yurchenco, P., Skubitz, A. P. N., Roskelley, C. & Bissell, M. J. (1995b). Laminin mediates tissue-specific gene expression in mammary epithelia. J. Cell Biol. 129, 591-603.

Streuli, C. H., Edwards, G. M., Delcommenne, M., Whitelaw, C. B. A., Burdon, T. G., Schindler, C. & Watson, C. J. (1995b). Stat5 as a target for regulation by extracellular matrix. J. Biol. Chem. 270, 21639-21644.

Streuli, C. H. & Edwards, G. M. (1998). Control of normal mammary epithelial phenotype by integrins. J. Mammary Gland Biol. and Neoplasia. 3, 151-163.

Superti-Furga, G. & Courtneidge, S. A. (1995). Structure-function relationships in Src family and related protein tyrosine kineses. Bioessays. 17, 321-330.

Symington, B. E. (1992). Fibronectin receptor modulates cyclin-dependent kinase activity. J. Biol. Chem. 267, 25744-25747.

Symington, B. E. (1995). Growth signaling through the α5β1 fibronectin receptor. Biochem. Biophys. Res. Comm. 208, 126-134.

Tanaka, S., Morishita, T., Hahimoto, Y., Hattori, S., Nakamura, S., Shibuya, M., Matuoka, K., Takenawa, T., Kurata, T. & Nagashima, K. (1994). C3G, a guanine nucleotide-releasing protein expressed ubiquitously, binds to the Src-homology 3 domains of Crk and Grb2/Ash proteins. Proc. Natl. Acad. Sci. 91, 3443-3447.

Tapley, P., Horowitz, A., Buck, C., Burridge, K., Duggan, K. & Rohrschneider, L. (1989). Integrins isolated from Rous sarcome virus-transformed chicken embryo fibroblasts. Oncogene. 4, 325-333.

Tapon, N. & Hall, A. (1997). Rho, Rac and Cdc42 GTPases regulate the organization of the actin cytoskeleton. Curr. Opin. Cell Biol. 9, 86-92.

Treisman, R. (1996). Regulation of transcription of MAP kinase cascades. Curr. Opin. Cell Biol. 8, 205-215.

Tucker, R. W., Meade-Cobun, K. & Ferris, D. (1990). Cell shape and increased free cytostolic calcium $[Ca^{2+}]i$ induced by growth factors. Cell Calcium. 11, 201-209.

Turner, C. E. (1994). Paxillin: a cytoskeletal target for tyrosine kineses. Bioessays. 16, 47-52.

Turner, C. E. & Miller, J. T. (1994). Primary sequence of paxillin contains putative SH2 and SH3 domain binding motifs and multiple LIM domains: identification of a vinculin and pp^{125FAK}-binding region. J. Cell Sci. 107, 1583-1591.

Vignoud, L., Usson, Y., Balzac, F., Tarone, G. & Block, M. R. (1994). Internalisation of the α5β1 integrin does not depend on NPXY signals. Biochem. Biophys. Res. Commun. 199, 603-611.

Vignoud, L., Albiges-Rizo, C., Frachet, P. & Block, M. R. (1997). NPXY motifs control the recruitment of the $\alpha5\beta1$ integrin in focal adhesions independently of the association of talin with the $\beta1$ chain. J. Cell Sci. 110, 1421-1430.

Vuori, K. & Ruoslahti, E. (1993). Activation of protein kinase C precedes $\alpha5\beta1$ integrin-mediated cell spreading on fibronectin. J. Biol. Chem. 268, 21459-21462.

Vuori, K. & Ruoslahti, E. (1994). Association of insulin receptor substrate-1 with integrins. Science. 266, 1576-1578.

Vuori, K., Hirai, H., Aizawa, S. & Ruoslahti, E. (1996). Induction of p130^{cas} signaling complex formation upon integrin-mediated cell adhesion: a role for Src family kineses. Mol. Cell. Biol. 16, 2606-2613.

Wachsstock, D. H., Wilkins, J. A. & Lin, S. (1987). Specific interactions of vinculin with α-actinin . Biochem. Biophys. Res. Comm. 146, 554-560.

Wary, K. K., Mainiero, F., Isakoff, S. J., Marcantonio, E. E. & Gianocotti, F. G. (1996). The adapter protein Shc couples a class of integrins to the control of cell cycle progression. Cell. 87, 733-743.

Weng, Z. G., Taylor, J. A., Turner, Ce, Brugge, J. S. & Seidel-Dugan, C. (1993). Detection of Src homology 3-binding proteins, including paxillin, in normal and v-Src-transformed Balb/c 3T3 cells. J. Biol. Chem. 268, 14956-14963.

Werb, Z., Tremble, P. M., Berhrendtsen, O., Crowley, E. & Damsky, C. H. (1989). Signal transduction through the fibronectin receptor induces collagenase and stomelysin gene expression. J. Cell Biol. 109, 877-889.

Wood, C. K., Turner, C. E., Jackson, P. & Critchley, D. R. (1994). Characterisation of the paxillin-binding site and the C-terminal focal adhesion targeting sequence in vinculin. J. Cell Sci. 107, 709-717.

Woods, A. & Couchman, J. R. (1992). Protein kinase C involvement in focal adhesion formation. J. Cell Sci. 101, 277-290.

Xing, Z., Chen, H. C., Nowlen, J. K., Taylor, S. J., Shalloway, D. & Guan, J. L. (1994). Direct interaction of v-Src with focal adhesion kinase mediated by the Src SH2 domain. Mol. Biol. Cell. 5, 413-421.

Xu, W., Harrison, S. C. & Eck, M. J. (1997). Three-dimensional structure of the tyrosine kinase c-Src. Nature. 385, 595-602.

Ylanne, J., Chen, Y., O'Toole, T. E., Loftus, J. C., Takada, Y. & Ginsberg, M. H. (1993). Distinct functions of integrin α and β subunit cytoplasmic domains in cell spreading and formation of focal adhesions. J. Cell Biol. 122, 223-233.

Yamada, K. M. & Miyamoto, S. (1995). Integrin transmembrane signaling and cytoskeletal control. Curr. Opin. Cell Biol. 7, 681-689.

Yurochko, A. D., Liu, D. Y., Eierman, D. & Haskill, S. (1992). Integrins as a primary signal transduction molecule regulating monocyte immediate-early gene induction. Proc. Natl. Acad. Sci. 89, 9034-9038.

Zhang, J., King, W. G., Dillon, S., Hall, A., Feig, L. & Rittenhouse, S. E. (1993). Activation of platelet phosphatidylinositol 3-kinase requires the small GTP-binding protein Rho. J. Biol. Chem. 268, 22251-22254.

Zhu, X. & Assoian, R. K. (1995). Integrin-dependent activation of MAP kinase: a link to shape-dependent cell proliferation. Mol. Biol. Cell. 6, 273-282.

Zhu, X., Ohtsubo, M., Bohmer, R. M., Roberts, J. M. & Assoian, R. K. (1996). Adhesion-dependent cell-cycle progression linked to the expression of cyclin D1, activation of cyclin E-cdk2, and phosphorylation of the retinoblastoma protein. J. Cell Biol. 133, 391-403.

Zigmond, S. H. (1996). Signal transduction and actin filament organisation. Curr. Biol. 8, 6673.

SIGNALING AND PLATELET ADHESION

Xiaoping Du and Mark H. Ginsberg

Advances in Molecular and Cell Biology
Volume 28, pages 269-301.

ISBN: 0-7623-0495-2

I. INTRODUCTION

Platelet adhesion is critical for response to vascular injury and formation of primary hemostatic thrombi. Platelets in normal circulation are in a resting, non-adherent state. At sites of vascular injury, platelets adhere to the exposed subendothelial matrix. This initial platelet adhesion triggers signals which, in concert with signals initiated by soluble platelet agonists, leads to the activation of platelets. The activated platelets firmly adhere and spread onto the subendothelial matrix, and aggregate (adhere to each other) to form a primary thrombus. Platelet aggregation leads to additional intracellular signaling and secondary platelet responses including secretion of granule contents, further platelet aggregation, release of procoagulant membrane particles, and clot retraction. Thus, the platelet response to vascular injury involves a cascade of amplified cycles of platelet adhesion and signaling. At the molecular level, adhesion and aggregation of platelets are mediated by the interaction of platelet adhesion receptors with their ligands. Ligand binding to platelet adhesion receptors can be regulated by intracellular signaling and can trigger intracellular signaling events. Thus, understanding the relationship between the platelet adhesion receptors and the intracellular signaling pathways has been and will be a focus of platelet biology.

II. PLATELET ADHESION RECEPTORS

Platelet adhesion receptors may be divided into two major groups: the initiating adhesion receptors that mediate initial platelet adhesion and activation and the responsive adhesion receptors that mainly function as a response to platelet activation.

A. Initiating Adhesion Receptors

The initiating adhesion receptors recognize insoluble adhesive proteins that are normally not exposed to the blood, but become available to platelets after disruption of endothelial cell barrier. The subendothelial adhesive proteins mediating platelet adhesion include collagen, subendothelial matrix-bound von Willebrand factor (vWF), fibronectin, and possibly other subendothelial matrix components. The initiating platelet adhesion receptors are constitutively active on the surface of circulating platelets and function immediately when exposed to the subendothelial matrix. These adhesion receptors include glycoprotein Ib-IX complex (GPIb-IX), which functions as a vWF receptor, the integrin $\alpha_2\beta_1$ which functions as a collagen receptor, and other platelet collagen receptors such as glycoprotein IV (GPIV) and glycoprotein VI (GPVI. p62), and may include other β_1 integrins such as $\alpha_5\beta_1$ as a fibronectin receptor, $\alpha_6\beta_1$ as a laminin receptor. It is known that GPIb-IX and collagen receptor(s) initiate transmembrane signals that leads to platelet activation.

B. Responsive Adhesion Receptors

This group of platelet adhesion receptors includes the integrin $\alpha_{IIb}\beta_3$ (GPIIb-IIIa), which is the most abundant adhesion receptor in platelets. The integrin $\alpha_{IIb}\beta_3$ functions as a receptor for fibrinogen, fibronectin, vitronectin, vWF and thrombospondin, and mediates platelet aggregation, firm adhesion as well as spreading. In addition, P-selectin (GMP140), which binds glycoprotein ligands such as PSGL-1 and mediate adhesion of activated platelets to white blood cells, appears on the platelet surface only after activation (for review of P-selectin, see Furie and Furie, 1995). The ligands for this group of adhesion receptors usually are present in circulating blood, but the interaction with their receptors are regulated either by altering receptor affinity (in the case of $\alpha_{IIb}\beta_3$) or by translocating the receptor to the platelet surface (P-selectin).

III. GLYCOPROTEIN Ib-IX COMPLEX-MEDIATED ADHESION AND SIGNALING

A. The Glycoprotein Ib-IX Complex

Platelet glycoprotein Ib-IX complex (GPIb-IX) consists of three glycosylated polypeptide chains, GPIbα, GPIbβ and GPIX (Berndt et al., 1983, 1985; Du et al., 1987; Phillips and Agin, 1977). GPIbα and GPIβ are linked by disulfide bonds. GPIX is non-covalently associated with GPIb (Berndt et al., 1985; Du et al., 1987). There has been controversy with regard to whether GPIX directly interact with GPIbα or GPIbβ (Lopez et al., 1994; Wu et al., 1996). GPIb-IX is associated with another platelet membrane glycoprotein, glycoprotein V (GPV) (Modderman et al., 1992). Each subunit of GPIb-IX is the product of a distinct gene (Lopez et al., 1987, 1988; Hickey et al., 1989; Hickey et al., 1993; Hickey and Roth, 1993). All of these subunits, however, shares homologous sequences characterized by one or more tandem repeats of leucine-rich sequences (Hickey et al., 1993; Lopez et al., 1990, 1987; Lopez et al., 1988). Leucine rich motifs are present in a variety of proteins and have been implicated in protein-protein interactions (Lopez, 1994). GPIbα is the largest subunit with 610 amino acid residues, including a 485 residue extracellular domain, a 29 residue transmembrane segment and an intracellular domain of 100 residues (Lopez et al., 1987). The central region of GPIbα extracellular domain is highly glycosylated. Carbohydrate side chains in GPIbα account for about 50% of its molecular weight, and are the major contributor of the so-called "glycocalyx" layer on the platelet surface (Okumura and Jamieson, 1976a). Using Rotary-shadowing electron microscopy technique, GPIb-IX appears to be a semi-flexible rod with one globular domain on each end (Fox et al., 1988). The rod-like region corresponds to heavily glycosylated region of GPIbα (macroglycopeptide region). The small globular domain represents the N-terminal ligand binding region of $GPIb_{\alpha,}$ and the larger globular domain corresponds to the membrane-proximal and cytoplasmic regions of GPIbα complexed with GPIbβ and GPIX. In the cytoplasmic domain of GPIbβ, serine residue 166 is phosphorylated by cAMP-dependent protein kinase (Fox et al., 1987; Wardell et al., 1989). This phosphorylation is associated with regulation of actin polymerization (Fox and Berndt, 1989). Both GPIbβ and GPIX are acylated with palmitic acid or myristic acid (GPIX) (Muszbek and Laposata, 1989; Schick and Walker, 1996). The cytoplasmic domain of GPIb-IX is attached to the membrane skeleton, which is the short actin filamental structure underlining the membrane (Fox, 1985). The association is mediated by the interaction between the actin-binding protein 250 (ABP-250, also called filamin) and the central region (Thr536-Phe568) of GPIbα cytoplasmic domain (Andrews and Fox, 1992). Association of GPIb-IX to the membrane skeleton may provide platelet membranes with anchoring site to the cytoskeleton and thus contribute to membrane stability and cell shape. This association may also render GPIb-IX-mediated platelet adhesion resistant to shear stress.

B. GPIb-IX as a vWF Receptor

The function of GPIb-IX as a vWF receptor and its role in platelet adhesion was first indicated in studies on the hereditary platelet functional disorder, Bernard-Soulier syndrome (Nurden and Caen, 1975). Bernard-Soulier platelets which lacks GPIb-IX fail to adhere to denuded subendothelial matrix and to respond to ristocetin which induces vWF binding to platelets and vWF-dependent platelet agglutination. This was confirmed by studies using proteases, anti-GPIb-IX antibodies, and in vitro binding of vWF to purified GPIb-IX (Berndt et al., 1988) (for reviews see Lopez, 1994; Ruggeri, 1991).

Normally, GPIb-IX on the platelet surface does not bind to soluble vWF in circulation. Binding of GPIb-IX occurs when vWF is deposited on the subendothelial matrix or other surface (Sakariassen et al., 1986; Savage et al., 1992; Weiss et al., 1974), under pathologically high shear rate conditions (Peterson et al., 1987), or in type IIB and platelet type von Willebrand disease caused by mutational changes in either vWF (Cooney et al., 1992; De Marco et al., 1985b, 1987; Randi et al., 1991; Ware et al., 1991) or GPIb-IX (Miller et al., 1991; Russell and Roth, 1993). As the latter conditions do not occur in normal circulation, exposure of subendothelial matrix in the case of vascular injury is probably the physiological triggering mechanism of the vWF-GPIb-IX interaction. In *in-vitro* studies, two modulators of vWF, ristocetin (Howard and Firkin, 1971) and botrocetin (Andrews et al., 1989; Read et al., 1989), are often used to induce vWF binding to GPIb-IX. These two modulators bind to vWF and induce a change that mimics the effect of vWF binding to subendothelial matrix (Andrews et al., 1989; Sugimoto et al., 1991; Berndt et al., 1992).

Several lines of evidence convincingly indicate that vWF binds to the N-terminal 45 kDa fragment of GPIbα (residues 1-290) (for reviews, see Ruggeri, 1991; Lopez, 1994). Further examination of vWF binding using synthetic peptides showed that synthetic peptides corresponding to Ser251-Glu279 (Vicente et al., 1990) and Asp235-Lys262 (Katagiri et al., 1990) were the most potent in inhibiting vWF binding to GPIb induced by ristocetin, indicating a vWF binding site. C-terminal to the ristocetin-dependent vWF binding site is the sequence containing a cluster of sulfated tyrosine residues (Tyr^{275}-Asp-Tyr-Tyr-Pro-Glu-Glu) (Dong et al., 1994). This region is critical for the vWF binding selectively induced by botrocetin but not ristocetin (Vicente et al., 1990; Ward et al., 1996). Sulfation on tyrosine residues enhances vWF binding (Dong et al., 1994; Marchese et al., 1995). In addition, studies on the variant Bernard-Soulier platelet expressing dysfunctional GPIb-IX with mutations at Leu57(to Phe) and Alal56 (to Val) suggest that N-terminal and leucine rich motifs are also critical for the vWF binding (Miller et al., 1992; Ware et al., 1993).

VWF binding to GPIb-IX mediates platelets adhesion to the subendothelial matrix. This vWF and GPIb-IX-dependent platelet adhesion is reversible, but effectively retains the fast flowing platelets onto the matrix surface (Savage et al., 1996; Weiss et al., 1986), in a way similar to leukocyte rolling on vascular wall. In this process, platelets becomes activated, and firmly adhere and spread onto the matrix via the integrin $\alpha_{IIb}\beta_3$ (Weiss et al., 1986; Savage et al., 1992; Savage et al., 1996). Platelet activation can also be induced in vitro when vWF is induced to bind to GPIb-IX by ristocetin and botrocetin, under high shear stress (Peterson et al., 1987), or desialylation (De Marco et al., 1985a; Gralnick et al., 1985). vWF from type IIb von Willebrand disease also induces platelet activation (De Marco et al., 1985b; De Marco et al., 1987). This indicates that vWF-GPIb-IX interaction initiates a platelet activation signal. GPIb-IX-dependent platelet adhesion to subendothelial matrix plays a major role under high shear rate conditions, which occurs in small arteries and capillaries where blood flow is fast and blood vessels are narrower (Baumgartner et al., 1980; Weiss et al., 1978). As GPIb-IX binding to surface-bound vWF is not dependent on the shear rate (Savage et al., 1992), it is likely that vWF and GPIb-IX-induced platelet activation rather than vWF-GPIb interaction per se requires high shear stress.

C. GPIb-IX as a Thrombin Receptor

GPIb-IX is also the major thrombin binding protein on the platelet surface (Okumura and Jamieson, 1976). The binding site for thrombin is also localized in the 45 kDa N-terminal fragment of GPIbα (Okumura and Jamieson, 1976b; Wicki and Clemetson, 1987). Studies using synthetic peptides and proteolytic cleavages indicate that region Tyr276-Glu282 (YDYYPEE) is critical for thrombin binding (De Marco et al., 1994; Ward et al., 1996). The region around YDYYPEE is highly anionic and aligns well with the thrombin binding sites of the heptahelical thrombin receptor and with the thrombin inhibitor, hirudin (De Marco et al., 1994; Lopez, 1994). Several monoclonal antibodies that inhibited thrombin binding to fixed platelets recognize this region of GPIbα, indicating that the C-terminal region of the 45-kDa fragment contains the thrombin binding site (Ward et al., 1996). The thrombin binding site overlaps with the sequence critical for botrocetin-induced vWF binding.

The role of GPIb-IX in thrombin-induced platelet activation is not clear. In addition to GPIb-IX, platelets express a heptahelical G-protein coupled thrombin receptor (Hung et al., 1992; Vu et al., 1991) that is involved in platelet activation. However, in GPIb-deficient platelets either from Bernard-Soulier syndrome patients (Jamieson and Okumura, 1978) or protease-treated (Cooper et al., 1982), the response to thrombin is impaired. Further, monoclonal antibodies against the N-terminal thrombin binding region of GPIbα block low dose thrombin-induced platelet aggregation (McGregor et al., 1983; Yamamoto et al., 1985). Thus, GPIb-IX seems to potentiate thrombin-induced signaling via the heptahelical receptor. Alternatively, it is also possible that thrombin binding to GPIb-IX may initiate an independent activation signal, as the thrombin-induced increase in intracellular calcium was completely inhibited only in the presence of both antibodies against GPIb and the heptahelical receptor (Greco et al., 1996). In this regard, mouse platelets lacking the heptahelical thrombin receptor are still responsive to thrombin (Connolly et al., 1996).

D. Signaling Mediated by GPIb-IX

Interaction of vWF and a snake venom protein alboaggregin (Andrews et al., 1996) with GPIb-IX induces platelet activation associated with a series of intracellular signaling events. These include the breakdown of phosphatidylinositol 4,5-bisphosphate, the production of phosphatidic acid, the activation of protein kinase C (PKC), increase of cytoplasmic calcium, the synthesis of thromboxane A2 (Kroll et al., 1991, 1993), increased tyrosine phosphorylation of intracellular proteins (Razdan et al., 1994) as well as the activation and translocation of PI-3 kinase and protein tyrosine kinase pp^{60src} (Jackson et al., 1994). The sequence of these signaling events has not been elucidated. Release of arachidonic acid by phospholipase A2 and production of thromboxane A2 appear to be early events as the breakdown of phosphatidylinositol 4,5-bisphosphate the production of phosphatidic acid, the activation of protein kinase C (PKC), increase of cytoplasmic calcium were shown to be inhibitable by indomethacin which inhibits cyclooxygenase in the arachidonic acid pathway (Kroll et al., 1991). However, the activation of PI-3 kinase and translocation of pp^{60src} was not inhibited by aspirin which also inhibits the cycloxygenase (Jackson et al., 1994), indicating that these latter events are likely to precede or to be on a pathway different from the arachidonic acid signaling pathway.

It is not clear how GPIb-IX is linked to intracellular signaling pathways. As the cytoplasmic domain of GPIbα is linked to the membrane skeleton, one possibility is that vWF binding to GPIb-IX may alter the GPIb-IX-associated membrane skeleton. Cunningham and co-workers (1996) showed that vWF binding to recombinant GPIb-IX transfected cells caused changes in cell morphology. Expression of GPIb-IX with truncated cytoplasmic domain of GPIbα (to abolish cytoskeleton interaction) induced a different morphological change. Truncation of GPIbα cytoplasmic domain, however, blocks not only the cytoskeleton interaction, but also the interaction of GPIb-IX with other intracellular molecules. We have found that GPIb-IX is associated with the ξ-form 14-3-3 protein (Du et al., 1994). The binding site for the ξ-form 14-3-3 protein is located in a 15 residue region at the C-terminus of GPIbα (Du et al., 1996). In addition, a phosphorylation-dependent 14-3-3ξ binding site is located in the cytoplasmic domain of GPIbβ (Andrews et al., 1998; Calverley et al., 1998). High affinity binding between 14-3-3ξ and GPIb-IX probably involves a dimerized 14-3-3ξ and both binding sites in GPIbα and GPIbβ (Andrews et al., 1998; Gu and Du, 1998).

The ξ-form 14-3-3 protein in platelets was first reported by Zupan and colleagues (Zupan et al., 1992) to be identical to a 30 kDa cytosolic phospholipase A2 (PLA2). However, this report has not been confirmed by other groups (Robinson et al., 1994). Recent studies revealed that the 14-3-3 protein interacts with many intracellular signaling molecules including C-Raf (Fanti et al., 1994; Freed et al., 1994; Fu et al., 1994; Irie et al., 1994; Li et al., 1995), protein kinase C (Toker et al., 1990; Aitken et al., 1992; Isobe et al., 1992; Tanji et al., 1994; Meller et al., 1996) Bcr kinase (Reuther et al., 1994) and PI-3 kinase (Bonnefoy et al., 1995), and the interaction appears to require phosphorylation of these molecules on serine residues (Muslin et al., 1996). However, no significant phosphorylation has been identified in the cytoplasmic domain of GPIbα (Wyler et al., 1986). A synthetic peptide corresponding to the 14-3-3 protein binding site in GPIbα bound to the 14-3-3 protein, suggesting that phosphorylation may not be required for GPIbα binding (Du et al., 1996). It is possible that the phosphorylation-independent binding of the 14-3-3 protein to GPIbα may serve as a mechanism to link the phosphorylated signaling molecules to the cytoplasmic domain of GPIbα. The 14-3-3 protein regulates the functions of Raf kinase (Fanti et al., 1994; Freed et al., 1994; Fu et al., 1994; Irie et al., 1994; Li et al., 1995), protein kinase C (Toker et al., 1990; Aitken et al., 1992; Isobe et al., 1992; Tanji et al., 1994; Meller et al., 1996), and PI-3 kinase (Bonnefoy et al., 1995). GPIb-mediated signals activate protein kinase C and PI-3 kinase (Jackson et al. 1994; Kroll et al., 1991, 1993). Raf is upstream in the mitogen activated protein (MAP) kinase pathway. Raf activates MAP kinase kinase which subsequently activates MAP kinase. Whether the 14-3-3 protein may link GPIb-IX to these intracellular pathways is yet to be examined. Recently, Dong and colleagues (1997) showed that deletion of GPIbα C-terminal 14-3-3ξ

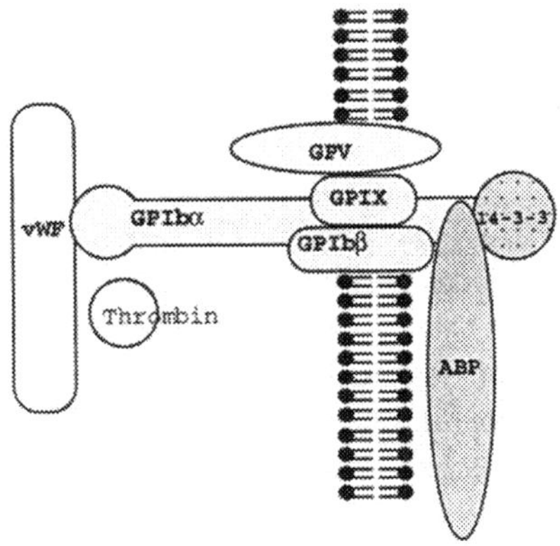

Figure 1. A schematic of the glycoprotein Ib-IX complex and its ligand interaction.

binding sequence increased lateral mobility of GPIb-IX on the membrane. As the lateral mobility of GPIb-IX is probably constrained by the actin filament network linked to GPIbα via ABP-250, it is possible that interaction between GPIb-IX and 14-3-3ξ is involved in the regulation of GPIb-IX interaction with the cytoskeleton. Interestingly, phosphorylation of GPIbβ by cAMP-dependent protein kinase is associated with inhibition of actin polymerization during platelet activation (Fox and Berndt, 1989). As phosphorylation of GPIbβ regulates 14-3-3 binding (Andrews et al., 1998; Calverley et al., 1998), it is possible that regulation of GPIb-IX-14-3-3 interaction is responsible for the inhibitory effect of cAMP-dependent kinase on actin polymerization during platelet activation.

IV. INTEGRIN-MEDIATED PLATELET ADHESION AND SIGNALING

A. Platelet Integrins

Integrins are a family of heterodimeric cell adhesion receptors (Hypes, 1992). Members of integrin family show strong similarity in structure and function. Platelet contains two major types of integrins, β1 and β3 integrins (Ginsberg et al., 1995). The β1 integrins expressed in platelets are $\alpha_2\beta_1$ (collagen receptor), $\alpha_5\beta_1$ (fibronectin receptor), and $\alpha_6\beta_1$ (laminin receptor). The β_3 integrins expressed in platelets are $\alpha_{IIb}\beta_3$, and $\alpha_v\beta_3$. $\alpha_{IIb}\beta_3$ is the most abundant integrin and the major responsive adhesion receptor on the platelet surface. $\alpha_{IIb}\beta_3$ and $\alpha_V\beta_3$ share common ligands such as fibrinogen, vWF, vitronectin, and fibronectin. Binding of β_1 integrins to their ligands immobilized in extracellular matrix (collagen, fibronectin, laminin) does not appear to require prior platelet activation, and thus appears to be regulated by exposure of subendothelial matrix.

B. Integrin $\alpha_{IIb}\beta_3$

Integrin $\alpha_{IIb}\beta_3$ is a cation-dependent heterodimer complex of two type I membrane glycoproteins, α_{IIb} and β_3. The α_{IIb} consists of a heavy chain (GPIIbα, 125 kDa) and a light chain (GPIIbβ, 25 kDa) linked by a disulfide bond (Phillips and Agin, 1977). The heavy and light chains are generated by proteolytic cleavage of a single polypeptide post-translationally (Poncz et al., 1987; Loftus et al., 1988). The α_{IIb} heavy chain has 871 residues and light chain 137 residues. The predicted trans-membrane domain and a 26 residue cytoplasmic domain are located within the light chain (Poncz et al., 1987). Features in α_{IIb}'s primary structure include four extracellular divalent cation binding motifs and an cytoplasmic GFFKR motif, both of which are highly homologous to other integrin α subunits, and are important to the regulation of ligand binding (Poncz et al., 1987; D'Souza et al., 1990, 1991; Hughes et al., 1995;). The β_3 subunit (95 kDa) consists of 762 amino acid residues with a 692-residue extracellular domain, ~25 residue transmembrane domain and a 45 residue cytoplasmic domain (Fitzgerald et al., 1987; Rosa et al., 1988). In its extracellular domain, there is a protease-resistant cysteine rich region with several intra-molecular disulfide bonds. A loop formed by disulfide bonding links the cysteine-rich region to the N-terminal region (Calvete et al., 1991b). The cytoplasmic domain of β_3 contains two NXXY motifs which are similar to the NPXY motif identified in LDL receptor as a internalization signal (Chen et al., 1990). There are four calpain cleavage sites flanks the two NXXY motifs in the cytoplasmic domain of β_3 (Du et al., 1995).

By rotary-shadowing electron microscopy (Carrell et al., 1985; Weisel et al., 1992), The $\alpha_{IIb}\beta_3$ complex appears to have a ~10 nm globular head domain which probably contains the complexed N-terminal regions of α_{IIb} and β_3, and two 18 nm long flexible tails which probably contain C-terminal portions of α_{IIb} (Weisel et al., 1992) and β_3 (Du et al., 1993) respectively. Stable complex between α_{IIb} and β_3 is mediated by the interaction of the N-terminal portions of both subunit (Lam, 1992; Wippler et al., 1994). There is evidence that the cytoplasmic domains of both subunits can also interact with each other in a divalent cation dependent manner (Haas and Plow, 1996).

C. Ligand Binding Function of Integrin $\alpha_{IIb}\beta_3$

The integrin $\alpha_{IIb}\beta_3$ recognizes adhesive proteins usually containing RGD sequence or mimetics of RGD sequence, including fibrinogen, vWF, vitronectin, fibronectin, and thrombospondin (for reviews see Ginsberg et al., 1993, 1995). Fibrinogen contains four RGD sites, one at N-terminal ($A\alpha^{95\text{-}97}$) and one at C-terminal ($A\alpha^{572\text{-}574}$) ends of each α-chain (Andrieux et al., 1991; Cheresh et al., 1989). Integrin $\alpha_{IIb}\beta_3$ also recognizes the fibrinogen γ-chain C-terminal 12 residues sequence, HHLGGAKQAGDV (γ400–411) (Kloczewiak et al., 1984), which probably functions as an RGD mimetic, and compete with RGD for the same binding site in $\alpha_{IIb}\beta_3$ (Lam et al., 1987). The γ-chain C-terminal sequence is probably primarily responsible for the interaction with $\alpha_{IIb}\beta_3$ in the process of platelet aggregation (Farrell et al., 1992; Farrell and Thiagarajan, 1994; Holmback et al., 1996), but may not be required for integrin interaction with fibrin during clot retraction (Holmback et al., 1996; Rooney et al., 1996). VWF contains an RGD sequence in its C-terminal region (residues 1744–1746), which is responsible for binding to $\alpha_{IIb}\beta_3$.

The N-terminal globular region of the $\alpha_{IIb}\beta_3$ complex contains the ligand binding sites, which involves both α_{IIb} and β_3 (Lam, 1992; Weisel et al., 1992; Wippler et al., 1994). In β_3, several lines of evidence suggest that residues 109–171 are likely to be a ligand binding site: (1) RGD was cross-linked to the residues 109–171 fragment (D'Souza et al., 1988); (2) Point mutations at residues 119, 121, and 123 abrogate the ligand binding to the integrin (mutation at D119 to Y naturally occurred in a variant thrombasthenic patient) (Loftus et al., 1990; Bajt and Loftus, 1994); (3) Antibodies that recognize this region inhibit ligand binding to $\alpha_{IIb}\beta_3$ (Andrieux et al., 1991; Calvete et al., 1991; D'Souza et al., 1994); and (4) A peptide derived from this region directly binds to fibrinogen (D'Souza et al., 1994). A second region of β_3 (residues 211–222) may be involved in fibrinogen binding. A synthetic peptide corresponding to this region of β_3 inhibited fibrinogen binding to $\alpha_{IIb}\beta_3$, and bound to fibrinogen (Charo et al., 1991). Point mutations within this region and antibodies against this region abolished ligand binding to $\alpha_{IIb}\beta_3$ (Charo et al., 1991; Bajt et al., 1992). However, the possibility that this region may have an alternative regulatory role was suggested by the observation that this peptide also bound to $\alpha_{IIb}\beta_3$ itself (Steiner et al., 1993). Fibrinogen γ-chain peptide was cross-linked to a fragment of α_{IIb} (residues 294-314) (D'Souza et al. 1990, 1991). Synthetic peptides from this region as well as antibodies against this peptide inhibited fibrinogen binding and platelet aggregation, indicating that it is also a ligand contact site (D'Souza et al., 1991; Taylor and Gartner, 1992). It is likely that these different ligand contact sites in $\alpha_{IIb}\beta_3$ form a ligand binding pocket.

Divalent cations probably directly participate the formation of the recognition site. The ligand contact region in β_3 (residues 109–131) can bind divalent cations (D'Souza et al., 1994). This region contains sequence similarity to the EF hand calcium binding region (Loftus et al., 1990). Mutation that substitute D119 to an alanine completely abolished the ligand binding function and also results in a conformation change in the integrin similar to the loss of cation binding (Loftus et al., 1990). Loftus and co-workers (Loftus et al., 1996) found that an α_V/α_{IIb} chimera with the N-terminal portion of α_V substituted with a 334 residue fragment of α_{IIb} reconstituted the ligand specificity of $\alpha_{IIb}\beta_3$, indicating that the N-terminal region including four calcium binding repeats probably forms the structural basis for ligand specificity. However, a predicted turn structure well conserved in the N-terminal regions of α_{IIb}, and other α subunits outside the calcium binding domains has been proposed to be important for ligand binding (Irie et al., 1995; Kamata et al., 1996).

The physiological role of $\alpha_{IIb}\beta_3$ was first indicated in the studies on a hereditary hemostatic disorder, Glanzmann's thrombasthenia (Nurden and Caen, 1974). Thrombasthenic platelets lacks the integrin and are defective in platelet adhesion, aggregation, fibrinogen binding, and hemostasis. Soluble fibrinogen binding to $\alpha_{IIb}\beta_3$ is required for platelet aggregation and normal hemostasis (Holmback et al., 1996). Soluble vWF binding to $\alpha_{IIb}\beta_3$ is important for platelet aggregation under high shear conditions (Peterson et al., 1987). The activated $\alpha_{IIb}\beta_3$ also interact with matrix bound vWF (Savage et al., 1992, 1996). This interaction is required for platelet spreading and firm adhesion on the subendothelial matrix under high shear conditions (Weiss et al., 1986; Savage et al., 1996).

$\alpha_{IIb}\beta_3$ primarily functions as a responsive adhesion receptor (Ginsberg et al., 1992; Phillips et al., 1991). In circulation, resting platelets are unable to bind either plasma fibrinogen or vWF with high affinity. Platelet activation either induced by soluble agonists or by initiating platelet adhesion to subendothelial matrix proteins induces high affinity binding of the ligands to $\alpha_{IIb}\beta_3$. The $\alpha_{IIb}\beta_3$ binding to matrix bound vWF also requires prior activation by initial interaction between the initiating platelet adhesion receptor GPIb-IX with vWF or by soluble platelet agonists (Savage et al., 1992). The tight regulation of the function of this integrin diminishes the probability of undesired platelet adhesion and aggregation in normal circulation but maintains quick hemostatic response of platelets to vascular injury. Under certain circumstances, however, integrin $\alpha_{IIb}\beta_3$ can also function as an initiating adhesion receptor. $\alpha_{IIb}\beta_3$ mediates platelet adhesion to immobilized fibrinogen and fibrin without prior platelet activation (Coller, 1980; Savage et al., 1992). Direct activation of the integrin by ligand recognition sequences is probably the underlying mechanism for this activation-independent binding (Du et al., 1991).

D. Activation of the Integrin $\alpha_{IIb}\beta_3$ by Inside-Out Signaling

Activation of the ligand binding function of the integrin $\alpha_{IIb}\beta_3$ requires intracellular signaling. Inhibitors of intracellular signaling pathways may inhibit the activation of $\alpha_{IIb}\beta_3$. Prostaglandin E_1 or other reagents that elevate intracellular cAMP levels may inhibit the integrin activation induced by many platelet agonists. Arachidonic acid-induced $\alpha_{IIb}\beta_3$ activation is inhibitable by aspirin; thrombin-induced $\alpha_{IIb}\beta_3$ activation is inhibitable by G-protein inhibitors such as pertussis toxin and GDPβS (Shattil and Brass, 1987). Reagents that mimic or activate the action of intracellular signaling molecules, such as thromboxane analogues, Ca^{2+} ionophores, GTPγS, and phorbal esters, activate $\alpha_{IIb}\beta_3$ (Shattil and Brass, 1987). The specific intracellular signaling pathway leading to integrin activation remain unresolved. As $\alpha_{IIb}\beta_3$ activation is the common consequence induced by all platelet agonists via diversified pathways, it is likely that different signaling pathways should converge to a common regulator of the integrin. One of the possibilities is that a protein kinase C pathway may provide a regulatory mechanism. Indeed, activation of protein kinase C and phosphorylation of its intracellular substrates precede the activation of the integrins. The cytoplasmic domain of β_3 can be phosphorylated by protein kinase C in activated platelets (Hillery et al., 1991; van Willigen et al., 1996). While Hillery and colleagues (1991) reported that only a small percentage of $\alpha_{IIb}\beta_3$ was phosphorylated, Van Willigen and co-workers (1996) reported that the stoichiometry of phosphorylation was 80%. However, the association of phosphorylation of the integrin and its activation is strictly correlative. In fact, inhibition of phosphoinositide 3-kinase (PI 3-K) by wortmannin, inhibited both thrombin- and phorbol ester–induced ligand binding to integrin $\alpha_{IIb}\beta_3$, suggesting that PI-3 kinase pathway is involved in the integrin activation pathway downstream of protein kinase C (Zhang et al., 1996). Two types of PI 3-K has been identified in platelets, PI 3-Kγ and p85/PI 3-K. The latter has been suggested to be preferentially involved in protein kinase C-stimulated integrin activation (Zhang et al., 1996). It is still not clear how PI-3 kinase relays signals that activate the integrin. PI-3 kinase regulate actin polymerization and cytoskeleton organization. However, the effect of rho and PI-3 kinase does not appear to be dependent on actin polymerization as inhibition of actin-polymerization by cytochalasin D does not affect ligand binding to $\alpha_{IIb}\beta_3$ per se (Addo et al., 1995), and Wortmannin did not affect actin polymerization during thrombin-induced platelet activation (Kovacsovics et al., 1995).

While the identity of the proximate intracellular regulator of $\alpha_{IIb}\beta_3$ function is unclear, it is known that the cytoplasmic domains of $\alpha_{IIb}\beta_3$ are critical for the activation of the integrin via the intracellular signaling. A single point mutation at the β_3 cytoplasmic domain (Ser_{752} to Pro) in a thrombasthenic patient resulted in abrogation of $\alpha_{IIb}\beta_3$ activation in platelets and bleeding diathesis (Chen et al., 1992, 1994). In the cytoplasmic domain of α_{IIb}, mutations deleting the highly conserved GFFKR motif resulted in expression of constitutively activated $\alpha_{IIb}\beta_3$, suggesting that this motif may be required to keep the integrin in a default resting conformation (O'Toole et al., 1990, 1994). In the membrane proximal region of the cytoplasmic domain of β_3, a negatively charged aspartic acid residue is spatially close to a positively charged lysine residue in the α_{IIb} GFFKR motif. Replacement of either one of these two charged residues resulted in activation of the ligand binding function of the integrin, indicating that a potential salt bridge between the GFFKR region of α_{IIb} and the membrane proximal region of β_3 may be involved in the maintenance of a resting state (Hughes et al., 1996). In an effort to test whether the integrin cytoplasmic domains may be regulated by possible intracellular elements, Chen and co-workers (1994) found that over-expression of a fusion protein containing β_3 cytoplasmic domain could reverse activated conformation of $\alpha_{IIb}\beta_3$. This suggests that the fusion protein containing β_3 cytoplasmic domain can compete with intact integrins for an intracellular regulatory factor. Thus, it is likely that intracellular signals may regulate the ligand binding function by interacting with the cytoplasmic domain of integrins. Several intracellular proteins have been suggested to interact with the cytoplasmic domain of integrins. Calreticulin binds to the GFFKR motif in the membrane proximal region of α_{IIb} in vitro (Rojiani et al., 1991), but recent studies have questioned the physiological relevance of this interaction (Opas et al., 1996). β3-endonexin binds to the β_3 cytoplasmic domain (Shattil et al., 1995). In addition, $\alpha_{IIb}\beta_3$ was shown to coimmunoprecipitate with membrane proteins, integrin-associated protein (IAP) (Brown et al., 1990) and CD9 (Indig et al., 1997).

Whether these proteins may be involved in the regulation of the integrin will be resolved in future investigations.

$\alpha_{IIb}\beta_3$ activation appears to be accompanied by a conformational change in the ligand binding domain. This is indicated by resonance energy transfer of fluorescently labeled antibodies (Sims et al., 1991) and by using monoclonal antibodies directed against the ligand binding site (Shattil et al., 1985). Recombinant $\alpha_{IIb}\beta_3$ expressed in Chinese Hamster Ovary (CHO) cells is intrinsically unable to bind soluble fibrinogen and can not be activated by physiological platelet agonists such as thrombin and ADP. However, it can be activated by the binding of certain anti-$\alpha_{IIb}\beta_3$ monoclonal antibodies (O'Toole et al., 1990). Furthermore, platelet $\alpha_{IIb}\beta_3$ can be solubilized and purified in a resting state, and subsequently activated using monoclonal antibodies or ligand-mimetic peptides that induce conformational changes of $\alpha_{IIb}\beta_3$ (Du et al., 1991; O'Toole et al., 1990). Thus, it is likely that activation of extracellular ligand binding function of $\alpha_{IIb}\beta_3$ results from intracellular signal-induced conformational change in the extracellular ligand binding sites.

E. Integrin $\alpha_{IIb}\beta_3$-Mediated Outside-In Signaling

Ligand interaction with $\alpha_{IIb}\beta_3$ induces a series of intracellular signaling events, including reorganization of the cytoskeleton (Jennings et al., 1981), elevation of intracellular Ca^{++} level (Fujimoto et al., 1991a, 1991b; Pelletier et al., 1992; Powling and Hardisty, 1985) and pH (Banga et al., 1986), hydrolysis of membrane lipids and phosphoinositide metabolism (Banga et al., 1986; Sultan et al., 1991), and activation of intracellular tyrosine kinases (Ferrell and Martin, 1989; Golden et al., 1990; Lipfert et al., 1992; Haimovich et al., 1993; Huang et al., 1993) as well as serine/threonine protein kinases such as protein kinase C (Haimovich et al., 1996). Thus, ligand binding to $\alpha_{IIb}\beta_3$ initiates transmembrane signal transduction (for reviews see Clark and Brugge, 1995; Schwartz et al., 1995). The integrin-mediated signaling leads to secondary platelet responses such as secretion of granule contents, second wave of platelet aggregation, release of procoagulant membrane vesicles, cytoskeleton reorganization and retraction of fibrin clot.

Integrin $\alpha_{IIb}\beta_3$ -mediated outside-in signals require the cytoplasmic domain of $\alpha_{IIb}\beta_3$. Ligand binding to mutant $\alpha_{IIb}\beta_3$ that lacks the cytoplasmic domain of β_3 significantly reduced outside-in signals as detected by tyrosine-phosphorylation of pp^{125FAK} (Leong et al., 1995). Mutant β_3 with deleted cytoplasmic domain or disrupted NPXY sequence in the cytoplasmic domain failed to mediated cell spreading and to be incorporated into focal adhesion complexes (Ylanne et al., 1993, 1995; Filardo et al., 1995). A proline substitution at S^{752} of β_3 not only abolished inside-out signaling but also reduced β_3-mediated cell spreading on immobilized fibrinogen, focal adhesion, and fibrin clot retraction, suggesting that outside-in signaling was also impaired (Chen et al., 1994b). It is thus possible that outside-in signals may be relayed by interaction of the cytoplasmic domain of $\alpha_{IIb}\beta_3$ with the intracellular molecules.

There are several possible intracellular signaling mechanisms and pathways involved in relaying the integrin signals

1. Formation of integrin-cytoskeleton signaling complex. Interaction of integrins with their ligands induces formation of focal adhesion complexes containing integrins and numerous cytoskeleton and signaling molecules. In nucleated cells, focal adhesions form at cell contact points with the extracellular matrix and the anchor points of intracellular stress fiber (actin filaments) (Burridge and Fath, 1989; Clark and Brugge, 1995). Integrin $\alpha_{IIb}\beta_3$-mediated platelet adhesion and aggregation also induces association of the integrin with cytoskeletal proteins analogous to those in the focal adhesion complex (Jennings et al., 1981; Fox, 1994). As a result, $\alpha_{IIb}\beta_3$ is incorporated into triton X-100-insoluble cytoskeleton (Jennings et al., 1981). Several important intracellular signaling molecules are also recruited to the detergent-insoluble structure in platelets and in other cell types. These include pp^{60src}, pp^{125FAK}, $pp^{62c\text{-}yes}$, pp^{21ras} (Fox et al., 1993), and PI3-kinase (Guinebault et al., 1995). As both the translocation of protein tyrosine kinases such as pp^{60Src} and pp^{125FAK} to cytoskeleton and integrin-dependent activation of these protein kinases are inhibited by cytochalasin D which inhibits actin polymerization, it is likely that formation of integrin-cytoskeleton-signaling molecule complex is a necessary step in the activation of these protein kinases and possibly other signaling molecules.

   It is still not clear how the integrin regulates and associates with the cytoskeleton. In fibroblasts, formation of stress fiber and focal adhesion sites involve the small G protein Rho A which regulates actin polymerization. Inhibition of Rho A by exoenzyme C, although it does not abolish integrin activation, inhibits platelet aggregation, adhesion, and spreading, suggesting a critical role for Rho A in integrin-mediated outside-in signaling (Morii et al., 1992; Leng et al., 1998). Focal adhesion proteins talin (Knezevic et al., 1996) and α-actinin (Otey et al., 1990) interact with the cytoplasmic domains of $\alpha_{IIb}\beta_3$ in vitro. Both the cytoplasmic domains of α_{IIb} and β_3 may be involved in the interaction with talin (Knezevic et al., 1996).

2. Activation of protein kinases. Ligand binding to integrin $\alpha_{IIb}\beta_3$ induces tyrosine phosphorylation of several intracellular proteins (Ferrell and Martin, 1988; Golden et al., 1990), including protein tyrosine kinases such as focal adhesion kinase (pp^{125FAK}) (Lipfert et al. 1992) and pp^{72syk} (Clark et al., 1994). Integrin-mediated tyrosine phosphorylation of these protein tyrosine kinases is associated with activation of their enzymatic activities (Lipfert et al., 1992; Clark et al., 1994). Tyrosine-phosphorylation of pp^{72syk} can be induced by fibrinogen binding to platelets without platelet activation by other agonists, and appeared to precede platelet aggregation process (Clark et al., 1994). Tyrosine phosphorylation of pp^{125FAK} is dependent upon platelet

aggregation (Shattil et al., 1994), requires co-stimulation of platelets by platelet agonists and is inhibitable by protein kinase C inhibitors (Shattil et al., 1994; Haimovich et al., 1996). Phosphorylation and activation of these protein kinases can be inhibited by cytochalasin D which inhibits actin polymerization (Clark et al., 1994; Lipfert et al., 1992) suggesting that activation of protein tyrosine kinases is probably downstream the integrin-induced formation of integrin-cytoskeleton-signaling complex.

Law and co-workers (Law et al., 1996) found that during integrin-mediated platelet aggregation, the cytoplasmic domain of β_3 becomes phosphorylated on the tyrosine residues within the two NXXY motifs. Tyrosine phosphorylated β_3 cytoplasmic domain peptides binds to SH2 domain containing proteins GRG2 and SHC. Wary and colleagues (Wary et al., 1996) reported that certain β_1 and β_3 integrins may interact with SHC and mediate SHC phosphorylation. This interaction, however, does not appear to be mediated by the cytoplasmic domains of either α or β subunits is rather specified by the transmembrane and extracellular domain of α subunits. SHC and GRB2 as adapter proteins plays a central role in linking the tyrosine-phosphorylated receptors to the MAP kinase pathways. Ligand binding to integrins has been shown to activate MAP kinase pathway in several cell types (Chen et al., 1994, 1996; Zhu and Assoian, 1995; Clark and Hynes, 1996; Renshaw et al., 1996; Wary et al., 1996), and this activation is mediated by SHC (Wary et al., 1996). However, as cytocalasin D and a dominant negative Rho A inhibited the integrin-mediated MAP kinase activation (Zhu and Assoian, 1995; Renshaw et al., 1996), it appears that activation of MAP kinase pathway also requires assembly of integrin-cytoskeleton complex.

3. Regulation of intracellular calcium. Ligand interaction with $\alpha_{IIb}\beta_3$ induces increased intracellular calcium. Blocking fibrinogen binding to $\alpha_{IIb}\beta_3$ by monoclonal anti-$\alpha_{IIb}\beta_3$ antibodies or by RGDS peptides in thrombin-stimulated platelets inhibits the Ca^{++} influx into platelets and reduces the probability of opening of Ca^{++} channel in subsequently prepared membrane vesicle (Fujimoto et al., 1991a; Powling and Hardisty, 1985). In cells expressing recombinant $\alpha_{IIb}\beta_3$, calcium oscillations can be induced by surface-bound fibrinogen or anti-$\alpha_{IIb}\beta_3$ antibody (Pelletier et al., 1992). While the signaling pathway to the increased intracellular calcium level is not clear, Rybak and co-workers (1988) and Fujimoto and colleagues (Fujimoto, 1991b) showed that $\alpha_{IIb}\beta_3$ preparation incorporated in liposomes or planar phospholipid bilayer had Ca^{++} channel activity, implicating that $\alpha_{IIb}\beta_3$ or an associated component may function as a calcium channel. Another study suggests that the integrin-associated protein (IAP) may be involved in the integrin-mediated calcium influx (Schwartz et al., 1993). Intracellular calcium as a second messenger

regulates activities of many aspects of platelet functions and activities of numerous intracellular proteins (Sage et al., 1993; Clapham, 1995).

The $\alpha_{IIb}\beta_3$-mediated platelet aggregation is responsible for the activation of calcium dependent protease calpain, presumably by increasing intracellular calcium level (Fox et al., 1993). Calpain is colocalized with integrins in the focal adhesion complex (Beckerle et al., 1987). Many identified calpain substrates are also colocalized with integrin in the focal adhesion complex and / or associated with integrin signaling. These include protein tyrosine phosphatase IB (Frangioni et al., 1993), pp^{60src} (Ode et al., 1993), pp^{125FAK}(Cooray et al., 1996), protein kinase C (Saido et al., 1991), talin (Fox et al., 1985), actin-binding protein (Fox et al., 1985), and the cytoplasmic domain of the integrin β_3 subunit (Du et al., 1995). Cleavage of the β_3 subunit at sites flanking two functionally important NXXY motifs are likely to disrupt integrin interaction with cytoskeleton and integrin signaling (Du et al., 1995).

F. Transmembrane Signaling Mechanisms of Integrin $\alpha_{IIb}\beta_3$

Ligand binding to integrin $\alpha_{IIb}\beta_3$ induces conformational changes in the integrin (Parise et al., 1987). The ligand-induced conformational changes can be identified by monoclonal antibodies recognizing ligand-induced binding sites (LIBS)

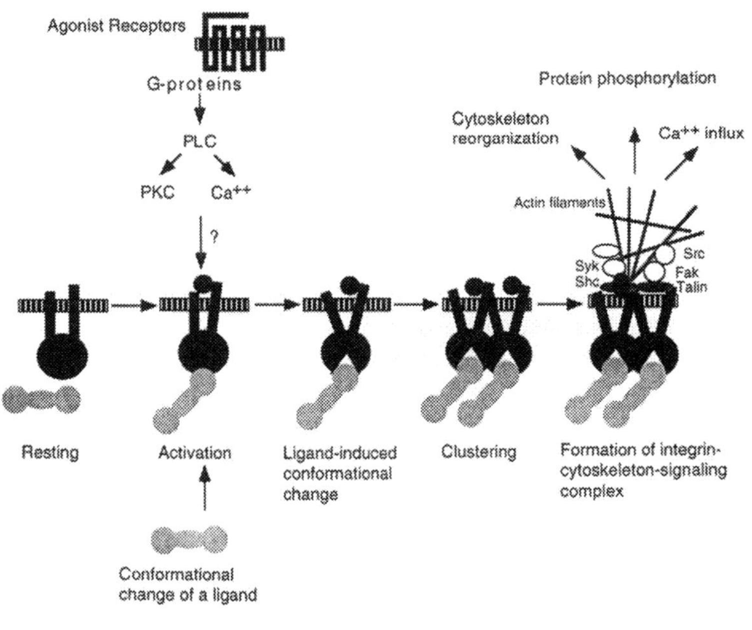

Figure 2. A schematic of integrin signaling.

(Frelinger et al., 1988, 1990, 1991; Kouns et al., 1990). The ligand binding induced-conformation of $\alpha_{IIb}\beta_3$ is an activated form with high affinity for fibrinogen (Du et al., 1991). Ligand binding-induced conformational change is not limited to the ligand binding region of $\alpha_{IIb}\beta_3$. This conformational change propagates through the entire 18 nm length of the $\alpha_{IIb}\beta_3$ molecule to the C-terminal tails of both α_{IIb} and β_3. This long range-propagated conformational change was reported by a monoclonal antibody, Anti-LIBS2, recognizing a ligand-induced binding site in the C-terminal extracellular domain of β_3, (Du et al., 1993), and recently by a monoclonal antibody directed against the ligand induced binding site in the cytoplasmic domain of α_{IIb} (Lam, personal communication).

On the other hand, although Anti-LIBS2 antibody has a low affinity interaction with the resting $\alpha_{IIb}\beta_3$, it induces ligand binding to $\alpha_{IIb}\beta_3$, indicating a retrograde propagated conformational change (from C-terminal region to the ligand binding site) (Du et al., 1993). This mimics physiological activation of the ligand binding function.

However, conformational changes of $\alpha_{IIb}\beta_3$ induced by soluble monomeric ligands are not sufficient to induce all platelet responses (Kouns et al., 1991). $\alpha_{IIb}\beta_3$-mediated platelet responses appear to require multivalent ligands. In fact, integrin mediated signaling can be initiated by certain multivalent antibodies (Pelletier et al., 1992), indicating that clustering of the integrin may be required for the integrin signaling. The clustering of the integrin molecules by itself, however, also may not be sufficient, as the binding of the monomeric RGD ligands, although not able to initiate platelet responses, can induce clustering of the integrin on the platelet surface (Isenberg et al., 1987, 1989). Thus, it appears that conformational changes, clustering of the integrin, and subsequent formation of integrin-focal adhesion like signaling complex may be important. Similar concepts have emerged from studies on the β_1 integrins (Miyamoto et al., 1995). When the integrin $\alpha_5\beta_1$ was bound to fibronectin-coated beads, actin, talin, vinculin α-actinin, tensin, and focal adhesion kinase (pp^{125FAK}) were recruited to the cytoplasmic side of the adhesion site. When integrins were clustered by non-inhibitory antibodies, however, only tensin and pp^{125FAK} were recruited. Adding RGD peptide to the antibody-clustered integrin resulted in recruitment of the full complement of cytoplasmic proteins. Thus, both ligand occupancy and receptor clustering are required for the full signaling complex (Miyamoto et al., 1995).

V. COLLAGEN RECEPTORS

Collagen is a principal structural component of blood vessel wall. When exposed to platelets, collagen not only mediates platelet adhesion, but also function as a strong platelet agonist, inducing platelet aggregation and secretion. It appears that collagen interacts with platelets by binding to multiple receptors. The reported collagen receptors include the integrin $\alpha_2\beta_1$, glycoprotein IV (GPIV, CD36), glycoprotein

VI (GPVI, p62), and a C1q receptor (Peerschke and Ghebrehiwet 1990). In addition, platelet adhesion to collagen under high shear conditions may be mediated by collagen-bound vWF via GPIb-IX (Houdijk et al., 1985). Involvement of the integrin $\alpha_{IIb}\beta_3$ in platelet interaction with collagen was also reported (Morton et al., 1994; Wu et al., 1996).

A. The Integrin $\alpha_2\beta_1$

The role of $\alpha_2\beta_1$ as a collagen receptor was identified in a patients with congenital platelet dysfunction. The platelets from the patient lacked the integrin $\alpha_2\beta_1$ (known also as GPIa-IIa), and were defective in collagen-induced platelet activation and in platelet adhesion to subendothelial matrix (Nieuwenhuis et al., 1985). $\alpha_2\beta_1$ binding to collagen and $\alpha_2\beta_1$-dependent platelet adhesion to collagen were also directly demonstrated (Santoro, 1986). One of the characteristics of the $\alpha_2\beta_1$-dependent platelet adhesion to collagen is its strong dependence on the Mg^{++} (Staatz et al., 1989) and inhibition by Ca^{2+}. The collagen binding site is located in the I-domain region of α_2 chain (Kamata et al., 1994; Kamata and Takada, 1994). I domain is a ~200 residue region in the a subunits of several integrins that is inserted possibly by exon shuffling, and is a ligand recognition site in these integrins (Corbi et al., 1988; Michishita et al., 1993; Kamata et al., 1994; Kern et al., 1994). I domains are homologous to A domains in vWF and certain adhesive proteins. A DxSxSx motif in the I domain is critical for both the ligand binding and divalent cation binding (Loftus et al., 1994).

$\alpha_2\beta_1$-mediated platelet adhesion initiates signals activating the integrin $\alpha_{IIb}\beta_3$, which leds to the firm adhesion and spreading of platelets (Wu et al., 1996). Similar to fibrinogen binding to $\alpha_{IIb}\beta_3$, collagen induces activation of pp^{72Syk} as an early event which was partially inhibited by anti-$\alpha_2\beta_1$ antibodies and totally inhibited by cytocalasin D (Asazuma et al., 1996; Keely and Parise, 1996). However, in an apparently conflicting report, one group suggested that collagen-induced pp^{72Syk} activation was not inhibited by cytocalasin D (Fuji et al., 1994). $\alpha_2\beta_1$ is not likely to be responsible for all of the intracellular signaling induced by collagen, as collagen-induced platelet activation and intracellular signaling may occur even when the integrin is inhibited with EDTA or antibodies (Asazuma et al., 1996; Keely and Parise, 1996; Morton et al., 1994, 1995).

B. GPIV (CD36)

GPIV as a collagen receptor was suggested by the study showing that purified GPIV binds to collagen and an antibody against the purified GPIV inhibited platelet activation and aggregation induced by collagen (Tendon et al., 1989). Monoclonal antibodies against GPIV were also reported to inhibit platelet adhesion to collagen in the absence of Mg^{++} (Matsuno et al., 1996). GPIV was also reported to be a receptor for thrombospondin, oxidized lipoprotein and Plasmodium falciparum-infected red cells (Tendon et al., 1991). The capacity of GPIV to bind collagen or thrombospondin appears to be regulated by phosphorylation in its ectodomain. Dephosphorylation results in loss of collagen binding but increased thrombospondin binding and vice versa (Asch et al., 1993). GPIV, however, is absent from a population of Japanese whose platelets function normally (Yamamoto et al., 1990, 1992), and show normal response to collagen (Daniel et al., 1994). However, GPIV-deficient platelets respond normally to collagen type I, II, III, and IV, but not type V (Kehrel et al., 1993). Thus, there may be redundancy in collagen receptors. The cytoplasmic domain of GPIV contains binding site for src family of protein kinases such as fyn, Lyn, and Yes (Huang et al., 1991), indicating a linkage to protein tyrosine kinase pathway.

C. GPVI (p62)

Evidence that GPVI functions as a collagen receptor includes patients whose platelets lack GPVI and are defective in response to collagen (Moroi et al.. 1989; Ryo et al., 1992; Arai et al., 1995). Antibodies against this protein inhibit platelet responses to collagen. Under flow conditions, GPVI-deficient platelets showed an adhesion deficiency similar to that caused by inhibitors of integrin $\alpha_{IIb}\beta_3$, suggesting that GPVI may be involved in the collagen-induced platelet signaling leading to activation of the integrin $\alpha_{IIb}\beta_3$ (Moroi et al., 1996). The cloning and sequencing of GPVI may provide insight into collagen-platelet interaction.

VI. CONCLUSIONS

As a cell specialized to adhere, a major function of platelet signaling machinery is to regulate adhesion receptors. Platelet adhesion receptors not only mediate adhesion but also function as triggers of intracellular signaling. Recent advances in both clinical and basic research have defined platelet adhesion receptors and their roles. This provides the means to develop and use adhesion receptor inhibitors in preventing and treating thrombosis.

ACKNOWLEDGMENTS

Xiaoping Du is an Established Investigator of the American Heart Association, and is supported by grant HL52547 from the National Institute of Health. Mark Ginsberg is supported by grants HL48728 and HL31950 from the National Institute of Health.

REFERENCES

Addo, J. B., Bray, P. F., Grigoryev, D., Faraday, N., & Goldschmidt, C. P. (1995). Surface recruitment but not activation of integrin alpha IIb beta 3 (GPIIb-IIIa) requires a functional actin cytoskeleton. Arterioscler. Thromb. Vasc. Biol. 15, 1466-1473.

Aitken, A., Amess, B., Howell, S., Jones, D., Martin, H., Patel, Y., Robinson, K., & Toker, A. (1992). The role of specific isoforms of 14-3-3 protein in regulating protein kinase activity in the brain. Biochem. Soc. Trans. 20, 607-611.

Andrews, R. K., Booth, W. J., Gorman, J. J., Castaldi, P. A., & Berndt, M. C. (1989). Purification of botrocetin from Bothrops jararaca venom. Analysis of the botrocetin-mediated interaction between von Willebrand factor and the human platelet membrane glycoprotein IL-IX complex. Biochemistry 28, 8317-8326.

Andrews, R. K. & Fox, J. E. (1992). Identification of a region in the cytoplasmic domain of the platelet membrane glycoprotein IL-IX complex that binds to purified actin-binding protein. J. Biol. Chem. 267, 18605-18611.

Andrews, R. K., Kroll, M. H., Ward, C. M., Rose, J. W., Scarborough, R. M., Smith, A. I., Lopez, J. A., & Berndt, M. C. (1996). Binding of a novel 50-kilodalton alboaggregin from Trimeresurus albolabris and related viper venom proteins to the platelet membrane glycoprotein IL-IX-V complex. Effect on platelet aggregation and glycoprotein Ib-mediated platelet activation. Biochemistry 35,12629-12639.

Andrews, R. K., Harris, S. J., McNally, T., & Berndt, M. C. (1998). Binding of purified 14-3-3 zeta signaling protein to discrete amino acid sequences within the cytoplasmic domain of the platelet membrane glycoprotein Ib-IX-V complex. Biochemistry 37, 638-647.

Andrieux, A., Rabiet, M. J., Chapel, A., Concord, E., & Marguerie, G. (1991). A highly conserved sequence of the Arg-Gly-Asp-binding domain of the integrin beta 3 subunit is sensitive to stimulation. J. Biol. Chem. 266, 14202-14207.

Arai, M., Yamamoto, N., Moroi, M., Akamatsu, N., Fukutake, K., & Tanoue, K. (1995). Platelets with 10% of the normal amount of glycoprotein VI have an impaired response to collagen that results in a mild bleeding tendency. Br. J. Haematol. 89, 124-130.

Asazuma, N., Yatomi, Y., Ozaki, Y., Qi, R., Kuroda, K., Satoh, K., & Kume, S. (1996). Protein-tyrosine phosphorylation and p72syk activation in human platelets stimulated with collagen is dependent upon glycoprotein Ia/IIa and actin polymerization. Thromb. Haemost. 75, 648-654.

Asch, A. S., Liu, I., Briccetti, F. M., Barnwell, J. W., Kwakye, B. F., Dokun, A., Goldberger, J., & Pernambuco, M. (1993). Analysis of CD36 binding domains: ligand specificity controlled by dephosphorylation of an ectodomain. Science 262, 1436-1440.

Bajt, M. L., Ginsberg, M. H., Frelinger, A. d., Berndt, M. C., & Loftus, J. C. (1992). A spontaneous mutation of integrin alpha IIb beta 3 (platelet glycoprotein IIb-IIIa) helps define a ligand binding site. J. Biol. Chem. 267, 3789-3794.

Bajt, M. L. & Loftus, J. C. (1994). Mutation of a ligand binding domain of beta 3 integrin. Integral role of oxygenated residues in alpha IIb beta 3 (GPIIb-IIIa) receptor function. J. Biol. Chem. 269, 20913-20919.

Banga, H. S., Simons, E. R., Brass, L. F., & Rittenhouse, S. E. (1986). Activation of phospholipases A and C in human platelets exposed to epinephrine: role of glycoproteins IIb/IIIa and dual role of epinephrine. Proc. Natl. Acad. Sci. U S A 83, 9197-9201.

Baumgartner, H. R., Tschopp, T. B., & Meyer, D. (1980). Shear rate dependent inhibition of platelet adhesion and aggregation on collagenous surface by antibodies to human factor VIII/von Willebrand factor. Brit. J. Haematol. 44, 127-139.

Beckerle, M. C., Burridge, K., DeMartino, G. N., & Croall, D. E. (1987). Colocalization of calcium-dependent protease II and one of its substrates at sites of cell adhesion. Cell 51, 569-577.

Berndt, M. C., Du, X. P., & Booth, W. J. (1988). Ristocetin-dependent reconstitution of binding of von Willebrand factor to purified human platelet membrane glycoprotein IL-IX complex. Biochemistry 27, 633-640.

Berndt, M. C., Gregory, C., Chong, B. H., Zola, H., & Castaldi, P. A. (1983). Additional glycoprotein defects in Bernard-Soulier syndrome: Confirmation of genetic basis by parental analysis. Blood 62, 800-807.

Berndt, M. C., Gregory, C., Kabral, A., Zola, H., Fournier, D., & Castaldi, P. A. (1985). Purification and preliminary characterization of the glycoprotein Ib complex in the human platelet membrane. Eur. J. Biochem. 151, 637-649.

Berndt, M. C., Ward, C. M., Booth, W. J., Castaldi, P. A., Mazurov, A. V., & Andrews, R. K. (1992). Identification of aspartic acid 514 through glutamic acid 542 as a glycoprotein IL-IX complex receptor recognition sequence in von Willebrand factor. Mechanism of modulation of von Willebrand factor by ristocetin and botrocetin. Biochemistry 31, 11144-11151.

Bonnefoy, B. N., Liu, Y. C., von, W. M., Sung, A., Wily, C., Mustelin, T., Yoshida, H., Ishizaka, K., & Altman, A. (1995). Inhibition of phosphatidylinositol 3-kinase activity by association with 14-3-3 proteins in T cells. Proc. Natl. Acad. Sci. U S A 92, 10142-10146.

Brown, E., Hooper, L., Ho, T., & Gresham, H. (1990). Integrin-associated protein: a 50-kD plasma membrane antigen physically and functionally associated with integrins. J. Cell. Biol. 111, 2785-2794.

Burridge, K. & Fath, K. (1989). Focal contacts: transmembrane links between the extracellular matrix and the cytoskeleton. Bioessays 10, 104-108.

Calverley, D. C., Kavanagh, T. J., & Roth, G. J. (1998). Human signaling protein 14-3-3 zeta interacts with platelet glycoprotein Ib subunits Ib alpha and Ib beta. Blood 91, 1295-1303.

Calvete, J. J., Arias, J., Alvarez, M. V., Lopez, M. M., Henschen, A., & Gonzalez-Rodriguez, J. (1991a). Further studies on the topograghy of the N-terminal region of human platelet glycoprotein IIIa: Localization of monoclonal antibody epitopes and the putative fibrinogen-binding sites. Biochem. J. 274, 457-463.

Calvete, J. J., Henschen, A., & Gonzalez, R. J. (1991b). Assignment of disulphide bonds in human platelet GPIIIa. A disulphide pattern for the beta-subunits of the integrin family. Biochem. J. 63-71.

Carrell, N. A., Fitzgerald, L. A., Steiner, B., Erickson, H. P., & Phillips, D. R. (1985). Structure of human platelet membrane glycoproteins IIb and IIIa as determined by electron microscopy. J. Biol. Chem. 260, 1743-1749.

Charo, I. F., Nannizzi, L., Phillips, D. R., Hsu, M. A., & Scarborough, R. M. (1991). Inhibition of fibrinogen binding to GPIIb-IIIa by a GP IIIa peptide. J. Biol. Chem. 266, 1414-1421.

Chen, Q., Kinch, M. S., Lin, T. H., Burridge, K., & Juliano, R. L. (1994). Integrin-mediated cell adhesion activates mitogen-activated protein kinases. J. Biol. Chem. 269, 26602-26605.

Chen, Q., Lin, T. H., Der, C. J., & Juliano, R. L. (1996). Integrin-mediated activation of mitogen-activated protein (MAP) or extracellular signal-related kinase kinase (MEK) and kinase is independent of Ras. J. Biol. Chem. 271, 18122-18127.

Chen, W. J., Goldstein, J. L., & Brown, M. S. (1990). NPXY, a sequence often found in cytoplasmic tails, is required for coated pit-mediated internalization of the low density lipoprotein receptor. J. Biol. Chem. 265, 3116-3123.

Chen, Y. P., Djaffar, I., Pidard, D., Steiner, B., Cieutat, A. M., Caen, J. P., & Rosa, J. P. (1992). Ser-752—Pro mutation in the cytoplasmic domain of integrin beta 3 subunit and defective activation of platelet integrin alpha IIb beta 3 (glycoprotein IIb-IIIa) in a variant of Glanzmann thrombasthenia. Proc. Natl. Acad. Sci. U S A 89, 10169-10173.

Chen, Y. P., O'Toole, T. E., Shipley, T., Forsyth, J., LaFlamme, S. E., Yamada, K. M., Shattil, S. J., & Ginsberg, M. H. (1994a). "Inside-out" signal transduction inhibited by isolated integrin cytoplasmic domains. J. Biol. Chem. 269, 18307-18310.

Chen, Y. P., O'Toole, T. E., Ylanne, J., Rosa, J. P., & Ginsberg, M. H. (1994b). A point mutation in the integrin beta 3 cytoplasm c domain (S752—P) impairs bidirectional signaling through alpha IIb beta 3 (platelet glycoprotein IIb-IIIa). Blood 84, 1857-1865.

Cheresh, D. A., Berliner, S. A., Vicente, V., & Ruggeri, Z. M. (1989). Recognition of distinct adhesive sites on fibrinogen by related integrins on platelets and endothelial cells. Cell 58, 945-953.

Clapham, D. E. (1995). Calcium Signaling. Cell 80, 259-268.

Clark, E. A. & Brugge, J. S. (1995). Integrins and signal tranduction pathways: the road taken. Science 268 233-239.

Clark, E. A. & Hynes, R. O. (1996). Ras activation is necessary for integrin-mediated activation of extracellular signal-regulated kinase 2 and cytosolic phospholipase A2 but not for cytoskeletal organization. J. Biol. Chem. 271, 14814-14818.

Clark, E. A., Shattil, S. J., Ginsberg, M. H., Bolen, J., & Brugge, J. S. (1994). Regulation of the protein tyrosine kinase pp72syk by platelet agonists and the integrin alpha IIb beta 3. J. Biol. Chem. 269, 28859-28864.

Coller, B. S. (1980). Interaction of normal, thrombasthenic, and Bernard-Soulier platelets with immobilized fibrinogen: Defective platelet-fibrinogen interaction in thrombasthenia. Blood 55, 169-178.

Connolly, A. J., Ishihara, H., Kahn, M. L., Farese, R. J., & Coughlin, S. R. (1996). Role of the thrombin receptor in development and evidence for a second receptor. Nature 381, 516-519.

Cooney, K. A., Lyons, S. E., & Ginsburg, D. (1992). Functional analysis of a type IIB von Willebrand disease missense mutation: increased binding of large von Willebrand factor multimers to platelets. Proc. Natl. Acad. Sci. U S A 89, 2869-2872.

Cooper, H. A., Bennett, W. D., White, G. C. I., & Wagner, R. H. (1982). Hydrolysis of human platelet membrane glycoproteins with a serratia marcescens metalloprotease: Effect on response to thrombin and von Willebrand factor. Proc. Natl. Acad. Sci. USA. 79,1433-1437.

Cooray, P., Yuan, Y., Schoenwaelder, S. M., Mitchell, C. A., Salem, H. H., & Jackson, S. P. (1996). Focal adhesion kinase (ppl25FAK) cleavage and regulation by calpain. Biochem. J. 318, 41-47.

Corbi, A. L., Kishimoto, T. K., Miller, L. J., & Springer, T. A. (1988). The human leukocyte adhesion glycoprotein Mac-1 (complement receptor type 3, CD11b) alpha subunit. Cloning, primary structure, and relation to the integrins, von Willebrand factor and factor B. J. Biol. Chem. 263, 12403-12411.

Cunningham, J. G., Meyer, S. C., & Fox, J. E. (1996). The cytoplasmic domain of the alpha-subunit of glycoprotein (GP) Ib mediates attachment of the entire GP IL-IX complex to the cytoskeleton and regulates von Willebrand factor-induced changes in cell morphology. J. Biol. Chem. 271, 11581-11587.

D'Souza, S. E., Ginsberg, M. H., Burke, T. A., Lam, S. C., & Plow, E. F. (1988). Localization of an Arg-Gly-Asp recognition site within an integrin adhesion receptor. Science 242, 91-93.

D'Souza, S. E., Ginsberg, M. H., Burke, T. A., & Plow, E. F. (1990). The ligand binding site of the platelet integrin receptor GPIIb-IIIa is proximal to the second calcium binding domain of its alpha subunit. J. Biol. Chem. 265, 3440-3446.

D'Souza, S. E., Ginsberg, M. H., Matsueda, G. R., & Plow, E. F. (1991). A discrete sequence in a platelet integrin is involved in ligand recognition. Nature 350, 66-68.

D'Souza, S. E., Haas, T. A., Piotrowicz, R. S., Byers, W. V., McGrath, D. E., Soule, H. R., Cièrniewski, C., Plow, E. F., & Smith, J. W. (1994). Ligand and cation binding are dual functions of a discrete segment of the integrin beta 3 subunit: cation displacement is involved in ligand binding. Cell 79, 659-667.

Daniel, J. L., Dangelmaier, C., Strouse, R., & Smith, J. B. (1994). Collagen induces normal signal transduction in platelets deficient in CD36 (platelet glycoprotein IV). Thromb Haemost. 71, 353-356.

De Marco, L., Girolami, A., Russell, S., & Ruggeri, Z. M. (1985a). Interaction of asialo von Willebrand factor with glycoprotein Ib induces fibrinogen binding to the glycoprotein IIb/IIIa complex and mediates platelet aggregation. J. Clin. Invest. 75, 1198-1203.

De Marco, L., Girolami, A., Zimmerman, T. S., & Ruggeri, Z. M. (1985b). Interaction of purified type IIB von Willebrand factor with the platelet membrane glycoprotein Ib induces fibrinogen binding to the glycoprotein IIb/IIIa complex and initiates aggregation. Proc. Natl. Acad. Sci. U S A 82, 7424-7428.

De Marco, L., Mazzucato, M., Masotti, A., & Ruggeri, Z. M. (1994). Localization and characterization of an alpha-thrombin-binding site on platelet glycoprotein Ib alpha. J. Biol. Chem. 269, 6478-6484.

De Marco, L., Mazzuccato, M., Grazia Del Ben, M., Budde, U., Federici, A. B., Girolami, A., & Ruggeri, Z. M. (1987). Type IIB von Willebrand factor with normal sialic acid content induces

platelet aggregation in the absence of ristocetin. Role of platelet activation, fibrinogen, and two distinct membrane receptors. J. Clin. Invest. 80, 475-482.
Dong, J. F., Li, C. Q., & Lopez, J. A. (1994). Tyrosine sulfation of the glycoprotein IL-IX complex: identification of sulfated residues and effect on ligand binding. Biochemistry 33, 13946-13953.
Dong, J. F., Li, C. Q., Sae, T. G., Hyun, W., Afshar, K. V., & Lopez, J. A. (1997). The cytoplasmic domain of glycoprotein (GP) Ib alpha constrains the lateral diffusion of the GP Ib-IX complex and modulates von Willebrand factor binding. Biochemistry 36, 12421-12427.
Du, X., Beutler, L., Ruan, C., Castaldi, P. A., & Berndt, M. C. (1987). Glycoprotein Ib and glycoprotein IX are fully complexed in the intact platelet membrane. Blood 69, 1524-1527.
Du, X., Fox, J. E., & Pei, S. (1996). Identification of the binding sequence for the ξ-form 14-3-3 protein in the cytoplasmic domain of the adhesion receptor, platelet glycoprotein Ibα. J. Biol. Chem. 271, 7362-7367.
Du, X., Gu, M., Weisel, J. W., Nagaswami, C., Bennett, J. S., Bowditch, R., & Ginsberg, M. H. (1993). Long range propagation of conformational changes in integrin $\alpha_{IIb}\beta_3$. J. Biol. Chem. 268, 23087-23092.
Du, X., Harris, S. J., Tetaz, T. J., Ginsberg, M. H., & Berndt, M. C. (1994). Association of a phospholipase A2 (14-3-3 protein) with the platelet glycoprotein IL-IX complex. J. Biol. Chem. 269, 18287-18290.
Du, X., Saido, T. C., Tsubuki, S., Indig, F. E., Williams, M. J., & Ginsberg, M. H. (1995). Calpain cleavage of the cytoplasmic domain of the integrin β_3 subunit. J. Biol. Chem. 270, 26146-26151.
Du, X. P., Plow, E. F., Frelinger, A. d., O'Toole, T. E., Loftus, J. C., & Ginsberg, M. H. (1991). Ligands "activate" integrin alpha IIb beta 3 (platelet GPIIb-IIIa). Cell 65, 409-416.
Fanti, W. J., Muslin, A. J., Kikuchi, A., Martin, J. A., MacNicol, A. M., Gross, R. W., & Williams, L. T. (1994). Activation of Raf-1 by 14-3-3 proteins. Nature 371, 612-614.
Farrell, D. H. & Thiagarajan, P. (1994). Binding of recombinant fibrinogen mutants to platelets. J. Biol. Chem. 269, 226-231.
Farrell, D. H., Thiagarajan, P., Chung, D. W., & Davie, E. W. (1992). Role of fibrinogen alpha and gamma chain sites in platelet aggregation. Proc. Natl. Acad. Sci. U S A 89, 10729-10732.
Ferrell, J. J. & Martin, G. S. (1988). Platelet tyrosine-specific protein phosphorylation is regulated by thrombin. Mol. Cell. Biol. 8, 3603-3610
Ferrell, J. J. & Martin, G. S. (1989). Tyrosine-specific protein phosphorylation is regulated by glycoprotein IIb-IIIa in platelets. Proc. Natl. Acad. Sci. U S A 86, 2234-2238.
Filardo, E. J., Brooks, P. C., Deming, S. L., Damsky, C., & Cheresh, D. A. (1995). Requirement of the NPXY motif in the integrin p3 subunit cytoplasmic tail for melanoma cell migration in vitro and in vivo. J. Cell Biol. 130, 441-450.
Fitzgerald, L. A., Steiner, B., Rall, S. C. J., Lo, S. S., & Phillips, D. R. (1987). Protein sequence of endothelial glycoprotein IIIa derived from a cDNA clone. Identity with platelet glycoprotein IIIa and similarity to "integrin". J. Biol. Chem. 262, 3936-3939.
Fox, J. E. (1994). Transmembrane signaling across the platelet integrin glycoprotein IIb-IIIa. Ann. N Y Acad. Sci. 714, 75-87.
Fox, J. E., Aggerbeck, L. P., & Berndt, M. C. (1988). Structure of the glycoprotein Ib.IX complex from platelet membranes. J. Biol. Chem. 263, 4882-4890.
Fox, J. E. & Berndt, M. C. (1989). Cyclic AMP-dependent phosphorylation of glycoprotein Ib inhibits collagen-induced polymerization of actin in platelets. J. Biol. Chem. 264, 9520-9526.
Fox, J. E., Goll, D. E., Reynolds, C. C., & Phillips, D. R. (1985). Identification of two proteins (actin-binding protein and P235) that are hydrolyzed by endogenous Ca2+-dependent protease during platelet aggregation. J. Biol. Chem. 260, 1060-1066.
Fox, J. E., Lipfert, L., Clark, E. A., Reynolds, C. C., Austin, C. D., & Brugge, J. S. (1993). On the role of the platelet membrane skeleton in mediating signal transduction. Association of GP IIb-IIIa, pp60c-src, pp62c-yes, and the p21ras GTPase-activating protein with the membrane skeleton. J. Biol. Chem. 268, 25973-25984.

Fox, J. E., Reynolds, C. C., & Johnson, M. M. (1987). Identification of glycoprotein Ib beta as one of the major proteins phosphorylated during exposure of intact platelets to agents that activate cyclic AMP-dependent protein kinase. J. Biol. Chem. 262, 12627-12631.

Fox, J. E., Taylor, R. G., Taffarel, M., Boyles, J. K., & Goll, D. E. (1993). Evidence that activation of platelet calpain is induced as a consequence of binding of adhesive ligand to the integrin, glycoprotein IIb-IIIa. J. Cell. Biol. 120, 1501-1507.

Fox, J. E. B. (1985). Linkage of a membrane skeleton to integral membrane glycoproteins in human platelets. Identification of one of the glycoproteins as glycoprotein Ib. J. Clin. Invest. 76, 1673-1683.

Frangioni, J. V., Oda, A., Smith, M., Salzman, E. W., & Neel, B. G. (1993). Calpain-catalyzed cleavage and subcellular relocation of protein phosphotyrosine phosphatase 1B (PTP-1B) in human platelets. Embo. J. 12, 4843-4856.

Freed, E., Symons, M., Macdonald, S. G., McCormick, F., & Ruggieri, R. (1994). Binding of 14-3-3 proteins to the protein kinase Raf and effects on its activation. Science 265, 1713-1716.

Frelinger, A. d., Du, X. P., Plow, E. F., & Ginsberg, M. H. (1991). Monoclonal antibodies to ligand-occupied conformers of integrin alpha IIb beta 3 (glycoprotein IIb-IIIa) alter receptor affinity, specificity, and function. J. Biol. Chem. 266, 17106-17111.

Frelinger, A. d., Lam, S. C., Plow, E. F., Smith, M. A., Loftus, J. C., & Ginsberg, M. H. (1988). Occupancy of an adhesive glycoprotein receptor modulates expression of an antigenic site involved in cell adhesion. J. Biol. Chem. 263, 12397-12402.

Frelinger, A. L. I., Cohen, I., Plow, E. F., Smith, M. A., Roberts, J., Lam, S. C., & Ginsberg, M. H. (1990). Selective inhibition of integrin function by antibodies specific for ligand-occupied receptor conformers. J. Biol. Chem. 265, 6346-6352.

Fu, H., Xia, K., Pallas, D. C., Cui, C., Conroy, K., Narsimhan, R. P., Mamon, H., Collier, R. J., & Roberts, T. M. (1994). Interaction of the protein kinase Raf-1 with 14-3-3 proteins. Science 266, 126-129.

Fujii, C., Yanagi, S., Sada, K., Nagai, K., Taniguchi, T., & Yamamura, H. (1994). Involvement of protein-tyrosine kinase p72syk in collagen-induced signal transduction in platelets. Eur. J. Biochem. 226 243-248.

Fujimoto, T., Fujimura, K., & Kuramoto, A. (1991a). Electrophysiological evidence that glycoprotein IIb-IIIa complex is involved in calcium channel activation on human platelet plasma membrane. J. Biol. Chem. 266, 16370-16375.

Fujımoto, T., Fujimura, K., & Kuramoto, A. (1991b). Functional Ca2+ channel produced by purified platelet membrane glycoprotein IIB-IIIA complex incorporated into planar phospholipid bilayer. Thromb. Haemost. 66, 598-603.

Furie, B. & Furie, B. C. (1995). The molecular basis of platelet and endothelial cell interaction with neutrophils and monocytes: role of P-selectin and the P-selectin ligand, PSGL-1. Thromb. Haemost. 74, 224-227

Ginsberg, M. H., Du, X., O'Toole, T. E., & Loftus, J. C. (1995). Platelet Integrins. Thrombosis and Haemostasis 74, 352-359.

Ginsberg, M. H., Du, X., & Plow, E. F. (1992). Inside-out integrin signalling. Curr. Opin. Cell Biol. 4, 766-771.

Ginsberg, M. H., Xiaopmg, D., O'Toole, T. E., Loftus, J. C., & Plow, E. F. (1993). Platelet integrins. Thromb. Haemost. 70, 87-93.

Golden, A., Brugge, J. S., & Shattil, S. J. (1990). Role of platelet membrane glycoprotein IIb-IIIa in agonist-induced tyrosine phosphorylation of platelet proteins. J. Cell Biol. 3117-3127.

Gralnick, H. R., Williams, S. B., & Coller, B. S. (1985). Asialo von Willebrand factor interactions with platelets. Interdependence of glycoproteins Ib and IIb/IIIa for binding and aggregation. J. Clin. Invest. 75,19-25.

Greco, N. J., Tandon, N. N., Jones, G. D., Kornhauser, R., Jackson, B., Yamamoto, N., Tanoue, K., & Jamieson, G. A. (1996). Contributions of glycoprotein Ib and the seven transmembrane domain receptor to increases in platelet cytoplasmic [Ca2+] induced by alpha-thrombin. Biochemistry 35, 906-914.

Gu, M. & Du, X. (1998). A novel ligand-binding site in the ξ-form 14-3-3 protein recognizing the platelet glycoprotein Iba and distinct from the c-Raf-binding site. J. Biol. Chem. 273, 33465-33471.

Guinebault, C., Payrastre, B., Racaud-Sultan, C., Mazarguil, H., Breton, M., Mauco, G., Plantavid, M., & Chap, H. (1995). Integrin-dependent translocation of phosphoinositide 3-kinase to the cytoskeleton of thrombin-activated platelets involves specific interactions of p85 alpha with actin filaments and focal adhesion kinase. J. Cell. Biol. 129, 831-842.

Haas, T. A. & Plow, E. F. (1996). The cytoplasmic domain of alphaIIb beta3. A ternary complex of the integrin alpha and beta subunits and a divalent cation. J. Biol. Chem. 271, 6017-6026.

Haimovich, B., Kaneshiki, N., & Ji, P. (1996). Protein kinase C regulates tyrosine phosphorylation of pp125FAK in platelets adherent to fibrinogen. Blood 87, 152-161.

Haimovich, B., Lipfert, L., Brugge, J. S., & Shattil, S. J. (1993). Tyrosine phosphorylation and cytoskeletal reorganization in platelets are triggered by interaction of integrin receptors with their immobilized ligands. J. Biol. Chem. 268, 15868-15877.

Hickey, M. J., Hagen, F. S., Yagi, M., & Roth, G. J. (1993). Human platelet glycoprotein V: characterization of the polypeptide and the related IL-V-IX receptor system of adhesive, leucine-rich glycoproteins. Proc. Natl. Acad. Sci. U S A 90, 8327-8331

Hickey, M. J. & Roth, G. J. (1993). Characterization of the gene encoding human platelet glycoprotein IX. J. Biol. Chem. 268, 3438-3443.

Hickey, M. J., Williams, S. A., & Roth, G. J. (1989). Human platelet glycoprotein IX: an adhesive prototype of leucine-rich glycoproteins with flank-center-flank structures. Proc. Natl. Acad. Sci. U S A 86, 6773-6777.

Hillery, C. A., Smyth, S. S., & Parise, L. V. (1991). Phosphorylation of human platelet glycoprotein IIIa (GPIIIa). Dissociation from fibrinogen receptor activation and phosphorylation of GPIIIa in vitro. J. Biol. Chem. 266 14663-14669.

Holmback, K., Danton, M. J., Suh, T. T., Daugherty, C. C., & Degen, J. L. (1996). Impaired platelet aggregation and sustained bleeding in mice lacking the fibrinogen motif bound by integrin alpha IIb beta 3. Embo. J. 15, 5760-5771.

Houdijk, W. P. M., Sakariassen, K. S., Nievelstein, P. F. E. M., & Sixma, J. J. (1985). Role of factor VIII-von Willebrand factor and fibronectin in the interaction of platelets in flowing blood with monomeric and fibrillar collagen types I and III. J. Clin. Invest. 75, 531-540.

Howard, M. A. & Firkin, B. G. (1971). Ristocetin—a new tool in the investigation of platelet aggregation. Thrombosis et Diathesis Haemorrhagica 26, 362-369.

Huang, M. M., Bolen, J. B., Barnwell, J. W., Shattil, S. J., & Brugge, J. S. (1991). Membrane glycoprotein IV (CD36) is physically associated with the Fyn, Lyn, and Yes protein-tyrosine kinases in human platelets. Proc. Natl. Acad. Sci. U S A 88, 7844-7848.

Huang, M. M., Lipfert, L., Cunningham, M., Brugge, J. S., Ginsberg, M. H., & Shattil, S. J. (1993). Adhesive ligand binding to integrin alpha IIb beta 3 stimulates tyrosine phosphorylation of novel protein substrates before phosphorylation of pp125FAK. J. Cell Biol. 122, 473-483.

Hughes, P. E., Diaz, G. F., Leong, L., Wu, C., McDonald, J. A., Shattil, S. J., & Ginsberg, M. H. (1996). Breaking the integrin hinge. A defined structural constraint regulates integrin signaling. J. Biol. Chem. 271, 6571-6574.

Hughes, P. E., O'Toole, T. E., Ylanne, J., Shattil, S. J., & Ginsberg, M. H. (1995). The conserved membrane-proximal region of an integrin cytoplasmic domain specifies ligand binding affinity. J. Biol. Chem. 270, 12411-12417.

Hung, D. T., Vu, T. K., Wheaton, V. I., Ishii, K., & Coughlin, S. R. (1992). Cloned platelet thrombin receptor is necessary for thrombin induced platelet activation. J. Clin. Invest. 89, 1350-1353.

Hynes, R. O. (1992). Integrins: versatility, modulation, and signaling in cell adhesion. Cell 69, 11-25.

Indig, F. E., Diaz, G. F., & Ginsberg, M. H. (1997). Analysis of the tetraspanin CD9-integrin alpha IIb beta 3 (GPIIb-IIIa) complex in platelet membranes and transfected cells. Biochem J. 327 (Pt. 1), 291-298.

Irie, A., Kamata, T., Puzon, M. W., & Takada, Y. (1995). Critical amino acid residues for ligand binding are clustered in a predicted beta-turn of the third N-terminal repeat in the integrin alpha 4 and alpha 5 subunits. Embo. J. 14, 5550-5556.

Irie, K., Gotoh, Y., Yashar, B. M., Errede, B., Nishida, E., & Matsumoto, K. (1994). Stimulatory effects of yeast and mammalian 14-3-3 proteins on the Raf protein kinase. Science 265 1716-1719.

Isenberg, W. M., McEver, R. P., Phillips, D. R., Shuman, M. A., & Bainton, D. F. (1989). Immunogold-surface replica study of ADP-induced ligand binding and fibrinogen receptor clustering in human platelets. Am. J. Anat. 185, 142-148.

Isenberg, W. M., McEver, R. P., Phillips, D. R., Shuman, M. A., & Bainton, D. F. (1987). The platelet fibrinogen receptor: an immunogold-surface replica study of agonist-induced ligand binding and receptor clustering. J. Cell. Biol. 104,1655-1663.

Isobe, T., Hiyane, Y., Ichimura, T., Okuyama, T., Takahashi, N., Nakajo, S., & Nakaya, K. (1992). Activation of protein kinase C by the 14-3-3 proteins homologous with Exo1 protein that stimulates calcium-dependent exocytosis. Febs. Lett. 308, 121-124.

Jackson, S. P., Schoenwaelder, S. M., Yuan, Y., Rabinowitz, I., Salem, H. H., & Mitchell, C. A. (1994). Adhesion receptor activation of phosphatidylinositol 3-kinase. von Willebrand factor stimulates the cytoskeletal association and activation of phosphatidylinositol 3-kinase and pp60c-src in human platelets. J. Biol. Chem. 269, 27093-27099.

Jamieson, G. A. & Okumura, T. (1978). Reduced thrombin binding and aggregation in Bernard-Soulier platelet. J. Clin. Invest. 61, 861-864.

Jennings, L. K., Fox, J. E., Edwards, H. H., & Phillips, D. R. (1981). Changes in the cytoskeletal structure of human platelets following thrombin activation. J. Biol. Chem. 256, 6927-6932.

Kamata, T., Irie, A., Tokuhira, M., & Takada, Y. (1996). Critical residues of integrin alphaIIb subunit for binding of alphaIIbbeta3 (glycoprotein IIb-IIIa) to fibrinogen and ligand-mimetic antibodies (PAC-1, OP-G2, and LJ-CP3). J. Biol. Chem. 271, 18610-18615.

Kamata, T., Puzon, W., & Takada, Y. (1994). Identification of putative ligand binding sites within I domain of integrin alpha 2 beta 1 (VLA-2, CD49b/CD29). J. Biol. Chem. 269, 9659-9663.

Kamata, T. & Takada, Y. (1994). Direct binding of collagen to the I domain of integrin alpha 2 beta 1 (VLA-2, CD49b/CD29) in a divalent cation-independent manner. J. Biol. Chem. 269, 26006-26010.

Katagiri, Y., Hayashi, Y., Yamamoto, K., Tanoue, K., Kosaki, G., & Yamazaki, H. (1990). Localization of von Willebrand factor and thrombin-interactive domains on human platelet glycoprotein Ib. Thromb. Haemost. 63,122-126.

Keely, P. J. & Parise, L. V. (1996). The $\alpha_2\beta_1$ integrin is a necessary co-receptor for collagen-induced activation of Syk and the subsequent phosphorylation of phospholipase Cγ2 in platelets. J. Biol. Chem. 271, 26668-26676.

Kehrel, B., Kronenberg, A., Rauterberg, J., Niesing-Bresch, D., Niehues, U., Kardoeus, J., Schwippert, B., Tschope, D., van de Loo, J., & Clemetson, K. J. (1993). Platelets deficient in glycoprotein IIIb aggregate normally to collagens type I and III but not to collagen type V. Blood 82, 3364-3370.

Kern, A., Briesewitz, R., Bank, I., & Marcantonio, E. E. (1994). The role of the I domain in ligand binding of the human integrin alpha 1 beta 1. J. Biol. Chem. 269, 22811-22816.

Kloczewiak, M., Timmons, S., Lukas, T. J., & Hawiger, J. (1984). Platelet receptor recognition site on human fibrinogen. Synthesis and structure-function relationship of peptides corresponding to the carboxy-terminal segment of the γ-chain. Biochemistry 23, 1767-1774.

Knezevic, I., Leisner, T. M., & Lam, S. (1996). Direct binding of the platelet integrin alphaIIbbeta3 (GPIIb-IIIa) to talin. Evidence that interaction is mediated through the cytoplasmic domains of both alphaIIb and beta 3. J. Biol. Chem. 271, 16416-16421.

Kouns, W. C., Fox, C. F., Lamoreaux, W. J., Coons, L. B., & Jennings, L. K. (1991). The effect of glycoprotein IIb-IIIa receptor occupancy on the cytoskeleton of resting and activated platelets. J. Biol. Chem. 266, 13891-13900.

Kouns, W. C., Wall, C. D., White, M. M., Fox, C. F., & Jennings, L. K. (1990). A conformation-dependent epitope of human platelet glycoprotein IIIa. J. Biol. Chem. 265, 20594-20601.

Kovacsovics, T. J., Bachelot, C., Toker, A., Vlahos, C. J., Duckworth, B., Cantley, L. C., & Hartwig, J. H. (1995). Phosphoinositide 3-kinase inhibition spares actin assembly in activating platelets but reverses platelet aggregation. J. Biol. Chem. 270, 11358-11366.

Kroll, M. H., Harris, T. S., Moake, J. L., Handin, R. I., & Schafer, A. I. (1991). von Willebrand factor binding to platelet GpIb initiates signals for platelet activation. J. Clin. Invest. 88, 1568-1573.

Kroll, M. H., Hellums, J. D., Guo, Z., Durante, W., Razdan, K., Hrbolich, J. K., & Schafer, A. I. (1993). Protein kinase C is activated in platelets subjected to pathological shear stress. J. Biol. Chem. 268, 3520-3524.

Lam, S. C. (1992). Isolation and characterization of a chymotryptic fragment of platelet glycoprotein IIb-IIIa retaining Arg-Gly-Asp binding activity. J. Biol. Chem. 267, 5649-5655.

Lam, S. C.-T., Plow, E. F., Smith, M. A. Andrieux, A., Ryckwaert, J.-J., Marguerie, G. A., & Ginsberg, M. H. (1987). Evidence that Arginyl-Glycyl-Aspartate peptides and fibrinogen gamma chain peptides share a common binding site on platelets. J. Biol. Chem. 262, 947-950.

Law, D. A., Nannizzi, A. L., & Phillips, D. R. (1996). Outside-in integrin signal transduction. $\alpha_{IIb}\beta_3$ -(GPIIb-IIIa) tyrosine phosphorylation induced by platelet aggregation. J. Biol. Chem. 271, 10811-10815.

Leng, L., Kashiwagi, H., Ren, X. D., & Shattil, S. J. (1998). RhoA and the function of platelet integrin alpha IIb beta 3. Blood 91, 4206-4215.

Leong, L., Hughes, P., Schwartz, M. A., Ginsberg, M. H., & Shattil, S. J. (1995). Integrin signaling: roles for the cytoplasmic tails of alpha IIb beta 3 in the tyrosine phosphorylation of ppl25FAK. J. Cell Sci. 108, 3817-3825.

Li, S., Janosch, P., Tanji, M., Rosenfeld, G. C., Waymire, J. C., Mischak, H., Kolch, W., & Sedivy, J. M. (1995). Regulation of Raf-1 kinase activity by the 14-3-3 family of proteins. Embo. J. 14, 685-696

Lipfert, L., Haimovich, B., Schaller, M. D., Cobb, B. S., Parsons, J. T., & Brugge, J. S. (1942). Integrin-dependent phosphorylation and activation of the protein tyrosine kinase ppl25FAK in platelets. J. Cell Biol. 119, 905-912.

Loftus, J. C., Halloran, C. E., Ginsberg, M. H., Feigen, L. P., Zablocki, J. A., & Smith, J. W. (1996). The amino-terminal one-third of alpha IIb defines the ligand recognition specificity of integrin alpha IIb beta 3. J. Biol. Chem. 271, 2033-2039.

Loftus, J. C., O'Toole, T. E., Plow, E. F., Glass, A., Frelinger, A. d., & Ginsberg, M. H. (1990). A beta 3 integrin mutation abolishes ligand binding and alters divalent cation-dependent conformation. Science 249, 915-918.

Loftus, J. C., Plow, E. F., Jennings, L. K., & Ginsberg, M. H. (1988). Alternative proteolytic processing of platelet membrane glycoprotein IIb. J. Biol. Chem. 263, 11025-11028.

Lopez, J. A. (1994). The platelet glycoprotein IL-IX complex. Blood Coagul Fibrinolysis 5, 97-119.

Lopez, J. A., Chung, D. W., Fujikawa, K., Hagen, F. S., Davie, E. W., & Roth, G. J. (1988). The alpha and beta chains of human platelet glycoprotein Ib are both transmembrane proteins containing a leucine-rich amino acid sequence. Proc. Natl. Acad. Sci. U S A 85, 2135-2139.

Lopez, J. A., Chung, D. W., Fujikawa, K., Hagen, F. S., Papayannopoulou, T., & Roth, G. J. (1987). Cloning of the alpha chain of human platelet glycoprotein Ib: a transmembrane protein with homology to leucine-rich alpha 2-glycoprotein. Proc. Natl. Acad. Sci. U S A 84, 5615-5619.

Lopez, J. A., Weisman, S., Sanan, D. A., Sih, T., Chambers, M., & Li, C. Q. (1994). Glycoprotein (GP) Ib beta is the critical subunit linking GP Ib alpha and GP IX in the GP Ib-IX complex. Analysis of partial complexes. J. Biol. Chem. 269, 23716-23721.

Marchese, P., Murata, M., Mazzucato, M., Pradella, P., De, M. L., Ware, J., & Ruggeri, Z. M. (1995). Identification of three tyrosine residues of glycoprotein Ib alpha with distinct roles in von Willebrand factor and alpha-thrombin binding. J. Biol. Chem. 270, 9571-9578.

Matsuno, K., Diaz, R. M., Montgomery, R. R., Aster, R. H., Jamieson, G. A., & Tandon, N. N. (1996). Inhibition of platelet adhesion to collagen by monoclonal anti-CD36 antibodies. Br. J. Haematol. 92, 960-967.

McGregor, J. L., Brochier, J., Wild, F., Follea, G., Trzeciad, M.-C., James, E., Dechavanne, M., McGregor, L., & Clemetson, K. J. (1983). Monoclonal antibodies against platelet membrane glycoproteins. Characterization and effect on platelet function. Eur. J. Biochem. 131, 427-436.

Meller, N., Liu, Y. C., Collins, T. L., Bonnefoy, B. N., Baier, G., Isakov, N., & Altman, A. (1996). Direct interaction between protein kinase C theta (PKC theta) and 14-3-3 tau in T cells: 14-3-3 overexpression results in inhibition of PKC theta translocation and function. Mol. Cell. Biol. 16, 5782-5791.

Michishita, M., Videm, V., & Arnaout, M. A. (1993). A novel divalent cation-binding site in the A domain of the beta 2 integrin CR3 (CDllb/CD18) is essential for ligand binding. Cell 72, 857-867.

Miller, J. L., Cunningham, D., Lyle, V. A., & Finch, C. N. (1991). Mutation in the gene encoding the alpha chain of platelet glycoprotein Ib in platelet-type von Willebrand disease. Proc. Natl. Acad. Sci. U S A 88, 4761-4765.

Miller, J. L., Lyle, V. A., & Cunningham, D. (1992). Mutation of leucine-57 to phenylalanine in a platelet glycoprotein Ib alpha leucine tandem repeat occurring in patients with an autosomal dominant variant of Bernard-Soulier disease. Blood 79, 439-46.

Miyamoto, S., Akiyama, S. K., & Yamada, K. M. (1995a). Synergistic roles for receptor occupancy and aggregation in integrin transmembrane function. Science 267, 883-885.

Miyamoto, S., Teramoto, H., Coso, O. A., Gutkind, J. S., Burbelo, P. D., Akiyama, S. K., & Yamada, K. M. (1995b). Integrin function: molecular hierarchies of cytoskeletal and signaling molecules. J. Cell Biol. 131, 791-805.

Modderman, P. W., Admiraal, L. G., Sonnenberg, A., von, dem, Borne, and Ae (1992). Glycoproteins V and IL-IX form a noncovalent complex in the platelet membrane. J. Biol. Chem. 267, 364-369.

Morii, N., Teru, u. T., Tominaga, T., Kumagai, N., Kozaki, S., Ushikubi, F., & Narumiya, S. (1992). A rho gene product in human blood platelets. II. Effects of the ADP-ribosylation by botulinum C3 ADP-ribosyltransferase on platelet aggregation. J. Biol. Chem. 267, 20921-20926.

Moroi, M., Jung, S. M., Okuma, M., & Shinmyozu, K. (1989). A patient with platelets deficient in glycoprotein VI that lack both collagen-induced aggregation and adhesion. J. Clin. Invest. 84, 1440-1445.

Moroi, M., Jung, S. M., Shinmyozu, K., Tomiyama, Y., Ordinas, A., & Diaz, R. M. (1996). Analysis of platelet adhesion to a collagen-coated surface under flow conditions: the involvement of glycoprotein VI in the platelet adhesion. Blood 88, 2081-2092.

Morton, L. F., Hargreaves, P. G., Farndale, R. W., Young, R. D., & Barnes, M. J. (1995). Integrin alpha 2 beta 1-independent activation of platelets by simple collagen-like peptides: collagen tertiary (triple-helical) and quaternary (polymeric) structures are sufficient alone for alpha 2 beta 1-independent platelet reactivity. Biochem. J. 306, 337-344.

Morton, L. F., Peachey, A. R., Zijenah, L. S., Goodall, A. H., Humphries, M. J., & Barnes, M. J. (1994). Conformation-dependent platelet adhesion to collagen involving integrin alpha 2 beta 1-mediated and other mechanisms: multiple alpha 2 beta 1-recognition sites in collagen type I. Biochem. J. 299, 791-797.

Muslin, A. J., Tanner, J. W., Allen, P. M., & Shaw, A. S. (1996). Interaction of 14-3-3 with signaling proteins is mediated by the recognition of phosphoserine. Cell 84, 889-897.

Muszbek, L. & Laposata, M. (1989). Glycoprotein Ib and glycoprotein IX in human platelets are acylated with palmitic acid through thioester linkages. J. Biol. Chem. 264, 9716-9719.

Nieuwenhuis, H. K., Akkerman, J. W., Houdijk, W. P., & Sixma, J. J. (1985). Human blood platelets showing no response to collagen fail to express surface glycoprotein Ia. Nature 318, 470-472.

Nurden, A. T. & Caen, J. P. (1974). An abnormal platelet glycoprotein pattern in three cases of Glanzmann's thrombasthenia. Brit. J. Haemat. 28, 253-260.

Nurden, A. T. & Caen, J. P. (1975). specific roles for platelet surface glycoproteins in platelet function. Nature 255, 720-722.

O'Toole, T. E., Katagiri, Y., Faull, R. J., Peter, K., Tamura, R., Quaranta, V., Loftus, J. C., Shattil, S. J., & Ginsberg, M. H. (1994). Integrin cytoplasmic domains mediate inside-out signal transduction. J. Cell Biol. 124. 1047-1059.

O'Toole, T. E., Loftus, J. C., Du, X. P., Glass, A. A., Ruggeri, Z. M., Shattil, S. J., Plow, E. F., & Ginsberg, M. H. (1990). Affinity modulation of the alpha IIb beta 3 integrin (platelet GPIIb-IIIa) is an intrinsic property of the receptor. Cell Regul. 1, 883-893.

Oda, A., Druker, B. J., Ariyoshi, H., Smith, M., & Salzman, E. W. (1993). pp60src is an endogenous substrate for calpain in human blood platelets. J. Biol. Chem. 268, 12603-12608.

Okumura, T. & Jamieson, G. A. (1976a). Platelet glycocalicin. II. Purification and characterization. J. Biol. Chem. 251, 5944-5949.

Okumura, T. & Jamieson, G. A. (1976b). Platelet glycocalicin: A single receptor for platelet aggregation induced by thrombin or ristocetin. Thromb. Res. 8, 701-706.

Opas, M., Szewczenko-Pawlikowski, M., Jass, G. K., & Mesaeli, N. (1996). Calreticulin modulates cell adhesiveness via regulation of vinculin expression. J. Cell Biol. 135, 1913-1923.

Otey, C. A., Pavalko, F. M., & Burridge, K. (1990). An interaction between alpha-actinin and the beta 1 integrin subunit in vitro. J. Cell Biol. 111, 721-729.

Parise, L. V., Helgerson, S. L., Steiner, B., Nannizzi, L., & Phillips, D. R. (1987). Synthetic peptides derived from fibrinogen and fibronectin change the conformation of purified platelet glycoprotein IIb-IIIa. J. Biol. Chem. 262, 12597-12602.

Peerschke, E. I. & Ghebrehiwet, B. (1990). Platelet C1q receptor interactions with collagen-and C1q-coated surfaces. J. Immunol. 145, 2984-2988.

Pelletier, A. J., Bodary, S. C., & Levinson, A. D. (1992). Signal transduction by the platelet integrin alpha IIb beta 3: induction of calcium oscillations required for protein-tyrosine phosphorylation and ligand-induced spreading of stably transfected cells. Mol. Biol. Cell 3, 989-998.

Peterson, D. M., Stathopoulos, N. A., Giorgio, T. D., Hellums, J. D., & Moake, J. L. (1987). Shear-induced platelet aggregation requires von Willebrand factor and platelet membrane glycoproteins Ib and IIb-IIIa. Blood 69, 625-628.

Phillips, D. R. & Agin, P. P. (1977). Platelet plasma membrane glycoproteins. Evidence for the presence of nonequivalent disulfide bonds using nonreduced-reduced two-dimensional gel electrophoresis. J. Biol. Chem. 252, 2121-2126.

Phillips, D. R., Charo, I. F., & Scarborough, R. M. (1991). GPIIb-IIIa: the responsive integrin. Cell 65, 359-362.

Poncz, M., Eisman, R., Heidenreich, R., Silver, S. M., Vilaire, G., Surrey, S., Schwartz, E., & Bennett, J. S. (1987). Structure of the platelet membrane glycoprotein IIb. Homology to the alpha subunits of the vitronectin and fibronectin membrane receptors. J. Biol. Chem. 262, 8476-8482.

Powling, M. J. & Hardisty, R. M. (1985). Glycoprotein IIb-IIIa complex and Ca2+ influx into stimulated platelets. Blood 66, 731-734.

Randi, A. M., Rabinowitz, I., Mancuso, D. J., Mannucci, P. M., & Sadler, J. E. (1991). Molecular basis of von Willebrand disease type JIB. Candidate mutations cluster in one disulfide loop between proposed platelet glycoprotein Ib binding sequences. J. Clin. Invest. 87, 1220-1226.

Razdan, K., Hellums, J. D., & Kroll, M. H. (1994). Shear-stress-induced von Willebrand factor binding to platelets causes the activation of tyrosine kinase(s). Biochem. J. 681-686.

Read, M. S., Smith, S. V., Lamb, M. A., & Brinkhous, K. M. (1989). Role of botrocetin in platelet agglutination: formation of an activated complex of botrocetin and von Willebrand factor. Blood 74,1031-1035.

Renshaw, M. W., Toksoz, D., & Schwartz, M. A. (1996). Involvement of the small GTPase rho in integrin-mediated activation of mitogen-activated protein kinase. J. Biol. Chem. 271, 21691-21694.

Reuther, G. W., Fu, H., Cripe, L. D., Collier, R. J., & Pendergast, A. M. (1994). Association of the protein kinases c-Bcr and Bcr-Abl with proteins of the 14-3-3 family. Science 266,129-133.

Robinson, K., Jones, D., Patel, Y., Martin, H., Madrazo, J., Martin, S., Howell, S., Elmore, M., Finnen, M. J., & Aitken, A. (1994). Mechanism of inhibition of protein kinase C by 14-3-3 isoforms. 14-3-3 isoforms do not have phospholipase A2 activity. Biochem. J. 299, 853-861.

Rojiani, M. V., Finlay, B. B., Gray, V., & Dedhar, S. (1991). In vitro interaction of a polypeptide homologous to human Ro/SS-A antigen (calreticulin) with a highly conserved amino acid sequence in the cytoplasmic domain of integrin alpha subunits. Biochemistry 30, 9859-9866.

Rooney, M. M., Parise, L. V., & Lord, S. T. (1996). Dissecting clot retraction and platelet aggregation. Clot retraction does not require an intact fibrinogen gamma chain C terminus. J. Biol. Chem. 271, 8553-8555.

Rosa, J. P., Bray, P. F., Gayet, O., Johnston, G. I., Cook, R. G., Jackson, K. W., Shuman, M. A., & McEver, R. P. (1988). Cloning of glycoprotein IIIa cDNA from human erythroleukemia cells and localization of the gene to chromosome 17. Blood 72, 593-600.

Roth, G. J., Church, T. A., McMullen, B. A., & Williams, S. A. (1990). Human platelet glycoprotein V: a surface leucine-rich glycoprotein related to adhesion. Biochem. Biophys. Res. Commun. 170, 153-161.

Ruggeri, Z. M. (1991). The platelet glycoprotein IL-IX complex. Prog. Hemost. Thromb. 10. 35-68.

Russell, S. D. & Roth, G. J. (1993). Pseudo-von Willebrand disease: a mutation in the platelet glycoprotein Ib alpha gene associated with a hyperactive surface receptor. Blood 81, 1787-1791.

Rybak, M. E., Renzulli, L. A., Bruns, M. J., & Cahaly, D. P. (1988). Platelet glycoproteins IIb and IIIa as a calcium channel in liposomes. Blood 72, 714-720.

Ryo, R., Yoshida, A., Sugano, W., Yasunaga, M., Nakayama, K., Saigo, K., Adachi, M., Yamaguchi, N., & Okuma, M. (1992). Deficiency of P62, a putative collagen receptor, in platelets from a patient with defective collagen-induced platelet aggregation. Am. J. Hematol. 39, 25-31.

Sage, S. O., Sargeant, P., Heemskerk, J. W., & Mahaut, S. M. (1993). Calcium influx mechanisms and signal organisation in human platelets. Adv. Exp. Med. Biol. 344, 69-82.

Saido, T. C., Mizuno, K., & Suzuki, K. (1991). Proteolysis of protein kinase C by calpain: effect of acidic phospholipids. Biomed. Biochim. Acta. 50, 485-489.

Sakariassen, K. S., Nievelstein, P. F., Coller, B. S., & Sixma, J. J. (1986). The role of platelet membrane glycoproteins Ib and IIb-IIIa in platelet adherence to human artery subendothelium. Br. J. Haematol. 63, 681-691.

Santoro, S. A. (1986). Identification of a 160,000 dalton platelet membrane protein that mediates the initial divalent cation-dependent adhesion of platelets to collagen. Cell 46, 913-920.

Savage, B., Saldivar, E., & Ruggeri, Z. M. (1996). Initiation of platelet adhesion by arrest onto fibrinogen or translocation on von Willebrand factor. Cell 84, 289-297.

Savage, B., Shattil, S. J., & Ruggeri, Z. M. (1992). Modulation of platelet function through adhesion receptors. A dual role for glycoprotein IIb-IIIa (integrin alpha IIb beta 3) mediated by fibrinogen and glycoprotein Ib-von Willebrand factor. J. Biol. Chem. 267, 11300-11306.

Schick, P. K. & Walker, J. (1996). The acylation of megakaryocyte proteins: glycoprotein IX is primarily myristoylated while glycoprotein Ib is palmitoylated. Blood 87, 1377-1384.

Schwartz, M. A., Brown, E. J., & Fazeli, B. (1993). A 50-kDa integrin-associated protein is required for integrin-regulated calcium entry in endothelial cells. J. Biol. Chem. 268, 19931-19934.

Schwartz, M. A., Schaller, M. D., & Ginsberg, M. H. (1995). Integrins: emerging paradigms of signal transduction. Ann. Rev. Cell Dev. Biol. 11, 549-599.

Shattil, S. J. & Brass, L. F. (1987). Induction of the fibrinogen receptor on human platelets by intracellular mediators. J. Biol. Chem. 262, 992-1000.

Shattil, S. J., Haimovich, B., Cunningham, M., Lipfert, L., Parsons, J. T., Ginsberg, M. H., & Brugge, J. S. (1994). Tyrosine phosphorylation of ppl25FAK in platelets requires coordinated signaling through integrin and agonist receptors. J. Biol. Chem. 269, 14738-14745.

Shattil, S. J., Hoxie, J. A., Cunningham, M., & Brass, L. F. (1985). Changes in the platelet membrane glycoprotein IIb.IIIa complex during platelet activation. J. Biol. Chem. 260, 11107-11114.

Shattil, S. J., O'Toole, T., Eigenthaler, M., Thon, V., Williams, M., Babior, B. M., & Ginsberg, M. H. (1995). Beta 3-endonexin, a novel polypeptide that interacts specifically with the cytoplasmic tail of the integrin beta 3 subunit. J. Cell Biol. 131, 807-816.

Sims, P. J., Ginsberg, M. H., Plow, E. F., & Shattil, S. J. (1991). Effect of platelet activation on the conformation of the plasma membrane glycoprotein IIb-IIIa complex. J. Biol. Chem. 266, 7345-7352.

Staatz, M. D., Rajpara, S. M., Wayner, E. A., Carter, W. G., & Santoro, S. A. (1989). The membrane glycoprotein Ia-IIa (VLA-2) complex mediates the Mg^{++}-dependent adhesion of platelets to collagen. J. Cell Biol. 108, 1917-1924.

Steiner, B., Trzeciak, A., Pfenninger, G., & Kouns, W. C. (1993). Peptides derived from a sequence within beta 3 integrin bind to platelet alpha IIb beta 3 (GPIIb-IIIa) and inhibit ligand binding. J. Biol. Chem. 268, 6870-6873.

Sugimoto, M., Mohri, H., McClintock, R. A., & Ruggeri, Z. M. (1991). Identification of discontinuous von Willebrand factor sequences involved in complex formation with botrocetin. A model for the regulation of von Willebrand factor binding to platelet glycoprotein Ib. J. Biol. Chem. 266, 18172-18178.

Sultan, C., Plantavid, M., Bachelot, C., Grondin, P., Breton, M., Mauco, G., Levy, T. S., Caen, J. P., & Chap, H. (1991). Involvement of platelet glycoprotein IIb-IIIa (alpha IIb-beta 3 integrin) in thrombin-induced synthesis of phosphatidylinositol 3',4'-bisphosphate. J. Biol. Chem. 266, 23554-23557.

Tandon, N. N., Kralisz, U., & Jamieson, G. A. (1989). Identification of glycoprotein IV (CD36) as a primary receptor for platelet-collagen adhesion. J. Biol. Chem. 264, 7576-7583.

Tandon, N. N., Ockenhouse, C. F., Greco, N. J., & Jamieson, G. A. (1991). Adhesive functions of platelets lacking glycoprotein IV (CD36). Blood 78, 2809-2813.

Tanji, M., Horwitz, R., Rosenfeld, G., & Waymire, J. C. (1994). Activation of protein kinase C by purified bovine brain 14-3-3: comparison with tyrosine hydroxylase activation. J. Neurochem. 63, 1908-1916.

Taylor, D. B. & Gartner, T. K. (1992). A peptide corresponding to GPIIb alpha 300-312, a presumptive fibrinogen gamma-chain binding site on the platelet integrin GPIIb/IIIa, inhibits the adhesion of platelets to at least four adhesive ligands. J. Biol. Chem. 267, 11729-11733.

Toker, A., Ellis, C. A., Sellers, L. A., & Aitken, A. (1990). Protein kinase C inhibitor proteins. Purification from sheep brain and sequence similarity to lipocortins and 14-3-3 protein. Eur. J. Biochem. 191, 421-429.

Tozer, E. C., Liddington, R. C., Sutcliffe, M. J., Smeeton, A. H., & Loftus, J. C. (1996). Ligand binding to integrin alphaIIbbeta3 is dependent on a MIDAS-like domain in the beta3 subunit. J. Biol. Chem. 271, 21978-21984.

van Willigen, G., Hers, I., Gorter, G., & Akkerman, J. W. (1996). Exposure of ligand-binding sites on platelet integrin alpha IIB/beta 3 by phosphorylation of the beta 3 subunit. Biochem. J. 314, 769-779.

Vicente, V., Houghten, R. A., & Ruggeri, Z. M. (1990). Identification of a site in the alpha chain of platelet glycoprotein Ib that participates in von Willebrand factor binding. J. Biol. Chem. 265, 274-280.

Vu, T. K., Hung, D. T., Wheaton, V. I., & Coughlin, S. R. (1991). Molecular cloning of a functional thrombin receptor reveals a novel proteolytic mechanism of receptor activation. Cell 64,1057-1068.

Ward, C. M. Andrews, R. K., Smith, A. I., & Berndt, M. C. (1996). Mocarhagin, a novel cobra venom metalloproteinase, cleaves the platelet von Willebrand factor receptor glycoprotein Ibalpha. Identification of the sulfated tyrosine/anionic sequence Tyr-276-Glu-282 of glycoprotein Ibalpha as a binding site for von Willebrand factor and alpha-thrombin. Biochemistry 35, 4929-4938.

Wardell, M. R., Reynolds, C. C., Berndt, M. C., Wallace, R. W., & Fox, J. E. (1989). Platelet glycoprotein Ib beta is phosphorylated on serine 166 by cyclic AMP-dependent protein kinase. J. Biol. Chem. 264, 15656-15661.

Ware, J., Dent, J. A., Azuma, H., Sugimoto, M., Kyrle, P. A., Yoshioka, A., & Ruggeri, Z. M. (1991). Identification of a point mutation in type IIB von Willebrand disease illustrating the regulation of von Willebrand factor affinity for the platelet membrane glycoprotein Ib-IX receptor. Proc. Natl. Acad. Sci. U S A 88, 2946-2950.

Ware, J., Russell, S. R., Marchese, P., Murata, M., Mazzucato, M., De, M. L., & Ruggeri, Z. M. (1993). Point mutation in a leucine-rich repeat of platelet glycoprotein Ib alpha resulting in the Bernard-Soulier syndrome. J. Clin. Invest. 92, 1213-1220.

Wary, K. K., Mainiero, F., Isakoff, S., Marcantonio, E. E., & Giancotti, F. G. (1996). The adaptor protein Shc couples a class of integrins to the control of cell cycle progression. Cell 87, 733-743.

Weisel, J. W., Nagaswami, C., Vilaire, G., & Bennett, J. S. (1992). Examination of the platelet membrane glycoprotein IIb-IIIa complex and its interaction with fibrinogen and other ligands by electron microscopy. J. Biol. Chem. 267, 16637-16643.

Weiss, H. J., Tschopp, T. B., Baumgartner, H. R., Sussman, I. I., Johnson, M. M., & Egan, J. J. (1974). Decreased adhesion of giant (Bernard-Soulier) platelets to subendothelium. Further implications on the role of von Willebrand factor in hemostasis. Am. T. Med. 57. 920-925.

Weiss, H. J., Turitto, V. T., & Baumgartner, H. R. (1978). Effect of shear rate on platelet interaction with subendothelium in citrated and native blood. I. Shear rate dependent decrease of adhesion in von Willebrand's disease and the Bernard-Soulier syndrome. J. Lab. Clin. Med. 92, 750-764.

Weiss, H. J., Turitto, V. T., & Baumgartner, H. R. (1986). Platelet adhesion and thrombus formation on subendothelium in platelets deficient in glycoproteins IIb-IIIa, Ib, and storage granules. Blood 67, 322-330

Wicki, A. N. & Clemetson, K. J. (1987). The glycoprotein Ib complex of human blood platelets. Eur. J. Biochem. 163, 43-50.

Wippler, J., Kouns, W. C., Schlaeger, E. J., Kuhn, H., Hadvary, P., & Steiner, B. (1994). The integrin alpha IIb-beta 3, platelet glycoprotein IIb-IIIa, can form a functionally active heterodimer complex without the cysteine-rich repeats of the beta 3 subunit. J. Biol. Chem. 269, 8754-8761.

Wu, C. C., Ko, F. N., Huang, T. F., & Teng, C. M. (1996a). Mechanisms-regulated platelet spreading after initial platelet contact with collagen. Biochem. Biophys. Res. Commun. 220, 388-393.

Wu, G., Meloni, F. J., & Shapiro, S. S. (1996b). Platelet glycoprotein (Gp) IX associates with GpIb alpha in the platelet membrane GpIb complex. Blood 87, 2782-2787.

Wyler, B., Bienz, D., Clemetson, K., J., & Luscher, E. F. (1986). Glycoprotein Ib_{β} is the only phosphorylated major membrane glycoprotein in human platelets. Biochem. J. 234, 373-379.

Yamamoto, N., Akamatsu, N., Yamazaki, H., & Tanoue, K. (1992). Normal aggregations of glycoprotein IV (CD36)-deficient platelets from seven healthy Japanese donors. Br. J. Haematol. 81, 86-92.

Yamamoto, N., Ikeda, H., Tandon, N. N., Herman, J., Tomiyama, Y., Mitani, T., Sekiguchi, S., Lipsky, R., Kralisz, U., & Jamieson, G. A. (1990). A platelet membrane glycoprotein (GP) deficiency in healthy blood donors: Naka- platelets lack detectable GPIV (CD36). Blood 76, 1698-1703.

Yamamoto, N., Kitagawa, H., Tanoue, K., & Yamazaki, H. (1985). Monoclonal antibody to glycoprotein Ib inhibits both thrombin- and ristocetin-induced platelet aggregations. Thromb. Res. 39, 751-759.

Ylanne, J., Chen, Y., O'Toole, T. E., Loftus, J. C., Takada, Y., & Ginsberg, M. H. (1993). Distinct functions of integrin alpha and beta subunit cytoplasmic domains in cell spreading and formation of focal adhesions. J. Cell Biol. 122, 223-233.

Ylanne, J., Huuskonen, J., O'Toole, T. E., Ginsberg, M. H., Virtanen, I., & Gahmberg, C. G. (1995). Mutation of the cytoplasmic domain of the integrin beta 3 subunit. Differential effects on cell spreading, recruitment to adhesion plaques, endocytosis, and phagocytosis. J. Biol. Chem. 270, 9550-9557.

Zhang, J., Zhang, J., Shattil, S. J., Cunningham, M. C., & Rittenhouse, S. E. (1996). Phosphoinositide 3-kinase gamma and p85/phosphoinositide 3-kinase in platelets. Relative activation by thrombin receptor or beta-phorbol myristate acetate and roles in promoting the ligand-binding function of alpha IIb beta 3 integrin. J. Biol. Chem. 271, 6265-6272.

Zhu, X. & Assoian, R. K. (1995). Integrin-dependent activation of MAP kinase: a link to shape-dependent cell proliferation. Mol. Biol. Cell. 6, 273-282.

Zupan, L. A., Steffens, D. L., Berry, C. A., Landt, M., & Gross, R. W. (1992). Cloning and expression of a human 14-3-3 protein mediating phospholipolysis. Identification of an arachidonoyl-enzyme intermediate during catalysis. J. Biol. Chem. 267, 8707-8710.

SIGNALING BY CELL ADHESION MOLECULES IN THE NERVOUS SYSTEM

John J. Hemperly

Advances in Molecular and Cell Biology
Volume 28, pages 303-320.

ISBN: 0-7623-0495-2

I. INTRODUCTION

The vertebrate nervous system is a paradigm for signaling between cells and it is through cellular contacts and interactions that signals arising both inside and outside of the body are integrated. However, most classical neuronal signaling involves the interaction of cells via synapses and diffusible neurotransmitters, where the cells are not directly in contact. This type of neuronal signaling will not be addressed in this review. Instead, this chapter will focus on the cellular signals that are transmitted directly via what are usually called cell adhesion molecules (CAMs). These molecules were originally identified on the basis of their ability to mediate direct cell adhesion in in vitro adhesion assays, or on their protein sequence similarity to other, previously identified CAMs (Edelman and Crossin, 1991; Ruoslahti and Obrink, 1996). Cell adhesion molecules have been classified into a number of families based on their structures and on the types of adhesion they mediate (Horwitz and Hunter, 1996). Molecules of all of the major classes of cell adhesion molecules are found in the brain, spinal cord, and peripheral tissues such as nerve and muscle. Probably the most prominent group of neuronal CAMs are those of the immunoglobulin superfamily (Brummendorf end Rathjen, 1995; Hemperly, 1996). Integrins are also found in the brain (Defreitas et al., 1995), as are certain of the cadherins (Redies and Takeichi, 1996). The selectins seem less prevalent in the brain, but there are many neural cell adhesions dependent on carbohydrates, particularly those involving the glucosaminoglycans found on proteoglycans (Margolis et al., 1996).

Although all of these molecules can mediate cellular adhesion in vitro and in vivo, it has been recognized that CAMs can also be major signaling molecules in their own right. In fact, the level of adhesion seen in vitro may be a secondary or minor component of their natural function. In this chapter, I will review briefly some of the neural cell adhesion molecules implicated in neuronal cell signaling and then describe some recent attempts to determine the signaling mechanisms they use.

II. SIGNALING PARADIGMS IN THE NERVOUS SYSTEM

To investigate signaling by adhesion receptors in the nervous system, it is obviously necessary to choose an experimental paradigm(s) that can be readily manipulated and analyzed at several levels (see Figure 1).

The model systems most commonly used for studying signaling mediated by CAMs are neurite outgrowth, biochemical analysis of protein phosphorylation and protein-protein interactions, cell migration and cell adhesion.

A. Neurite Outgrowth

It has long been recognized that both developing and regenerating neuronal processes must recognize and find their ways through a network of environmental cues

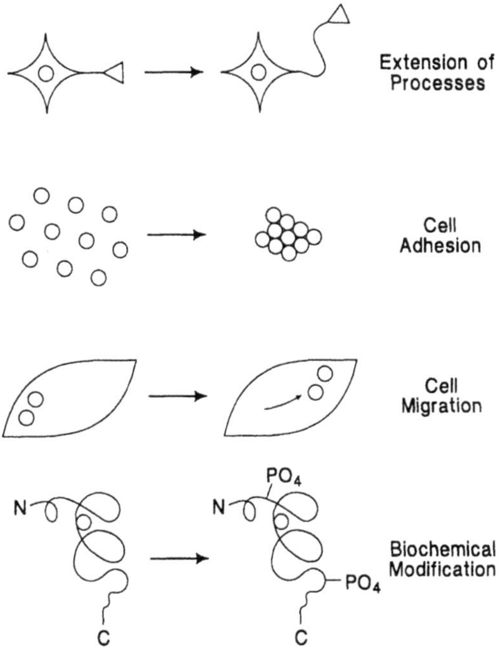

Figure 1. To assay the effects of CAM-mediated cellular signaling, a number of experimental paradigms have been devised. These include process extension, cell adhesion, cell migration, and biochemical modification.

(Tessier-Lavigne and Goodman, 1996; Varela-Echavarria and Guthrie, 1997). These cues may comprise soluble diffusible substances as well as the more firmly localized cues from neighboring cells and from the extracellular matrix. Because the elaboration of processes is such a highly visible and measurable response, the production of processes in vitro is often used in studies of cell signaling. In these measurements, the number of cell processes as well as the length of the neurites, either as the total length of all processes or as the length of the longest process, have been used. In its simplest form, a ligand for a cell adhesion molecule is attached to a tissue culture dish in vitro, cells capable of forming processes are plated, and the effects observed morphometrically after a fixed time. Alternatively, the growth of processes can be captured by video microscopy and the changes observed directly. This is particularly important for studies looking at the shape of the growth cones, specialized structures at the tips of neuron processes (Burdengulley et al., 1995). The effects of various pharmacological agents or blocking antibodies are then simply measured by adding them to the growth medium to look for increases or decreases in the extent of neurite outgrowth.

A very commonly used way of attaching adhesion ligands to the plastic substrate is by mixing them with nitrocellulose (Lagenaur and Lemmon, 1987). This type of assay is also convenient and relatively easy to do. However, as with many model systems, there can be problems. For example, the protein absorbed to a plastic surface consists not only of molecules in their native conformation, but also denatured and improperly oriented protein as well. To avoid some of these complications, another signaling paradigm is to measure the outgrowth of CAM-containing cellular processes over a monolayer of cells containing the CAM ligand. The ligand can be a molecule occurring naturally in the cell monolayer, for example, integrins on fibroblasts, or it may be added experimentally using recombinant DNA techniques. This method has been used particularly for studying the outgrowth of neurites on CAM-transfected cells by Walsh and Doherty and their colleagues as discussed below. It should be noted particularly, that the parameter being measured in this type of assay is not adhesion per se, (that is, the binding of cells which are capable of putting out processes to the monolayer) but instead is the stimulated growth of processes. Therefore, the growth is dependent on adhesion, but it is also dependant on signals generated within the cell.

B. Protein Phosphorylation and Binding Interactions

Another commonly used measurement of cell signaling is the phosphorylation and/or dephosphorylation of cellular proteins as a result of the "triggering" of adhesion molecules at the neuronal cell surface. The triggering usually uses the addition of cells bearing CAM ligands, the ligands themselves, or more recently chimeric proteins comprising the ligand fused to the Fc portion of an antibody molecule. Antibodies, both monoclonal and polyclonal, have also been used as triggering reagents. In this type of assay, one can measure the phosphorylation level of the cell extract as a whole or the phosphorylation level of defined proteins by immunoprecipitation. In any case, the levels of phosphorylation presumably reflect the modulation of the activities of specific kinases and phosphatases. In some cases, it is possible to measure the specific activity of the kinases and phosphatases directly, and a major effort has been to determine particular enzymes involved. A more recent approach to studying the biochemical signaling resulting from the binding to adhesion ligands is to generate transgenic mice deficient in particular kinases and phosphatases.

In addition to looking at changes in protein phosphorylation levels, it has sometimes been possible to directly look for new protein–protein interactions that result from CAM signaling. This usually involves immunoprecipitation of the CAM to see if new proteins immunoprecipitate as a result of ligand engagement, and if phosphoproteins are detected, what is their level of phosphorylation.

C. Cell Migration

Another assay for the cellular signaling that results from CAM signaling is cell migration, that is, the increased movement of cells through a membrane filter as a result of CAM binding. There have been fewer studies of this type and it is difficult to know if cell migration in neuronal cells is simply a reflection of the same type of processes involved in neurite outgrowth. There is also the complication of chemotaxis, chemokinesis, and haptotaxis to sort out. However, there are antibody reagents such as monoclonal antibody NCAM16 which can cause the migration of rat primary motor neurons and human cell lines through a membrane (Ackley et al., 1997). As an example of cell migration in vivo, when cells leave the neural crest during development, they regulate their level of CAM expression. It is interesting to speculate what signaling pathways may be effected during the migration as a result of these changing CAM levels.

D. Cell Adhesion

Cell adhesion is also used as a measure of cell signaling, and has been particularly useful in studying the signaling mediated by integrins. These molecules exist in conformationally active and inactive states that can be switched either endogenously via signals from inside the cell (inside-out signaling) or by antibodies. In turn, the integrins can also signal from the outside to the inside of the cell (outside-in signaling) to lead to actual changes in cell adhesion

III. ADHESION MOLECULES OF THE NERVOUS SYSTEM

Using the paradigms described above, a large number of molecules in the nervous system have been identified that can mediate cell adhesion. Most of the molecules fall into broad classes based on their structural motifs (see Table 1).

The members of these classes have been well described in other chapters of this volume, so I will only review a few selected molecules which have been implicated in signaling.

Table 1. Selected Neuronal Adhesion Receptors Implicated in Cellular Signaling

Family	*Example*	*Membrane Anchor*	*Signaling Pathway*
Ig superfamily	N-CAM (180 kD, 140 kD)	transmembrane	fyn
	N-CAM (120 kD)	GPI	
	L1	transmembrane	src, fyn
	contactin/F3/F1 l	GPI	fyn
Integrins	$a_v\beta_3$	transmembrane	phosphotyrosine
Cadherins	N-cadherin	transmembrane	catenins and cytoskeleton; wnt1
	R-cadherin	transmembrane	

A. Immunoglobulin Superfamily

Structural Motifs

Probably the largest number of cell adhesion molecules in the nervous system belong to the immunoglobulin (Ig) superfamily. It was once estimated that most of the cell surface molecules of the brain belong to this family. In general, Ig superfamily members have a number of extracellularly disposed domains which are similar to structural domains found in immunoglobulins. For many molecules, this similarity has been deduced solely on primary sequence comparisons. However, this comparison is not based on extensive areas of sequence identity, but rather on the conservation of a small number of amino acid residues adjacent to a central pair of cysteine residues spaced approximately 50 residues apart (see Brummendorf and Rathjen, 1995). For example, the first of the neural cell adhesion molecules, N-CAM, was postulated to be a member of the immunoglobulin superfamily based on just this type of analysis (Hemperly et al., 1986). Recently, the three dimensional structure of an extracellular domain of N-CAM has been determined, bearing out this prediction (Thomsen et al., 1996).

In addition to the immunoglobulin-like domains, many neuronal CAMs contain fibronectin type III repeats. This motif was originally identified in the extracellular matrix protein fibronectin. More recent three-dimensional structure analysis suggests that fibronectin type III- repeats also have a domain structure which is similar to that of an immunoglobulin domain (Leahy et al., 1996). Together, the immunoglobulin-like and fibronectin type III repeats probably form a structural scaffold of compact domains. It would then be the particular amino acids within the domains that mediate particular and specific bimolecular interactions. For example, although N-CAM comprises five extracellular immunoglobulin-like domains and two fibronectin type III domains, it is largely the second Ig-like domain that is involved in the interaction of N-CAM with heparin sulfate proteoglycans (Cole and Akeson, 1989) and largely the third Ig-like domain that is involved in homophilic binding of one N-CAM molecule to another (Rao et al., 1994; Ranheim et al., 1996).

Binding Partners

Although all of the Ig superfamily members appear able to mediate homophilic adhesion, particularly if expressed in relatively large amounts, it is clear that binding to other molecules (heterophilic binding) is also quite important. For example, although Ng-CAM can mediate binding, another molecule, axonin-1, is necessary for optimal neurite outgrowth, and binding is correlated with the activity of cellular kinases (Kunz et al., 1996).

In addition to binding other cell surface molecules, many, if not all, CAMs of the immunoglobulin superfamily can bind extracellular matrix proteins. For example, it has been reported that N-CAM can bind collagen (Probstmeier et al., 1992) as well as neuronal proteoglycans such as phosphacan and neurocan (Grumet et al., 1996).

A particularly exciting recent development is the observation that contactin/F3/F11, which has six immunoglobulin-like domains and five fibronectin type III repeats is capable of binding the receptor tyrosine phosphatase RTPβ, resulting in neurite outgrowth and demonstrating clearly that CAMs can interact with receptor tyrosine phosphatases as well as kinases (Sakurai et al., 1997)

As will be discussed below, CAMs of the immunoglobulin superfamily can bind intracellular ligands. Originally it was reported that N-CAM, and in particular the larger, 180 kD form of N-CAM can bind to spectrin, a component of the cytoskeleton (Pollerberg et al, 1987). However. these observations have not been extended significantly since then. Although still an exciting hypothesis, there is not extensive evidence for direct CAM binding to the cytoskeleton

More recently, multiple laboratories have demonstrated interactions of CAMs with intracellular kinases and with extracellular domains of protein kinases and phosphatases (see below). It is these observations that have also raised the possibility of intracellular signaling via these molecules.

Membrane Association

The majority of cell adhesion receptors in the nervous system are transmembrane proteins comprising an extracellular domain, a single hydrophobic membrane spanning region, and a cytoplasmic carboxyl terminal region which may or may not have functional enzymatic activity. These molecules are readily released in soluble form by non-ionic detergents such as NP40 and Triton X-100. The extracellular regions, when expressed using recombinant DNA techniques or released by proteolytic digestion, are relatively soluble and exhibit their expected ligand binding activities. There have also been reports that the larger variants of N-CAM, which presumably retain this single transmembrane segment, can be found in aqueous brain extracts and in cerebrospinal fluid (Bock et al., 1987; Poltorak et al., 1995). These polypeptides are soluble in the sense of not being pelleted by centrifugation at 100,000 *g*. It is not known if these soluble forms retain associated membrane lipids as well.

In contrast, a number of cell adhesion receptors are linked to the cell surface by linkage to the lipid phosphatidylinositol. It is thought that a precursor form of the CAM comprising the majority of the mature molecule linked to a hydrophobic tail is transferred in the endoplasmic reticulum via a transamination reaction to a preformed anchoring moiety (Takeda and Kinoshita, 1995). The anchor itself comprises the phospholipid phosphatidylinositol linked through glucosamine and a variable number of sugars to ethanolamine. It is the ethanolamine that becomes linked to the polypeptide with release of the nascent hydrophobic tail as well as eight to ten non-conserved residues (Vaughan, 1996).

Somewhat surprisingly, the lipid linked molecules of this type are not solubilized very well by the cold NP-40 or Triton X-100 used to release the transmembrane forms of cell adhesion receptors. This is particularly clear in the case of N-CAM, which exists as three major isoforms of 180 kD, 140 kD, and 120 kD. The former polypeptides are readily solubilized by NP-40, whereas the 120 kD GPI-linked isoform is largely insoluble. How these lipid anchored CAMs, which do not transverse the membrane, communicate with the inside of the cell to transmit signals is as yet unknown.

B. Neuronal Integrins

The integrins found in the nervous system appear to be similar to those found in other tissues in terms of their primary sequence. In particular, neuronal integrins are also comprised of two polypeptides, the α and β chains, that interact to form homodimers. As in other tissues, neuronal integrins seem to transmit both inside-out and outside-in signals. It has also been found recently, that the $\alpha V\beta 3$ integrins are ligands for the L1 antigen (Montgomery et al., 1996; Kadmon and Altevogt, 1997). Therefore, integrins are capable not only of interacting with the neuronal extracellular matrix but also with cell surface CAMs. It is possible. however. that some of the integrin signaling investigated in fibroblasts, may not be precisely the same in neurons. For example, a major protein $p125^{FAK}$, which is phosphorylated in fibroblasts upon engagement of integrins with the extracellular matrix, is quite differently distributed in neurons and does not associate with focal adhesions.

C. Cadherins

Cadherins are a family of cell adhesion molecules which mediate calcium-dependent binding of cells to each other. This is a large family of molecules some of which, such as N-cadherin and R-cadherin, were originally identified in the nervous system (Redies and Takeichi, 1996). One of the most investigated cadherins is E-cadherin, which is found on many epithelia. It is quite clear that E-cadherin is a structural component of adherens junctions and is found in complexes with another type of proteins called the catenins. The most recent interest in the cadherins involves (1) the down regulation of E-cadherin in a large number of carcinomas and (2) their signaling interactions with members of the wnt family of signaling molecules (Gumbiner, 1995). In carcinomas, the loss of E-cadherin seems to correlate with a decreased adhesion of the tumor cells with each other and an increased ability to metastasize to distant sites. (Similar correlations are found with the loss of catenins, either situation presumably leading to the loss of functional adhesion.)

Somewhat surprisingly, N-cadherin does not seem strictly analogous to E-cadherin. In fact, N-cadherin, although interacting with its own set of catenin molecules, does not seem as robust in mediating adhesion or localizing to adherens junctions. Also, in clinical samples, the ability of the cells to metastasize is not correlated with a loss of N-cadherin, and in fact seems more related to increased levels of N-cadherin (Shinoura et al., 1995). The differences in the biology of these cadherins, despite their homologous amino acid sequences and domain organization, is still being investigated.

D. Receptor Tyrosine Kinases and their Ligands

Growth Factor Receptor Kinases and their Ligands

Signaling in axon guidance undoubtedly involves the balanced actions of multiple tyrosine kinases and phosphatases, and these activities are highly conserved from insects to higher mammals (Desai et al., 1997). The major growth factor receptor tyrosine kinase implicated in adhesive signaling in the nervous system is the fibroblast growth factor receptor (FGFR). A long standing series or experiments by Patrick Doherty, Frank Walsh, and their colleagues have demonstrated that the FGFR is capable of interacting with a number of cell adhesion molecules, including N-CAM, L I, and N-cadherin. The results of their studies are discussed below, but it should be noted that the FGFR actually comprises a number of related molecules, all of which comprise extracellular, ligand-binding domains with variable numbers of immunoglobulin-like repeats which arise by alternative splicing of RNA. Moreover, the various FGF receptors are able to bind multiple forms of fibroblast growth factor (Green et al., 1996). The extracellular domains of the FGFR are followed by a single, membrane-spanning hydrophobic segment and intracellular tandemly-repeated tyrosine kinase domains.

Eph Kinases and their Ligands

A particularly exciting area of investigation has been the discovery of receptor tyrosine kinases with ligand bound receptors that are distributed in gradients in the developing nervous system (Harris and Holt, 1995; Tessier-Lavigne, 1995; Tessier-Lavigne and Goodman, 1996; Nieto, 1996). These so-called eph kinases form the largest family of receptor tyrosine kinases. The corresponding eph receptor ligands seem to fall into two sets, one linked to the cell by GPI anchors and the other containing transmembrane segments. As with the FGFR, the eph kinases consist of an extracellular segment capable of binding to appropriate ligands, a transmembrane segment, and an intracellular kinase domain.

The ligands for the eph receptors were isolated under a variety of names such as AL-1, RAGS, and ELF-1 using quite diverse systems, and the members and significance of these ligands remain to be elucidated. However, it is quite clear that the balance of these ligand-receptor pairs may be critical for the proper routing of neuronal projections to their appropriate targets. How the intracellular signals produced by this binding result in proper guidance is a balance of growth stimulation and growth repelling signals with the neuron itself.

IV. ADHESION RECEPTOR SIGNALING MECHANISMS

As mentioned above, cell adhesion molecules on one cell can directly bind to molecules on adjacent cells to mediate adhesion. Moreover, this binding can also activate signaling pathways in the participating cells that lead to changes in cellular behavior and ultimately to changes in gene expression. A number of recent studies have addressed how these signals are mediated at the molecular level, particularly in the area of neurite outgrowth (see Figure 2). These studies have begun in many ways.

Occasionally, they have asked the basic pharmacological questions: What changes in intracellular signals can be detected (Schuch et al., 1989; Von Bohlen et al., 1992)? Also, what drugs can interfere with the process of neurite extension, and hence what intracellular pathways may be involved? This latter approach led to the

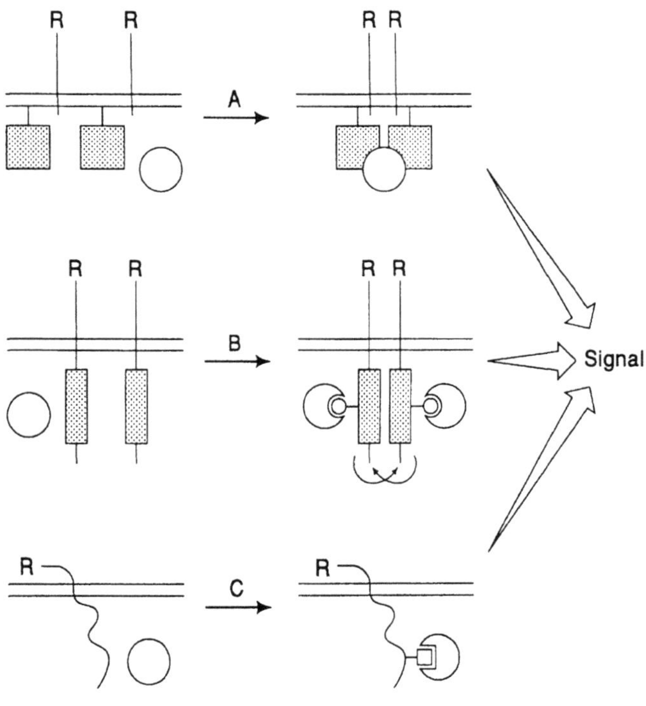

Figure 2. A number of possible signaling mechanisms involve the rearrangement of ligands at the cell surface. These include (**A**) the clustering of receptors (R) with non-receptor effector molecules (shaded boxes) and the recruitment of additional intracellular messenger molecules (open circles), and (**B**) the association of combined receptor/effectors with cross modification and subsequent recruitment of messengers. (**C**) The direct modification of receptors to generate new messenger binding sites is also possible.

hypothesis that CAMs may signal through the classical fibroblast growth factor receptor (FGFR) to activate the pathways utilized by cytokine signaling in other cell types. In the case of neurons or neuronal cell lines, these signals result in process outgrowth.

Another approach began with the observation that bona fide signaling molecules such as non-receptor tyrosine kinases, are found in neuronal growth cones. These are the specialized structures at the tips of growing neurites. It is these structures that are presumably of paramount importance in sensing the environmental cues that determine the path of the neurite.

Another well known mechanism of intracellular signaling involves the attachment of phosphate groups on serine and threonine residues to effect enzymatic activities. This has led some workers to investigate whether cell adhesion molecules are subject to differential phosphorylation themselves and if this phosphorylation can modulate cellular behavior.

A final mechanism that is still not well understood is how molecules that are attached to the cell membrane through lipid anchors can signal. These molecules do not connect directly with the inside of the cell, and hence must meditate their effects in combination with other transmembrane adaptor molecules

Because the signaling via integrins in other cell types is well covered in this volume and because the particulars of signaling by integrins in the nervous system is not yet well understood, the focus of this section will be the signaling mediated by members of the Ig superfamily.

A. Nonreceptor Tyrosine Kinases

Nonreceptor Tyrosine Kinases in the Nervous System

Nonreceptor tyrosine kinases of the src family are widely distributed in the nervous system where they can interact directly with adhesion receptors. One of the first studied of these interactions was the interaction of N-CAM and L1 antigen with the kinases src and fyn.

When it was found that a neuronal form of src was associated with the growth cones of developing neurons (Atashi et al., 1992; Bixby and Jhabvala, 1993), it was clear that levels of phosphorylation in the growth cone might be involved in axon outgrowth and guidance (Klinz et al., 1995). It was soon found that triggering of cell adhesion receptors such as N-CAM and L 1 antigen could affect the level of phosphorylation in growth cone preparations with the level of phosphorylation of the alpha and beta forms of tubulin being the primary measure (Atashi et al., 1992).

The availability of transgenic mice lacking specific nonreceptor tyrosine kinases made it possible to determine which receptors are involved in the signaling. For example, neurons deficient in the src kinase lose their ability to grow neurites over L1 (Ignelzi et al., 1994). Conversely, the loss of fyn leads to poor outgrowth on N-CAM (Beggs et al., 1994).

More direct evidence for the interaction of CAMS with non-receptor tyrosine kinases has been the direct biochemical detection of the L1 with fyn using immunoprecipitation (Beggs et al., 1997). In these experiments, N-CAM association with fyn can be triggered at the cell surface of transfected cells carrying both N-CAM and fyn using antibodies to N-CAM. Immunoprecipitation of the N-CAM as a brij 96 detergent extract also precipitated fyn (but not src). Moreover, it was the 140 kD transmembrane form of N-CAM which was most effective in interacting with fyn. (Note that this is somewhat in contrast to the GPI-signaling results discussed below, where fyn is associated with the GPI-linked CAM contactin/F3/F11, but not with the 140 kD form of N-CAM. The reasons for this discrepancy will need to be determined and the actual experimental systems will need to be compared.)

B. Receptor Tyrosine Kinases and Phosphatases

The signaling of neural adhesion receptors through receptor tyrosine kinases is an area of much recent investigation.

Signaling through the FGF Receptor

As mentioned above, there exists quite a large body of work on the binding of cell adhesion receptors to the FGFR through the work of Doherty, Walsh, and their colleagues (Hall et al., 1996; Doherty and Walsh, 1996). In their experimental paradigm, cells capable of extending neuronal processes are plated in vitro on cell monolayers. After a given time, the cells are fixed and assayed for the extent of process outgrowth.

In initial experiments, the cells extending processes were usually primary cerebellar or dorsal root gangial neurons or the PC12 pheochromacytoma cells line. The cell monolayer consisted of fibroblasts transfected with adhesion molecules and their isoforms. It was early appreciated that the level of adhesion of the neuronal cells to the monolayer was largely independent of the transfected "adhesion molecules," and is most likely due to adhesion though other endogenous molecules such as integrins. However, neurite outgrowth was clearly dependent on the form of the transfected molecule and on such factors as the presence of a cytoplasmic domain (Saffell et al., 1995) and the molecule remaining attached to the monolayer surface. For example, GPI-linked isoforms of N-CAM supported increased neurite outgrowth, but the effect was lost if the cells were treated with phospholipase C to release the N-CAM into the growth medium. Similarly, isoforms of N-CAM containing a variably spiced insert in the fourth immunoglobulin domain (Hall et al., 1996) did not support outgrowth as well as the VASE-less isoform (Walsh et al., 1992). This is similar to the situation observed in vivo, where the VASE isoform is absent during periods of extensive neurite projection, and is then expressed later in development when process outgrowth is reduced (Walsh et al., 1992).

Using the neurite outgrowth assay and pharmacological approaches, it was possible to determine a biochemical signaling pathway leading from cell surface binding to calcium influxes resulting in neurite outgrowth. For example, there seemed to be a receptor tyrosine kinase involved which leads to the activation of phospholipase C gamma to generate diacylglycerol (Hall et al., 1996). The DAG could then be converted to arachadonic acid, which in turn could activate the influx of calcium into the cell via calcium channels (Williams et al., 1994). This calcium influx could in turn mediate Ca^{2+}/calmodulin kinase and entirely account for process outgrowth (Doherty and Walsh, 1996).

A somewhat surprising result of these experiments was that three quite distinct molecules, N-CAM and L1 (which are both members of the immunoglobulin superfamily) and N-cadherin (a member of the cadherin family, which support calcium dependent adhesion) can both activate the same pathways. These studies, and the observation that all of these CAMs show a similarity with a segment of the FGF receptor, led to the proposal that these CAMs may be mediating their effect through activation of this tyrosine kinase. In fact, peptides composing this region of the FGFR can inhibit the neurite-promoting effect (Williams et al., 1994).

In more recent studies, it has been shown that a dominant-negative form of the FGFR can inhibit neurite outgrowth in the neuronal cell directly (Saffell et al., 1997). Similar to the situation where all three CAMs are introduced into the fibroblast monolayer, the dominant-negative FGFR can inhibit process outgrowth on all three molecules. The mutant FGFR, however, does not prevent outgrowth if it is stimulated by a down stream effector such as calcium influx stimulated by membrane depolarization. In all cases, the appropriate controls seem in place, suggesting that CAMs can indeed signal by this mechanism. It is interesting to see that the in vivo studies continue to support this pathway. For example, adult songbird neurons show Ng-CAM mediated calcium fluxes (Goldman et al., 1996) and the developing embryo utilizes these same pathways (Brittis et al., 1996).

Eph Kinases and their Ligands

As for signaling via the FGFR, investigators are now beginning to sort out the extracellular pathways activated by signaling via the eph receptor tyrosine kinases. Eck kinase, for example, can bind phosphatidylinositol-3-kinase (Pandey et al., 1994). It is also possible for eph ligands to phosphorylate other cell adhesion molecules. For example, the cek5 kinase can phosphorylate the L1 antigen (Zisch et al., 1997). Also, because a number of the transmembrane ligands for the eph kinases are themselves substrates for tyrosine phosphorylation, this opens the possibility that engagement of eph ligands can give rise to changes in both cellular partners and hence two-way signaling as a result of adhesion interactions. Moreover, it is also possible to activate cytoskeletal-mediated signaling by this kinase family. For example, members of the Elk subfamily of eph kinase can bind to pl20gap and to the nonreceptor tyrosine kinases fyn and yes (Zisch et al., 1997) leading to the rho and nck pathways for reorganization of the cytoskeleton. A major area for future investigation is sorting out the downstream aspects of signaling by this group of molecules, and as discussed below, understanding how the neuronal cell integrates all of the signals impinging upon it to respond appropriately.

C. Serine/Threonine Phosphorylation of CAMs

One of the major pathways for intracellular signaling is the phosphorylation of proteins on the amino acids threonine, serine, and/or tyrosine. Because of this, an immediate question is whether neural adhesion molecules are indeed subject to phosphorylation, and if so, on which amino acids and by which kinases are they phosphorylated.

It was reported early after N-CAM was discovered that it could be phosphorylated on serine and threonine, but there was little tyrosine phosphorylation in the chick. These observations led to the purification of an associated protein kinase. However, it is still not clear if there is a functional significance in vivo to this phosphorylation. The phosphorylation of L1 antigen has also been described, and an associated protein kinase was co-isolated with L1 (Sadoul et al., 1989; Heiland et al., 1996). Other recent studies have shown that both the p90rsk kinase (Wong et al., 1996b) and casein kinase II (Wong et al., 1996a) can phosphorylate L1 on serine and threonine. It also appears that L1 may be phosphorylated on tyrosine in vivo and in vitro (Zisch et al., 1997).

Another interesting facet of L1 phosphorylation is that the L1 can consist of isoforms either containing or lacking a tetrapeptide in the cytoplasmic region of the molecule proximal to the membrane spanning region. This tetrapeptide, RSLE, is alternatively spliced, and the RSLE-containing form seems more prevalent in muscle and cell-lines derived from muscle (Reid and Hemperly, 1992). Moreover, the presence or absence of this sequence changes the way that neurite outgrowth is supported extracellularly (Takeda et al., 1996).

D. Integrin-Mediated Mechanisms

Integrin-mediated signaling is an extremely well studied area. An example that relates particularly to signaling by neuronal integrins is that the $\alpha_v\beta_3$ integrin is itself an extracellular ligand for the L1 antigen (Montgomery et al., 1996; Kadmon and Altevogt, 1997). Here is an example where a signaling molecule on one cell could directly interact with a different type of signaling molecule on another cell.

E. Signaling via Lipid Anchors

In addition to signaling via proteins containing transmembrane polypeptide segments, there can also be signaling via receptors which are linked to the cell surface by phosphatidylinositol glycans. Probably the prototype molecule of this type is Thy-1 antigen, which was originally described as a marker of T-cells of the immune system, but is also a major brain antigen. Cross-linking of Thy-l can lead to cell activation. It is now known that a number of neuronal cell adhesion molecules are GPI-linked (see Table 1), and the study of their signaling is relatively new (for a recent review, see Vaughan, 1996). It has recently been reported that the GPI-linked CAM contactin/F3/F11 can bind with the non-receptor tyrosine kinase fyn and with L1, possibly through a detergent-insoluble lipid complex (Olive et al., 1995; Zisch et al., 1995).

V. INTEGRATION OF CELLULAR SIGNALS

One of the concepts that emerges as a result of these studies on signaling by adhesive receptors of the nervous system and elsewhere is that the cell is in fact sensing and responding to multiple cues, both soluble and substrate- or cell-bound. It is of course the challenge now, not only to understand the detailed biochemical and cell biological mechanisms at work with each of the molecules discussed above, but also to determine how the cell can integrate all of the signals to respond appropriately to its environment. Clearly this integration is of importance during development. Particularly exciting is the recent demonstration that CAMs may be directly involved in learning and memory through synaptic stabilization (Fields and Itoh, 1996; Martin and Kandel, 1996). In addition, this integration of signals arising from cell adhesive events is likely to be very relevant to disease and to finding new mechanisms for diagnosing and treating human suffering.

REFERENCES

Ackley, R. L., Madison, R. D., Archibald, S. J., & Hemperly, J. J. (1997). Monoclonal antibody interaction with the third immunoglobulin-like domain of N-CAM is sufficient to cause cell migration. Mol. Cell. Neurosci. 10, 117-129.

Atashi, J. R., Klinz, S. G., Ingraham, C. A., Matten, W. T., Schachner, M., & Maness, P. F. (1992). Neural cell adhesion molecules modulate tyrosine phosphorylation of tubulin in nerve growth cone membranes. Neuron 8, 831-842.

Beggs, H. E., Baragona, S. C., Hemperly, J. J., & Maness, P. F. (1997). NCAM140 interacts with the focal adhesion kinase pl25(fak) and the SRC-related tyrosine kinase p59(Fyn). J. Biol. Chem. 272, 8310-8319.

Beggs, H. E., Soriano, P., & Maness, P. F. (1994). NCAM-dependent neurite outgrowth is inhibited in neurons from Fyn-minus mice. J. Cell Biol. 127, 825-833.

Bixby, J. L. & Jhabvala, P. (1993). Tyrosine phosphorylation in early embryonic growth cones. J. Neurosci. 18, 3421-3432.

Bock, E., Edvardsen, K., Givson, A., Linnemann, D., Lyles, J. M., & Nybroe, O. (1987) Characterization of soluble forms of NCAM. FEBS Lett. 225, 33-36.

Brittis, P. A., Silver, J., Walsh, F. S., & Doherty, P. (1996). Fibroblast growth factor receptor function is required for the orderly projection of ganglion cell axons in the developing mammalian retina. Mol. Cell. Neurosci. 8, 120-128.

Brummendorf, T. & Rathjen, F. G. (1995). Cell adhesion molecules .1. Immunoglobulin superfamily. Protein Profile 2, 963-1058.

Burdengulley, S. M., Payne, H. R., & Lemmon, V. (1995). Growth cones are actively influenced by substrate-bound adhesion molecules. J. Neurosci. 15, 4370-4381.

Cole, G. J. & Akeson, R. (1989) Identification of a heparin binding domain of the neural cell adhesion molecule N-CAM using synthetic peptides. Neuron 2, 1157-1165.

Defreitas, M. F., Yoshida, C. K., Frazier, W. A., Mendrick, D. L., Kypta, R. M., & Reichardt, L. F. (1995). Identification of integrin alpha(3)beta 1 as a neuronal thrombospondin receptor mediating neurite outgrowth. Neuron 15, 333-343

Desai, C. J., Sun, Q., & Zinn, K. (1997). Tyrosine phosphorylation and axon guidance: Of mice and flies. Curr. Opin. Neurobiol. 7, 70-74.

Doherty, P. & Walsh, F. S. (1996). CAM-FGF receptor interactions: A model for axonal growth. Mol. Cell. Neurosci. 8, 99- 111.

Edelman, G. M. & Crossin, K. L. (1991). Cell adhesion molecules: Implications for a molecular histology. Ann. Rev. Biochem. 60, 155-190.

Fields, R. D. & Itoh, K. (1996). Neural cell adhesion molecules in activity-dependent development and synaptic plasticity. Trends Neurosci. 19, 473-480.

Goldman, S. A., Williams, S., Barami, K., Lemmon, V., & Nedergaard, M. (1996). Transient coupling of Ng-CAM expression to NgCAM-dependent calcium signaling during migration of new neurons in the adult songbird brain. Mol. Cell. Neurosci. 7, 29-45.

Green, P. J., Walsh, F. S., & Doherty, P. (1996). Promiscuity of fibroblast growth factor receptors. Bioessays 18, 639-646.

Grumet, M., Friedlander, D. R., & Sakurai, T. (1996). Functions of brain chondroitin sulfate proteoglycans during development: Interactions with adhesion molecules. Perspectives on Dev. Neurobiol. 3, 319-330.

Gumbiner, B. M. (1995). Signal transduction of beta-catenin. Curr. Opin. Cell Biol. 7, 634-40.

Hall, H., Walsh, F. S., & Doherty, P. (1996). A role for the FGF receptor in the axonal growth response stimulated by cell adhesion molecules? Cell Adhesion and Communication 3, 441.

Hall, H., Williams, E. J., Moore, S. E., Walsh, F. S., Prochiantz, A., & Doherty, P. (1996). Inhibition of FGF-stimulated phosphatidylinositol hydrolysis and neurite outgrowth by a cell-membrane permeable phosphopeptide. Current Biology 6, 580-587.

Harris, W. A. & Holt, C. E. (1995). From Tags to RAGS: chemoaffinity finally has receptors and ligands. Neuron 15, 241-244.

Heiland, P. C., Hertlein, B., Traub, O., Griffith, L. S., & Schmitz, B. (1996). The neural adhesion molecule L1 is phosphorylated on tyrosine and serine residues. Neuroreport 7, 26752678.

Hemperly, J. J. (1996). Neural cell adhesion molecules of the immunoglobulin superfamily. In: Advances in Molecular and Cell Biology (Colman, D., Ea.) Vol. 16, pp. 137-157.

Hemperly, J. J., Murray, B. A., Edelman, G. M., & Cunningham, B. A. (1986) Sequence of a cDNA clone encoding the polysialic acid-rich and cytoplasmic domains of the neural cell adhesion molecule N-CAM. Proc. Natl. Acad. Sci. U. S. A. 83, 3037-3041.

Horwitz, A. F. & Hunter, T. (1996). Cell adhesion: integrating circuitry. Trends in Cell Biology 6, 460-461.

Ignelzi, M. A., Miller, D. R., Soriano, P., & Maness, P. F. (1994). Impaired neurite outgrowth of src-minus cerebellar neurons on the cell adhesion molecule L1. Neuron 12, 873-884.

Kadmon, G. & Altevogt, P. (1997). The cell adhesion molecule L1: Species- and cell-type-dependent multiple binding mechanisms. Differentiation 61, 143-150.

Klinz, S. G., Schachner, M., & Maness, P. F. (1995). L1 and N-CAM antibodies trigger protein phosphatase activity in growth cone-enriched membranes. J. Neurochem. 65, 84-95.

Kunz, S., Ziegler, U., Kunz, B., & Sonderegger, P. (1996). Intracellular signaling is changed after clustering of the neural cell adhesion molecules axonin-1 and NgCAM during neurite fasciculation. J. Cell Biol. 135, 253-267.

Lagenaur, C. & Lemmon, V (1987). An L1-like molecule, the 8D9 antigen, is a potent substrate for neurite extension. Proc. Natl. Acad. Sci. USA 84, 7753-7757.

Leahy, D. J., Aukhil, I., & Erickson, H. P. (1996). 2.0 angstrom crystal structure of a four-domain segment of human fibronectin encompassing the RGD loop and synergy region. Cell 84, 155-164.

Margolis, R. K., Rauch, U., Maurel, P., & Margolis, R. U. (1996). Neurocan and phosphacan: Two major nervous tissue-specific chondroitin sulfate proteoglycans. Perspec. Dev. Neurobio. 3, 273-290.

Martin, K. C. & Kandel, E. R. (1996). Cell adhesion molecules, CREB, and the formation of new synaptic connections. Neuron 17, 567-570.

Montgomery, A. M. P., Becker, J. C., Siu, C. H., Lemmon, V. P., Cheresh, D. A., Pancook, J. D., Zhao, X. N., & Reisfeld, R. A. (1996). Human neural cell adhesion molecule L1 and rat homologue NILE are ligands for integrin alpha(v) beta(3). J. Cell Biol. 132, 475-485.

Nieto, M. A. (1996). Molecular biology of axon guidance. Neuron 17, 1039-1048

Olive, S., Dubois, C., Schachner, M., & Rougon, G. (1995). The F3 neuronal glycosylphosphatidylinositol-linked molecule is localized to glycolipid-enriched membrane subdomains and interacts with L1 and fyn kinase in cerebellum. J. Neurochem. 65, 2307-2317.

Pandey, A., Lazar, D. F., Saltiel, A. R., & Dixit, V. M. (1994). Activation of the Eck receptor tyrosine kinase stimulates phosphatidyl-inositol 3-kinase activity. J. Biol. Chem. 269, 30154-30157.

Pollerberg, G. E., Burridge, K., Krebs, K. E., Goodman, S. R., & Schachner, M. (1987) The 180kD component of the neural cell adhesion molecule N-CAM is involved in cell-cell contacts and cytoskeleton-membrane interactions. Cell Tissue Res. 250, 227-236.

Poltorak, M., Khoja, I., Hemperly, J. J., Williams, J. R., Elmallakh, R., & Freed, W. J. (1995). Disturbances in cell recognition molecules (N-CAM and L1 antigen) in the CSF of patients with schizophrenia. Exp. Neurol. 131, 266-272.

Probstmeier, R., Fahrig T, Spiess E, and Schachner M. (1992). Interactions of the neural cell adhesion molecule and the myelin-associated glycoprotein with collagen type I: Involvement in fibrillogenesis. J. Cell Biol. 116, 1063-1070

Ranheim, T. S., Edelman, G. M., & Cunningham, B. A. (1996). Homophilic adhesion mediated by the neural cell adhesion molecule involves multiple immunoglobulin domains. Proc. Natl. Acad. Sci. U S A 93, 4071-4075.

Rao, Y., Zhao, X. N., & Siu, C. H. (1994). Mechanism of hemophilic binding mediated by the neural cell adhesion molecule NCAM - Evidence for isologous interaction. J. Biol. Chem. 269, 27540-27548.

Redies, C. & Takeichi, M. (1996). Cadherins in the developing central nervous system: An adhesive code for segmental and functional subdivisions. Dev. Biol. 180, 413-423.

Reid, R. A. & Hemperly, J. J. (1992) Variants of human L1-cell adhesion molecule arise through alternate splicing of RNA. J. Mol. Neurosci. 3, 127- 135.

Ruoslahti, E. & Obrink, B. (1996). Common principles in cell adhesion. Exp. Cell Res. 227, 111

Sadoul, R., Kirchhoff, F., & Schachner, M. (1989) A protein kinase activity is associated with and specifically phosphorylates the neural cell adhesion molecule L1. J. Neurochem. 53, 1471-1478.

Saffell, J. L., Doherty, P., Tiveron, M. C., Morris, R. J., & Walsh, F. S. (1995). NCAM requires a cytoplasmic domain to function as a neurite outgrowth-promoting neuronal receptor. Mol. Cell. Neurosci. 6, 521-531.

Saffell, J. L., Williams, E. J., Mason, I. J., Walsh, F. S., & Doherty, P. (1997). Expression of a dominant negative FGF receptor inhibits axonal growth and FGF receptor phosphorylation stimulated by CAMs. Neuron 18, 231 -242.

Sakurai, T., Lustig, M., Nativ, M., Hemperly, J. J., Schlessinger, J., Peles, E., & Grumet, M. (1997). Induction of neurite outgrowth through contactin and Nr-CAM by extracellular regions of glial receptor tyrosine phosphatase beta. J. Cell Biol. 136, 907-918.

Schuch, U., Lohse, M. J., & Schachner, M. (1989) Neural cell adhesion molecules influence second messenger systems. Neuron 3, 1-20.

Shinoura, N., Paradies, N. E., Warnick, R. E., Chen, H., Larson, J. J., Tew, J. J., Simon, M., Lynch, R. A., Kanai, Y., Hirohashi, S., Hemperly, J. J., Menon, A. G., & Brackenbury, R. (1995). Expression of N-cadherin and alpha-catenin in astrocytomas and glioblastomas. Br J. Cancer 72, 627-633.

Takeda, J. & Kinoshita T (1995). GPI-anchor biosynthesis. Trends Biochem. Sci. 20, 367-71.

Takeda, Y., Asou, H., Murakami, Y., Miura, M., Kobayashi, M., & Uyemura, K. (1996). A nonneuronal isoform of cell adhesion molecule L1: Tissue-specific expression and functional analysis. J. Neurochem. 66, 2338-2349.

Tessier-Lavigne, M. (1995). Eph receptor tyrosine kinases, axon repulsion, and the development of topographic maps. Cell 82, 345-348.

Tessier-Lavigne, M., & Goodman, C. S. (1996). The molecular biology of axon guidance. Science 274, 1123-1133.

Thomsen, N. K., Soroka, V., Jensen, P. H., Berezin, V., Kiselyov, V. V., Bock, E., & Poulsen, F. M. (1996). The three-dimensional structure of the first domain of neural cell adhesion molecule. Nature Struc. Biol. 3, 581-585.

Varela-Echavarria, A., & Guthrie, S. (1997). Molecules making waves in axon guidance. Genes & Development 11, 545-557.

Vaughan, L. (1996). Signal transduction by GPI-anchored receptors in the nervous system. Seminars in the Neurosciences 8, 397-403.

Von Bohlen und Halbach, F., Taylor, J., & Schachner, M. (1992). Cell type-specific effects of the neural adhesion molecules L1 and N-CAM on diverse second messenger systems. Eur. J. Neurosci. 4, 896-909.

Walsh, F. S, Furness J, Moore SE, Ashton S, and Doherty P (1992). Use of the neural cell adhesion molecule VASE exon by neurons is associated with a specific down-regulation of neural cell adhesion molecule-dependent neurite outgrowth in the developing cerebellum and hippocampus. J. Neurochem. 59, 1959-62.

Williams, E. J., Furness, J., Walsh, F. S., & Doherty, P. (1994). Activation of the FGF receptor underlies neurite outgrowth stimulated by L I, N-CAM, and N-cadherin. Neuron 13, 583-594.

Wong, E. V., Schaefer, A. W., Landreth, G., & Lemmon, V. (1996a). Casein kinase II phosphorylates the neural cell adhesion molecule ,1. J. Neurochem. 66, 779-786.

Wong, E. V., Schaefer, A. W., Landreth, G., & Lemmon, V. (1996b). Involvement of p90(rsk) in neurite outgrowth mediated by the cell adhesion molecule L1. J. Biol. Chem. 271, 18217-18223.

Zisch, A. H., D'Alessandri, L., Amrein, K., Ranscht, B., Winterhalter, K. H., & Vaughan, L. (1995). The glypiated neuronal cell adhesion molecule contactin/F11 complexes with src-family protein tyrosine kinase Fyn. Mol. Cell. Neurosci. 6, 263-279.

Zisch, A. H., Stallcup, W. B., Chong, L. D., DahlinHuppe, K., Voshol, J., Schachner, M., & Pasquale, E. B. (1997). Tyrosine phosphorylation of L1 family adhesion molecules: Implication of the Eph kinase Cek5. J. Neurosci. Res. 47, 655-665.

PART IV

ADHESIVE PROCESSES

VASCULAR ENDOTHELIAL CELL ADHESION MOLECULES AND THE CONTROL OF LEUKOCYTE TRAFFIC IN CUTANEOUS INFLAMMATION

Dorian O. Haskard, Justin C. Mason, and Julie McHale

Advances in Molecular and Cell Biology
Volume 28, pages 323-344.

ISBN: 0-7623-0495-2

I. INTRODUCTION

Inflammation is the physiological response of living tissues to injury, and serves to localize and eliminate the mediating agent, to limit tissue injury, and then to restore the tissue towards normality. The four cardinal signs of inflammation were first described by Celsus in the first century AD,. viz.: rubor, tumor, calor, and dolor (redness, swelling, heat and pain). The fifth clinical sign, loss of function, was later added by Virchow. Pathophysiologically the clinical signs can be explained by (i) changes in vascular caliber and increased blood flow, leading to erythema (rubor) and warmth (calor); (ii) increased vascular permeability to plasma proteins, leading to edema (tumor); and (iii) escape of leukocytes from blood vessels into extravascular tissue spaces, leading to the release of free radicals and proteases. While neutrophils tend to dominate the early phase of inflammatory responses, monocytes and lymphocytes become the major cell types as the lesion becomes chronic, particularly in T lymphocyte-mediated inflammatory reactions such as delayed hypersensitivity.

A. Leukocytes

Over the last few years, there have been major advances in our understanding of the molecular mechanisms responsible for the traffic of different leukocyte populations into and within the tissues during inflammation. Interactions between adhesion molecules on the surface of leukocytes and counter-structures expressed by resident tissue cells or by extracellular matrix are fundamental to the control of leukocyte movement and function, and consequently have become a large field of research. There is now a substantial literature on the expression and functional importance of different adhesion molecules, both in animal models of inflammation and in human inflammatory diseases (Mason and Haskard, 1994). In this chapter we will focus on interactions between leukocytes and vascular endothelial cells (EC), with particular emphasis on changes in adhesion molecule expression by EC that stimulate neutrophil and lymphocyte recruitment into simple models of inflammation in skin.

B. The Adhesion Cascade

Early in vivo studies of leukocyte interactions with EC using intra-vital microscopy had shown that the immobilization of the circulating leukocyte on the endothelial surface of blood vessels is preceded by their initial tethering and rolling on the EC surface (Atherton and Born, 1972). This process is now known to be due to the sequential involvement of distinct families of adhesion molecules and is referred to as the "adhesion cascade" (Butcher, 1991; Springer, 1994). The series of adhesion events is initiated by reversible binding between selectins and their ligands, resulting in the rolling of leukocytes on the surface of endothelium. As a consequence of this, leukocytes are exposed to local activation signals (e.g., chemokines) capable of upregulating the affinity and/or avidity of surface β_2 (e.g. LFA-1, Mac-1) or α_4 (e.g. VLA-4) integrins, and thereby stimulating the adhesion of integrins to their EC ligands (e.g., ICAM-1, ICAM-2, VCAM-1). This results in stable attachment and immobilization of leukocytes on the endothelial surface, and is followed by their transmigration through endothelium into the perivascular tissue. The molecular mechanisms responsible for transmigration are not yet fully established but include the involvement of β_2 integrins and PECAM-1 (CD31).

Although the concept of the adhesion cascade acts as an excellent frame-work with which to approach leukocyte-EC interactions, it is important to realize that the precise adhesion mechanisms governing leukocyte recruitment are likely to vary between different vascular beds and in response to different inflammatory stimuli (Bullard et al., 1995). Furthermore, in pathological situations where blood flow and shear rates are critically reduced, leukocytes may not need or be able to use selectins, and adhesion to endothelium may instead be initiated by integrins (Gaboury and Kubes, 1994; Finger et al., 1996).

A key feature of the "adhesion cascade" is that it allows for diversity between different forms of inflammation in the types of leukocyte that are recruited into the tissues. Thus, whether or not a particular leukocyte emigrates into the tissues during an inflammatory reaction depends upon it fulfilling each of the steps in the cascade by making use of the adhesion ligands and chemoattractants that are locally available. This process governing selectivity of leukocyte traffic has been termed the "area code" model (Springer, 1994). The implication of the model is that mechanisms that differentially regulate expression of adhesion molecules with a selective capacity to bind particular leukocyte populations may have a major impact on the types of leukocyte entering an inflammatory lesion. In this regard we have been particularly interested in understanding how expression of VCAM-1 is differentially regulated, since this molecule is likely to promote the selective migration of monocytes, eosinophils and lymphocytes through acting as a ligand for their α_4 integrins.

II. ENDOTHELIAL ACTIVATION

Experiments with cultured cells have greatly advanced our understanding of how adhesion molecule expression by EC is controlled. While resting EC are poorly adhesive for leukocytes, leukocyte adhesion is enhanced by EC activation. The increased adhesion of leukocytes to activated EC is due not only to an increase in surface expression of adhesion ligands but also to the coordinated release or surface expression by EC of factors capable of stimulating the function of leukocyte integrins (e.g. platelet activating factor (PAF), chemokines). Defining different forms of EC activation in vitro has been very helpful in interpreting the phases of leukocyte traffic that can be distinguished in different inflammatory responses in vivo.

A. Rapid Activation

Probably the earliest adhesive change on the luminal surface of EC in inflammation is due to the translocation of intracellular secretary granules, known as Weibel-Palade bodies, to the plasma membrane, resulting in the surface expression of presynthesised P-selectin. This response occurs within seconds to minutes of stimulation, and can be induced by a number of agonists including histamine, thrombin, phorbol esters, C5a, and, C5b-C9 membrane attack complexes (Hattori et al., 1989a, 1989b; Geng et al., 1990; Foreman et al., 1994). In addition to upregulating P-selectin expression, many of the same agonists stimulate synthesis of PAF, thereby providing an EC-derived stimulus for the integrin-mediated immobilisation of leukocytes rolling on P-selectin (Zimmerman et al., 1990). In vitro this form of EC activation tends to be transient, with expression of P-selectin falling back to base-line by 30 minutes after stimulation.

B. Subacute Activation

Stimulation of cultured EC with interleukin-1 (IL-1) α or β or with tumor necrosis factor (TNF) α or β results in a dramatic increase in adhesiveness of cultured EC for most leukocyte populations (Bevilacqua et al.,1985.; Gamble et al., 1985; Cavender et al., 1986). This is attributable to the orchestrated transcriptional activation of a number of primary and secondary response genes, mediated in part through the nuclear translocation of the transcription factor NFκB (Pober and Cotran, 1990; Collins et al., 1995). Many of these genes are also directly activated by endotoxin, although in vivo the effects of endotoxin on EC are amplified through the generation of IL-1 and/or TNF. Among molecules that are regulated on EC by IL-1 and TNF are the adhesion molecules E-selectin, ICAM-1, MAdCAM-1, and VCAM-1 and a number of chemokines including IL-8, MCP-1, and fractalkine (Pober and Cotran, 1990; Bazan et al., 1997). Although there are differences between each of these molecules in the precise kinetics of expression following stimulation by IL-1 or TNF, expression is usually maximal between 2 and 24 hours.

Expression of E-selectin by cultured EC in response to IL-1 or TNF is first detectable by about one hour and is maximal by 4–6 hours, after which expression tends to decline (Pober et al., 1986; Wellicome et al., 1990; Tsang et al., 1995). Compared to E-selectin, VCAM-l expression by cultured EC shows a number of differences which may be relevant to the traffic of mononuclear cells in chronic inflammation: (i) the peak of VCAM-1 expression occurs at 8-10 hours, which is later than that of E-selectin (Wellicome et al., 1990; Tsang et al., 1994) (Figure 1). This relative delay in VCAM-1 expression probably contributes to the observation that TNF leads to a more prolonged increase in EC adhesiveness for lymphocytes than for neutrophils in simple static adhesion assays (Thornhill et al., 1990; Tsang et al., 1994); (ii) while TNF and IL-1 are similar in their capacity to stimulate E-selectin

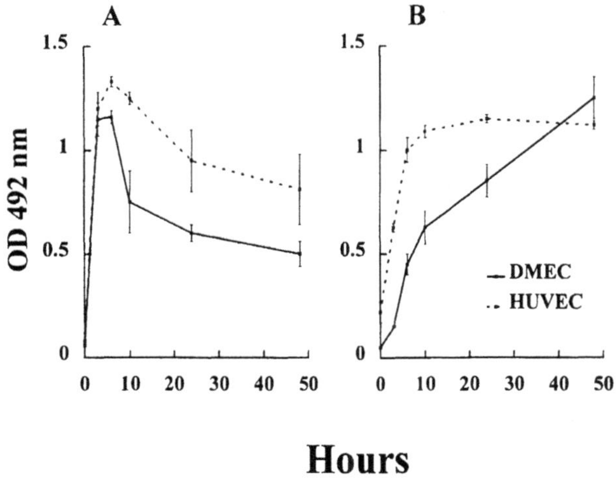

Figure 1. Kinetics of expression of E-selectin and VCAM-1 on EC stimulated with TNF. Dermal microvascular EC and HUVEC were plated at confluence in 96 well plates and stimulated with human recombinant TNFα (10 ng/ml) for the times shown. The expression of E-selectin (**A**) and VCAM-1 (**B**) were then measured by ELISA. The results show mean ± SD of triplicate wells. The figure shows the delayed expression of VCAM-1 in relation to E-selectin, and the more persistent expression of E-selectin on DMEC compared to HUVEC.

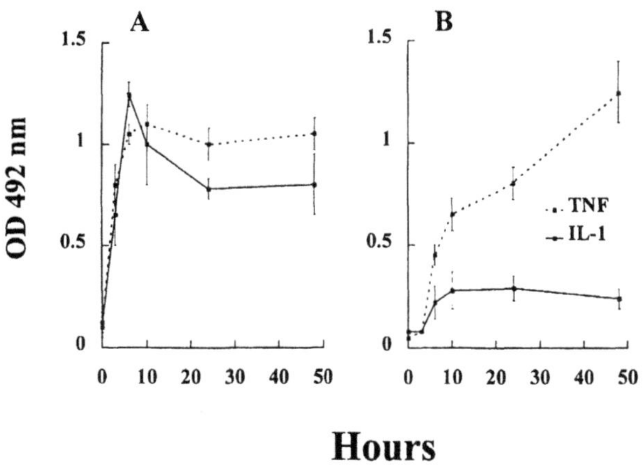

Figure 2. Expression of E-selectin (**A**) and VCAM-1 (**B**) by dermal microvascular EC (DMEC) in response to IL-1 and TNF. DMEC monolayers were cultured with human recombinant IL-1β (10 ng/ml) or human recombinant TNFα (10 ng/ml) for varying times and adhesion molecule expression measured by ELISA. The figure shows that IL-1 and TNF have similar effects on E-selectin expression but that TNF is more effective than IL-1 at inducing VCAM-1 expression.

expression, TNF is a much more potent stimulus for VCAM-1 (Figure 2) expression than IL-1. This is particularly true for dermal EC, which have been found to contain a transcriptional repressor of IL-1α-mediated VCAM-1 expression (Swerlick et al., 1992; Gille et al., 1996); and (iii) expression of VCAM-1 in response to TNFα can be selectively enhanced by IL-4 (Figure 3) (Thornhill and Haskard, 1990), which is now known to stabilize VCAM-1 mRNA (Iademarco et al., 1995). Thus the release of IL-4, or of similarly acting lymphokines, in immune-mediated inflammation might augment VCAM-1 expression, and consequently amplify mononuclear cell and/or eosinophil recruitment.

C. Chronic Activation

In vitro experiments provide only limited information on the regulation of EC adhesion molecule expression in established chronic inflammation, since in practice it is difficult to maintain cultures of stimulated EC for longer than 3–4 days. Endothelial cells may become adapted to chronic stimulation both by down-regulating the synthesis of adhesion molecules which are maximally expressed during the sub-acute phase, and by expressing further adhesion ligands not seen earlier in the response (e.g. ligands for L-selectin; Michie et al., 1993). Active down-regulation of EC activation during on-going inflammatory responses is likely to be important for preventing excessive leukocyte influx and thereby in protecting tissues from leukocyte-mediated injury.

In the case of E-selectin there is evidence that expression can be inhibited by the actions of IL-4 or transforming growth factor-β (TGFβ) (Thornhill and Haskard,

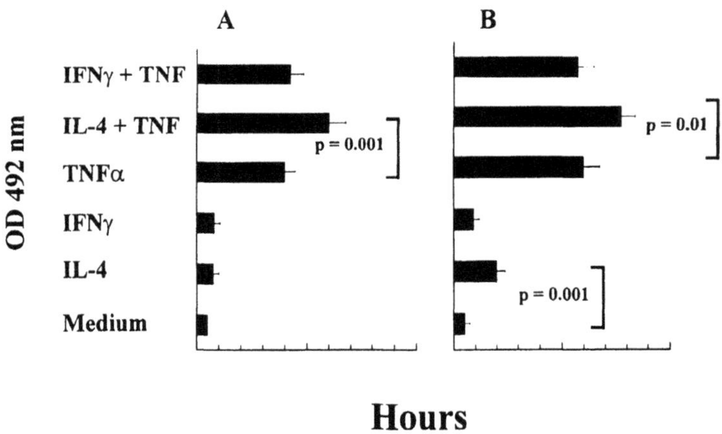

Figure 3. IL-4 synergizes with TNF to upregulate VCAM-1 expression. DMEC (**A**) and HUVEC (**B**) monolayers were cultured with human recombinant TNFα (10 ng/ml), human recombinant IL-4 (20 ng/ml) or human recombinant IFNγ (500 u/ml) alone or in combination for 24 hours, after which VCAM-1 was measured by ELISA.

1990; Gamble et al., 1993). A further mechanism of particular importance is the capacity of EC to autoregulate their synthesis of adhesion molecules in response to IL-1 or TNF through the active inhibition of transcriptional activity. One of the principal mechanisms responsible for the down-regulation of the response to IL-1 and TNF is the induction by the same cytokines of the inhibitory factor IκBα, which binds NFκB in the cytoplasm and prevents its translocation to the nucleus (De Martin et al., 1993; Read et al., 1994). The degree to which this mechanism is operative varies between adhesion molecules and between types of EC, and is particularly marked in the case of E-selectin expression by HUVEC.

III. LEUKOCYTE RECRUITMENT INTO MODELS OF INFLAMMATION IN THE SKIN

The skin provides excellent opportunities for studying EC adhesion molecules experimentally, since it is fully accessible for stimulation, and the large surface area permits the study of separate internally-controlled inflammatory sites at the same time. For example, in the pig it is possible to distribute up to 60 small inflammatory sites over the abdomen of a single animal. Using the cutaneous models described below it has been possible to provide a link between the cell biology of cultured EC and the dynamics and functional importance of adhesion molecule expression by EC in inflammation in vivo.

A. Technical Considerations

Quantifying Leukocyte Traffic

There are a number of ways by which leukocyte traffic into inflamed tissues can be measured. Histological techniques quantify the steady state of leukocytes in the tissue, but fail to account for the rates of leukocyte entry into the tissues, leukocyte cell death, and the emigration of leukocytes from the tissue into afferent lymphatics. Techniques based on harvesting leukocytes (for example, skin windows, skin blisters, sub-cutaneous sponges) selectively measure leukocytes that are migrating rather than adherent. More exact information on the dynamics of leukocyte entry into the skin at different times during inflammatory responses can be gained by counting the local uptake of intravenously administered radiolabeled leukocytes. This approach was developed by studies investigating neutrophil traffic in rabbits (Issekutz and Movat, 1980), monocyte traffic in rabbits (Issekutz et al., 1981), and lymphocyte traffic in pigs (Binns et al., 1990). Our group has adapted these techniques in order to simultaneously analyze the uptake into skin of differentially radiolabelled leukocyte populations, bearing in mind that differences in the whole-body distribution of leukocyte subsets influence the absolute number of injected cells that migrate into skin lesions. For example, neutrophils remain largely intravascular, while a large proportion of lymphocytes home rapidly to lymphoreticular tissues which support constitutive lymphocyte trafficking (Pabst and Binns, 1989).

Quantifying Expression of EC Adhesion Molecule Expression

Although immunocytochemistry can readily detect expression of adhesion molecules by EC, quantification of immunostaining is notoriously problematic. Moreover, immunocytochemistry does not readily distinguish the proportion of antigen that is surface-expressed rather than cytoplasmic and therefore available to interact with leukocytes. For these reasons we have developed a novel method for quantifying expression of adhesion molecules on the lumenal surface of EC in vivo, utilizing radiolabelled monoclonal antibodies (mAb) injected in tracer amounts that are insufficient to significantly interfere with leukocyte traffic (Keelan et al., 1994). By counting the uptake into individual skin lesions of mAb directed against an EC adhesion molecule (e.g. E-selectin) in relation to the uptake of a differentially radiolabelled isotype-matched control antibody, it has been possible to gain detailed information on the kinetics of adhesion molecule expression in vivo with a sensitivity not previously obtained (Keelan et al., 1994). Furthermore, by simultaneously administering the radiolabelled anti-adhesion molecule mAb with differentially radiolabelled neutrophils and/or lymphocytes, we have been able to assess to what degree expression of the individual adhesion molecule relates to the uptake of specific leukocyte populations at different time-points during inflammatory responses (Binns et al., 1996a, 1996b; Chapman et al., 1996).

B. Inflammatory Lesions Induced by IL-1 or TNF

Early studies using radiolabelled leukocytes to study leukocyte traffic into cutaneous inflammation tended to concentrate on the effects of injecting chemoattractants such as PAF, leukotriene B4, α-casein, C5a, or IL-8 (Colditz and Movat, 1984; Colditz et al., 1989). These agents typically stimulate an early and transient influx of neutrophils that occurs within two hours of injection. The exact changes in endothelial cell adhesion molecule expression which occur during chemoattractant-stimulated inflammation are not yet clear.

Injection of IL-I or TNF into the skin results in the recruitment of a mixed population of leukocytes, with an onset of action that is delayed by 30–60 minutes compared to the effects of injecting chemoattractants (Cybulsky et al., 1988; Wankowicz et al., 1988; Cybulsky et al., 1989; Issekutz and Stolz, 1989; Issekutz, 1990, Binns et al., 1992, Issekutz and Issekutz, 1993). Our studies in the pig have shown that, although IL-1 and TNF stimulate the uptake of both neutrophils and lymphocytes, lymphocyte uptake follows that of neutrophils. Furthermore, although IL-1 and TNF stimulate an equivalent amount of neutrophil uptake into skin, TNF is a much more potent stimulus for promoting lymphocyte traffic (Binns et al., 1996a). Thus, the uptake of neutrophils into skin stimulated with either IL-1

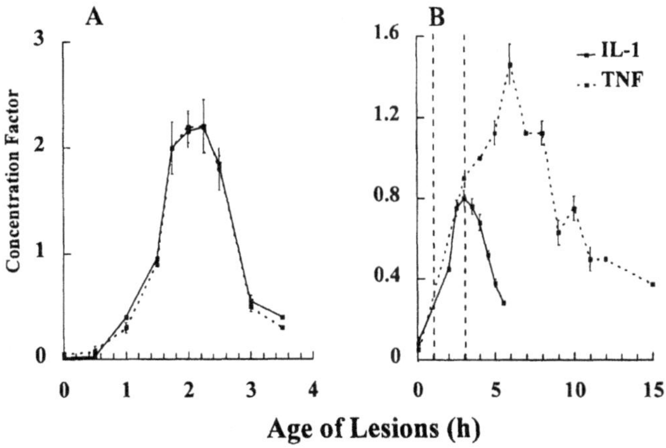

Figure 4. Kinetics of neutrophil (**A**) and lymphocyte (**B**) uptake in skin lesions stimulated by injection of IL-lα (3000 U/site for neutrophils, 2,250 U/site for lymphocytes) or TNF a (2,250 U/site) for neutrophils (**A**) and lymphocytes (**B**). The Concentration factor was calculated as the sample-specific cpm/gram divided by injected cpm/body weight in grams. In (**B**) the duration between the parallel dotted lines represents the period of optimal neutrophil uptake shown in (**A**). These graphs are compiled from data shown as Figures 4 and 5 in Binns et al., (1996a). (Reproduced by kind permission of the Editor of the American Journal of Physiology).

or TNF starts at around 1 hour, peaks at around 2 hours, and then falls sharply and is nearly over by 3 hours. IL-1 stimulates a modest influx of lymphocytes which peaks at 3 hours, while TNF stimulated lymphocyte recruitment peaks later, at 5–6 hours, and is more substantial in amount (Figure 4).

Whereas the uptake of leukocytes due to injection of chemoattractants is unaffected by the protein synthesis inhibitor cycloheximide, IL-1 or TNF-mediated leukocyte recruitment is protein synthesis dependent (Rampart and Williams, 1988; Cybulsky et al., 1989). The kinetics and protein synthesis dependence of the response to IL-1 or TNF are therefore in keeping with a direct action of these cytokines on adhesion molecule gene induction in EC, as shown with cultured EC (see above). Furthermore, the capacity of IL-1 or TNF to induce EC expression of adhesion molecules was shown immunocytochemically using organ culture of human skin (Messadi et al., 1987), and by direct injection of cytokines into the skin of subhuman primates (Munro et al., 1989; Briscoe et al., 1992) and humans (Groves et al., 1992).

E-Selectin

Our initial in vivo studies were focused on demonstrating the feasibility of measuring the lumenal expression of an EC adhesion molecule in vivo by counting the differential uptake of iv-injected ^{99m}Tc-labelled mAb 1.2B6 (anti-E-selectin) and ^{111}In-labelled MOPC 21 (control IgG1) into pig skin injected with human recombinant IL-lα. We were able to show that specific uptake of the anti-E-selectin mAb commenced at about 45 minutes after IL-lα injection, peaked at 2-3 hours and then declined, consistent with the known kinetics of E-selectin expression in vitro (Keelan et al., 1994). Further experiments showed that the kinetics of anti-E-selectin uptake were similar in response to TNFα, although slightly more prolonged (Figure 5). In contrast, phorbol myristate acetate (PMA) and phytohaemagglutinin (PHA) had more delayed effects on the uptake of anti-E-selectin into the skin, possibly due to the actions of these agents being mediated indirectly through the release of IL-1 and/or TNF locally in the skin. The conclusion that we drew from this initial study was that the specific uptake of anti-E-selectin mAb was a sensitive measurement of E-selectin expressed luminally by EC during inflammation in the skin.

Having defined the kinetics of E-selectin induction following injection of IL-1 or TNF, we were able to examine the relationship between E-selectin expression

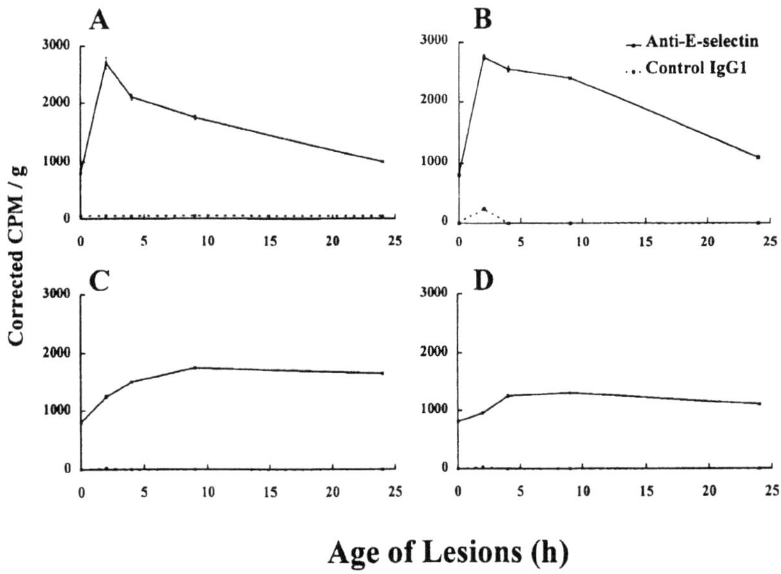

Figure 5. Uptake of ^{99m}Tc-labelled mAb 1.2B6 (anti-E-selectin) and ^{111}In-labelled MOPC 21 (control IgG_1) into skin sites prepared by injection of IL-lα (2,250 U/site; **A**), TNFα (2250 U/site; **B**), phytohaemagglutinin A (PHA, 25 mg/site; **C**), or PMA (10 ng/site; **D**) 24, 9, 4, or 2 hours before exsanguination. Labeled antibodies were intravenously injected simultaneously into animals 10 minutes before exsanguination, and lesions were then excised for counting of localized radioactivity. CPM/g were corrected for body weight and specific activities of the isotopes. Values are means ± SEM of 4 animals, with 4 sites in each animal (Keelan et al., 1994). (Reproduced by kind permission of the Editor, American Journal of Physiology).

and the timing of the recruitment of leukocytes into the IL-1- or TNF-induced lesions. Since E-selectin is capable in vitro of binding neutrophils, monocytes, and a sub-population of lymphocytes (Bevilacqua et al., 1989, Carlos et al., 1991: Picker et al., 1991), how are the kinetics of expression of this adhesion molecule related to the separate phases of recruitment of different leukocyte populations? We found that the onset of E-selectin expression in response to either IL-1 or TNF is closely paralleled by the onset of neutrophil recruitment (Binns et al., 1996a). Furthermore, E-selectin expression was necessary for neutrophil uptake since the administration of a saturating quantity of anti-E-selectin mAb led to a significant inhibition of neutrophil uptake in the period from 95–105 min following injection of IL-1 or TNF. Surprisingly, when the relationship between the time-course of E-selectin uptake and neutrophil recruitment was examined in detail, we found that neutrophil uptake dropped off after 105 min in response to both IL-1 and TNF, while uptake of anti-E-selectin continued to increase and peaked later at about 120 minutes (Figure 6). The reason for the drop in the uptake of neutrophils at 105 min despite increasing expression of E-selectin is not clear, but could include a reduction in expression of another adhesion molecule(s) acting in concert with E-selectin (e.g. P-selectin), reduction in expression of a critical chemoattractant. or the appearance of an inhibitor

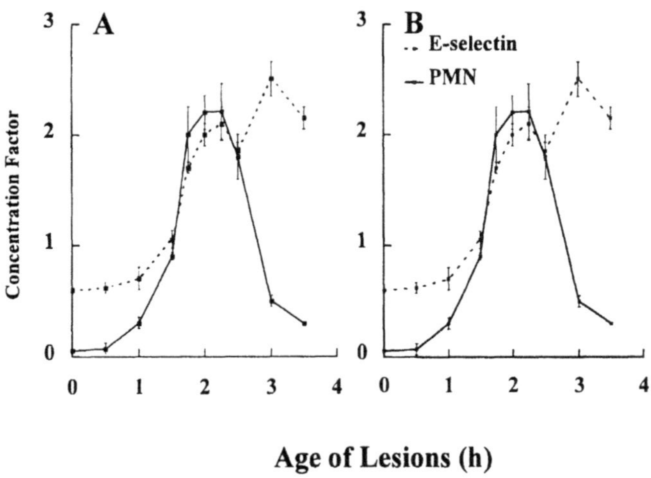

Figure 6. Relationship in the time-course of uptake of neutrophils to the expression of E-selectin following injection into the skin of IL-lα (3,000 U/site) (**A**) or TNFα (2,250 U/site) (**B**). ^{99m}Tc-labeled-anti-E-selectin and ^{111}In-labeled neutrophils were intravenously injected simultaneously into animals 7.5 and 6 minutes respectively before exsanguination, and lesions were then excised for counting of localized radioactivity. Note that the duration of neutrophil traffic is relatively brief and that expression of E-selectin, as shown by uptake of labeled anti-E-selectin, is increasing at the time neutrophil uptake starts to drop (Binns et al., 1996a). (Reproduced by kind permission of the Editor, American Journal of Physiology.)

of neutrophil uptake. This phenomenon of neutrophil uptake falling off before the peak in expression of E-selectin could also be observed in the inflammatory response to sodium urate crystals, which probably induce inflammation in part through stimulating the local release of IL-1 and TNF (Chapman et al., 1996, 1997).

As discussed above, the onset of lymphocyte entry into the skin following injection of IL-1 or TNF follows that of neutrophil recruitment, with peak accumulation of lymphocytes at about 3 hours after IL-1 and 5-6 hours after TNF. As E-selectin expression precedes lymphocyte recruitment, this molecule is unlikely to explain the onset of the lymphocyte traffic. However E-selectin clearly plays a role in the uptake of lymphocytes as well as neutrophils, since lymphocyte uptake at both 90 and 180 minutes was significantly inhibited by saturating levels of anti-E-selectin mAb. Thus, in this model, E-selectin facilitates optimal recruitment of lymphocytes as well as neutrophils. but other mechanisms are clearly required to determine the specific leukocyte populations that are recruited at different time-points during the period in which E-selectin is expressed.

VCAM-1

Expression of VCAM-1 should particularly favor lymphocyte and monocyte recruitment since VCAM-1 acts as a ligand for α_4 integrins which are expressed at much greater density on monocytes and lymphocytes than on neutrophils. By measuring the uptake of intravenously injected ^{111}In-mAb 10.2C7 (anti-VCAM-1), ^{125}I-mAb 1.2B6 (anti-E-selectin), and ^{99m}Tc-MOPC-21 (control IgG1), it was possible to determine the relative kinetics of expression of E-selectin and VCAM-1 in the same developing lesions (Harrison et al., 1997). Both IL-1 and TNF stimulated VCAM-1 expression, but with an onset that followed the peak of E-selectin by approximately 6 hours, as observed in vitro (see above). Also as observed in vitro, there was significantly greater VCAM-1 expression in response to TNF than in response to IL-1. Although we have not yet performed blocking experiments with anti-α_4 or anti-VCAM-1 antibodies to establish that the VCAM-1 expression in this context is functionally important, it is reasonable to suppose that the delayed expression of VCAM-1 compared to E-selectin contributes to the relative delay in lymphocyte compared to neutrophil uptake following cytokine injection, and that the greater effect on VCAM-1 expression of TNF compared to IL-1 contributes to the relatively greater effect of TNF on lymphocyte recruitment.

Previous studies had shown that the traffic of neutrophils into IL-1-stimulated skin lesions in rabbits was strongly inhibited by anti-β_2 integrin mAb (Rampart and Williams, 1988). We found that the anti-β_2 integrin mAb 6.5E almost completely inhibited neutrophil uptake into IL-1 and TNF-stimulated lesions in the pig model and also significantly inhibited the uptake of lymphocytes. The capacity of both anti-E-selectin and anti-β_2 integrin mAb to inhibit leukocyte traffic indicates a lack of redundancy of these molecules and is consistent with their acting sequentially in the adhesion cascade. The greater dependency of neutrophils than lymphocytes on

β_2 integrins obviously may be due to the contribution of α_4 integrins in supporting lymphocyte traffic into the skin.

C. Delayed-Type Hypersensitivity

Delayed-type hypersensitivity reactions (DTH) are useful models with which to study T lymphocyte-mediated inflammatory responses, since they are simple to establish and have predictable kinetics of lesion development. It seems likely that most DTH reactions involve an initial phase over the first few hours during which antigen-specific CD4+ T-cells are recruited to the site of antigen challenge (Van Loveren and Askenase, 1984; Herzog et al., 1990).T cells that have been recruited into the tissues become activated locally by antigen presenting cells and release lymphokines that then perpetuate the response. Most recent work investigating the immune and inflammatory mechanisms underlying delayed hypersensitivity have focused on contact reactions elicited by painting oxazalone or dinitrofluorobenzene (DNFB) on the ear of sensitized mice and have measured ear thickness as the read-out of the response.

A number of immunocytochemical studies have demonstrated induction or upregulation of EC adhesion molecules for leukocytes (e.g., E-selectin, ICAM-1, VCAM-1) during the course of DTH reactions in a variety of species (Cotran et al., 1986; Norris et al., 1991; Waldorf et al., 1991; Silber et al., 1994a: Hanseleit et al., 1996; Sunderkotter et al., 1996). Furthermore, studies with neutralizing mAb or soluble adhesion molecule fusion proteins have demonstrated the importance in the elicitation phase of DTH responses of L-selectin (Watson et al., 1991; Dawson et al., 1992), P-selectin (Tipping et al., 1996), E-selectin (Silkier et al., 1994b), PSGL-1 (Borges et al., 1997), α_L (Scheynius et al., 1993, Issekutz, 1993), α_M (Nielson et al., 1994), α_4 (Issekutz, 1993; Chisolm et al., 1993; Ferguson and Kupper, 1993), ICAM-1 (Scheynius et al., 1993), VCAM-1 (Silber et al., 1994b), and CD44 (Camp et al., 1993).

The advent of genetically-engineered mice has allowed a further exploration of the role of individual adhesion molecules in inflammatory responses, while avoiding possible artefactual effects that antibodies can have on leukocyte distribution (Hynes and Wagner, 1996). Interestingly, while contact hypersensitivity was found to be depressed in mice deficient in L-selectin (Tedder et al., 1995; Xu et al., 1996), or P-selectin (Subramaniam et al., 1995), or mice doubly deficient in both P and E-selectin (Staite et al., 1996), E-selectin deficient mice have been reported to have a normal contact hypersensitivity response, suggesting that, among the selectins, E-selectin is redundant in this model in the mouse (Labow et al., 1994). The DTH response is also reduced in mice deficient in ICAM-1 (Sligh et al., 1993).

By studying the uptake of ^{111}In-labelled neutrophils and ^{51}Cr-labelled lymphocytes, it was possible for us to demonstrate two phases of cell traffic in DTH responses in the pig, either when elicited by injection of purified protein derivative (PPD) in animals sensitized to Bacille Calmette-Guérin (BCG) vaccine, or when elicited by contact challenge of pigs sensitized by topical application of DNFB (Binns et al., 1996). The first phase of leukocyte traffic into PPD-stimulated DTH reactions in pig skin was detectable by 3 hours and involved the uptake of both neutrophils and lymphocytes (Figure 7). In the case of neutrophils, uptake was equivalent in naïve and sensitized animals, reflecting a non-specific inflammatory response to the injection of PPD. In contrast, lymphocyte uptake was significantly greater during the first 3-8 hours in sensitized pigs, indicating an immune component to lymphocyte recruitment even at this early stage. After 24 hours, very few neutrophils or lymphocytes are taken up into the skin in animals that had not been sensitized, while in sensitized pigs, uptake of leukocytes was found to enter a second phase, peaking at 24 hours and predominated by lymphocytes rather than neutrophils. Neutrophil and lymphocyte uptake in naïve and sensitized pigs in response to DNFB was broadly similar to the PPD response, although more prolonged and with some subtle differences compared to the PPD response in the proportion of neutrophils to lymphocytes entering the lesions at the different stages.

Taken together, these data are consistent with there being at least three mechanisms contributing to leukocyte traffic in DTH: (i) a proportion of the early phase, which occurs within three hours, is a non-specific inflammatory response to antigen challenge, and is also seen in naïve animals, (ii) superimposed on the early non-specific phase is an immune-dependent mechanism that is demonstrated by

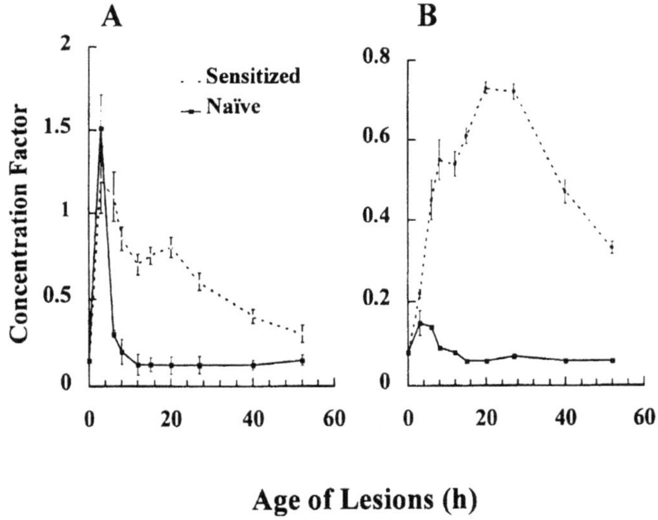

Figure 7. Kinetics of neutrophil (**A**) and lymphocyte (**B**) uptake in skin lesions stimulated by injection of PPD (2,250 U) in pigs that had been previously sensitized by injection of bacillus Calmette-Guérin or in naïve pigs. The data shows mean + SEM in three pigs (Binns et al., 1996b). (Reproduced by kind permission of the Editor, Journal of Immunology.)

increase in uptake of lymphocytes but not neutrophils in sensitized compared to naïve pigs, and (iii) the two components of the early response are followed by the delayed phase of leukocyte traffic over the second 24-hour period. The delayed phase is immune-specific and is characterised by a gradual increase in lymphocyte uptake to a peak at 24–36 hour coincident with a gradual decrease in neutrophil uptake into the same lesions.

E-Selectin

Using the techniques outline above, we have been able to relate the expression of E-selectin in DTH to the traffic of neutrophils and lymphocytes (Binns et al., 1996b; Harrison et al., 1997). In naïve pigs, an increase in E-selectin expression was detectable by 3 hours after challenge but had returned to base-line by 24 hours, consistent with the transient non-specific inflammatory response (Figure 8). Sensitized pigs showed a biphasic expression of E-selectin, comprising an early phase that was identical to that

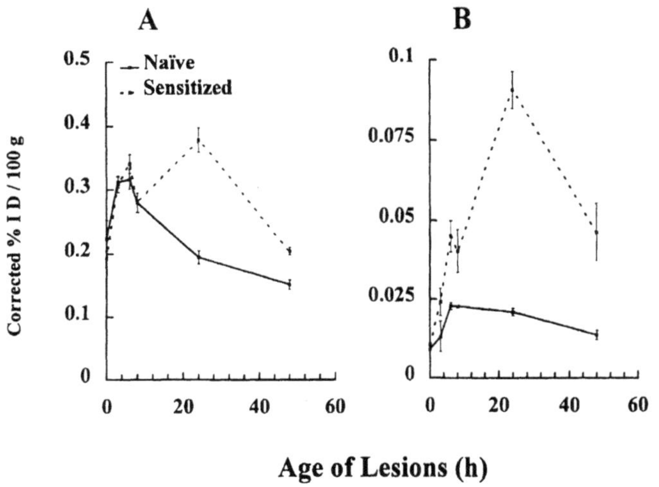

Figure 8. Kinetics of expression of E-selectin (**A**) and VCAM-1 (**B**) in skin lesions stimulated by injection of PPD (2250 V) in pigs that had been sensitized by injection of bacillus Calmette-Guérin or in naïve pigs. Expression of E-selectin and VCAM-1 were measured by counting the uptake of ^{125}I-labelled mAb 1.2B6 and ^{111}In-labelled mAb 10.2C7 respectively. The percentage injected dose (ID)/100 g of lesion has been corrected by subtraction of the uptake (% ID/100 gram) of ^{99m}Tc-labelled control IgG1. Data points are mean + SEM of data a total of 18–24 skin lesions from three pigs. During the first 24 hours of the response, E-selectin expression is similar in sensitized and naïve animals. In contrast VCAM-1 is only significantly expressed during the first 24 hours in sensitized pigs. Both E-selectin and VCAM-1 show a second phase of expression which is maximal at 24 hours and is only seen in the sensitized animals (Harrison et al., 1997). (Reproduced by kind permission of the Editor, Journal of Immunology.)

seen in naïve pigs, followed by a delayed phase which peaked at 24 hours. Blocking studies with neutralizing concentrations of anti-E-selectin mAb demonstrated highly significant inhibition of uptake of both neutrophils and lymphocytes during the delayed phase of the response, either following challenge with PPD or DNFB.

The conclusions to be drawn from these experiments looking at E-selectin in DTH responses are similar to those coming out of the investigation of E-selectin expression in response to injection of IL-1 or TNF and indicate that this molecule is necessary for optimal uptake of neutrophils and lymphocytes both in the non-specific inflammatory and in the immune-specific phases of the response. The dependence of both neutrophils and lymphocytes on E-selectin in this model is surprising considering the apparently normal DTH responses that have been observed in E-selectin deficient mice (Labow et al., 1994), and raises the possibility that there may be species differences in the relative importance of the three selectins.

VCAM-1

In contrast to E-selectin, there was very little expression of VCAM-1 during the non-specific inflammatory response to PPD in naïve animals (Figure 8). In sensitized pigs, VCAM-1 was expressed both during the early immune phase and, to a greater degree, during the delayed phase of the response. In view of the greater effects of TNF than IL-1 on VCAM-1 expression, both in vitro and when injected into the skin in vivo, we tested the effect of administering neutralizing anti-TNFα mAb before challenge with PPD. This abolished the expression of VCAM-1 during the early immune phase and significantly reduced VCAM-1 expression in the delayed phase, emphasizing the importance of TNF for VCAM-1 expression (Figure 9). On the other hand, although there was slight reduction of uptake of anti-E-selectin in the anti-TNFα treated pigs when compared with untreated sensitized animals, this difference was not statistically significant.

The greater effects of TNF compared to IL-1 on VCAM-1 expression allow us to start to build up a model of how endothelium becomes activated to express adhesion molecules during the DTH response. The following tentative proposals can be made: (i) the endothelial activation seen in the non-specific response shown by naïve animals is unlikely to involve TNFα, since only E-selectin and not VCAM-1 is expressed and is probably due to the release of IL-1 from stores in keratinocytes and perhaps other cell types; (ii) VCAM-1 expression in the early immune phase requires TNFα, since it was strongly abrogated by anti-TNFα antibodies. Judging from previous work, TNFα may be released from mast cells soon after challenge with antigen (Klein et al., 1989; Walsh et al., 1991); (iii) TNFα is also involved in activating EC during the delayed immune phase, since anti-TNFα antibodies partially inhibited VCAM-1 expression. However, the lack of inhibition of E-selectin expression by anti-TNFα during the delayed phase argues that EC activation is also maintained by factors. These might include cytokines released either from resident cells or from leukocytes that have migrated into the lesion, or possibly also contact-dependent activation of EC by emigrating monocytes and lymphocytes (Damle et al., 1991; Rainger et al., 1996); (iv) Although we have not yet established that VCAM-1 is functionally involved in this model, it seems likely that expression of VCAM-1 is responsible for the predominance of mononuclear cells entering the tissue in sensitized animals, and that this is attributable in turn to the actions of TNFα.

IV. FUTURE WORK

The studies that have been outlined in this chapter demonstrate an approach that can now be extended to understand in detail how the regulated expression of adhesion molecules on vascular endothelium relates to the traffic of particular leukocyte populations during inflammatory responses. It will now clearly be important to extend these studies in order to determine how other important adhesion molecules such as P-selectin and ICAM-1 behave in relation to E-selectin and VCAM-1, and

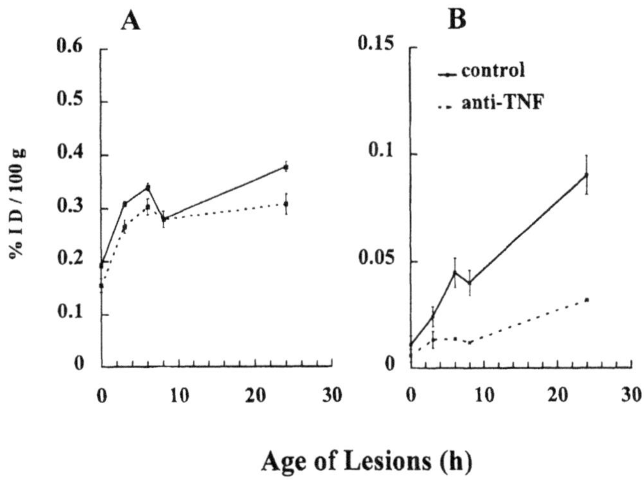

Figure 9. Effect of anti-TNFα on expression of E-selectin and VCAM-1 during the DTH response to PPD in pigs that had been previously sensitized by injection of bacillus Calmette-Guérin. Pigs treated with anti-TNFα received mAb BAY X1351 (5 mg. kg) intravenously two hours before starting a series of intra-cutaneous injections of PPD (2250 U) timed to give lesions of 24, 8, 6, and 3 h duration at the end of the experiment. Expression of E-selectin and VCAM-1 were measured as in Figure 8. In anti-TNFα-treated pigs there was no significant uptake of anti-VCAM-1 at 3, 8, and 8 h and uptake into 24-hour lesions was significantly less than in untreated control pigs. In contrast, although there was slight reduction in uptake of anti E-selectin in anti-TNFα-treated pigs, this was not statistically significant (Harrison et al., 1997). (Reproduced by kind permission of the Editor, Journal of Immunology.)

to assess the relative contribution that individual adhesion molecules and chemoattractants make to the selective recruitment of different leukocyte populations.

ACKNOWLEDGMENTS

The authors gratefully acknowledge the major contributions that several highly valued colleagues have made to this work. These include Drs Richard Binns, Peter Chapman, Andrew Harrison, Ted Keelan, and Martyn Robinson.

REFERENCES

Atherton, A. & Born, G. V. R. (1972). Quantitative investigations of the adhesiveness of circulating polymorphonuclear leukocytes to blood vessel walls. J. Physiol. 222, 447-474.

Bazan, J. F., Bacon, K. B., Hardiman, G., Wang, W., Soo, K., Rossi, D., Greaves, D. R., Zlotnik, A., & Schall, T. J. (1997). A new class of membrane-bound chemokine with a CX3C motif. Nature 385, 640-644.

Bevilacqua, M. P., Pober, J. S., Wheeler, M. E., Cotran, R. S., & Gimbrone, M. A. (1985). Interleukin 1 acts on cultured human vascular endothelium to increase the adhesion of polymorphonuclear leukocytes, monocytes, and related cell lines. J. Clin. Invest. 76, 2003-2011.

Bevilacqua, M. P., Stengelin, S., Gimbrone, M. A., & Seed, B. (1989). Endothelial leukocyte adhesion molecule 1: an inducible receptor for neutrophils related to complement regulatory proteins and lectins. Science 243, 1160-1164

Binns, R. M., Licence, S. T., & Wooding, S. T. (1990). Phytohemagglutinin induces major short-term protease-sensitive lymphocyte traffic involving high endothelium venule-like blood vessels in acute delayed-type hypersensitivity-like reactions in skin and other tissues. Eur. J. Immunol. 20, 1067-1071.

Binns, R. M., Licence, S. T., Wooding, F. B. P., & Duffus, W. P. H. (1992). Active lymphocyte traffic induced in the periphery by cytokines and phytohemagglutinin: three different mechanisms? Eur. J. Immunol. 22, 2195-2203.

Binns, R. M., Licence, S. T., Harrison, A. A., Keelan, E. T. D., Robinson, M. K., & Haskard, D. O. (1996a). in vivo E-selectin upregulation correlates with early infiltration of PMN, later with PBL-entry: mAbs block both. Am. J. Physiol. 270, H183-193.

Binns, R. M., Whyte, A., Licence, S. T., Harrison, A. A., Tsang, Y., Haskard, D. O., & Robinson, M. K. (1996b). The role of E-selectin in lymphocyte and polymorphonuclear cell recruitment into cutaneous delayed hypersensitivity reactions in sensitised pigs. J. Immunol. 157, 4094-4099.

Borges, E., Tietz, W., Steegmaier, M., Moll, T., Hallmann, R., Hamann, A., & Vestweber, D. (1997). P-selectin glycoprotein ligand-1 (PSGL-1) on T helper 1 but not on T helper 2 cells binds to P-selectin and supports migration into inflamed skin. J. Exp. Med. 185, 573-578.

Briscoe, D. M., Cotran, R. S., & Pober, J. S. (1992). Effects of tumor necrosis factor, lipopolysaccharide, and IL-4 on the expression of vascular cell adhesion molccule-1 in vivo: Correlation with CD3+ T cell infiltration. J. Immunol. 149, 2954-2960.

Bullard, D. C., Qin, L., Lorenzo, I., Quinlin, W. M., Doyle, N. A., Bosse, R., Vestweber, D., Doershuk, C. M. & Beaudet, A. L. (1995) P-selectin/ICAM-1 double mutant mice: acute emigration of neutrophils into peritoneum is completely absent but is normal into pulmonary alveoli. J. Clin. Invest. 95, 1782-1788.

Butcher, E. C. (1991). Leukocyte-endothelial cell recognition: Three (or more) steps to specificity and diversity. Cell 67, 1033-1036.

Carlos, T., Kovach, N., Schwartz, B., Rosa, M., Newman, B., Wayner, E., Benjamin, C., Osborn, L., Lobb, R., & Harlan, J. (1991). Human monocytes bind to two cytokine-induced adhesive ligands on cultured human endothelial cells: Endothelial-leukocyte adhesion molecule-1 and vascular cell adhesion molecule-1. Blood 77, 2266-2271.

Camp, R. L., Scheynius, A., Johansson, C., & Pure, E. (1993). CD44 is necessary for optimal contact allergic responses but is not required for normal leukocyte extravasation. J. Exp. Med. 178, 497-507.

Cavender, D. E., Haskard, D. O., Joseph, B., & Ziff, M. (1986). Interleukin-1 increases the binding of human B and T lymphocytes to endothelial cell monolayers. J. Immunol. 136, 203-207.

Chapman, P. T., Jamar, F., Harrison, A. A., Schofield, J. B., Peters, A. M., Binns, R. M., & Haskard. D. O. (1996). Characterization of E-selectin expression, leukocyte traffic and clinical sequelae in urate crystal-induced inflammation: an insight into gout. Brit. J. Rheumatol. 35, 323-334.

Chapman, P. T., Yarwood, H., Harrison, A. A., Stocker, C. J., Jamar, F., Gundel, R. H., Peters, A. M., & Haskard, D. O. (1997). Endothelial activation in monosodium urate monohydrate crystal-induced inflammation: in vitro and in vivo studies on the roles of tumor necrosis factor-alpha and interleukin-1. Arthritis Rheum 40, 955-965.

Chisolm, P. L., Williams, C. A., & Lobb, R. R. (1993). Monoclonal antibodies to integrin α_4 subunit inhibit the murine contact hypersensitivity response. Eur. J. Immunol. 23, 682-688.

Colditz, I. G. & Movat, H. Z. (1984). Kinetics of neutrophil accumulation in acute inflammatory lesions induced by chemotaxins and chemotaxinigens. J. Immunol. 133, 2169-2173.

Colditz, I., Zwahlen, R., Dewald, B., & Baggiolini, M. (1989). in vivo inflammatory activity of neutrophil-activating factor, a novel chemotactic peptide derived from human monocytes. Am. J. Pathol. 134, 755-760.

Collins, T., Read, M. A., Neish, A. S., Whitley, M. Z., Thanos, D., & Maniatis, T. (1995). Transcriptionall regulation of endothelial cell adhesion molecules: NF-κB and cytokine-inducible enhancers. FASEB J. 9, 899-909.

Cotran, R. S., Gimbrone, M. A., Bevilacqua, M. P., Mendrick, D. L., & Pober J. S. (1986). Induction and detection of a human endothelial activation antigen in vivo. J. Exp. Med. 164, 661-666.

Cybulsky, M. I., Chan, M. K. W., & Movat, H. Z. (1988). Acute inflammation and microthromosis induced by endotoxin, interleukin-1, and tumour necrosis factor and their implication in Gram-negative infection. Lab. Invest. 58, 365-378.

Cybulsky, M. I., McComb, D. J., & Movat, H. Z. (1989). Protein synthesis dependent and independent mechanisms of neutrophil emigration. Am. J. Pathol. 135, 227-237.

Damle, N. K., Eberhardt, C., & Van der Vieren, M. (1991). Direct interaction with primed CD4+ CD45R0+ memory T lymphocytes induces expression of endothelial leukocyte adhesion molecule- 1 and vascular cell adhesion molecule-1 on the surface of vascular endothelial cells. Eur. J. Immunol. 21, 2915-2923

Dawson, J., Sedgwick, A. D., Edwards, J. C. W., & Lees, P. (1992). The monoclonal antibody MEL-14 can block lymphocyte migration into a site of chronic inflammation. Eur. J. Immunol. 22, 1647-1650.

De Martin, R., Vanhove, B., Cheng, Q., Hofer, E., Csizmadia, V., Winkler, H., & Bach, F. H. (1993). Cytokine-inducible expression in endothelial cells of an IkappaBa-like gene is regulated by NFκB. EMBO J. 12, 2773-2779.

Ferguson, T. A. & Kupper, T. S. (1993). Antigen-independent processes in antigen-specific immunity. J. Immunol. 150, 1172-1182.

Finger, E. B., Puri, K. D., Alon, R., Lawrence, M. B., Von Andrian, U. H., & Springer, T. A. (1996). Adhesion through L-selectin requires a threshold hydrodynamic shear. Nature 379, 266-269.

Foreman, K. E., Vaporciyan, A. A., Bonish, B. K., Jones, M. L., Johnson, K. J., Glovsky, M. M., Eddy, S. M., & Ward, P. A. (1994). C5a-induced expression of P-selectin in endothelial cells. J. Clin. Invest. 94, 1147-1155

Gaboury, J. P. & Kubes, P. (1994). Reductions in physiologic shear rates lead to CD11/CD18-dependent, selectin-independent leukocyte rolling in vivo. Blood 83, 345-350.

Gamble, J. R., Harlan, J. M., Klebanoff, S. J., & Vadas, M. A. (1985). Stimulation of the adherence of neutrophils to umbilical vein endothelium by human recombinant tumour necrosis factor. Proc. Natl. Acad. Sci. USA 82, 8667-8671.

Gamble. J. R., Khew-Goodall, Y., & Vadas, M. A. (1993). Transforming growth factor-β inhibits E-selectin expression on human endothelial cells. J. Immunol. 150, 4494-4503.

Geng, J.-G., Bevilacqua, M. P., Moore, K. L., McIntyre, T. M., Prescott, S. M., Kim, J. M., Bliss, G. A., Zimmerman, G. A., & McEver, R. P. (1990). Rapid neutrophil adhesion to activated endothelium mediated by GMP-140. Nature 343, 757-760

Gille, J., Swerlick, R. A., Lawley, T. J., & Caughman, S. W. (1996). Differential regulation of vascular cell adhesion molecule-1 gene transcription by tumour necrosis factor alpha and interleukin-1 alpha in dermal microvascular endothelial cells. Blood 87, 211-217.

Groves, R. W., Ross, E., Barker, J. N. W. N., Ross, J. S., Camp, R. D. R., & MacDonald, D. M. (1992). Effect of in vivo interleukin-1 on adhesion molecule expression in normal human skin. J. Invest. Dermatol. 98, 384-387.

Harrison, A. A., Stocker, C. J., Chapman P. T., Tsang, Y. T., Heuhns, T. Y., Gundel, R. H., Peters, A. M., Davies, K. A., George, A. J., Robinson, M. K., & Haskard, D. O. (1997). Expression of VCAM-1 by vascular endothelial cells in immune- and non-immune inflammatory reactions in the skin. J. Immunol. 159, 4546-4554.

Hattori, R., Hamilton, K. K., Fugate, R. D., McEver, R. D., & Sims, P. J. (1989a). Stimulated secretion of endothelial von Willebrand factor is accompanied by rapid redistribution to the cell surface of the intracellular granule membrane protein GMP-140. J. Biol. Chem. 264, 7768-7771.

Hattori, R., Hamilton, K. K., McEver, R. P., & Sims, P. J. (1989b). Complement proteins C5b-9 induce secretion of high molecular weight multimers of endothelial von Willebrand factor and translocation of granule membrane protein GMP-140 to the cell surface. J. Biol. Chem. 264, 9053-9060.

Henseleit, U., Steinbrink, K., Goebeler, M., Roth, J., Vestweber, D., Sorg, C., & Sunderkotter, C. (1996). E-selectin expression in experimental models of inflammation in mice. J. Pathol. 180, 317-325.

Herzog, W.-R., Millet, I., Ferreri, N. R., Ramabhadran, R., Schreurs, J., & Askenase, P.W. (1990) An antigen-specific DTH-initiating cell clone. J. Immunol. 144, 3667-3676.

Hynes, R. O. & Wagner, D. D. (1996) Genetic manipulation of adhesion molecules in mice. J. Clin. Invest. 98, 2193-5.

Iademarco, M. F., Barks, J. L., & Dean, D. C. (1995). Regulation of vascular cell adhesion molecule-1 expression by IL-4 and TNFα in cultured endothelial cells. J. Clin. Invest. 95, 264-271

Issekutz, T. B. (1990). Effects of six different cytokines on lymphocyte adherence to microvascular endothelium and in vivo lymphocyte migration in the rat. J. Immunol. 144, 2140-2146.

Issekutz, T. B. (1993). Dual inhibition of VLA-4 and LFA-1 maximally inhibits cutaneous delayed-type hypersensitivity-induced inflammation. Am J. Pathol. 143, 1286-1293.

Issekutz, A. C. & Issekutz, T. B. (1993). Quantitation and kinetics of blood monocyte migration to acute inflammatory reactions, and IL-lα, tumor necrosis factor-α, and IFN-γ. J. Immunol. 151, 2105-2115.

Issekutz, T. B., Issekutz, A. C., & Movat, H. Z. (1981). The in vivo quantification and kinetics of monocyte migration into acute inflammatory tissue. Am. J. Pathol. 103, 47-55.

Issekutz, A. C. & Movat, H. Z. (1980). The in vivo quantification and kinetics of rabbit neutrophil leukocyte accumulation in the skin in response to chemotactic agents and Escherichia coli. Lab. Invest. 42, 310-317

Issekutz, T. B. & Stoltz, J. M. (1989). Stimulation of lymphocyte migration by endotoxin, tumour necrosis factor, and interferon. Cell. Immunol. 120, 165-173.

Keelan, E. T. M., Licence, S. T., Peters, A. M., Binns, R. M., & Haskard, D. O. (1994). Characterization of E-selectin expression in vivo using a radiolabelled monoclonal antibody. Am. J. Physiol. 266, H279-290

Klein, L. M., Lavker, R. M., Matis, W. L., & Murphy, G. F. (1989). Degranulation of human mast cells induced an endothelial antigen central to leukocyte adhesion. Proc. Natl. Acad. Sci. USA 86, 8972-8976.

Labow. M. A., Norton, C. R., Rumberger, J. M., Lombard-Gillooly, K. M., Shuster, D. J., Hubbard, J., Bertko, R., Knaack, P. A., Terry, R. W., Harbison, P. A., Kontgen, F., Stewart, C. L., McIntyre, K. W., Will, P. C., Burns, D. K., & Wolitzky, B. A. (1994). Characterization of E-selectin-deficient mice: demonstration of overlapping function of the endothelial selecting. Immunity 1, 709-720.

Mason. J. C. & Haskard, D. O. (1994). The clinical importance of leukocyte and endothelial cell adhesion molecules in inflammation. Vasc. Med. Rev. 5, 249-275.

Messadi, D. V., Pober, J. S., Fiers, W., Gimbrone, M. A., & Murphy, G. F. (1987). Induction of an activation antigen on postcapillary venular endothelium in human skin organ culture. J. Immunol. 139, 1557-1562.

Michie, S. A., Streeter, P. R., Bolt, P. A., Butcher, E. C., & Picker, L. J. (1993). The human peripheral lymph node vascular addressin: an inducible endothelial antigen involved in lymphocyte homing. Am. J. Pathol. 143, 1688-1698.

Munro, J. M., Pober, J. S., & Cotran, R. S. (1989). Tumor necrosis factor and interferon-y induce distinct patterns of endothelial activation and associated leukocyte accumulation in skin of Papio Anubis. Am. J. Pathol. 135, 121-133.

Norris, P., Poston, R. N., Thomas, D. S., Thornhill, M., Hawk, J., & Haskard, D. O. (1991). The expression of endothelial leukocyte adhesion molecule-1 (ELAM-1), Intercellular adhesion molecule-1 (ICAM-1) and vascular cell adhesion molecule-1 (VCAM-1) in experimental cutaneous inflammation: a comparison of ultraviolet- B erythema and delayed hypersensitivity. J. Invest. Dermatol. 96, 763-770.

Nielson, H. V., Christensen, J. P., Anderson, E. C., Marker, O., & Thomsen, A. R. (1994). Expression of Type 3 complement receptor on activated CD8+ T cells facilitates homing to inflammatory sites. J. Immunol. 153, 2021-2028.

Pabst, R. & Binns, R. M. (1989). Heterogeneity of lymphocyte homing physiology: several mechanisms operate in the control of migration to lymphoid and non-lymphoid organs in vivo. Immunological reviews 108, 83-109.

Picker, L. J., Kishimoto, T. K., Smith, C. W., Warnock, R. A., & Butcher, E. C. (1991). ELAM-1 is an adhesion molecule for skin-homing T cells. Nature 349, 796-799.

Pober, J. S. & Cotran, R. S. (1990). Cytokines and endothelial cell biology. Physiological Reviews 70, 427-451.

Pober, J. S., Bevilacqua, M. P., Mendrick, D. L., Lapierre, L. A., Fiers, W., & Gimbrone, M. A. (1986). Two distinct monokines, interleukin 1 and tumor necrosis factor, each independently induce the biosynthesis and transient expression of the same antigen on the surface of cultured human vascular endothelial cells. J. Immunol. 136, 1680-1687.

Rainger, G. E., Wautier, M. P., Nash, G. B., & Wautier, J.-L. (1996). Prolonged E-selectin induction by monocytes potentiates the adhesion of flowing neutrophils to cultured endothelial cells. Br. J. Haematol. 92, 192-199.

Rampart, M. & Williams, T.J. (1988). Evidence that neutrophil accumulation induced by interleukin-1 requires both local protein biosynthesis and neutrophil CD18 antigen expression in vivo. Br. J. Pharmacol. 94, 1143-1148.

Read, M. A., Whitley, M. Z., Williams, A. J., & Collins, T. (1994). NFκB and IκBα: an inducible regulatory system in endothelial activation. J. Exp. Med. 179, 503-512.

Scheynius, A., Camp, R. L., & Puré, E. (1993). Reduced contact sensitivity reactions in mice treated with monoclonal antibodies to leukocyte function-associated molecule-1 and intercellular adhesion molecule-1. J. Immunol. 150, 655-663.

Silber, A., Newman, W., Reimann, K. A., Hendricks, E., Walsh, D., & Ringler, D. J. (1994a). Kinetic expression of endothelial adhesion molecules and relationship to leukocyte recruitment in two cutaneous models of inflammation. Lab. Invest. 70, 163-175.

Silber, A., Newman, W., Sasseville, V. G., Pauley, D., Beall, D., Walsh, D. G., & Ringler, D. J. (1994b). Recruitment of lymphocytes during cutaneous delayed hypersensitivity in non-human primates is dependent on E-selectin and vascular cell adhesion molecule-l. J. Clin. Invest. 93, 1554-1563.

Sligh, J. E., Jr., Ballantyne, C. M., Rich, S. S., Hawkins, H. K., Smith, C. W., Bradley, A., & Beaudet, A. L. (1993). Inflammatory and immune responses are impaired in mice deficient in intercellular adhesion molecule 1. Proc. Natl. Acad. Sci. USA 90, 8529-8533.

Springer, T. A. (1994). Traffic signals for lymphocyte recirculation and leukocyte emigration: the multistep paradigm. Cell 76, 301-314.

Staite, N. D., Justen, J. M., Sly, L. M., Beaudet, A. L., & Bullard, D. C. (1996). Inhibition of delayed-type contact hypersensitivity in mice deficient in both E-selectin and P-selectin. Blood 88, 2973-2979.

Subramaniam, M., Saffaripour, S., Watson, S. R., Mayadas, T. N., Hynes, R. O., & Wagner, D. D. (1995). Reduced recruitment of inflammatory cells in a contact hypersensitivity response in P-selectin-deficient mice. J. Exp. Med. 181, 2277-2282.

Sunderkotter, C., Steinbrink, K., Henseleit, U., Bosse, R., Schwartz, A., Vestweber, D., & Sorg, C. (1996). Activated T cells induce expression of E-selectin *in vitro* and in an antigen-dependent manner in vivo. Eur. J. Immunol. 26, 1571-1579.

Swerlick, R. A., Lee, K. H., Li, L.-J., Sepp, N. T., Caughman, S. W., & Lawley, T. J. (1992). Regulation of vascular cell adhesion molecule 1 on human dermal microvascular endothelial cells. J. Immunol. 149, 698-705.

Tedder, T. F., Steeber, D. A., & Pizcueta, P. (1995). L-selectin deficient mice have impaired leukocyte recuitment into inflammatory sites. J. Exp. Med. 181, 2259-2264.

Thornhill, M. H. & Haskard, D. O. (1990). IL-4 regulates endothelial activation by IL-1, tumor necrosis factor or IFNγ. J. Immunol. 145, 865-872.

Thornhill, M., Kyan-Aung, U., Lee, T. L., & Haskard, D. O. (1990). T cells and neutrophils exhibit differential adhesion to cytokine stimulated endothelial cells. Immunology 69, 287-292.

Tipping, P. G., Huang, X. R., Berndt, M. C., & Holdsworth, S. R. (1996). P-selectin directs T lymphocyte-mediated injury in delayed-type hypersensitivity responses: studies in glomerulonephritis and cutaneous delayed-type hypersensitivity. Eur. J. Immunol. 26, 454-460.

Tsang, Y. T. M., Haskard, D. O., & Robinson, M. K. (1994). Cloning and expression kinetics of porcine vascular cell adhesion molecule. Biochem. Biophys. Res. Commun. 201, 805-812.

Tsang, Y., Stevens, P. E., Licence, S. T., Haskard, D. O., Binns, R. M., & Robinson, M. K. (1995). Porcine E-selectin: cloning and functional characterization. Immunology 85, 140-145.

Van Loveren, H. & Askenase, P. W. (1984) Delayed-type hypersensitivity is mediated by a sequence of two different T cell activities. J. Immunol. 133, 2397-2401.

Waldorf, H. A., Walsh, L. J., Schechter, N. M., & Murphy, G. F. (1991). Early cellular events in evolving cutaneous delayed hypersensitivity in humans. Am. J. Pathol. 138, 477-486.

Walsh, L. J., Trinchieri, G., Waldorf, H. A., Whitaker, D., & Murphy, G. F. (1991). Human dermal mast cells contain and release tumor necrosis factor a, which induces endothelial leukocyte adhesion molecule 1. Proc. Natl. Acad. Sci. USA 88, 4220-4224.

Wankowicz, Z., Megyeri, P., & Issekutz, A. (1988). Synergy between tumour necrosis factor alpha and interleukin-1 in the induction of polymorphonuclear leukocyte migration during inflammation. J. Leukoc. Biol. 43, 349-356.

Watson, S. R., Fennie, C., & Lasky, L. A. (1991). Neutrophil influx into an inflammatory site inhibited by a soluble homing receptor-IgG chimaera. Nature 349, 164-167.

Wellicome, S. M., Thornhill, M. H., Pitzalis, C., Thomas, D. S., Lanchbury, J. S. S., Panayi, G. S., & Haskard, D. O. (1990). A monoclonal antibody that detects a novel antigen on endothelial cells that is induced by tumor necrosis factor, IL-1 or lipopolysaccharide. J. Immunol. 144, 2558-2565.

Xu, J., Grewal, I. S., Geba, G. P., & Flavell, R. A. (1996). Impaired primary T cell responses in L-selectin-deficient mice. J. Exp. Med. 183, 589.

Zimmerman, G. A., McIntyre, T. M., Mehra, M., & Prescott, S. M. (1990). Endothelial Cell-associated Platelet-activating Factor: A novel mechanism for signaling intercellular adhesion. J. Cell. Biol. 110, 529-540.

THE ROLE OF ADHESION IN METASTASIS:
POTENTIAL MECHANISMS AND MODULATION OF INTEGRIN ACTIVITY

John F. Marshall and Ian R. Hart

Advances in Molecular and Cell Biology
Volume 28, pages 345-365.

ISBN: 0-7623-0495-2

I. THE BIOLOGICAL PROCESSES OF METASTASIS

Metastasis, which is defined as the transfer of malignant tumor cells from one organ or part to another not directly connected with it, is responsible (directly or indirectly) for the majority of cancer deaths. Though approximately half of all patients suffering from solid cancers will have metastatic deposits already by the time of initial presentation (Sugarbaker et al., 1982) there is considerable interest in studying the pathogenesis of the process in the hope that increased understanding will lead to improved interventional strategies.

For secondary tumor growth to occur, malignant cells must invade and infiltrate normal host tissue surrounding the primary mass, there must be penetration of small vascular channels or lymphatics, individual cells or emboli must be released into the circulation and should be able to survive in this milieu, the disseminating cells must arrest in the capillary beds of distant organs, and they must extravasate by penetrating the walls of the arresting vessel before growing in their secondary location.

This metastatic process can be defined in terms of a series of events involving adhesion and de-adhesion (see Figure 1). Normal tissue and, presumably, primary tumor growth requires a combination of both growth-factor and adhesion dependent signaling. The down-regulation of intercellular cohesion is a prerequisite for the dissociation of malignant cells away from the primary neoplasm. Migration through the local extracellular-matrix (ECM) or on the surface of other cells (e.g., stromal fibroblasts, smooth muscle cells) requires that cells are able to develop

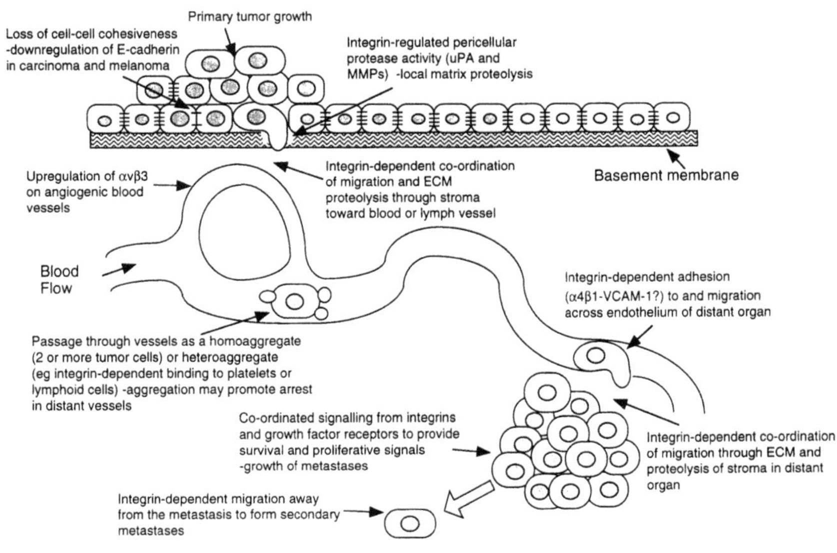

Figure 1. Potential integrin-dependent mechanisms in metastasis.

adhesive interactions at one pole of the cell which will provide sufficient traction to enable the cell to propel itself forward but that these adhesive interactions will not be so great as to prevent cell movement (Palecek et al., 1997). Transmigration of individual cells or emboli into the vasculature or lymphatics allows their distribution to distant organs or lymph nodes where the cells may arrest. The arrest may be promoted by tumor cell adhesion to lymphoid cells or platelets that facilitate mechanical entrapment of multicellular aggregates (Gasic et al., 1973). Extravasation will be followed by secondary tumor growth if a permissive local environment is present (that is, appropriate ECM, growth factors, oxygen supply, neovascularisation) and it is believed that the growth-factor and adhesion dependent signaling involved in primary tumor growth must be duplicated at such distant sites.

One of the major molecules responsible for cell–cell cohesion in epithelial cells is E-cadherin. Loss or loss of function of these calcium-dependent homotypic adhesion molecules is strongly correlated with progression toward metastasis in many types of carcinoma (Birchmeier et al., 1995) as well as in tumors of neuroectodermal origin such as metastatic melanoma (Danen et al., 1996; Seline et al., 1996). Since the cadherins are described elsewhere in this volume (Chapters 2 and 10) we will confine our review of the role of adhesive events in tumor spread to the other major families of adhesion molecules—the selectins, the immunoglobulin superfamily and the integrins. Most of our attention, however, will focus upon the integrins responsible for adhesion to, and thus migration through, the ECM.

Integrins are transmembrane, non-covalently associated αβ heterodimers which bind to ECM glycoproteins or cell surface-molecules of the immunoglobulin superfamily (or as in the case of αIELβ7 to E-cadherin; Cepek et al., 1994). The specificity of integrins is determined partly by the particular association of α and β subunits and, to date, over 20 different αβ combinations have been described (see Table 1). Almost all nucleated mammalian cells express at least some integrins and, given the multiple potential roles for adhesion in the metastatic cascade, it seems highly likely that integrins play a major part in the process of tumor spread.

Having disengaged from the cohesive restraints of the primary tumor mass, the disseminating cancer cell migrates and it seems likely that successful spread will be enhanced by an active migratory ability. Palecek and colleagues (1997) have shown that α5β1-and αIIbβ3-dependent motility was determined by the relative contribution of three factors: concentration of substrate, concentration of integrin receptor and the activation state of the integrin (see Figure 2). These factors interact to ensure that maximum motility occurred at an intermediate level of adhesiveness. Tumor cells appear to be subject to migratory control via these different components. Thus Mould and colleagues (1994) reported that an excess of VCAM-1 substrate reduced the α4β1-dependent migration of the A375M melanoma cell line and Giancotti and Ruoslahti (1990) determined that a high expression of α5 in CHO cells reduced their migration while suppressing the tumorigenic phenotype.

Table 1.

Integrin	*ECM Glycoprotein*	*Cell-Membrane-Associated Glycoprotein*
αIβ1	Lm, Coll	
α2β1	Lm, Coll, Tn	
α3β1	Lm, Coll, Fn, Epi, Ent	
α4β1	Fn (CS1), Fn (CS5), Fn (HepII), Fn (ROD)	VCAM-1
α5β1	Fn (RGD), Fgen	
α6β1	Lm, Mer, Kal	
α7β1	Lm	
α8β1	Tsp, Vn, Fn	
α9β1	Tn	
αLβ2		ICAM-1, ICAM-2, ICAM-3
αMβ2	iC3b, Fgen, Factor X	ICAM-1
αXβ2	iC3b, Fgen	
αDβ2		ICAM-3
αIIbβ3	Fgen, Fn(RGD), vWF, Vn	
αvβ1	Fn, Vn, Fgen, OP	
αvβ3	Vn, Fgen, vWF, Fn, Lm, Tn, Coll(denat.), OP, TSP, Perl	CD31
αvβ5	Vn, OP	
αvβ6	Fn	
αvβ8	Vn	
α6β4	Lm, Kal	
α4β7	Fn (CS1),	VCAM-1, MADCAM
αIELβ7		E-cadherin

Abbreviations: Fn (fibronectin), Vn (vitronectin), Fgen (fibrinogen), Coll (collagen), Tn (tenascin), Epi (epiligrin), Ent (entactin), Mer (merosin), Kal (kalinin), Lm (laminin), OP (osteopontin), TSP (thrombospondin), Perl (perlecan), vWF (von Willebrand factor), ICAM (intercellular adhesion molecule), VCAM (vascular cell adhesion molecule), MADCAM (mucosal addressin cell adhesion molecule), denat. (denatured)

Integrins may also influence metastasis by providing growth and survival signals (see Schwartz, 1994; Meredith and Schwartz, 1997, for references) in addition to providing the physical link between the ECM and the cell cytoskeleton which enables motility, and thus migration, to occur. The fact that metastatic tumor cells are able to continue proliferating when they find themselves in a foreign environment, and presumably are receiving some positional information through integrin receptors suggests that defects in the growth/survival pathways may be an important aspect of secondary cancer development. However, it is not possible to review all the evidence associating changes in integrin function with tumor progression and metastasis and the reader is referred to several recent reviews (Juliano and Varner, 1993; Giancotti and Mainiero, 1994; Pignatelli and Stamp, 1995; Marshall and

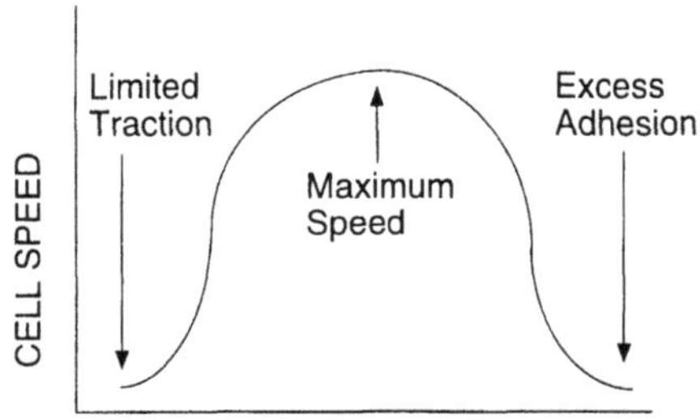

Figure 2. Maximum motility requires intermediate adhesiveness-A balance of adhesion and de-adhesion. (Based upon Palecek et al., 1997.)

Hart, 1996; Varner and Cheresh, 1996). This chapter will focus upon recent evidence concerning the role of integrins in regulating proteinases, in mediating cross-talk with other adhesion molecules, and in inhibiting apoptosis and in promoting angiogenesis.

II. CADHERIN/INTEGRIN CROSSTALK

For E-cadherin to mediate strong cell–cell adhesion it must associate with the cytoskeleton via the catenins. Disruption of this coordinated multi-molecular complex enhances the ability of malignant cells to detach from the primary tumor mass and invade surrounding tissue. The various mechanisms that lead to cadherin/catenin disruption are described elsewhere in this volume (Chapters 2 and 10). It is sufficient to note here that it is not only disturbances in cell adhesion but also disruptions in adhesion-related signaling especially those which regulate the activity of β-catenin (Hinck et al., 1994; Hoschuetzky et al., 1994; Peifer, 1995; Shibata et al., 1996), which might have profound implications for cancer evolution (Korinek et al., 1997; Morin et al., 1997; Rubinfeld et al., 1997). The feature to be considered in this chapter is the possibility that the deregulation of homotypic cell adhesion is linked somehow to changes in cell-matrix interactions. Certainly the activity of certain oncogenes have been implicated in the deregulation of both cadherins and integrins. For example, transfection of the MTSV1-7 breast line with c-erbB2 cDNA resulted in a dose-dependent loss of α2β1 expression (D'Souza et al., 1993) and it also has been shown that c-erbB2 can phosphorylate β-catenin, (Shibata et al., 1996) which could lead to a disruption of the E-cadherin/catenin complex (Behrens et al., 1993; Hoschuetzky et al., 1994). The increased expression of c-erbB2, which is associated with a poor prognosis in breast cancer (Lofts and Gullick, 1992), may be partly responsible for the loss of $\alpha2\beta1$ expression which is a common finding in carcinomas (Pignatelli et al., 1990; Hall et al., 1991; Koukolis et al., 1991; Pignatelli et al., 1991). Moreover, inhibition of the c-erbB-2/β-catenin association suppressed the invasive potential of a human gastric carcinoma cell line (Shibata et al., 1996). In addition to these data, however, accumulating evidence suggests that there is another type of coordinated regulation or "cross-talk" which may exist between cell adhesion molecules from different families such that changes in the functional capacity of one class of receptors (e.g. cadherins) has an impact on the activity of another class of receptors (e.g. integrins).

The epithelial morphotype of the rat bladder cell line, NBT-II, is maintained partly by Ca^{2+}-independent cell adhesion molecules (C-CAM) (Hunter et al., 1994) but undergoes an $\alpha2\beta1$-dependent epithelial-to-mesenchymal transition when plated on a collagen substrate (Valles et al., 1996). Weaver and colleagues (1997) noted that tumorigenic variants of the human breast line HMT-3522 could be induced to re-express functional E-cadherin/catenin complexes and lose tumorigenicity by incubation with antibodies which blocked β1-integrins. The mechanisms underlying these observations may be related to the work of Hannigan and colleagues. Overexpression of the integrin-linked kinase ($p59^{ILK}$), which was detected by application of the yeast 2-hybrid analysis using the β1-cytoplasmic tail as bait (Hannigan et al., 1996), resulted in transition of the rat intestinal epithelial line IEC-18 from a cobblestone epithelial morphology to a more stellate morphology with the cells failing to form close cell–cell adhesions. This transition was associated with loss of E-cadherin activity. The mechanism may involve the observed upregulation of members of the Lef-1/Tcf transcription family in the $p59^{ILK}$ overexpressing cell lines (Novak et al., 1998). Since Lef-1/Tcf proteins bind β-catenin prior to translocation to the nucleus (Behrens et al., 1996; Korinek et al., 1997), these transcription proteins may competitively inhibit β-catenin availability for binding to E-cadherin (see Figure 3).

There is also evidence of coordinated regulation of cadherins and integrins in normal cells. Hodivala and Watt (1994) reported that loss of integrin expression in differentiating keratinocytes correlated with an increase in E- and P-cadherin. Introduction of a dominant-negative E-cadherin mutant into such cells not only reduced E-cadherin expression but also reduced significantly the level of expression of $\alpha2$ and $\alpha3$ integrins and increased cell motility (Zhu and Watt, 1996).

Taken together, these data would seem to indicate that alterations in E-cadherin levels of expression/function can have an impact on integrin expression/function, suggesting that an, as yet, imperfectly understood co-ordinate regulation exists between these two families of adhesion molecules (see Figure 3). While such a mechanism is likely to be required during embryonic migration of cells there seems no doubt that recapitulation of this phenomenon could have a substantial impact upon the process of metastasis.

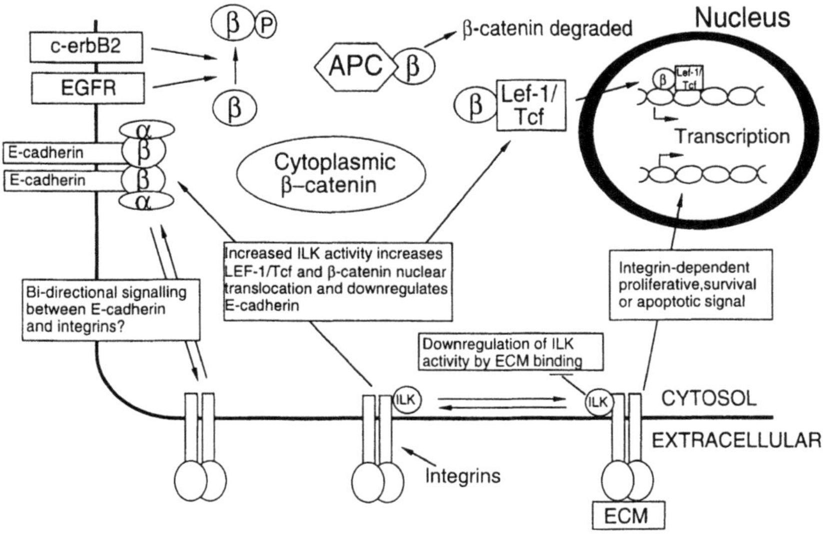

Figure 3. Potential relationships between integrins and E-cadherin in cancer cells.

III. INTEGRINS AND PERICELLULAR PROTEASE ACTIVITY

Destruction and degradation of the ECM are frequently observed phenomena associated with tumor invasion. In large part, such destructive activity appears to arise from the release and activation of proteolytic enzymes, either by the neoplastic cells themselves or by adjacent stromal cells (Basset et al., 1990; Poulsom et al., 1992; Newell et al., 1993). However, in order to modify the ECM in a regulated fashion the most efficient mechanism would be that tumor cells should sequester and activate matrix-degrading proteins at the cell surface and, ideally, such activity should be localized to sites of maximum interaction with the underlying stroma, for example at focal adhesion plaques or regions of integrin accumulation.

There is increasing evidence that the activity of two of the major protease systems involved in cancer progression—the matrix-metalloproteinase (MMP) and the urokinase-type plasminogen activator (uPA) systems—are controlled partly in such a manner by the integrins (see Figure 4).

Ligation of integrins can lead to expression of MMPs. Thus $\alpha 2\beta 1$-dependent binding has been shown to regulate the expression of collagenase type 1 (MMP1) (Langholz et al., 1995). Binding of antibodies to $\alpha v\beta 3$, or addition of vitronectin, increased MMP2 production by the A375M melanoma cell line and caused a concomitant increase in invasion through Matrigel (Seftor et al., 1992). Adhesion of rabbit synovial fibroblasts, via $\alpha 5\beta 1$, to the RGD region of fibronectin increased MMP9 mRNA and protein (Huhtala et al., 1995). Interestingly, if the $\alpha 4\beta 1$ integrin on the same cells was bound simultaneously to the CS1 domain of fibronectin the level of MMP9 production was reduced significantly (Huhtala et al., 1995). Thus the functions mediated by one integrin may be modulated by the activity of a different integrin.

The first evidence of a direct physical interaction between integrins and MMP2 was provided by Brooks and colleagues (1996) who reported that recombinant MMP2 could bind to αvβ3 expressed by the melanoma cell line CS-1. Moreover, the bound MMP2 was in the activated form. Thus, there is potential for a situation where the production and cell membrane sequestration of activated MMP2 could involve the activity of αvβ3 in malignant melanoma (Seftor et al., 1992; Brooks et al., 1996) and it perhaps is not surprising, therefore, that increasing malignant progression in this tumor type has been associated with increased expression of αvβ3 (see Marshall and Hart, 1996 for references).

Once a spatially-regulated destruction of the ECM has occurred it may have a profound effect upon the subsequent interactions with other integrins. Collagen for example is not a ligand for αvβ3, until the collagen is denatured, either by heat or the activity of a collagenase (Davis, 1992; Montgomery et al., 1994) and this denaturation also reduces the ability of the major collagen receptor, α2β1, to bind to the ligand (Davis, 1992; Montgomery et al., 1994). Therefore degradation of the constituents of the ECM may shift the repertoire of involved integrins whereby cells interact with their substrata. The implications of how such dynamic processes may affect the abilities of the various receptors to cross-talk to one another will be dealt with below.

It would be wrong to imply that integrins are the only, or even the major, membrane bound molecules capable of restricting MMP activity to the cell surface. Thus, the recently described membrane-bound protease MT1-MMP (Sato et al., 1994) serves as a major regulatory enzyme in controlling activation of MMP-2 at the cell surface (Sato et al., 1996) as does the broad-spectrum plasmin, arising from the activity of plasminogen activator (De Clerck and Imren, 1994; De Clerk and Lang, 1996).

Several reports have suggested that integrins may be involved in modulating the cellular location and activity of the urokinase type plasminogen activator (uPA) and its receptor (uPAR). Based on co-immunoprecipitation or immunocolocalization, uPAR appears to be closely associated with several different integrins including β2 integrins (Bohuslav et al., 1995; Simon et al., 1996; Wei et al., 1996), β1 and β3 integrins (Wei et al., 1996; Xue et al., 1997) as well as αvβ5 (Reinartz et al., 1995; Yebra et al., 1996). However, in non-lymphoid cells uPAR has been most commonly associated with αv integrins.

Antibody crosslinking of αvβ3 on melanoma cells, for example, doubled the expression of uPAR mRNA (Nip et al., *1995*). Some rather preliminary evidence has suggested that expression of integrins and uPAR genes is, at least in certain cell lines, co-ordinately regulated. Thus Nip and co-workers (1995) showed that antisense-mediated down-regulation of αv mRNA was associated with an equal down-regulation of uPAR mRNA. The proteins encoded for by these genes often appear to share a close topological association. Thus co-localization of uPAR to focal adhesions rich in αvβ3 has been reported for glioblastoma cells (Gladson et al., 1995), fibroblasts and rhabdomyosarcoma cells (Myohanen et al., 1993). However, co-localization of uPAR to focal adhesions is not always found (Stahl and Mueller, 1994; Yebra et al., 1996). Since vitronectin is the major binding protein for plasminogen activator inhibitor type 1 (PAI-1), which is an inhibitor of uPA activity (Cubellis et al., 1990), and because uPAR, particularly when associated with uPA, is a high affinity receptor for vitronectin (Waltz and Chapman, 1994; Wei, 1994) there is the possibility of accumulation of all these components of the uPA/uPAR/PAI-1 complex at the sites of vitronectin receptor expression which is suggestive of the possibility of co-ordinated interactions (see Figure 4). It has been reported that αvβ-dependent migration of smooth muscle cells on a vitronectin substrate was inhibited by the addition of PAI-1 (Stefansson and Lawrence, 1996). As PAI-1 binds to the N-terminal, somatomedin domain (residues 1-44) of vitronectin and these residues immediately precede the RGD cell adhesion site (residues 45-47) the authors suggested that this inhibition of cell motility was a consequence of steric hindrance (Stefansson and Lawrence, 1996).

Modulation of uPAR/uPA activity has been shown to affect tumor cell migration directly. Addition of exogenous uPA increased the migration of the metastatic melanoma cell line M24-met toward Matrigel, whereas antibody to uPA or addition of the inhibitor PAI-2 significantly reduced the ability of M24-met cells to invade Matrigel (Stahl and Mueller, 1994). The mechanisms involved may be related to those detailed in the report by Yebra and colleagues (1996) who showed that αvβ5-dependent migration of the FG pancreatic carcinoma cell line on vitronectin required activation of uPA. Previously this group had documented that activation of PKC (by EGF, insulin or TPA) was required for αvβ5-dependent migration on vitronectin (Klemke, 1994). They extended these observations by reporting that increasing PKC activity in FG cells, with PMA or TGFα, caused an increase in levels of uPA and uPAR mRNA and protein (Yebra et al., 1996). Inhibition of uPA function, by antibody-blockade of uPA binding to uPAR or by addition of the uPA inhibitor PAI-2, abrogated or nullified the αvβ5-dependent migration on vitronectin (Yebra et al., 1996). The uPA/uPAR effect was specific for αvβ5-dependent migration on vitronectin and did not affect either αvβ3-dependent migration on vitronectin nor α2β1-dependent migration on collagen. These results indicated the specificity of these interactions. Though the mechanisms of uPA/uPAR/αvβ5 co-operation were not determined one possible pathway was suggested by the work of Wei and co-workers (1996). These authors showed that uPAR modulation of integrin function may occur via a direct physical association between the molecules. Transfection of uPAR cDNA into 293 kidney cells (uPAR-293) significantly increased their binding to vitronectin but decreased their binding to fibronectin relative to the binding phenotype of the parental 293 cells. The authors proceeded to use a phage display library to identify a specific 25 amino-acid oligopeptide (peptide-

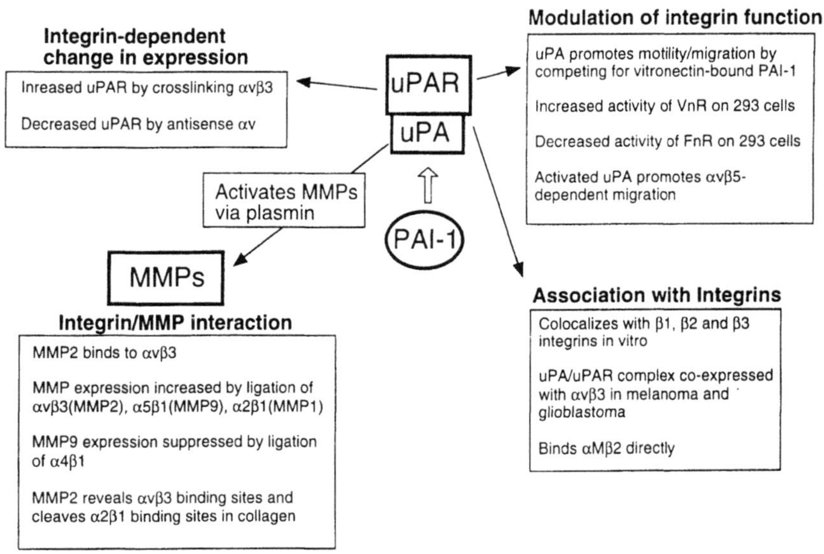

Figure 4. Interaction of integrins with the uPA/uPAR and MMP protease systems.

25) which bound to uPAR but which did not inhibit uPA binding or binding of uPAR to vitronectin. Addition of peptide-25 to uPAR-293 cells restored the parental binding phenotype (that is, strong binding to fibronectin and poor binding to vitronectin). The possibility that the uPAR in 293 cells was influencing integrins by binding directly to them was shown in further experiments where recombinant soluble uPAR (suPAR) bound to immobilized purified integrin, αMβ2 (Wei et al., 1996). The binding of suPAR to αMβ2 was enhanced by both antibody-(mAb KIM127) and cation (manganese)-dependent activation of integrins. Thus, modulation of integrin activity by uPAR appears not always to be either cell type or integrin specific and it may involve physical association with different integrins, particularly those which are in an activated state (Wei et al., 1996).

As discussed above, Yebra and colleagues (1996) showed that although αvβ5-dependent migration was uPA/uPAR dependent, αvβ3-dependent migration was independent of uPA/uPAR. Thus, tumors which down-regulate αvβ5 and upregulate αvβ3 or shift to expression of an αvβ3 heterodimer might be expected to exhibit a more migratory, and thus more invasive, phenotype in the presence of the uPA/uPAR/PAI-1 complex. Metastatic melanoma is one cancer which exhibits such a phenotype with many reports correlating an increase in αvβ3 (McGregor et al., 1989; Albelda et al., 1990; Danen et al., 1994; Si and Hersey, 1994; Danen et al., 1995; Natali et al., 1995) and a decrease in αvβ5 expression (Danen et al., 1995) with increased tumor progression.

The relationship between integrins and uPA/uPAR/PAI-1 is not restricted solely to experimental models but has been noted clinically. Thus uPA, uPAR, PAI-1, and

PAI-2 were not detected in benign naevi and early melanomas using immunohistochemistry but were found in advanced primary melanomas and metastases (de Vries et al., 1993). This pattern of expression closely resembles the expression of $\alpha v\beta 3$ in melanoma (McGregor et al., 1989; Albelda et al., 1990; Danen et al., 1994; Si and Hersey, 1994; Danen et al., 1995; Natali et al., 1995). uPA was detected both in stromal and tumor cells at the invasive front, whereas uPAR and PAI-2 were located on the tumor cell membranes (de Vries et al., 1995). Increased expression of uPA and PAI-1 has been shown to be associated with a poor prognosis in breast cancer (Bouchet et al., 1994; Duffy et al., 1994), which generally is $\alpha v\beta 3$-negative (Zutter et al., 1990; Gui et al., 1995) suggesting that if similar mechanisms exist in breast cancer alternative integrins may associate with, regulate, or be regulated by, uPAR/uPA complexes.

IV. INTEGRIN ACTIVATION STATUS DURING TUMOR CELL METASTASIS

While *de novo* expression (Giancotti and Ruoslahti, 1990; Chan et al., 1991), loss of (Danen et al., 1995), or changes in the level of expression (Plantefaber and Hynes, 1989) of integrins all have been associated with changes in tumor progression both clinically and experimentally, it also is likely that loss or gain of integrin function which occurs without change in levels of expression may also contribute to tumor development. Palecek (1997) showed that an increase in the activation state of $\alpha IIb\beta 3$ increased the motility of CHO cells on permissive concentrations of fibrinogen. Variation in the activation status of different integrins could then have profound effects upon the biological behaviour of neoplastic cells, which could relate directly to their metastatic propensity.

The $\alpha 4\beta 1$ heterodimer mediates heterophilic interactions between lymphocytes and activated endothelial cells as a prelude to the extravasation of immunocytes at areas of inflammation (Lobb and Hemler, 1994). However the cell surface expression of integrins is not necessarily indicative of functional capacity (Masumoto and Hemler, 1992) and changes in the affinity status of such molecules can have profound effects upon the binding behavior of the expressing cell types. Thus, while $\alpha 4\beta 1$ exists in a relatively non-functional state in cells of hematopoietic origin, its binding capacity can be increased dramatically following activation with a variety of treatments (Masumoto and Hemler, 1992). The interactions of lymphocytes and activated endothelium closely resemble those involved in the extravasation of disseminating tumor cells (see Figure 1) and it is highly likely that molecules common to both normal and transformed cells will regulate both processes. Indeed melanoma cells appear to bind to endothelial cells via $\alpha 4\beta 1$-VCAM-1 receptor-ligand interaction (Rice and Bevilacqua, 1989). It has been shown that human melanoma cells produce and secrete cytokines capable of activating endothelium and causing the up-regulation of a range of adhesion molecules (Burrows et al., 1991) so that the activation status of the $\alpha4\beta1$ heterodimer expressed by these tumor cells is a matter of some interest. It has been proposed that the ligand binding of integrins in tumor cells may be constitutively high with little capacity for further up-regulation (Mattila, 1992; Danen et al., 1994) and this certainly appears to be the case with $\alpha4\beta1$ expressed on malignant melanoma (Saini et al., 1997). Pre-treatment of mice with cytokines, such as TNFα (Okahara et al., 1994) or with IL-1 (Garofalo et al., 1995), which upregulate VCAM-1 expression on endothelial cells has resulted in an increase in lung tumor burden subsequent to systemic delivery of melanoma cells. Antibody-blockade of either the $\alpha4\beta1$ on the melanoma cells (Okahara et al., 1994; Garofalo et al., 1995) or the upregulated endothelial VCAM-1 (Okahara et al., 1994) inhibited this cytokine-dependent enhancement of metastasis. This positive correlation between $\alpha4\beta1$/VCAM-1 interactions and increasing malignancy however has not been a universal observation since it also has been reported that, rather than assisting metastatic formation, de novo expression of $\alpha4\beta1$ by melanoma cells inhibited the formation of spontaneous metastases from a subcutaneous tumor (Qian et al., 1994).

V. INTEGRINS, GROWTH FACTORS, AND APOPTOSIS

Having achieved a secondary location, it is necessary for disseminating cancer cells to grow at distant sites (see Figure 1) and again the integrins play a pivotal role in regulating this process. Schwartz and Ingber (1994) have suggested that normal cell homeostasis (proliferation, differentiation, apoptosis) is a consequence of combinations of signals generated by both integrin-ligand interactions and of soluble growth factors thus suggesting that regulated control may be lost at specific points. Mutations or disruptions prior to the convergence points of these two pathways would lead to either anchorage independent growth or growth factor independence, whereas lesions at or downstream of the convergence point should lead to both anchorage and growth factor independence (Schwartz and Ingber, 1994; Meredith and Schwartz, 1997). If this is so then perhaps metastatic cancer cells have lesions at or downstream of the convergence point since they are likely to be growth-factor and anchorage-independent. Moreover, since programmed cell death (apoptosis) may be considered a safety mechanism whereby cells which are in an inappropriate ECM environment undergo self-destruction (Raff, 1992), it should be expected that at least some part of the metastatic phenotype should be a failure to engage the apoptotic pathways. It also is possible, however, that metastatic cells either fail to respond to integrin-mediated apoptosis or, possibly by expressing constitutively active integrins, continually generate anti-apoptotic signals. Thus, expression of constitutively activated FAK in MDCK cells blocked their entrance into apoptosis when they where placed into suspension and even induced tumorigenicity (Frisch et al., 1996). In addition, increased migration and increased metastatic activity have been correlated with FAK overexpression (Weiner et al., 1993; Owens et al., 1995).

Different integrins appear to mediate different survival signals (see Meredith and Schwartz, 1997 for references). Thus in CHO cells expressing either of the fibronectin receptors αvβ1 or α5β1, only the α5β1 expressors survived when plated onto a fibronectin substrate in a serum-free environment; the αvβ1 expressors treated in a comparable fashion underwent apoptosis (Zhang et al., 1995). Such variation could well explain why different patterns of integrins are expressed by different malignant tumors. The melanoma cell line M21, for example, was protected from apoptosis in serum-free conditions in collagen gels by expression of αvβ3, which was presumed to bind to MMP2-degraded collagen (Montgomery et al., 1994). This might explain then why the upregulation of αvβ3 in melanoma cells is associated with enhanced survival in a collagen-rich dermal environment (Marshall and Hart, 1996).

VI. INTEGRINS AND ANGIOGENESIS

In order for tumors to expand beyond approximately 2mm in three-dimensional growth, a size at which they are undetectable and unlikely to metastasise, there is a requirement for angiogenesis or neovascularisation (Folkman, 1989). This provision of new capillary blood vessels is important not only for providing the growing mass with nutrients but, because of the "leaky" nature of the newly developing tumor vasculature, also for providing cells with ready access to the vasculature. It probably is for this reason that the degree of angiogenesis, as assessed by the simple expedient of counting microvessels, has been correlated with metastatic disease in a wide range of malignancies (Weidner et al., 1992, 1993). New angiogenic vessels arise from preexisting capillaries or venules, never from arteries, arterioles or veins, and they represent the ingrowth of columns of aligned endothelial cells (Mahadevan and Hart, 1990). Adjacent columns of endothelial cells contact to form loops which, subsequent to the development of a patent lumen, permit blood flow to begin. Recently, it has been shown that endothelial cells in these developing blood vessels express αvβ3 whereas this heterodimer is low or undetectable on the cells of quiescent, established vessels (Brooks et al., 1994). Blockade of αvβ3 with specific antibodies or peptides abrogated the angiogenesis generated in the chick chorioallantoic membrane (CAM) model by pieces of human tumor or filter-paper soaked with angiogenic factors such as bFGF (Brooks et al., 1994a, 1994b).

Pieces of human foreskin grafted onto SCID mice have formed anastamoses between the mouse and human blood vessels thus allowing survival of the xenograft. An inoculum of αvβ3-negative human breast cells into the graft tissue was inhibited from growing by administration of an αvβ3-blocking antibody (Brooks et al., 1995). Since the antibody was specific only for human cells, and as the human tumor cell line was αvβ3-negative, the authors concluded that the inhibited tumor growth occurred by the specific blocking of angiogenesis (Brooks et al., 1995).

The anti-endothelial cell activity of these antibodies, or indeed the synthetic peptides used to inhibit the angiogenic process, was related to their ability to induce apoptosis in the endothelial cells (Brooks et al., 1994b). Such angiogenesis may not be exclusively $\alpha v\beta 3$-dependent. Thus, whereas bFGF or TNFα treatment stimulated $\alpha v\beta 3$-dependent angiogenesis, treatment with PMA, VEGF or TGFα induced an $\alpha v\beta 5$-dependent angiogenic response (Friedlander et al., 1995). Presumably the nature of the angiogenic peptides released by the cancer cells can have a profound effect upon the nature of the integrin involvement in endothelial cell motility and this might have significant implications for anti-angiogenic therapies based upon the utilization of $\alpha v\beta 3$-blocking agents.

VII. THE SELECTINS

Selectins, a family of Ca^{2+}-dependent carbohydrate-binding proteins, mediate the initial attachment of leukocytes to the endothelium on the blood vessel walls during the rolling step of leukocyte extravasation in inflammation (Springer, 1990). Selectins recognise fucosylated carbohydrate ligands, especially structures containing Sialyl-Lewis^X (sLe^X) and Sialyl-Lewis^a(sLe^a) (Vestweber, 1993). These selectin/carbohydrate interactions permit leukocytes to roll along the vascular endothelium in the direction of blood flow as a prelude to integrin-mediated adhesion. The selectins, both their form and function, are covered in greater detail elsewhere in this volume, but the demonstration that carbohydrate ligands—such as sLe^a, which is a ligand for E-selectin, are expressed on some tumor cells but are not usually found on leukocytes—has underscored the belief that this interaction between E-selectin and sLe^a is more relevant to tumor metastases than to leukocyte trafficking (Mannori et al., 1995). For example, E-selectin has been found to regulate adhesion of human colon cancer cells to endothelium by binding to sLe^a and sLe^X carbohydrate ligands (Dejana et al., 1992; Mannori et al., 1995).

While it is tempting to dwell on the parallels between tumor cell attachment during metastasis and leukocyte attachment as a prelude to extravasation in inflammation such comparisons may be misleading. It is appreciated that tumor cells can secrete cytokines, which activate endothelium and thus can cause an increase in the expression of adhesion molecules on these cells (Burrows et al., 1991), but the mode of interaction between ligand and receptor may be very different between the two processes. Thus, leukocytes travelling under flow conditions need to initiate tethering and rolling as a prelude to integrin-mediated consolidation of attachment (Springer, 1990). In contrast, neoplastic cells adhere primarily to the endothelial cells of small capillaries (Fidler et al., 1978) where they often have to force their way through vessels of a diameter which is smaller than the disseminating cancer cell (Weiss, 1992). There is therefore no need for such neoplastic cells to mimic the attachment behavior of the substantially smaller leukocytes.

VIII. IMMUNOGLOBULIN SUPERFAMILY ADHESION MOLECULES

Though not all immunoglobulin-superfamily adhesion molecules mediate cell-cell interactions many which belong to this family do function as such adhesion receptors, for example, several of the intercellular adhesion molecules- (ICAM-1, ICAM-2, ICAM-3) and vascular cell adhesion molecule-1 (VCAM-1).

Cytolytic conjugates between cytotoxic T lymphocytes and tumor cells, presumed to be such an important means of regulating tumor spread, are stabilized by interactions between the LFA-1 integrin on the lymphocytes and the ICAM-1 ligand expressed by the target cell (Anichini et al., 1990). Therefore. it might be expected that enhanced expression of ICAM-1 would be associated with decreased metastatic proclivity. Paradoxically, this does not appear to be the case and, in malignant melanoma, ICAM-1 expression has been correlated with increasing malignancy (Johnson et al., 1989). It might be that the formation of larger intravascular emboli as a consequence of these interactions results in more efficient entrapment and thus greater disseminated tumor burden.

Elevated levels of soluble ICAM-1, which still retains the binding site for LFA-1 in its extracellular domain, have been found in the serum of melanoma and ovarian cancer patients (Gearing et al., 1992; Giavazzi et al., 1992). It has been suggested that this active form of ICAM-1 can inhibit binding sites on cytotoxic T-cell and natural killer cells and that, in its presence, tumor cells might better be able to escape immune destruction (Johnson, 1991). Interestingly, in this respect it has been shown that P-selectin expressed on platelets regulates their binding to tumor cells and that this phenomenon can prevent tumor cell destruction by protecting from immune mechanisms owing to the formation of multicellular aggregates (Honn and Tang, 1992).

IX. SUMMARY

In 1986, when the term integrin was coined, it was predicted that integrins probably played a major role in modulating the tumorigenic and metastatic phenotype of cells (Tamkun et al., 1986; Hynes, 1987) since they mediate cell matrix interactions. In the subsequent 10 years, the complexity of intracellular signals and accessory molecules which participate in integrin-dependent promotion of the tumorigenic/metastatic phenotype (that is, proliferation, migration, survival) has come to be appreciated. Coupled with this greater understanding of the role of integrins in tumor spread has been an appreciation that other adhesion molecules also play a significant part in determining cancer metastasis. Recently, it has appeared likely that the involvement of these different molecules is not due to entirely independent events but represents the outcome of cross-talk between the various factions. Understanding how these interactions are regulated should throw considerable light on the mechanistic basis of cancer spread.

ACKNOWLEDGMENT

We would like to thank Dr Shoukat Dedhar for supplying unpublished data.

REFERENCES

Albelda, S. M., Mette, S. A., Elder, D. E., Stewart, R., Damjanovch, L., Herlyn, M., & Buck, C. A. (1990). Integrin distribution in malignant melanoma: Association of the β3 subunit with tumor progression. Cancer Res. 50, 6757-6764.

Anichini, A., Mortarini, R., Supino, R., & Parmiani, G. (1990). Human melanoma cells with high susceptibility to cell-mediated lysis can be identified on the basis of ICAM-1 phenotype, VLA profile and invasive ability. Int. J. Cancer 46, 508-515.

Basset, P., Bellocq, J., Wolf, C., Stoll, I., Hutin, P., Limacher, J., Podhajcer, O., Chenard, M., Rio, M., & Chambon, P. (1990). A novel metalloproteinase gene specifically expressed in stromal cells of breast carcinomas. Nature 348, 699-704.

Behrens, J., Vakaet, L., Friis, R., Winterhager, E., Van Roy, F., Mareel, M., & Birchmeir, W. (1993). Loss of epithelial differentiation and gain of invasiveness correlates with tyrosine phosphorylation of the E-cadherin/beta-catenin complex in cells transformed with a temperature-sensitive v-SRC. J. Cell Biol. 120, 757-766.

Behrens, J., von Kries, J., Kuhl, M., Bruhn, L., Wedlich, D., Grosschedl, R., & Birchmeier, W. (1996). Functional interaction of β-catenin with the transcription factor LEF-1. Nature 382, 638-642.

Birchmeier, W., Hulsken, J., & Behrens, J. (1995). Adherens junction proteins in tumour progression in Cancer Surveys: Cell Adhesion and Cancer, Hart, I., & Hogg, N. (eds.), Cold Spring Harbor Laboratory Press 24, 129-140.

Bohuslav, J., Horejsi, V., Hansmann, C., Stockl, J., Weidle, U., Majdic, O., Bartke, I., Knapp, W., & Stockinger, H. (1995). Urokinase-plasminogen activator receptor, beta-2-integrins and src-kinases within a single receptor complex of human monocytes. J. Exp. Med. 181, 1381-1390.

Bouchet, C., Spyratos, F., Martin, P., Hacene, K., Gentile, A., & Oglobine, J. (1994). Prognostic value of urokinase-type plasminogen-activator (upa) and plasminogen -activator inhibitors pai-1 and pai-2 in breast carcinoma. Br. J. Cancer 69, 398-405.

Brooks, P., Stromblad, S., Klemke, R., Visscher, D., Sarkar, F., & Cheresh, D. (1995). Antiintegrin αvβ3 blocks human breast cancer growth and angiogenesis in human skin. J. Clin. Invest. 96, 1815-1822.

Brooks, P., Stromblad, S., Sanders, L., von Schlasa, T., Aimes, R., Stetler-Stevenson, W., Quigley, J., & Cheresh, D. (1996). Localization of matrix-metalloproteinase MMP-2 to the surface of invasive cells by interaction with integrin αvβ5. Cell 85, 683-693.

Brooks, P. C. Clark, R. A. F., & Cheresh, D. A. (1994a). Requirement of vascular integrin αvβ3 for angiogenesis. Science 264, 569-571.

Brooks, P. C., Montgomery, A. M. P., Rosenfeld, M., Reisfeld, R. A., Hu, T., Klier, G., & Cheresh, D. A. (1994b). Integrin αvβ3 antagonists promote tumour regression by inducing apoptosis of angiogenic blood vessels. Cell 79, 1157-1164.

Burrows, F., Haskard, D., Hart, I., Marshall, J., Selkirk, S., Poole, S., & Thorpe, P. (1991). Influence of tumor-derived interleukin 1 on melanoma-endothelial cell interactions in vitro. Cancer Res. 51, 4768-4775.

Cepek, K. L., Shaw, S. K., Parker, C., Russel, G. J., & Brenner, M. (1994). Adhesion between epithelial cells and T lymphocytes mediated by E-cadherin and the αEβ7 integrin. Nature 372, 190-193.

Chan, B. M. C., Matsuura, N., Takada, Y., Zetter, B. R., & Hemler, M. E. (1991). In vitro and in vivo consequences of VLA-2 expression on rhabdomyosarcoma cells. Science 251, 1600-1602.

Cubellis, M. V., Wun, T.-C., & Blasi, F. (1990). Receptor mediated internalization and degradation of urokinase is caused by its specific inhibitor PAI-1. EMBO J. 9, 1079-1085.

D'Souza, B., Berdichevsky, F., Kyprianou, N., & Taylor-Papadimitriou, J. (1993). Collagen-induced morphogenesis and expression of the alpha 2-integrin subunit is inhibited in c-ERB2-transfected human mammary epithelial cells. Oncogene 8, 1797-1806.

Danen, E., Devries, T., Morandini, R., Ghanem, G., Ruiter, D., & Vanuijen, G. (1996). E-cadherin expression in human melanoma. Melanoma Res. 6, 127-131.

Danen, E. H. J., Jansen, K. F. J., Van Kraats, A. A., Cornelissen, M. H. A., Ruiter, D. J., & Muijen, G. N. P. (1995). αv-integrins in human melanoma: Gain of αvβ3 and loss of αvβ5 are related to tumour progression in situ but not to metastatic capacity of cell lines in nude mice. Int. J. Cancer 61, 491-496.

Danen, E. H. J., ten Berge, P. J. M., van Muijen, G. N. P., van Thofgrootenboer, A. E., Brocker, E. B., & Ruiter, D. J. (1994). Emergence of α5β1 fibronectin receptor and αvβ3 vitronectin-receptor expression in melanocytic tumour progression. Histopath. 24, 249-256.

Davis, G. E. (1992). Affinity of integrins for damaged extracellular matrix: αvβ3 binds to denatured collagen type I through RGD sites. Biochem. Biophys. Res. Commun. 182, 1025-1031.

De Clerck, Y. A. & Imren, S. (1994). Protease inhibitors: Role and potential therapeutic use in human cancer. Eur. J. Cancer 30 2170-2180.

De Clerk, Y. & Lang, W. (1996). Cooperation between matrix metalloproteinases and the plasminogen activator-plasmin system in tumour progression. Enzyme Protein 49, 72-84.

de Vries, T. J., Mooy, C. M., Van _Balken, M. R., Luyten, G. B. M., Quax, P. H. A., Verspaget, H. W., Weidle, U. H., Ruiter, D. J., & van Muijen, G. N. P. (1995). Components of the plasminogen activation system in uveal melanoma: a clinicopathological study. J. Path. 175, 59-67.

de Vries, T. J., Quax, P. H. A., Denijn, M., Verrijp, K. N., Verheijen, J. H., Verspaget, H. W., Weidle, U. H., Ruiter, D. J., & van Muijen, G. N. P. (1993). Plasminogen activators, their inhibitors and urokinase receptor emerge in late stages of melanocytic tumor progression. Am. J. Pathol. 144, 70-81.

Dejana, E., Martin-Padura, I., Lauri, D., Bernasconi, S., Bani, M., Garofalo, A., Giavazzi, R., Magnani, J., Mantovani, A., & Menard, S. (1992). Endothelial leukocyte adhesion molecule-1-dependent adhesion of colon carcinoma cellls to vascular endothelium is inhibited by antibody to Lewis fucosylated type 1 carbohydrate chain. Lab. Invest. 66, 324-330.

Duffy, M., Reilly, D., McDermott, E., Ohiggins, N., Fennelly, J., & Andreasen, P. (1994). Urokinase plasminogen-activator (uPA) as a prognostic marker in different subgroups of patients with breast-cancer. Cancer 74, 2276-2280.

Fidler, I., Gersten, D., & Hart, I. (1978). The biology of cancer invasion and metastasis. Adv. Cancer Res. 28, 149-250.

Folkman, J., Watson, K., Ingber, D., & Hanahan, D. (1989). Induction of angiogenesis during the transition from hyperplasia to neoplasia. Nature 339, 58-61.

Friedlander, M., Brooks, P. C., Shaffer, R. W., Kincaid, C. M., Varner, J. A., & Cheresh, D. A. (1995). Definition of two angiogenic pathways by distinct αv integrins. Science 270, 1500-1502.

Frisch, S., Vuori, K., Ruoslahti, E., & Chan-Hui, P. (1996). Control of adhesion-dependent cell survival by focal adhesion kinase. J. Cell Biol. 134, 793-799.

Garofalo, A., Chirivi, R., Foglieni, C., Pigott, R., Mortarini, R., Martin-Padura, I., Anichini, A., Gearing, A., Sanchez-Madrid, F., Dejana, E., & Giavazzi, R. (1995). Involvement of the very late antigen 4 on melanoma in interleukin 1-augmented experimental metastasis. Cancer Res. 55, 414-419.

Gasic, G. J., Gasic, T. B., Galanti, N., Johnson, T., & Murphy, S. (1973). Platelet tumor interactions in mice. The role of platelets in the spread of malignant disease. Int. J. Cancer 11, 704-718.

Gearing, A., Hemingway, I., Piggot, R., Hughes, J., Ress, A., & Cashman, A. (1992). Soluble forms of vascular adhesion molecules E-selectin, ICAM-1 and VCAM-1: pathological significance. Ann. N.Y. Acad. Sci. 667, 324-331.

Giancotti, F. G. & Mainiero, F. (1994). Integrin-mediated adhesion and signaling in tumourigenesis. Biochim. Biophys. Acta 1198, 47-64.

Giancotti, F. G. & Ruoslahti, E. (1990). Elevated levels of the α5β1 fibronectin receptor suppress the transformed phenotype of Chinese hamster ovary cells. Cell 60, 849-859.

Giavazzi, R., Chirivi, R., Garofalo, A., Rambaldi, A., Hemingway, I., Pigott, R., & Gearing, A. (1992). Soluble intercellular adhesion molecule 1 is released by human melanoma cells and is associated with tumour growth in nude mice. Cancer Res. 52, 2628-2630.

Gladson, C. L., Pijuan-Thompson, V., Olman, M. A., Gillespie, G. Y., & Yacoub, I. Z. (1995). Up-regulation of urokinase and urokinase-receptor genes in malignant astrocytoma. Am. J. Path. 146, 1150-1160.

Gui, G., Wells, C., Browne, P., Yeomans, P., Jordan, S., Puddefoot, J., Vinson, G., & Carpenter, R. (1995). Integrin expression in primary breast cancer and its relation to axillary nodal status. Surgery 117, 102-108.

Hall, P., Coates, P., Lemoine, N., & Horton, M. (1991). Characterisation of integrin chains in normal and neoplastic human pancreas. J. Pathol. 165, 33-41.

Hannigan, G., Leung-Hagesteijn, C., Fitz-Gibbon, L., Coppolino, M., Radeva, G., Filmus, J., Bell, J., & Dedhar, S. (1996). Regulation of cell adhesion and anchorage-dependent growth by a new β1-integrin-linked protein kinase. Nature 379, 91-96.

Hinck, L., Nathke, I.S., Papkoff, J., & Nelson, W.J. (1994). β-catenin: a common target for the regulation of cell adhesion by Wnt-1 and Src signaling pathways. Trends Biol. Sci. 19, 538-542.

Hodivala, K. J. & Watt, F. M. (1994). Evidence that cadherins play a role in the downregulation of integrin expression that occurs during keratinocyte terminal differentiation. J. Cell Biol. 124, 589-600.

Honn, K. & Tang, D. (1992). Adhesion molecules and tumour cell interaction with endothelium and subendothelial matrix. Cancer Metastasis Rev. 11, 353-375.

Hoschuetzky, H., Aberle, H., & Kemler, R. (1994). β-catenin mediates the interaction of the cadherin-catenin complex with epidermal growth factor receptor. J. Cell Biol. 127. 1375-1380.

Huhtala, P., Humphries, M. J., McCarthy, J. B., Tremble, P. M., Werb, Z., & Damsky, C. H. (1995). Cooperative signaling by α5β1 and α4β1 integrins regulates metalloproteinase gene expression in fibroblasts adhering to fibronectin. J. Cell Biol. 129, 867-879.

Hunter, I., Lindh, M., & Obrink, B. (1994). Differential regulation of c-cam isoforms in epithelial-cells. J. Cell Sci. 107, 1205-1216.

Hynes, R. O. (1987). Integrins: A family of cell surface receptors. Cell 48, 549-554.

Johnson, J. P. (1991). The role of ICAM-1 in tumor development. Chem. Immunol. 50, 143-63.

Johnson, J. P., Stade, B. G., Holzmann, B., Schwable, W., & Rietmuller, G. (1989). De novo expression of cell adhesion molecule ICAM-1 in melanoma and increased risk of metastasis. Proc. Natl. Acad. Sci. USA 86, 641-644.

Juliano, R. L. & Varner, J. A. (1993). Adhesion molecules in cancer: the role of integrins. Curr. Opin. Cell Biol. 5, 812-818.

Klemke, R. L., Yebra, M., Bayna, E.M., & Cheresh, D. A. (1994). Receptor tyrosine kinase signaling required for integrin αvβ5-directed cell motility but not adhesion on vitronectin. J. Cell Biol. 127, 859-866.

Korinek, V., Barker, N., Morin, P., de Weger, R., Kinzler, K., Voglestein, B., & Clevers, H. (1997). Constitutive transcriptional activation by a β-catenin-Tcf complex in APC-/- colon carcinoma. Science 275, 1784-1787.

Koukolis, G., Virtanen, I., Korhonen, M., Laitinen, L., Quarant, V., & Gould, V. (1991). Immunohistochemical localization of integrins in the normal hyperplastic and neoplastic breast. Am. J. Pathol. 139, 787-799.

Langholz, O., Rockel, D., Mauch, C., Kozlowska, E., Bank, I., Krieg, T., & Eckes, B. (1995). Collagen and collagenase gene expression in three-dimensional collagen lattices are differentially regulated by α1β1 and α2β1 integrins. J. Cell Biol. 131, 1903-1915.

Lobb, R. & Hemler, M. (1994). The pathophysiologic role of α4 integrins in vivo. J. Clin. Invest. 94, 1722-1728.

Lofts, F. & Gullick, W. (1992). c-erbB2 amplification and overexpression in human tumours. Cancer Treat. Res. 61, 161-179.

Mahadevan, V. & Hart, I. (1990). Angiogenesis and metastasis. Annals Oncology 3, 97-103.

Mannori, G., Crottet, P., Cecconi, O., Hanasaki, K., Aruffo, A., Nelson, R., Varki, A., & Bevilacqua, M. (1995). Differential colon cancer cell adhesion to E-, P- and L-selectin: role of mucin-type glycoproteins. Cancer Res. 55, 4425-4431.

Marshall, J. & Hart, I. (1996). The role of αv-integrins in tumour progression and metastasis. Seminars Cancer Biol. 7, 129-138.

Masumoto, A. & Hemler, M. E. (1992). Multiple activation states of VLA-4: mechanistic differences between adhesion to CS1/fibronectin and to VCAM-1. J. Biol. Chem. 268, 228-234.

Mattila, P., Majuri, M-L., & Renkonen, R. (1992). VLA-4 integrin on sarcoma cell lines recognises endothelial VCAM-1 differential regulation of the VLA-4 avidity on various sarcoma cell lines. Int. J. Cancer 52, 98-923.

McGregor, B. C., McGregor, J. L., Weiss, L. M., Wood, G. S., Hu, C.-H., Boukerche, H., & Warnke, R. A. (1989). Presence of cytoadhesins (IIb-IIIa-like glycoproteins) on human metastatic melanomas but not on benign melanocytes. Am. J. Clin. Pathol. 92, 495-499.

Meredith Jr., J. & Schwartz, M. (1997). Integrins, adhesion and apoptosis. Trends Cell Biol. 7, 146-150.

Montgomery, A. M. P., Reisfeld, R. A., & Cheresh, D. A. (1994). Integrin αvβ3 rescues melanoma cells from apoptosis in three-dimensional dermal collagen. Proc. Natl. Acad. Sci 91, 8856-8860.

Morin, P., Sparks, A., Korinek, V., Barker, N., Clevers, H., Voglestein, B., & Kinzler, K. (1997). Activation of β-catenin-Tcf signaling in colon cancer by mutations in β-catenin or APC. Science 275, 1787-1790.

Mould, P., Askari, J., Craig, S., Garrat, A., Clements, J., & Humphries, M. (1994). Integrin α4β1-mediated melanoma cell adhesion and migration on vascular cell adhesion molecule-1 (VCAM-1) and the alternatively spliced IIICS region of fibronectin. J. Biol. Chem. 269, 27224-27230.

Myohanen, H. T., Stephens, R. W., Hedman, K., Tapiovaara, H., Ronne, E., Hoyerhansen, G., Dano, K., & Vaheri, A. (1993). Distribution and lateral mobility of the urokinase-receptor complex at the cell-surface. J. Histochem. Cytochem. 41, 1291-1301.

Natali, P. G., Nicotra, M. R., Difillipo, F., & Bigotti, A. (1995). Expression of fibronectin, fibronectin isoforms and integrin receptors in melanocytic lesions. Br. J. Cancer 71, 12431247.

Newell, A., Witty, J., Rodgers, W., & Matrisian, L. (1993). Expression and localization of matrix-degrading metalloproteinases during colorectal tumorigenesis. Molecular Carcinogenesis 10, 199-206.

Nip, J., Rabbani, S. A., Shibata, H. R., & Brodt, P. (1995). Coordinated expression of the vitronectin receptor and the urokinase-type plasminogen-activator receptor in metastatic melanoma-cells. J. Clin. Invest. 95, 2096-2103.

Novak, A., Hsu, S.-C., Leung-Hagesteijn, C., Radeva, G., Papkoff, J., Montesano, R., Roskelley, C., Grosschedl, R., & Dedhar, S. (1998). Cell adhesion and the integrin-linked kinase regulate the LEF-1 and β-catenin signaling pathways. Proc. Natl. Acad. Sci., 95, 4374-4379.

Okahara, H., Yagita, H., Miyake, K., & Okumura, K. (1994). Involvement of very late activation antigen 4 (VLA4) and vascular cell adhesion molecule-1 (VCAM-1) in tumor necrosis factor α enhancement of experimental metastasis. Cancer Res. 54, 3233-3236.

Owens, V., Xu, L., Craven, J., Dent, A., Weiner, M., Kornberg, L., Liu, T., & Cance, G. (1995). Overexpression of the focal adhesion kinase (p125FAK) in invasive human tumours Cancer Res. 55, 2752-2755.

Palecek, S., Loftus, J., Ginsberg, M., Lauffenburger, D., & Horwitz, A. (1997). Integrin-ligand binding properties govern cell migration speed through cell-substratum adhesiveness. Nature 385, 537-540.

Peifer, M. (1995). Cell adhesion and signal transduction: the armadillo connection. Trends Cell Biol. 5, 224-229.

Pignatelli, M., Hanby, A., & Stamp, G. (1991). Low expression of beta 1, alpha 2 and alpha 3 subunits of VLA integrins in malignant mammary tumours. J. Pathol. 165, 25-32.

Pignatelli, M., Smith, M. E. F., & Bodmer, W. F. (1990). Low expression of collagen receptors in moderate and poorly differentiated colorectal adenocarcinomas. Br. J. Cancer 61, 636-638.

Pignatelli, M. & Stamp, G. (1995). Integrins in tumour development and spread in Cell Adhesion and Cancer. I. Hart and N. Hogg (eds.), Cold Spring Harbour Laboratory Press. 24, 113-127.

Plantefaber, L. C. & Hynes, R. O. (1989). Changes in integrin receptors on oncogenically transformed cells. Cell 56, 281-290.

Poulsom, R., Pignatelli, M., Stetler-Stevenson, W., Liotta, L., Wright, P., Jeffrey, R., Longcroft, J., Rogers, L., & Stamp, G. (1992). Stromal expression of 72 kda type IV collagenase(MMP-2) and TIMP-2 mRNAs in colorectal neoplasia. Am. J. Pathol. 141, 389-396.

Qian, F., Vaux, D., & Weissman, I. (1994). Expression of the integrin $\alpha 4\beta 1$ can inhibit the invasive stage of metastasis formation. Cell 77, 335-347.

Raff, M.C. (1992). Social controls on cell survival and cell death. Nature 356, 397-400.

Reinartz, J., Schafer, B., Batrla, R., Klein, C., & Kramer, M. (1995). Plasmin abrogates $\alpha v\beta 5$-mediated adhesion of a human keratinocyte cell-line (HaCat) to vitronectin. Exp. Cell Res. 220, 274-282.

Rice, G. E. & Bevilacqua, M. P. (1989). An inducible cell surface glycoprotein mediates melanoma adhesion. Science 246, 1303-1306.

Rubinfeld, B., Robbins, P., El-Gamil, M., Albert, I., Porfiri, E., & Polakis, P. (1997). Stabilisation of β-catenin by genetic defects in melanoma cell lines. Science 275, 1790-1792.

Saini, A., Seller, Z., Davies, D., Marshall, J., & Hart, I. (1997). Activation status and function of the VLA-4 ($\alpha 4\beta 1$) integrin expressed on human melanoma cell lines. Int. J. Cancer 73, 264-270.

Sato, H., Takino, T., Kinoshita, T., Imai, K., Okada, Y., & Stetler-Stevenson, W. (1996). Cell surface binding and activation of gelatinase A induced by expression of membrane-type-1-matrix metalloproteinase (MT1-MMP). FEBS Lett. 385, 238-240.

Sato, H., Takino, T., Okada, Y., Cao, J., Shinagawa, A., Yamamoto, E., & Seiki, M. (1994). A matrix metalloproteinase expressed on the surface of invasive cells. Nature 370, 61-65.

Schwartz, M. A. & Ingber, D. E. (1994). Integrating with integrins. Mol. Biol. Cell 5, 389-393.

Seftor, R. E. B., Seftor, E. A., Gehlsen, K. R., Stetler-Stevenson, W. G., Brown, P. D., Ruoslahti, E., & Hendrix, M. J. C. (1992). Role of the $\alpha v\beta 3$ integrin in human melanoma cell invasion. Proc. Natl. Acad. Sci. USA 89, 1557-1561.

Seline, P., Norris, D., Horikawa, T., Fujita, M., Middleton, M., & Morelli, J. (1996). Expression of e-cadherin and p-cadherin by melanoma cells decreases in progressive melanomas following ultraviolet-radiation. J. Investig. Dermat. 106, 1324-1324.

Shibata, T., Ochiai, A., Kanai, Y., Akimoto, S., Gotoh, M., Yasui, N., Machinami, R., & Hirohashi, S. (1996). Dominant negative inhibition of the association between β-catenin and c-erbB-2 by N-terminally deleted β-catenin suppresses the invasion and metastasis of cancer cells. Oncogene 13, 883-889.

Si, Z. & Hersey, P. (1994). Immunohistological examination of the relationship between metastatic potential and expression of adhesion molecules and 'selectins' on melanoma cells. Pathology 26, 6-15.

Simon, D., Rao, N., Xu, H., Wei, Y., Majdic, O., Ronne, E., Kobzik, L., & Chapman, H. (1996). Mac-1(CD11b/CD18) and urokinase receptor (CD87) form a functional unit on monocytic cells. Blood 88, 3185-3194.

Springer, T. A. (1990). Adhesion receptors of the immune system. Nature 346, 425-434.

Stahl, A. & Mueller, B. M. (1994). Binding of urokinase to its receptor promotes migration and invasion of human-melanoma cells in vitro. Cancer Res. 54, 3066-3071.

Stefansson, S., & Lawrence, D. (1996). The serpin PAI-1 inhibits cell migration by blocking integrin $\alpha v\beta 3$ to vitronectin. Nature 383, 441-443.

Sugarbaker, E., Weingrad, D., & Roseman, J. (1982). Observations on cancer metastasis in man. In: *Tumour Invasion and Metastasis.* (Liotta, L. and Hart, I., Eds.) pp. 427-465. Martins Nijhoff, The Hague.

Tamkun, J. W., DeSimone, D. W., Fonda, D., Patel, R. S., Buck, C., Horwitz, A. F., and Hynes, R. O. (1986). Structure of integrin, a glycoprotein involved in the transmembrane linkage between fibronectin and actin. Cell 46, 271-282.

Valles, A., Boyer, N., Tarone, G. and Thiery, J. (1996). Alpha-2-beta-1 integrin is required for collagen and FGF-1–induced cell dispersion in a rat bladder-carcinoma cell-line. Cell Adhes. Comm. 4, 187-199.

Vestweber, D. (1993). The selectins and their ligands. Curr. Top. Microbiol. Immunol. 184, 65-75.

Waltz, D. A. & Chapman, H. A. (1994). Reversible cellular adhesion to vitronectin linked to urokinase receptor occupancy. J. Biol. Chem. 269, 14746-14750.

Weaver, V., Petersen, O., Laarabell, C., Briand, P., Damsky, C., & Bissell, M. (1997). Reversion of the malignant phenotype of human breast cells in three-dimensional culture and in vivo by integrin blocking antibodies. J. Cell Biol. 137, 231-245.

Wei, Y., Waltz, D. A., Rao, N., Drummond, R. J., Rosenberg, S., & Chapman, H.A. (1994). Identification of the urokinase receptor as an adhesion receptor for vitronectin. J. Biol. Chem. 269, 32380-32388.

Wei, Y., Lukashev, M., Simon, D., Bodary, S., Rosenberg, S., Doyle, M., & Chapman, H. (1996). Regulation of integrin function by the urokinase receptor. Science 273, 1551-1555.

Weidner, N., Carroll, P. R., Flax, J., Blumenfeld, W., & Folkman, J. (1993). Tumor angiogenesis correlates with metastasis in invasive prostate carcinoma. Am J Pathol 143, 401-9.

Weidner, N., Folkman, J., Pozza, F., Bevilacqua, P., Allred, E. N., Moore, D. H., Meli, S., & Gasparini, G. (1992). Tumor angiogenesis: a new significant and independent prognostic indicator in early-stage breast carcinoma. J. Natl. Cancer Inst. 84, 1875-87.

Weiner, T. M., Liu, E. T., & Craven, R. J. (1993). Expression of focal adhesion kinase gene and invasive cancer. The Lancet 342, 1024-1026.

Weiss, L. (1992). Biomechanical interactions of cancer cells with the microvasculature during hematogenous metastasis. Cancer Metastasis Rev. 11, 227-235.

Xue, W., Mizukami, I., Todd, R., & Petty, H. (1997). Urokinase-type plasminogen-activator receptors associate with β1 and β3 integrins of fibrosarcoma cells- dependence on extracellular-matrixcomponents. Cancer Res. 57, 1682-1689.

Yebra, M., Parry, G., Stromblad, S., Mackman, N., Rosenberg, S., Mueller, B., & Cheresh, D. (1996). Requirement of receptor-bound urokinase-type plasminogen activator for integrin αvβ5-directed cell migration. J. Biol. Chem. 271. 29393-29399.

Zhang, Z. H., Vuori, K., Reed, J. C., & Ruoslahti, E. (1995). The α5β1 integrin supports survival of cells on fibronectin and up-regulates bcl-2 expression. Proc. Natl. Acad. Sci. USA 92, 6161-6165.

Zhu, A. & Watt, F. (1996). Expression of a dominant negative cadherin mutant inhibits proliferation and stimulates terminal differentiation of human epidermal keratinocytes. J. Cell Sci. 109, 3013-3023.

Zutter, M. M., Mazouijio, G., & Santoro, S. A. (1990). Decreased expression of integrin adhesive receptors in adenocarcinoma of the breast. J. Pathol. 137, 863-870.

INTEGRIN ADHESION IN CELL MIGRATION

Sean P. Palecek, Elisabeth A. Cox, Anna Huttenlocher, Douglas A. Lauffenburger, and Alan F. Horwitz

I. INTRODUCTION

Cell migration is essential for normal embryonic development and the maintenance of health throughout life, but it also may lead to pathologic conditions that contribute to our death. Migrations are first apparent in early embryonic development in morphogenic processes where different cells come together and segregate to form

Advances in Molecular and Cell Biology
Volume 28, pages 367-388.

ISBN: 0-7623-0495-2

diverse tissues and organs. These morphogenic migrations are intricate, highly specific, and reproducible. In the adult, cellular migrations are required for the development of an immune response whereby immune cells migrate out of the vasculature and into areas of inflammation. While this process is required for normal immune function, it may also contribute to the pathogenesis of inflammatory disorders such as arthritis or asthma. Cell migrations also contribute to diseases such as cancer, in which migration of cells from the primary tumor mass to distant sites is the hallmark of the conversion of a benign to a malignant tumor. Thus, a molecular understanding of migration is central to our understanding of basic organismal biology. It is also critical to developing therapeutic strategies against several important pathologic conditions including vascular disease, cancer, and chronic and acute inflammation. Finally, it promises to aid efforts to accelerate wound repair and the production and utilization of replacement tissues.

The complex process of cell migration can be usefully analyzed as comprised of several distinct, dynamic events. These events include membrane protrusion and extension, formation of stable but dynamic contacts between the cell and its substratum, translocation of cell cortex, and release of cell-substratum adhesions at the rear of the cell (see Figure 1). This parsing of migration into distinct events underscores the spatially distributed nature of the component processes, which therefore require coordination—temporal as well as spatial—to accomplish locomotion. Understanding the mechanisms of cell migration, along with their regulation and coordination, is especially challenging because of this central role for physical integration. Different proportions of integration likely give rise to qualitatively different movement appearances among particular cell types or under various environmental conditions. For instance, effective coordination of lamellipod protrusion, cytoskeletal contraction, and cell-substratum de-adhesion may result in highly efficient "gliding" motion, as seen in keratocytes and neutrophils on substrata of relatively low adhesion. More distinct extension and retraction processes yields locomotion by relatively discrete "steps," as observed in fibroblasts and endothelial cells; at the same time, keratocyte and neutrophil movement behavior on highly adhesive substrata can become more saltatory and resemblent of fibroblasts. In a comparable vein, the process rate-limiting for movement may differ among cell types or across a span of conditions. For example, under some circumstances the rate of cell locomotion can be governed by events at the cell front (Wessels et al., 1994). However, cell speed is generally not limited by the rate of lamellipod protrusion (Abercrombie et al., 1970; Felder and Elson, 1990; Condeelis, 1993). Sometimes lamellipodia are not able to form stable attachments to the substratum (Bard and Hay, 1975) and often cell/substratum adhesions at the cell rear are determinative, particularly on highly adhesive substrata. It is conceivable, therefore, that the fundamental molecular mechanisms underlying the processes comprising migration are similar among different cell types despite apparent phenomenological variation.

The focus of this overview is the adhesive events that contribute to locomotion. This includes the formation and release of adhesions during migration and the role

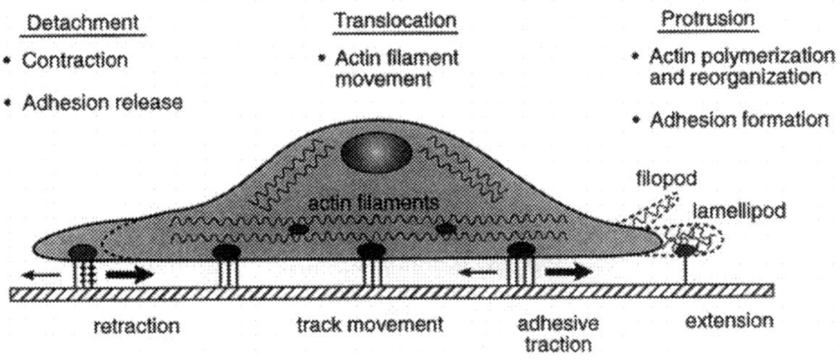

Figure 1. A model for cell migration. Protrusion, or extension, at the cell front involves actin polymerization and reorganization. These extensions are stabilized by nascent adhesions, which also serve to generate traction. Translocation of the cell cortex occurs after protrusion and involves actin filament movements mediated by myosin motors (see Figure 2). Following cortical translocation, contraction of the actin cytoskeleton and release of adhesions allows detachment to occur at the cell rear.

of adhesive strength in generating traction forces and governing migration properties. We emphasize that our focus represents our belief only that adhesion-related processes are of great importance in cell migration, but not exclusively so, along with the other processes listed above.

II. FORMATION OF NEW ADHESIONS

To initiate migration, cells must extend membrane protrusions and form new attachments to the substratum at their front. Nascent attachments may subsequently become highly organized structures that are connected to the actin cytoskeleton. Protrusions at the leading edge include broad, flat, veil-like protrusions called lamellipodia and needle-like projections called filopodia (for reviews of actin dynamics at cell front see Stossel, 1993; Mitchison and Cramer, 1996). These adhesive sites generate forward-based traction necessary for migration. Recent studies have contributed fresh insights into the basic mechanisms of adhesion formation.

Formation of integrin-containing adhesive complexes during migration has been examined using low-light level video microscopy and immunofluorescence (Regan and Horwitz, 1992) and laser optical trapping and nanometer level motion analysis (Schmidt et al., 1993). In migrating cells, nascent adhesions containing aggregated integrins, vinculin, focal adhesion kinase (FAK), and various phosphoproteins, form preferentially along the periphery of lamellipodia and the tips of filipodia (Nobes and Hall, 1995). On the ventral (lower) surface, these aggregates tend to remain fixed to the substratum, increasing in size and organization as the cell moves over them until they reach the rear (Regen and Horwitz, 1992). Thus the motile process actively removes integrins from the leading edge and concentrates them at the rear. This raises an important question. How are integrins replenished at the cell front so new adhesions can form? One potential mechanism is provided by polarized vesicle trafficking in which receptors on the cell surface, especially those at the rear, are endocytosed and delivered toward the cell front where they can supply integrins for nascent adhesions (Lawson and Maxfield, 1995; Bretscher, 1996). This appears to be an important mechanism for neutrophils, but its relative contribution in other cell types is less clear. Another mechanism, which may complement the endocytic mechanism just discussed, involves molecular motors carrying integrins to the cell front, thus having a similar replenishing effect. Experimental evidence supporting this has been obtained for integrins through tracking studies in which colloidal gold particles attached to small β1 integrin aggregates on the dorsal (upper) surface at the cell front were found to translocate to the leading edge (Schmidt et al., 1993).

In addition to preferential integrin localization to the cell front, mechanisms also exist to deliver actin, and perhaps other adhesion related molecules, to the leading edge. The mRNA for β-actin localizes to the lamellipodia of motile cells (for a review, see Bassell and Singer, 1997). Furthermore, inhibiting β-actin mRNA localization or its translation into protein, reduces motility in fibroblasts (Kislauskis et al., 1997). In summary, it is now clear that various mechanisms ensure that the cell front is supplied with the building blocks for nascent adhesions and actin protrusion. Precise determination of the mechanisms involved and the importance of these delivery modes among different cell types, as well as discovery of new mechanisms will give more insights into how the molecular constituents concentrate at the leading edge.

The mechanisms that govern adhesion formation and actin organization at the leading edge are not yet fully understood. Recent progress has elucidated some of the signaling molecules and mechanisms which regulate adhesion and cytoskeletal organization during migration. These include tyrosine phosphorylation, rho family proteins, and growth factor signaling pathways.

Tyrosine phosphorylation has a prominent role in adhesion formation and migration (however, it is also implicated in adhesive release as discussed below). Inhibition of tyrosine phosphorylation abrogates formation of adhesions, cell spreading and decreases motility (Burridge et al., 1992). Though the phosphoproteins responsible for initiating and assembling adhesions have not been clearly defined, the (FAK) has received much attention in this context. FAK associates with the cytoplasmic domains of integrins, is present in nascent adhesions and has a well-defined role in cell migration (Schaller and Parsons, 1994). FAK deficient mice are embryonic lethals that exhibit general mesodermal defects on the eighth day of development (Ilic et al., 1995). Fibroblast-like cells cultured from FAK deficient mice have a more rounded morphology, an increased number of focal adhesions and dramatically reduced motility compared to analogous cells derived from wild type mice.

Further evidence supporting a role for this kinase in regulating migration is provided by CHO cells that overexpress FAK. These cells show an enhanced motility on fibronectin, although the mechanism contributing to this phenotype has not been elucidated (Cary et al., 1996).

Another group of molecules that appear to be key regulators of adhesion formation during migration are those belonging to the rho family and in particular, rec. The rho family of small GTPases direct adhesive complex formation and also serve as actin architects, sculpting actin networks and linking them to these adhesive complexes. The most intensely studied members of this family include cdc42, rac, and rho. G-protein linked and growth factor receptors activate the rho family proteins (for review of rho family regulation see Machesky and Hall, 1996). Like ras, these proteins act as molecular switches controlling downstream signaling cascades. They are inactive when bound to GDP and are active when bound to GTP, which seems to occur concurrently with membrane localization. Three groups of regulatory molecules control the activity of these proteins: guanine nucleotide exchange factors (GEFs), which catalyze exchange of GDP for GTP; GTPase activating proteins (GAPs), which catalyze the hydrolysis of GTP to GDP; and GDP dissociation inhibitors (GDIs), which stabilize the inactive state.

Microinjection studies demonstrate that cdc42 induces filopod formation, rac orchestrates lamella extension and membrane ruffling, while rho coordinates stress fiber and focal adhesion assembly (Ridley and Hall, 1992; Kozma et al., 1995; Nobes and Hall, 1995). Cdc42 and rac also induce formation of adhesion complexes which have the same components as rho induced focal adhesions (clustered integrin, talin, vinculin, phosphorylated FAK and paxillin) but are smaller in size (Nobes and Hall, 1995). The rho family proteins are linked in a hierarchal cascade; cdc42 activates rac which activates rho, a temporal order consistent with the chronology of actin polymerization for membrane protrusion followed by attachment.

Motile cells have membrane ruffles and lamellipodia at the leading edge and diminished stress fibers relative to non-motile cells. Rac, which stimulates formation of membrane ruffles and lamellipodia, appears to have a paramount role in migration. For example, the expression of rac or Tiam1, a rac GEF, attributes an invasive phenotype to T-lymphoma cells (Michiels et al., 1995). Furthermore, active rac is essential for nerve growth factor (NGF) and phobol myristate acetel (PMA) induced motility of PC12 cells (Altun-Gultekin and Wagner, 1996). Rho, on the other hand, stimulates formation of strong adhesions and stress fibers through generating contractility (Chrzanowska-Wodnicka and Burridge, 1996), and its role in migration is less well defined. Some evidence exists that downregulation of rho is necessary for migration to occur. Microinjection of activated rho inhibits scatter factor/hepatocyte growth factor induced migration of MDCK cells, while inhibition of rho does not affect motility (Ridley et al., 1995). Alternatively, rho-stimulated contractility may be important in release of adhesions at the cell rear during migration (see below). The role of cdc42 in migration has not been examined, though this molecule is likely to be important as it directs filopodial formation. The relationship between these short-term ruffling events and longer-term sustained locomotion, however, is not clear at present. A plausible hypothesis may be that these short-term events are involved in initial reorganization of cytoskeleton from that of sessile cells to that required for effective force transmission in locomoting cells. Alternatively, the rho family-mediated events could remain important for continuing rounds of lamellipod/filopod extension during sustained migration. This issue merits further investigation.

Recently, a great deal of progress has been made towards elucidating the downstream effectors of the rho proteins (for review see Tapon and Hall, 1997). This has given insights into how these proteins form adhesions and stimulate actin polymerization. One of the most intensely studied downstream effectors of rac and rho is phosphatidylinositol 5-kinase (PI 5-kinase), which elevates phosphatidyinositol 4,5-bisphosphate (PIP_2) levels (Chong et al., 1994; Tolias et al., 1995; Hartwig et al., 1995). This is interesting because PIP_2 has many activities that may contribute to adhesion formation and cytoskeletal architecture. In particular, PIP_2 unmasks cryptic actin and talin binding sites on vinculin which is necessary for adhesion complex assembly (Gilmore and Burridge, 1996). It also modulates actin polymerization through interactions with numerous actin binding proteins including gelsolin, capZ, α255-actinin, and profilin (Stossel, 1993). PIP_2 also apparently has a role in activating ERM (ezrin-radixin-moesin) proteins, which act as cross-linkers between plasma membranes and the actin cytoskeleton (Hirao et al., 1996; Niggli et al., 1995). Highlighting the cumulative effect of these activities, microinjection of PIP_2 antibodies inhibits stress fiber and focal adhesion assembly (Gilmore and Burridge, 1996).

Another important downstream effector of rac and cdc42 is phosphatidylinositol 3-kinase (PI 3-kinase), whose activity is regulated by rho (Zhang et al., 1993, 1994; Kumagai et al., 1993). This is interesting because PI 3-kinase is implicated in growth factor mediated motogenic signaling pathways (see below). How PI 3-kinase influences motility is not understood but it is known to regulate integrin receptor avidity (Kovacsovics et al., 1995). Furthermore, PI 3-kinase catalyzes the synthesis of D3-phosphorylated inositides which activate various protein kinase C (PKC) family isoforms (Nakanishi et al., 1993; Toker et al., 1994). Activation of PKC has been shown to stimulate focal adhesion formation (Woods and Couchman, 1992).

Finally, growth factor receptors and certain downstream effectors have been implicated as regulators of adhesion in motile cells. Studies of chemotaxis towards growth factors have given insights into motogenic signaling pathways. Many growth factor receptors are tyrosine kinases whose activities are stimulated upon ligand binding, resulting in receptor autophosphorylation. Upon phosphorylation, several molecules associate with the growth factor receptors via SH2 domains, including PI 3-kinase, phospholipase C-γ (PLC-γ), rho GTPase activating protein (rhoGAP), and the GRB2-SOS complex. The motogenic and mitogenic pathways stimulated by growth factors are distinct; PI 3-kinase and PLC-γ are implicated in generating motogenic signals. The ability of platelet derived growth factor (PDGF) to stimulate chemotaxis correlates positively with PI 3-kinase activity (Wennstrom et al., 1994). Furthermore, tyrosine phosphorylation sites on the PDGF receptor that bind PI 3-kinase are required for chemotaxis (Wennstrom et al., 1994). In another study, both PI 3-kinase and PLC-γ were found to be essential in the motile response mediated by PDGF receptor, while rasGAP had an antagonistic effect (Kundra et al., 1994). PLC-γ is also important in epidermal growth factor (EGF) receptor initiated motility (Chen et al., 1994). Various inhibitors of PLC-γ abrogate the ability of EGF receptor to signal motogenically (Chen et al., 1994).

The manner in which PI 3-kinase and PLC-γ stimulate locomotion is unknown. Both these molecules can alter actin organization. PI 3-kinase is discussed above as being a downstream effector of rho proteins. In addition, this lipid kinase has been found to mediate PDGF induced activation of rac (Hawkins et al., 1995). The EGF receptor stimulates motility at least partially through PLC-γ mediated gelsolin release (Chen et al., 1996). PLC-γ hydrolyzes PIP_2 releasing membrane bound gelsolin, which can function as an actin-severing protein, freeing actin monomers for polymerization. Consistent with this activity, membrane extension rates in fibroblasts are increased via a PLC-γ pathway upon stimulation by EGF, proportional to but not entirely commensurate with enhanced migration speed (Wells et al., 1998). Both speed and directional persistance are affected by this signaling (Ware et al., 1998).

Much progress has been made towards discerning the mechanisms that orchestrate adhesion formation during migration—in particular the role of rho family proteins and interactions among focal adhesion components. Despite this however, much remains unknown concerning the upstream regulators and targets of the rho family regulatory proteins

III. RELEASE OF ADHESIONS AT THE CELL REAR

While there has been much progress in recent years toward understanding adhesive events at the front of motile cells, less is known about events at the cell rear. Although the rate of rear release determines the migration rate under many conditions (Chen, 1981), the mechanisms regulating this process are poorly understood. Understanding the dynamics of cell-substratum adhesions at the rear of a migrating cell may help contribute insight into release mechanisms. Previous studies suggest that under some conditions distraction of membrane components occurs at the rear of migrating cells both in vitro and in vivo (Chen, 1981; Hay, 1985). Using anti-integrin antibodies conjugated with fluorescent derivatives, the fates of integrin receptors at the cell rear have been characterized in fibroblasts. Studies using migrating fibroblasts demonstrate that a substantial fraction of integrin receptors can be released from the cell surface and remain on the substratum after detachment (Regen and Horwitz, 1992; Palecek, et al. 1996). A smaller fraction of the integrin receptor remains on the cell surface after fibroblasts detach from the substratum.

Integrin remaining on the cell surface may subsequently be taken up into endocytic vesicles which accumulate near the nucleus in fibroblasts. Alternatively, integrin may also remain on the cell surface as an aggregate which can move along the cell edge and reattach to the surface, or it may disperse along the cell surface.

The factors contributing to adhesive release at the cell rear are likely to involve a combination of contractile forces and biochemical regulatory mechanisms. There is increasing evidence supporting a role for contraction in the breakdown of adhesions at the rear of the cell. Several lines of evidence support a role for myosin II based contraction (Jay et al., 1995). Amoebae deficient in myosin II have decreased migration rates which are exaggerated on more adhesive substrata (Weasels et al., 1988, Jay et al., 1995). Myosin II also localizes to the rear of migrating fibroblasts (Conrad et al., 1993), suggesting a functional role in that area. Furthermore, injection of antibodies to myosin light chain kinase inhibits macrophage migration (Wilson et al., 1991). Using a permeabilized system to study mechanisms of adhesive release, additional evidence supports a role for contraction in cell detachment. Addition of exogenous ATP to the permeabilized cells promotes contraction and a breakdown of focal adhesions (Crowley and Horwitz, 1995). The importance of contraction and actin-myosin interactions is also demonstrated by the inhibition of focal adhesion breakdown by peptides that inhibit actin-myosin interactions. Externally-induced signaling pathways could elicit appropriate contractile activity for purposes of regulating detachment. For instance, MAP kinase activity stimulated by EGF has been found to cause disassembly of focal adhesions in fibroblasts with corresponding decrease in adhesion strength (Xie et al., 1998). The potential importance of contractility in migration is further suggested by a recent study which showed that MAP kinase enhances migration through activating myosin light chain (MLC) kinase activity, thus increasing MLC phosphorylation (Klemke et al., 1997). A particular locus for effects of MLC kinase activity, has not been identified; it could operate over the entire cell for cytoskeletal reorganization, or possibly primarily at the cell rear to facilitate detachment.

Upon application of force, cell-substratum linkages can be broken at two different loci: the integrin-ligand bond or the integrin-cytoskeletal bond. In general, the comparative probabilities of breakage at two alternative loci are inversely proportional to the square root of the comparative bond affinities (Saterbak and Lauffenburger, 1996). In migrating fibroblasts, the majority of integrin remains on the substratum suggesting preferential severing of the integrin-cytoskeletal bond at the rear of migrating fibroblasts. Direct evidence for this is demonstrated by studies showing that the integrin-cytoskeletal linkage is weaker at the rear of migrating fibroblasts (Schmidt et al., 1993). Optical trapping studies using integrin antibodies show that there is a large differential in cytoskeletal linkage between the cell front and rear, with at least a four fold lower probability of an integrin linking with the cytoskeleton at the cell rear.

As discussed below, the application of force to cell-substratum attachments can either strengthen or weaken the cytoskeletal linkages. Thus, if linkages at the cell rear need to be preferentially broken while those at the front need to be preferentially strengthened, some biochemical mechanism must exist to provide a differential effect. Accordingly, recent studies suggest that enzymatic weakening of cytoskeletal linkages may occur at the rear of migrating cells. These studies suggest that the calcium-dependent protease calpain may play an important role in the release of adhesions on both fibronectin and vitronectin (Huttenlocher et al., 1997; Palecek et al., 1998). Calpain localizes to focal adhesions, cleaves many focal adhesion proteins including FAK, talin, and the integrin cytoplasmic domain (Beckerle et al., 1987; Du et al., 1996; Cooray et al., 1996), and therefore is a likely candidate to play a role in severing the integrin-cytoskeletal linkage during migration. Inhibitor and genetic studies both support a role for calpain in adhesive release in migrating CHO cells. CHO cells treated with calpain inhibitors or cell lines expressing low levels of endogenous μ-calpain show lamellipodial protrusion but inhibited detachment from the substratum. Calpain inhibitors also alter the fate of integrin receptors at the rear of migrating cells; calpain inhibitors significantly reduce the amount of integrin that rips from the rear of migrating cells and remains on the substratum. This suggests that calpain inhibitors act by stabilizing the integrin-cytoskeletal linkages at the cell rear thereby inhibiting cell detachment.

In addition to mechanisms altering integrin-cytoskeletal linkages during rear retraction, alteration in the strength of integrin-ligand interactions may also play a regulatory role during rear release. It is possible, for example, that integrins at the cell front have a high affinity while those at the rear have a lower affinity, producing an adhesive asymmetry within the cell. The weaker adhesions at the rear would favor release while the stronger adhesions at the front would support adhesion. To date there is no direct evidence supporting a role for this kind of gradient in integrin affinity. However, the affinity of integrin receptors does contribute to migration. In immune cell migration, for example, transient activation is required for the transmigration of immune cells. The cells neither adhere nor migrate with their β2 receptors in their low affinity state (Springer, 1995). However locking β1, β2, or β3 integrin receptors in a high affinity state inhibits cell migration in several cell types (Kuijpers et al., 1993, Huttenlocher et al., 1996). The high integrin affinity inhibits migration by inhibiting release at the cell rear, as revealed by video imaging. Lowering the substrate ECM concentration can compensate for increased integrin affinity to result in efficient migration (Huttenlocher et al., 1996; Palecek et al., 1997). This is in accord with the notion that migration is inhibited because of a reduction in the rate of rear release with high affinity integrin.

Regulation of cell detachment likely involves signal transduction pathways, although the role of these pathways are not yet defined. Gradients of calcium, with higher concentrations seen at the cell rear, have been demonstrated in migrating eosinophils and are implicated in release of adhesions (Brundage et al., 1991). Buffering calcium inhibits migration of neutrophils on both fibronectin and vitronectin by inhibiting the release of adhesions at the rear (Maxfield, 1993). One target for calcium appears to be the calcium/calmodulin-dependent phosphatase calcineurin. Addition of a peptide inhibitor of calcineurin inhibits neutrophil detachment on vitronectin, suggesting that calcineurin is a calcium-dependent release mechanism specifically for the $\alpha v\beta 3$ interaction with vitronectin (Hendey et al., 1992).

The small GTP-binding protein rho is among signal transduction pathways that are implicated in rear release. Inhibitors of rho induce cell rounding and detachment (Paterson et al., 1990; Miura et al., 1993). Tyrosine phosphorylations also appear to play a role in cell detachment. Infection of cells with transforming viruses such as Rous sarcoma virus promote a rounded morphology and less organized adhesive structures as a result of increased pp60src tyrosine kinase expression (Burridge et al., 1988). Further evidence for a role for tyrosine phosphorylation is supported by studies using a permeabilized system which implicate tyrosine phosphorylation in the breakdown of focal adhesions. Although the evidence suggests that tyrosine phosphorylation is involved in adhesive release, the pathway involved is unknown.

IV. FORCES AS MECHANICAL AND MOLECULAR INTEGRATORS

While the formation and release of adhesions are essential during migration, the function of adhesions, once formed, is to transmit intracellularly generated forces to its environment. Adhesions serve primarily to provide the traction necessary for force transmission and cortical movement. Cellular forces also serve mechanistic roles in the release and formation of adhesions mentioned previously. At least two types of forces, protrusive and contractile, are required for cell locomotion. Protrusive forces are required to extend lamellipodia or filopodia. These forces must overcome the drag of the membrane and the deformation of the surrounding environment. Actin polymerization is the most likely origin of protrusive forces. Actin-binding and crosslinking proteins have been implicated in regulating actin polymerization and membrane protrusion. Contractile forces are required to move the cell body forward and to detach cell-substratum adhesions at the rear. Contractile forces presumably depend on muscle-like actin-myosin interactions and on myosin motors. Protrusive forces at the cell front have been extensively studied and reviewed recently (Mitchison and Cramer, 1996). Therefore, we will concentrate on contractile forces, whose roles are less well understood than those of protrusive forces.

At least two mechanisms have been suggested for generating the contractile forces which move the cell body forward (see Figure 2). Actin filament contraction, in which filaments of opposite polarity slide past each other (as in muscle sarcomeres), could generate tension aiding cortical translocation or rear retraction. Alternately, myosin motors could function to move one population of actin filaments forward over a track of stationary actin filaments on the ventral surface of migrating cells. A recent study of actin filament polarity in migrating cells supports both of

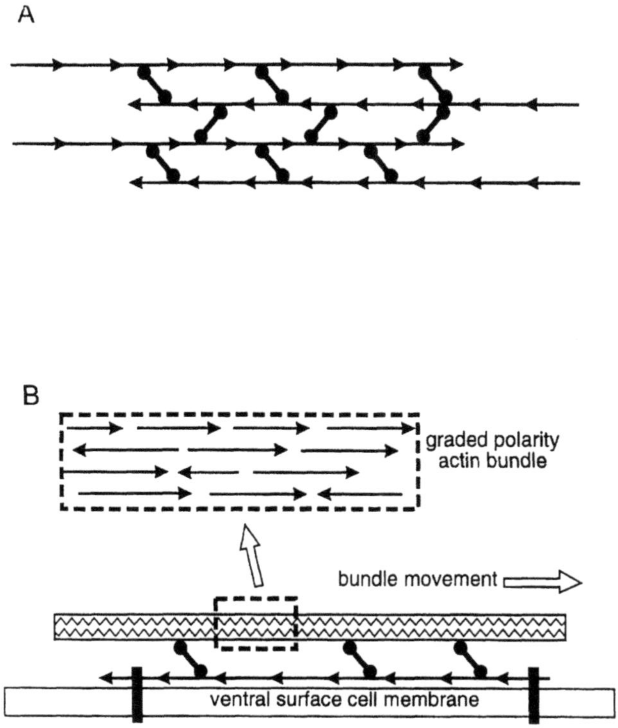

Figure 2. Models for contractile force generation during motility. (**A**) Individual filaments in actin bundles alternate in polarity (arrows point toward pointed end). Myosin II molecules (line with two balls) slide the actin filaments across each other to contract the bundle, as in a muscle sarcomere. This is the actin arrangement along the inner surface of the plasma membrane. (**B**) In the bundle/track mechanism, actin filaments are anchored (filled rectangles) to the substratum through the cell membrane and remain stationary with respect to the cell surface. Bundles of actin filaments are attached to myosins that transport the bundles along the filament track. The actin bundles are composed of filaments with graded polarity. Near the cell front, actin barbed ends tend to face forward, whereas near the cell rear, the actin pointed ends tend to face forward. This actin organization occurs along the entire length of the ventral cell surface.

these models (Cramer et al., 1997). A small population of actin filaments are organized in bundles which alternate in polarity. This is consistent with their role as mediators of contractile events. Another, larger population of actin filament bundles is organized in ventrally located bundles in which the polarity gradually changes along the length of the bundle. This arrangement is expected for filaments involved in track migration. Recently, observations on migrating cells demonstrate that one fraction of actin filaments in the cell body remains fixed with respect to the substratum and another moves forward at the rate of cell body translocation (Cramer et al.,

1997) This suggests that track-like movement generates cortical translation. Earlier observations point to contractile events during release at the cell rear, which is consistent with the actin filament arrangement predominantly found there (Conrad et al., 1993).

Contractile forces are transmitted from the a tin cytoskeleton to the extracellular environment through adhesive complexes. The transmission of these contractile forces to the environment leads to the exertion of a traction force on the surface. Likewise, the surface exerts an equal but opposite force on the cell. The magnitude of the traction force is obviously related to the magnitude of the tension generated by forces discussed above; but it is also influenced by properties of the cytoskeleton-adhesion receptor-ECM linkage. Measuring local contractile forces is technically challenging, but they have been measured on deformable silicone rubber substrata for migrating fibroblasts (Harris et al., 1980) and keratocytes (Lee et al., 1994), and by a micropatterned force-transducing surface for fibroblasts (Galbraith and Sheetz, 1997). Faster moving cells which glide over a substrate, such as keratocytes, tend to be less adherent and exert lower traction forces on the substrate than slower cells, like fibroblasts. Traction forces range from about 2×10^{-8} N for keratocytes to about 2×10^{-7} N for fibroblasts? with neutrophils intermediate. Migrating keratocytes do not exert detectable traction forces at the front of the cell where actin polymerization is thought to provide protrusive forces. Keratocytes exert inward-directed traction forces at the cell periphery near the cell rear. Fibroblasts also do not exert traction forces at the front of the cell. Behind the leading edge, however, fibroblasts exert large, inward directed forces, throughout the entire cell, that are oriented perpendicular to the peripheral contour of the cell. The differences in magnitude of these traction forces reflect the differences in cell-substratum adhesiveness among the cell types. The force required to overcome viscous drag on a migrating cell is about 3×10^{-13} N (Lee et al., 1994), much lower than the contractile force exerted by any cell type. The large contractile forces are instead required to facilitate detachment of cell-substratum adhesions. As mentioned above, the linkage can be disrupted either between the ECM ligand and adhesion receptor or between intracellular cytoskeletal bonds. As the force on a bond increases, the probability of bond dissociation increases (Aloe et al., 1995). The differences in the direction of exerted traction forces in fibroblasts and keratocytes possibly reflects the differences in coordination between extension and retraction in keratocytes and fibroblasts. Keratocytes must be very coordinated and move very quickly while fibroblasts do not need to move as quickly but must be able to generate larger forces for processes such as wound retraction.

Contractile forces serve not only to disrupt cell-substratum linkages during migration, but they also function to strengthen their interactions with cytoskeletal elements. Application of mechanical stress through integrins increases cytoskeletal stiffness proportionally with the applied stress (Wang et al., 1993). The stiffening requires an intact cellular architecture of actin filaments, microtubules, and intermediate filaments, suggesting that structural rearrangement of the cytoskeleton contributes to the mechanically induced stiffening (Wang and Ingber, 1994). Using a laser optical trap to exert forces on fibronectin-coated beads on a cell surface, Choquet and colleagues (1997) also demonstrated that reinforcement of integrin-cytoskeletal linkages is proportional to the applied force. The linkage strengthening requires occupancy of the ligand binding site and is inhibited by a phosphatase inhibitor. This phenomenon may be involved in cell guidance since linkages on rigid substrates will be stronger than linkages on compliant substrates, steering cells toward rigidity (Choquet et al., 1997).

Cell migration requires that the intracellular y generated forces be transmitted to the substratum in a spatially asymmetric manner. This front versus rear asymmetry in contractility or cell-substratum adhesiveness can be satisfied in a number of ways. The spatial distribution of adhesion receptors may lead to higher adhesiveness or reduced transmission of contractile forces at the cell front than at the cell rear. The triangular shape of many cells, with more contact area at the cell front than rear, may produce such a polarity. Asymmetries in adhesion receptor-ECM ligand affinity could also result in an adhesive gradient in the cell. As discussed above, locking integrins in a high-affinity state inhibits eosinophil migration (Kuijpers et al., 1993); this could be due to a superoptimal overall adhesiveness or loss of front-versus-rear gradient in adhesiveness. Also, locking β2 integrin in a high-affinity conformation for ICAM prevents lymphocyte locomotion (Dustin et al., 1997). Adhesion receptor-cytoskeleton linkage avidity appears to exhibit a spatial asymmetry in the cell; integrin linkages to the cytoskeleton in migrating NIH 3T3 fibroblasts are more prevalent in the cell front than at the cell rear (Schmidt et al., 1993). Spatial concentrations of signaling molecules may also contribute to the asymmetry. For example, intracellular Ca^{+2} or phophoinositide levels may regulate contractile forces or the strength of adhesive linkages. In migrating leukocytes, calcium concentration, and myosin-II based contractility, increases from the front to rear of the cell (Hahn et al., 1992). Also, intracellular calcium transients are required for detachment of the cell rear during neutrophil migration (Maxfield, 1993). Kinases and/or phosphatases may regulate the activity or affinity of adhesion complex components.

Contractility and adhesiveness must cooperate during the rear retraction phase of migration so that the front of the cell remains anchored while the rear detaches from the substratum. According to a theoretical model for migration (DiMilla et al., 1991), the magnitude of the front versus rear asymmetry governs the ranges of contractile forces and cell-substratum adhesiveness promoting locomotion. Cells with a large asymmetry are predicted to exhibit migration over a wide range of adhesiveness while cells with a low asymmetry are predicted to migrate only over a narrow range of adhesiveness. Therefore, adhesive and contractile asymmetries are candidates for explaining why some cell types (e.g. neutrophils) are able to migrate through many different environments while others (e.g. fibroblasts) only migrate under more specific conditions.

V. REGULATION OF MIGRATION BY ADHESIVENESS

As discussed above, the major function of cell substrate adhesion is to generate the traction required for translocation of the cell cortex. Consequently, the adhesive strength between a cell and its substratum affects migration speed (see Figure 3). At low adhesiveness sufficient traction cannot be generated with the substrate to efficiently move the cell while at high adhesiveness the cell cannot generate enough contractile force to fracture attachments at the rear. At intermediate adhesiveness the cell can both efficiently form new attachments at the front and release attachments at the rear. A quantitative model of cell migration confirms this simple analysis and suggests principles governing the dependence of cell migration speed on integrin-l CM ligand binding and cell mechanical properties (DiMilla et al., 1991). In this model, maximum migration speed occurs at an intermediate ratio of cell-substratum adhesiveness to intracellular contractile force. Considerable experimental evidence in many different cell systems supports the prediction that cell-substratum adhesiveness regulates cell migration speed, which is maximal at intermediate adhesiveness. Cell-substratum adhesiveness can be altered in many ways, including changes in ECM protein composition or concentration, integrin expression levels or spatial distribution, integrin-ligand affinity, cytoskeletal protein expression, or integrin-cytoskeletal affinity. Measurements of cell migration speed and short-term cell-substratum detachment force of CHO cells demonstrate that cell migration speed is regulated by cell-substratum adhesiveness (Palecek et al.,

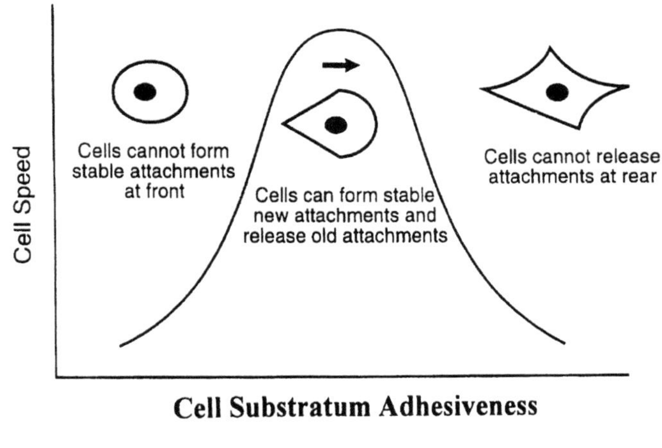

Figure 3. Cell-substratum adhesiveness regulates cell migration speed. At low adhesiveness, cells cannot form stable attachments to the substratum and generate traction forces required for migration. At high adhesiveness, adhesions at the cell rear are very strong and rear detachment limits cells speed. Cell speed is maximum at intermediate adhesiveness where the cells can efficiently form adhesions at the cell front and detach adhesions at the cell rear.

1997). As ECM substrate concentration, integrin expression level, or integrin-ECM ligand affinity change, migration speed is a constant. biphasic function of cell-substratum adhesiveness.

While not studied as systematically, observations in other cell types support a similar relationship, including in vivo (Ho et al., 1997). Human smooth muscle cells and murine myoblasts migrate maximally at intermediate concentrations of fibronectin and collagen, and laminin respectively (DiMilla et al., 1993; Goodman et al., 1989). Tenascin enhances glioma cell migration on fibronectin substrates presumably by modulating cell-substratum adhesiveness (Deryugina and Bourdon, 1996). Furthermore, addition of fibronectin to collagen type I gels inhibits migration of polymorphonuclear leukocytes through the gels due to increased cell adhesion (Kuntz and Saltzman, 1997).

Changes in adhesion receptor expression can also have varied effects on cell migration. Cells speed is maximal on collagen at intermediate levels of $\alpha2$ integrin subunit expression (Keely et al, 1995) and on fibronectin at intermediate levels of $\alpha5\beta1$ expression (Giancotti and Ruoslahti, 1990; Bauer et al., 1992; Palecek, et al., 1997). Neurons can actually modulate expression of $\alpha6\beta1$ integrin based on laminin concentration (Condic and Letourneau, 1997). At low laminin levels, $\alpha6\beta1$ integrin expression and cell adhesiveness increases, allowing the neurite to continue growing. This integrin regulation may allow the growth cone to respond to its environment and maintain a constant rate of motility over a variety of laminin concentrations.

Other receptors may interact with integrins to regulate integrin function and also promote adhesion and migration. For example, the urokinase-type plasminogen activator receptor (uPAR) interacts with $\alpha v\beta3$ integrin in human embryonic kidney 293 cells to enhance adhesion and migration on vitronectin (Wei et al., 1996). uPAR binds to vitronectin and probably transmits forces between the ECM and cytoskeleton via an integrin-mediated linkage.

Changes in cytoskeletal protein expression lead to changes in cell adhesion and migration. A decrease in vinculin or α-actinin expression in 3T3 cells increases motility (Fernandez et al., 1993; Gluck and Ben-Ze'ev, 1994) while a decrease in vinculin expression in PC12 cells diminishes neurite outgrowth by decreasing lamellipod stability (Varnum-Finney and Reichardt, 1994).

Integrin/ECM ligand affinity also regulates cell adhesion and migration. Integrins may be activated into a high affinity state by cytoplasmic domain alterations in either the α or β chains which presumably disrupt specific interactions between the α and β subunits (Hughes et al., 1996). Integrins may also be activated by specific antibodies which alter integrin conformation (Freilinger et al., 1991). CHO cells transfected with $\alpha IIb\beta3$ integrin migrate maximally on intermediate concentrations of fibrinogen, with the concentration promoting maximum migration decreasing as $\alpha IIb\beta3$/fibrinogen affinity increases (Huttenlocher et al., 1996). Neural crest cells are able to form focal adhesions and migrate on antibodies to the $\beta1$ integrin subunit (Duband et al., 1991). The antibody concentration required for optimal cell migration cells is lower for cells on high-avidity polyclonal antibodies than for cells on lower avidity monoclonal antibodies.

Altered integrin-cytoskeleton interactions can result in altered cell-substratum adhesion and motility. CHO cells transfected with different α4 cytoplasmic domain chimeras demonstrate a maximum in migration speed at an intermediate adhesiveness (Kassner et al., 1995). Truncation of the β3 cytoplasmic domain alters integrin-cytoskeletal interactions and inhibits focal adhesion formation (Huttenlocher et al., 1996). CHO cells transfected with this variant show maximum motility at higher fibrinogen concentrations than wild-type cells. Maximum migration speed does not change in a random migration assay but haptotaxis is inhibited in the truncation mutant. These results suggest that haptotaxis may be regulated differently than random motility.

Transmembrane or secreted matrix metalloproteinases (MMPs) are also involved in regulation of cell migration. For example, integrin signaling induces expression of collagenases in fibroblasts, keratinocytes, and melanoma cells (Werb et al., 1989; Larjava et al., 1993; Seftor et al., 1993). Expression of Kuzbanian, a metalloproteinase/disintegrin of the ADAM family, is required for axonal extension in *Drosophila* (Fambrough et al., 1996). Matrix metalloproteinase MMP-2 can directly bind αvβ3 integrin in angiogenic blood vessels and melanoma cells in vivo (Brooks et al., 1996). Clustering CD3 and αLβ2, α4β1, or α5β1 integrin in T lymphocytes leads to induction of uPAR transcription and translation (Bianchi et al., 1996). These proteinases can regulate cell locomotion by a number of different mechanisms. At the cell leading edge they can degrade the matrix, reducing the force required to extend membrane protrusions through the surrounding environment. At the cell rear, the matrix-degradation activity of MMPs may disrupt cell/substratum bonds, thereby decreasing cell/substratum adhesiveness. Understanding how these very complicated proteinases help govern migration will require devotion of quantitative and physical approaches similar to those recently brought to bear on integrin- and growth-factor-mediated effects.

VI. CONCLUSION

Research on cell migration is beginning to provide molecular explanations for a complex and highly integrative cellular process. While not reviewed here, considerable recent progress has been made in understanding cytoskeletal dynamics during migration. Although understanding of adhesive dynamics lags behind that of the cytoskeleton, it now shows great promise for major progress in the near future. However the challenge of understanding the integrative aspects of migration lies just up the road and is formidable. Some questions that need to be addressed are: What are the signal transduction pathways that regulate lamellipodial extension, formation and release of adhesions, and translocation? How are these processes separated spatially to allow them to occur in the same cell? How are these coordinated temporally so they can occur simultaneously? What are the roles of non biochemical mechanisms—forces—in communicating and coordinating migratory behavior across the cell? How do cells make directed migrations or respond to chemotactic stimuli? How does the cell interpret cues from external factors or through contact with other cells to initiate and terminate migration? The challenge of answering these questions lies just around the corner; however initial explorations are already being made. The next few years promise major progress in all areas of migration research.

ACKNOWLEDGMENTS

This work was supported by NIH grant GM53905 to A. F. Horwitz and D. A. Lauffenburger, an Arthritis Foundation grant to A. Huttenlocher, and a Whitaker Foundation Graduate Fellowship to S. P. Palecek.

REFERENCES

Abercrombie, M., Heaysman, J. E. M., & Pegrun, S. M. (1970). The locomotion of fibroblasts in culture. I. Movements of the leading edge. Exp. Cell Res. 59, 393-398.

Alon, R., Hammer, D. A., & Springer, T. A. (1995). Lifetime of the P-selectin-carbohydrate bond and its response to tensile force in hydrodynamic flow. Nature. 374. 539-542.

Altun-Gultekin, Z. F., & Wagner, J. A. (1996). Src, ras, and rac mediate the migratory response elicited by NGF and PMA in PC12 cells. J. Neurosci. Res. 44, 308-327.

Bard, J. B. L. & Hay, E. D. (1975). The behavior of fibroblasts from the developing avian cornea. J. Cell Biol. 67, 400-418.

Bassell, G. & Singer, R. H. (1997). mRNA and cytoskeletal filaments. Curr. Opin. Cell Biol. 9, 109-115.

Bauer, J. S., Schreiner, C., Giancotti, F. G., Ruoslahti, E., & Juliano, R. L. (1992). Motility of fibronectin receptor-deficient cells on fibronectin and vitronectin: collaborative interactions among integrins. J. Cell Biol. 116, 477-487.

Beckerle, M. C., Burridge, K., DeMartino, G. N., & Croall, D. E. (1987). Colocalization of calcium-dependent protease II and one of its substrates at sites of cell adhesion. Cell. 51, 569-577.

Bianchi, E., Ferrero, E., Fazioli, F., Mangili, F., Wang, J., Bender, J. R., Blasi, F., & Pardi, R. (1996). Integrin-dependent induction of functional urokinase receptors in primary T lymphocytes. J. Clin. Invest. 98, 1133-1141.

Bretscher, M. S. (1996). Getting membrane flow and the cytoskeleton to cooperate in moving cells. Cell. 87, 601-606.

Brooks, P. C., Stromblad, S., Sanders, L. C., von Schalscha, T. L., Aimes, R. T., Stetler-Stevenson, W. G., Quigley, J. P., & Cheresh, D. A. (1996). Localization of matrix metalloproteinase MMP-2 to the surface of invasive cells by interaction with integrin $\alpha v \beta 3$. Cell. 85, 683-693.

Brundage, R. A., Fogarty, K. E., Tuft, R. A., & Fay, F. S. (1991). Calcium gradients underlying polarization and chemotaxis of eosinophils. Science. 254, 703-706.

Burridge, K., Fath, K., Kelly, T., Nuckolls, G., & Turner, C. (1988). Focal adhesions: transmembrane junctions between the extracellular matrix and the cytoskeleton. Annu. Rev. Cell Biol. 4, 487-525.

Burridge, K., Turner, C. E., & Romer, L. H. (1992). Tyrosine phosphorylation of paxillin and ppl25FAK accompanies cell adhesion to extracellular matrix: a role in cytoskeletal assembly. J. Cell Biol. 119, 893-903.

Cary, L., Chang, J., & Guan, J.-L. (1996). Stimulation of cell migration by overexpression of focal adhesion kinase and its association with Src and Fyn. J. Cell Sci. 109, 1787-1794.

Chen, P., Xie, H., Sekar, M. C., Gupta, K., & Wells, A. (1994). Epidermal growth factor receptor-mediated cell motility: phospholipase C activity is required, but mitogen-activated protein kinase activity is not sufficient for induced cell movement. J. Cell Biol. 127, 847-857.

Chen, P., Murphy-Ullrich, J. E., & Wells, A. (1996). A role for gelsolin in actuating epidermal growth factor receptor-mediated cell motility. J. Cell Biol. 134, 689-698.

Chen, W. -T. (1981). Mechanism of retraction of the trailing edge during fibroblast movement. J. Cell Biol. 90, 187-200.

Chong, L. D., Traynor-Kaplan, A., Bokoch, G. M., & Schwartz, M. A. (1994). The small GTP-binding protein rho regulates a phosphatidylinositol 4-phosphate 5-kinase in mammalian cells. Cell. 79, 507-513.

Choquet, D., Felsenfeld, D. P., & Sheetz, M. P. (1997). Extracellular matrix rigidity causes strengthening of integrin-cytoskeleton linkages. Cell. 88, 39-41.

Chrzanowska-Wodnicka, M., & Burridge, K. (1996). Rho-stimulated contractility drives the formation of stress fibers and focal adhesions. J. Cell Biol. 133, 1403-1415.

Condeelis, J. (1993). Life at the leading edge. Annul Rev. Cell Biol. 9, 411-444.

Condic, M. L. & Letourneau, P. C. (1997). Ligand-induced changes in integrin expression regulate neuronal adhesion and neural outgrowth. Nature. (in press).

Conrad, P. A., Giuliano, K. A., Fisher, G., Collins, K., Matsudaira, P. T., & Taylor, D. L. (1993). Relative distribution of actin, myosin I, and myosin II during the wound healing response of fibroblasts. J. Cell Biol. 120, 1381-1391.

Cooray, P., Yuan, Y., Schoenwaelder, S. M., Mitchell, C. A., Salem, H. H., & Jackson, S. P. (1996). Focal adhesion kinase (pp125Fak) cleavage and regulation by calpain. Biochem. J. 318, 41 -47.

Cramer, L. P., Siebert, M., & Mitchison, T. J. (1997). Identification of novel graded polarity actin filament bundles in locomoting heart fibroblasts: implications for the generation of motile force. J. Cell Biol. 136, 1287-1305.

Crowley, E. & Horwitz, A. F. (1995). Tyrosine phosphorylation and cytoskeletal tension regulate the release of fibroblast adhesions. J. Cell Biol. 131, 525-537.

Deryugina, E. I. & Bourdon, M. A. (1996). Tenascin mediates human glioma cell migration and modulates cell migration on fibronectin. J. Cell Sci. 109, 643-652.

DiMilla, P. A., Barbee, K., & Lauffenburger, D. A. (1991). Mathematical model for the effects of adhesion and mechanics on cell migration speed. Biophys. J. 60. 15-37.

DiMilla, P. A., Stone, J. A., Quinn, J. A., Albeda, S. M., & Lauffenburger, D. A. (1993). Maximal migration of human smooth muscle cells on fibronectin and type IV collagen occurs at an intermediate attachment strength. J. Cell Biol. 122, 729-737.

Du, X., Saido, T. C., Tsubuki, S., Indig, F. E., Williams, M. J., & Ginsberg, M. H. (1996). Calpain cleavage of the cytoplasmic domain of the integrin β3 subunit. J. Biol. Chem. 270, 26146-26151.

Duband, J. L., Dufour, S., Yamada, S. S., Yamada, K. M., & Thiery, J. P. (1991). Neural crest cell locomotion induced by antibodies to β1 integrins. A tool for studying the roles of substratum molecular avidity and density in migration. J. Cell Sci. 98, 517-532.

Dustin, M. L., Bromley, S. K., Kan, Z., Peterson, D. A., & Unanue, E. R. (1997). Antigen receptor engagement delivers a stop signal to migrating T lymphocytes. Proc. Natl. Acad. Sci. USA. 94, 3909-3913

Fambrough, D., Pan, D., Rubin, G. M., & Goodman, C. S. (1996). The cell surface metalloprotease/disintegrin Kuzbanian is required for axonal extension in Drosophila. Proc. Natl. Acad. Sci. USA. 93, 13233-13238.

Felder, S. & Elson, E. L. (1990). Mechanics of fibroblast locomotion: quantitative analysis of forces and motions at the leading lamellas of fibroblasts. J. Cell Biol. 111, 2527-2536.

Fernandez, J. L. R., Geiger, B., Salomon, D., & Ben-Ze'ev, A. (1993). Suppression of vinculin expression by antisense transfection confers changes in cell morphology, motility, and anchorage-dependent growth of 3T3 cells. J. Cell. Biol. 122, 1285-1294.

Frelinger, A. L., Du, X., Plow, E. F., & Ginsberg, M. H. (1991). Monoclonal antibodies to ligand-occupied conformer of integrin $\alpha IIb\beta 3$ (glycoprotein IIb-IIa) alter receptor affinity, specificity and function. J. Biol. Chem. 266, 17106-17111.

Galbraith, C. G., & Sheetz, M. P. (1997). A micromachined device provides a new bend on fibroblast traction forces. Proc. Natl. Acad. Sci, USA. 94, 9114-9118.

Giancotti, F. G. & Ruoslahti, E. (1990). Elevated levels of the $\alpha 5\beta 1$ fibronectin receptor suppress the transformed phenotype of Chinese hamster ovary cells. Cell. 60, 849-859.

Gilmore, A. P. & Burridge, K. (1996). Regulation of vinculin binding to talin and actin by phosphatidyl-inositol-4-5-bisphosphate. Nature. 381, 531-535.

Gluck, U. & Ben-Ze'ev, A. (1994). Modulation of α-actinin levels affects cell motility and confers tumorigenicity on 3T3 cells. J. Cell Sci. 107, 1773-1782.

Goodman, S. L., Risse, G., & von der Mark, K. (1989). The E8 subfragment of laminin promotes locomotion of myoblasts over extracellular matrix. J. Cell Biol. 109, 799-809.

Hahn, K., DeBiasio, R., & Taylor, D. L. (1992). Patterns of elevated free calcium and calcineurin activation in living cells. Nature. 359, 736-738.

Harris, A. K., Wild, P., & Stopak, D. (1980). Silicone rubber substrata: a new wrinkle in the study of cell locomotion. Science. 208, 177-179.

Hartwig, J. H., Bokoch, G. M., Carpenter, C. L., Janmey, P. A., Taylor, L. A., Toker, A., & Stossel, T. P. (1995). Thrombin receptor ligation and activated rac uncap actin filament barbed ends through phosphoinositide synthesis in permeabilized human platelets. Cell. 82, 643-653.

Hawkins, P. T., Equinoa, A., Qui, R.-G., Stokoe, D., Cooke, F. T., Walters, R., Wennstrom, S., Claesson-Welsh, L., Evans, T., Symons, M., & Stephens, L. (1995). PDGF stimulates and increase in GTP-rac via activation of phosphoinositide 3-kinase. Curr. Biol. 5, 393-403.

Hay, E. D. (1985). Interaction of migrating embryonic cells with extracellular matrix. Exp. Biol. Med. 10, 174-193.

Hendey, B., Klee, C. B., & Maxfield, F. R. (1992). Inhibition of neutrophil chemotaxis on vitronectin by inhibitors of calcineurin. Science. 258, 296-299.

Hirao, M., Sato, N., Kondo, T., Yonemura, S., Monden, M., Sasaki, T., Takai, Y., Tsukita, Sh., & Tsukita, Sa. (1996). Regulation mechanism of ERM protein/plasma membrane association: possible involvement of phosphatidylinositol turnover and rho-dependent signaling pathway. J. Cell Biol. 135, 37-52.

Ho, W.-C., Heinemann, C., Hangan, D., Uniyal, S., Morris, V. L., & Chan, B. M. C. (1997). Modulation of in vivo migratory function of $\alpha 2\beta 1$ integrin in mouse liver. Molec. Biol. Cell 8, 1863-1875.

Huttenlocher, A, Ginsberg, M. H., & Horwitz, A. F. (1996). Modulation of cell migration by integrin-mediated cytoskeletal linkages and ligand-binding affinity. J. Cell Biol. 134, 1551-1562.

Huttenlocher, A., Palecek, S. P., Lu, Q., Zhang, W., Mellgren, R. L., Lauffenburger, D. A., Ginsberg, M. H., & Horwitz, A. F. (1997). Regulation of cell migration by the calcium-dependent protease calpain. J. Biol. Chem. 272, 32719-32722.

Hughes, P. E., Diaz-Gonzalez, F., Leong, L., Wu, C., McDonald, J. A., Shattil, S., & Ginsberg, M. H. (1996). Breaking the integrin hinge: a defined structural constraint regulates integrin signaling. J. Biol. Chem. 271, 6571-6574.

Ilic, D., Furuta, Y., Kanazawa, S., Takeda, N., Sobue, K., Nakatsuji, N., Nomura, S., Fujimoto, J., Okada, M., Yamamoto, T., & Aizawa, S. (1995). Reduced cell motility and enhanced focal adhesion contact formation in cells from FAK-deficient mice. Nature. 377, 539-544.

Jay, P. Y., Pham, P. A., Wong, S. A., & Elson, E. L. (1995). A mechanical function of myosin II in cell motility. J. Cell Sci. 108, 387-393.

Kassner, P. D., Alon, R., Springer, T. A., & Hemler, M. E. (1995). Specialized functional properties of the integrin Cc4 cytoplasmic domain. Mol. Biol. Cell 6, 661-674.

Keely, P. J., Fong, A. M., Zutter, M. M., & Santoro, S. A. (1995). Alteration of collagen-dependent adhesion, motility, and morphogenesis by the expression of antisense α2 integrin mRNA in mammary cells. J. Cell Sci. 108, 595-607.

Kislauskis, E. H., Zhu, X-C., & Singer, R. H. (1997). β-actin messenger RNA localization and protein synthesis augment cell motility. J. Cell Biol. 136, 1263-1270.

Klemke, R. K., Cai, S., Giannini, A. L., Gallagher, P. J., de Lanerolle, P., & Cheresh, D. A. (1997). Regulation of cell motility by mitogen-activated protein kinase. J. Cell Biol. 137, 481-492.

Kovacsovics, T. J., Bachelot, C, Toker, A., Viahos, C. J., Duckworth, B., Cantley, L. C., & Hartwig, J. H. (1995). Phosphoinositide 3-kinase inhibition spares actin assembly in activating platelets but reverses platelet aggregation. J. Biol. Chem. 270, 11358-11366.

Kozma, R., Ahmed, S., Best, A., & Lim, L. (1995). The ras-related protein cdc42Hs and bradykinin promote formation of peripheral actin microspikes and filopodia in Swiss 3T3 fibroblasts. Mol. Cell. Biol. 15, 1942-1952.

Kuijpers, T. W., Mul, E. P. J., Blom, M., Kovach, N. L., Gaeta, F. C. A., Tollefson, V., Elices, M. J., & Harlan, J. M. (1993). Freezing adhesion molecules in a state of high-avidity binding blocks eosinophil migration. J. Exp. Med. 178. 279-284.

Kumagai, N., Morii, N., Fujisawa, K., Nemoto, Y., & Narumiya, S. (1993). ADP-ribosylation of rho p21 inhibits lysophosphatidic acid-induced protein tyrosine phosphorylation and phosphatidylinositol 3-kinase activation in cultured Swiss 3T3 cells. J. Biol. Chem. 268, 24535-24538.

Kundra, V., Escodedo, J. A., Kaziauskas, A., Kim, H. K., Rhee, S. G., Williams, L. T., & Zetter, B. R. (1994). Regulation of chemotaxis by the platelet-derived growth factor receptor-β. Nature. 367, 474-476.

Kuntz, R. M. & Saltzman, W. M. (1997). Neutrophil motility in extracellular matrix gels: mesh size and adhesion affect speed of migration. Biophys. J. 72, 1472-1480.

Larjava, H., Lyons, J. G., Salo, T., Makela, M., Koivisto, L., Birkedal-Hansen, H., Akiyama, S. K., Yamada, K. M., & Heino, J. (1993). Anti-integrin antibodies induce type IV collagenase expression in keratinocytes. J. Cell. Physiol. 157, 190-200.

Lawson, M. A. & Maxfield? F. R. (1995). Ca^{2+}-and calcineurin-dependent recycling of an integrin to the front of migrating neutrophils. Nature. 377, 75-79.

Lee, J., Leonard, M., Oliver, T., Ishihara, A., & Jacobson, K. (1994). Traction forces generated by locomoting keratocytes. J. Cell Biol. 127, 1957-1964.

Machesky, L. M. & Hall, A. (1996). Rho: a connection between membrane signalling and the cytoskeleton. Trends Cell Biol. 6, 304-310.

Maxfield, F. R. (1993). Regulation of leukocyte locomotion by Ca^{2+} Trends Cell Biol. 3, 386-391.

Michiels, F., Habets, G. G. M., Stam, J. C., van der Kammer, R. A., & Collard, J. G. (1995). A role for rac in tiam1-induced membrane ruffling and invasion. Nature. 375, 338-340.

Mitchison, T. J. & Cramer, L. P. (1996). Actin-based cell motility and cell locomotion. Cell. 84, 371-379.

Miura, Y., Kikuchi, A., Musha, T., Kuroda, S., Yaku, H., Sasaki, T., & Takai, Y. (1993). Regulation of morphology by rho p21 and its inhibitory GDP/GTP exchange protein (rho GDI) in Swiss 3T3 cells. J. Biol. Chem. 268, 510-515

Nakanishi, H., Brewer, K. A., & Exton, J. H. (1993). Activation of the ξ isozyme of protein kinase C by phosphatidylinositol 3,4,5-triphosphate. J. Biol. Chem. 268, 13-16.

Niggli, V., Andreoli, C., Roy, C., & Mangeat, P. (1995). Identification of a phoshatidylinositol-4,5-bisphosphate-binding domain in the N-terminal region of ezrin. FEBS Lett. 376, 172-176.

Nobes, C. D. & Hall, A. (1995). Rho, rac, and cdc42 GTPases regulate the assembly of multimolecular focal complexes associated with a tin stress fibers, lamellipodia, and filopodia. Cell. 81. 53 -62.

Palecek, S. P., Schmidt, C. E., Lauffenburger, D. A., & Horwitz, A. F. (1996). Integrin dynamics on the tail region of migrating fibroblasts. J. Cell Sci. 109, 941-952.

Palecek, S. P., Loftus, J. C., Ginsberg, M. H., Lauffenburger, D. A., & Horwitz, A. F. (1997). Integrin-ligand binding properties govern cell migration speed through cell-substratum adhesiveness. Nature. 385 537-540.

Palecek, S. P., Huttenlucher, A., Horwitz, A. F., & Lauffenburger, D. A. (1998). Physical and biochemical regulation of integrin release during rear detachment of migrating cells. J. Cell Sci. 111, 929-940.

Paterson, H. F., Self, A. J., Garrett, M. D., Just, I., Aktories, K., & Hall, A. (1990). Microinjection of recombinant p21 rho induces rapid changes in cell morphology. J. Cell Biol. 111, 1001-1007.

Regen, C. M. & Horwitz, A. F. (1992). Dynamics of β1 integrin-mediated adhesive contacts in mottle fibroblasts. J. Cell. Biol. 119, 1347-1359.

Ridley, A. J. & Hall, A. (1992). The small GTP-binding protein rho regulates the assembly of focal adhesion and actin stress fibers in response to growth factors. Cell. 70, 389-399.

Ridley, A. J., Comoglio, P. M., & Hall, A. (1995). Regulation of scatter factor/hepatocyte growth factor responses by ras, rac, and rho in MDCK cells. Mol. Cell Biol. 15, 1110-1122.

Saterbak, A. & Lauffenburger, D. A. (1996). Adhesion mediated by bonds in series. Biotechnol. Prog. 12, 682-699.

Schaller, M. D. & Parsons, J. T. (1994). Focal adhesion kinase and associated proteins. Curr Opin. Cell Biol. 6, 705-710

Schmidt, C., Horwitz, A. F., Lauffenburger, D. A., & Sheetz, M. P. (1993). Integrin/cytoskeleton interactions in migrating fibroblasts are dynamic, asymmetric, and regulated. J. Cell. Biol. 123, 977-991.

Seftor, R. E., Seftor, E. A., Stetler-Stevenson, W. G., & Hendrix, M. J. (1993). The 72 kDa type IV collagenase is modulated via differential expression of αvβ3 and α5β1 integrins during human melanoma cell invasion. Cancer Res. 53. 3411-3415.

Springer, T. A. (1995). Traffic signals on endothelium for lymphocyte recirculation and leukocyte emigration. Annu Rev. Physiol. 57, 827-872.

Stossel, T. P. (1993). On the crawling of animal cells. Science. 260, 1086-1094.

Tapon, N. & Hall, A. (1997). Rho, rac and cdc42 GTPases regulate the organization of the actin cytoskeleton. Curr. Opin. Cell Biol. 9, 86-92.

Toker, A., Meyer, M., Reddy, K. K., Falck, J. R., Aneja, R., Aneja, S., Parra, A., Burns, D. J., Ballas, L. M., & Cantley, L. C. (1994). Activation of protein kinase C family members by the novel polyphosphoinositides PtdIns-3,4-P2 and PtdIns-3,4,5-P3. J. Biol. Chem. 269, 32358-32367.

Tolias, K. F., Cantley, L., & Carpenter, C. L. (1995). Rho family GTPases bind to phosphoinositide kinases. J. Biol. Chem. 270, 17656-17659.

Varnum-Finney, B., & Reichardt, L. F. (1994). Vinculin-deficient PC12 cell lines extend unstable lamellipodia and filopodia and have a reduced rate of neurite outgrowth. J. Cell. Biol. 127, 1071-1084.

Wang, N., Butler, J. P., & Ingber, D. E. (1993). Mechanotransduction across the cell surface and through the cytoskeleton. Science. 260, 1124-1127.

Wang, N. & Ingber, D. E. (1994). Control of cytoskeletal mechanics by extracellular matrix, cell shape, and mechanical tension. Biophys. J. 66, 2181 -2189.

Ware, M. F., Wells, A., & Lauffenburger, D. A. (1998). Epidermal growth factor alters fibroblast migration speed and directional persistence reciprocally and in a matrix-dependent manner. J. Cell Sci. 111, 2423-2432.

Wei, Y., Lukashev, M., Simon, D. I., Bodary, S. C., Rosenberg, S., Doyle, M. V., & Chapman, H. A. (1996). Regulation of integrin function by the urokinase receptor. Science. 273, 1551-1555.

Wells, A., Gupta, K., Chang, P., Swindle, S., Glading, A., & Shiraha, H. (1998). Epidermal growth factor receptor-mediated motility in fibroblasts. Microsc. Res. Tech. 43, 395-411.

Wennstrom, S., Siegbahn, A., Yokote, K., Arvidsson, A. -K., Heldin, C. -H., Mori, S., & Claesson-Welsh, L. (1994). Membrane ruffling and chemotaxis transduced by the PDGF β-receptor require the binding site for phosphatidylinositol 3' kinase. Oncogene. 9, 651-660.

Werb, Z., Tremble, P. M., Behrendtsen, O., Crowley, E., & Damsky, C. H. (1989). Signal transduction through the fibronectin receptor induces collagenase and stromelysin gene expression. J. Cell Biol. 109, 877-889.

Wessels, D., Soll, D. R., Knecht, D., Loomis, W. F., DeLozanne, A., & Spudich, J. (1988). Cell motility and chemotaxis in Dictyostelium amoebae lacking myosin heavy chain. Dev. Biol. 128, 164-177.

Wessels, D., Vawter-Hugart, H., Murray, J., & Soll, D. R. (1994). Three-dimensional dynamics of pseudopod formation and the regulation of turning during the motility cycle of Dictyostelium. Cell Motil. Cytoskeleton. 27, 1-12.

Wilson, A. K., Gorgas, G., Claypool, W. D., & de Lanerolle, P. (1991). An increase or decrease in myosin II phosphorylation inhibits macrophage motility. J. Cell Biol. 114, 277-283.

Woods, A. & Couchman, J. R. (1992). Protein kinase C involvement in focal adhesion formation. J Cell Sci. 101, 277-290.

Xie, H., Pallero, M. A., Gupta, K., Chang, P., Ware, M. F., Witke, W., Kwiatkowski, D. J., Lauffenburger, D. A., Murphy-Ulrich, J. A., & Wells, A. (1998). EGF receptor regulation of cell motility: EGF induces disassembly of focal adhesions independently of the motility-associated PLC-γ signaling pathway. (submitted for publication).

Zhang, J., King, W. C., Dillon, S., Hall, A., Feig, L., & Rittenhouse, S. E. (1993). Activation of platelet phosphatidylinositide 3-kinase requires the small GTP-binding protein rho. J. Biol. Chem. 268, 22251-22254.

Zheng, Y., Bagrodia, S., & Cerione, R. A. (1994). Activation of phosphoinositide 3-kinase activity by cdc42Hs binding to p85. J. Biol. Chem. 269, 18727-18730.

ADHESION RECEPTORS:
CRITICAL EFFECTORS OF TROPHOBLAST DIFFERENTIATION DURING IMPLANTATION AND PLACENTATION

Caroline H. Damsky, Yan Zhou, Olga Genbacev,
Jay Cross, and Susan J. Fisher

Advances in Molecular and Cell Biology
Volume 28, pages 389-408.

ISBN: 0-7623-0495-2

I. NORMAL PLACENTATION: A UNIQUE COLLABORATION BETWEEN EMBRYONIC AND MATERNAL TISSUES

During the early stages of mammalian embryogenesis, the conceptus devotes most of its effort towards segregating, expanding, and promoting the differentiation of extraembryonic cell lineages, the extraembryonic endoderm, and trophoblast. As a result, an intimate relationship develops between fetal trophoblast cells and the maternal uterine vasculature that is essential for the nourishment and maintenance of the fetus to term. In a parallel fashion, the most urgent task for the embryo proper is to develop an intact, patent circulatory system. The critical role for the developing placenta during the first half of pregnancy is to engineer the functional interface between the maternal and fetal circulations such that nutrients can be transported to the fetus, and waste and gases efficiently exchanged. This must occur in order to cope with the rapid growth of the fetus during the second half of pregnancy. This chapter describes the critical roles that cell–extracellular matrix and cell–cell adhesion receptors play in this process.

The trophoblast lineage has two very distinct functional roles in early mammalian development (see Figures 1 and 2). Its first role, during the preimplantation period, is to form a polarized, transporting epithelial monolayer, the trophectoderm, around an inner core of cells (the inner cell mass), which will go on to form the embryo proper. As a result of the ability of trophectoderm to transport fluid, the blastocoele cavity forms. The formation of E-cadherin-based cell–cell contacts is essential for the organization of the intercellular junctional complex in the outer trophectoderm layer, because blastulation fails to occur in embryos deficient in E-cadherin (Larue et al., 1994; Riethmacher et al., 1995).

In the mouse, transition from trophectoderm to trophoblast begins shortly after the blastocyst hatches from the zone pellucida (E3.5-4). Immediately after hatching, the apical surface of the trophoblast (the external surface of the embryo) is quiescent and non-adhesive. Several hours later, it displays extensive protrusive and adhesive activity (Sutherland et al., 1988) and becomes attachment competent. In vitro, blastocysts are able to bind to, and grow out on, several defined

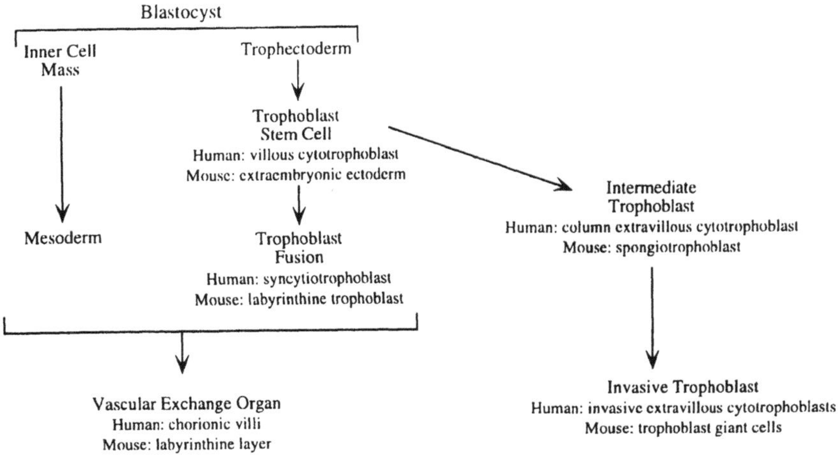

Figure 1. Schematic comparison of trophoblast derivatives in mouse and human placenta.

ECM ligands—including fibronectin, collagen and laminin—in a process that is dependent on the integrin family of ECM receptors (Sutherland et al., 1988, 1993; Schultz and Armant, 1995). During the transition from trophectoderm to trophoblast and the development of the ability to attach and migrate on ECM, trophoblast cells modulate expression of their integrin repertoire extensively: the $\alpha V\beta 3$ integrin is expressed constitutively, but $\alpha 6\beta 1$, which is expressed on trophectoderm, is downregulated on migrating trophoblast, whereas $\alpha 7\beta 1$ and $\alpha 1\beta 1$ are sequentially upregulated on the hatched blastocyst and migrating trophoblast, respectively (Damsky et al., 1993; Sutherland et al., 1993). Antibodies against $\alpha V\beta 3$ block trophoblast outgrowth on fibronectin, while antibodies against $\alpha 7$ partially block trophoblast outgrowth on laminin (A. Sutherland, unpublished experiments). Thus far, deletion of single integrin subunits in mice has not generated an obvious implantation phenotype. However, the trophoblast expresses a total of 7 different integrin complexes, representing members of both the αV and $\beta 1$ families, at the time it starts to invade the uterine stroma (E5). Therefore, it is likely that the remaining integrins compensate sufficiently for the absence of some integrins.

After the blastocyst implants in vivo (starting at E4.5-5) in the mouse, further expansion and differentiation of the trophoblast takes place in order to form the functional placenta. Trophoblast cells not in contact with the inner cell mass of the implanting embryo (mural trophoblast) undergo rapid terminal differentiation to primary giant cells and promote the initial invasion of the conceptus into the decidualized uterine stroma. Trophoblast cells in contact with the inner cell mass (polar trophoblast) remain proliferative and in the mouse, as so-called extraembryonic ectoderm cells, they go on to generate the chorion and ectoplacental cone. By day E7

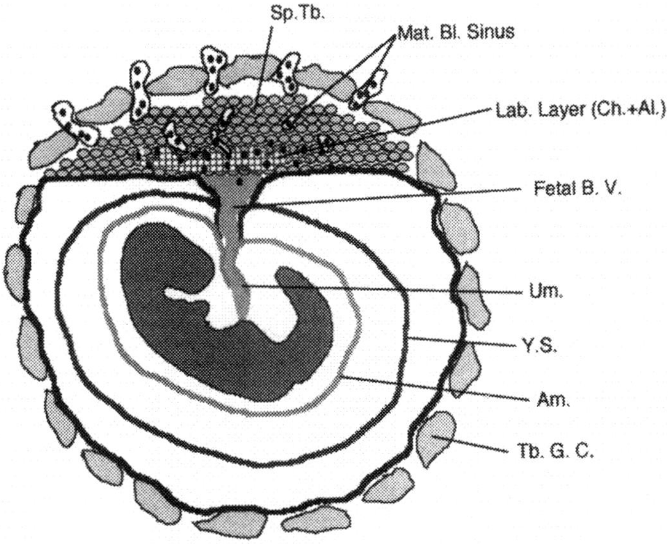

Figure 2. Schematic drawing of the definitive mouse chorioallantoic placenta at embryonic day 9. Fusion of the allantois and chorion to form the labyrinthine layer (Lab. Layer) brings the embryonic circulation (Fetal B.V.) in close proximity to trophoblast-lined maternal blood sinusoids (Mat. B1. Sinus). Sp. Tb., spongiotrophoblast layer; Tb. G. C., trophoblast giant cells; Am, amnion; ys, yolk sac; Um, umbilicus. For additional diagrams and explanation, see Cross and co-workers (1994).

in the mouse, an intimate relationship between maternal vasculature and trophoblast is first observed. This takes the form of maternal blood sinusoids lined by primary and secondary trophoblast giant cells. Thus, embryonic cells (trophoblast) come in direct contact with the maternal blood—the hallmark of hemochorial placentation.

Meanwhile, as a consequence of gastrulation, extraembryonic mesoderm forms and is the site of vasculogenesis within the yolk sac and allantois of the embryo. The chorioallantoic (definitive) placenta becomes functional by E8.5, when the extraembryonic ectoderm of the chorion (chorionic ectoderm) fuses with the allantois (within which umbilical vessels form). This process brings the trophoblast-modified maternal circulation and fetal circulation into sufficiently close proximity that nutrient transport and gas and waste exchange can take place successfully (see Figure 2; see Cross et al., 1994 for further description and diagrams). The interaction of the $\alpha 4\beta 1$ integrin on the chorionic plate with its counterreceptor, vascular cell adhesion molecule-1 (VCAM-1), on the tip of the allantois is important for chorioallantoic fusion, in that deletion of either molecule by gene targeting renders this process defective in about 50% of embryos, which then die at about E9.5 (Gurtner et al., 1995; Kwee et al., 1995; Yang et al., 1995). When these connections are fully developed (E9 in the mouse), the mouse placenta consists of several layers. Starting from the fetal side, these are the labyrinthine layer, the spongiotrophoblast, and trophoblast giant cells (see Figure 2; Cross et al., 1994).

Information is limited about the early post-implantation phase of trophoblast development in the human. However, extensive analysis of human placental tissues, starting from six weeks of gestation, have revealed a process of placentation (post-implantation morphogenesis at the fetal-maternal interface) functionally parallel to that seen in the mouse (see Figure 1; see Figure 3A), although histologically the placentas of the two species look quite different. In the human, a stem cell population of epithelial cytotrophoblasts (CTB), analogous to the mouse extraembryonic ectoderm, can differentiate along one of two pathways. In one pathway, these cells fuse to form a syncytium (analogous to the labyrinthine trophoblast layer of the mouse placenta: see Figure 2) that covers the surface of floating chorionic villi. These floating villi form the fetal compartment of the placenta and represent the bulk of the placental tissue that is expelled at birth. These villi are responsible for the gas and nutrient exchange between the maternal blood that bathes the intervillous space and the fetal circulation that resides in the villus stromal cores. In the second pathway, a subset of CTB in anchoring villi begin a remarkable transformation in which they become migratory and invasive, leave the fetal compartment and invade the decidualized maternal stroma (interstitial invasion) and uterine vasculature (endovascular invasion). These invasive CTB in the human are analogous to the giant cells of the implanting mouse embryo. By invading the uterine stroma and vascular network, these CTB serve to anchor the embryo/fetus to the mother (interstitial CTB) and to modify the maternal uterine circulation (endovascular CTB) so that blood efficiently perfuses the intervillous space that bathes the floating villi and nourishes the fetus.

The endovascular invasion process in human placentation is particularly remarkable. Studies in non-human primates (Blankenship et al., 1993a, 1993b) suggest the following scenarios. Initially, CTB enter the superficial arteriolar and venous segments of uterine vessels in the decidualized endometrium. They then migrate upstream in arterioles, but not veins, replacing the maternal endothelium on the walls of the spiral arterioles as far as the first third of the myometrium (see Figure 3A and 3B and Zhou et al., 1997a, 1997b). Utilizing their extensive proteolytic activity, the endovascular CTB remodel the muscular tunica media of the arteriole vessel walls they occupy, converting them to large-bore, low-resistance vessel segments. Thus, a hybrid spiral arteriole vasculature forms in the maternal compartment of the human placenta permitting enhanced perfusion of the intervillous space by well-oxygenated maternal blood. Trophoblast cells in the mouse are also invasive and line maternal vessel segments but the extent of endovascular invasion by trophoblast is less extensive than in the human. However, there is evidence in the mouse for extensive chemotaxis and neovascularization on the part of maternal uterine vasculature, driven in part by proliferin (Jackson and Linzer, 1997). Together, these observations lead to the hypothesis that a hybrid vasculature forms in

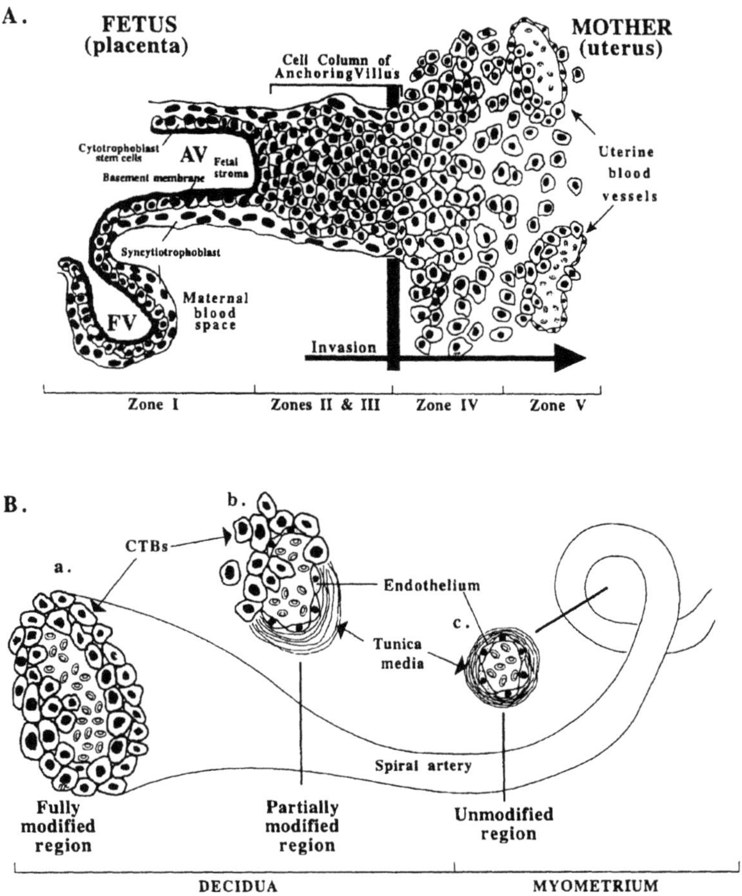

Figure 3. Diagrammatic representation of a first trimester human placental anchoring villus. (**A**) A subset of the cytotrophoblast (CTB) stem cells in anchoring villi forms a column of nonpolarized cells. These CTB go on to invade the uterine interstititum (interstitial invasion) and maternal vasculature (endovascular invasion). (**B**). Diagram of a spiral arteriole segment in which endovascular invasion is in progress (10-18 wks gestation). Decidual and then myometrial segments of the arteriole are modified progressively. In totally modified regions (a) the vessel diameter is large, and CTB are present in the lumen and occupy the entire surface of the vessel wall. In deeper, partially modified vessel segments (b), maternal endothelial cells and CTB occupy discrete regions of the vessel wall. In areas of intersection, CTB lie deep to the endothelium and in contact with the vessel wall (Zhou et al., 1997b). When the endovascular invasion process is complete, CTB will have taken over and remodeled the spiral arteriole wall as far as the first third of the myometrium. In deeper unmodified myometrial segments (c), maternal endothelium remains and the vessels retain their normal structure. (Reproduced from J. Clin. Investig. (1997) 99 (9): A, B: Figure 1, Zhou et al., 1997a, by copyright permission of The American Society for Clinical Investigation.)

all species that develop a hemochorial placenta, but that in some species (e.g. human) trophoblast invasion into the maternal vascular compartment is a primary mechanism for its creation, whereas in other species (e.g. the mouse) expansion of the maternal vasculature towards the conceptus and invading trophoblasts may play a more important role.

II. STEPWISE DIFFERENTIATION OF CYTOTROPHOBLASTS ALONG THE INVASIVE PATHWAY IN HUMAN PLACENTATION

Those cytotrophoblasts (CTB) that differentiate along the invasive pathway lose their ability to proliferate, and they form aggregates of non-polarized cells that become migratory and, finally, highly invasive. Several classes of molecules show altered expression patterns during this multistep process. These include hormones, proteinases and their inhibitors, and class I histocompatibility antigens (reviewed in Cross et al., 1994). However, when examining the morphology of the fetal-maternal interface in human first trimester placental bed biopsies, we became particularly interested in the regulation of adhesion molecule expression. Our initial hypothesis was that the repertoire of adhesion receptors expressed by CTB would have to change as they make their journey from the fetal to the maternal compartment. This was supported by our findings of adhesion molecule switching in trophoblast during the post-implantation period in mouse embryos (Sutherland et al., 1993). We also predicted that this adhesion molecule switching would have functional consequences for both interstitial and endovascular invasion by CTB and for normal human placentation. We tested this hypothesis by examining expression patterns of adhesion receptors in biopsy samples of the fetal-maternal interface, and by isolating cells from the fetal stem cell compartment and modeling their differentiation and acquisition of an invasive phenotype in vitro. Finally, we compared our morphological and functional findings in normal placental tissue with those in placental tissues from pregnancies complicated by preeclampsia, a disorder characterized by hypertension and proteinurea in the mother, intrauterine growth retardation in the fetus, and poor perfusion of the fetus by maternal blood due to shallow interstitial invasion of the uterine wall by CTB and their failure to modify maternal uterine arterioles (endovascular invasion).

We did indeed find that CTB modulate their adhesion molecule repertoire extensively as they differentiate along the invasive pathway. Our most surprising finding was that CTB transform their cell adhesion receptor phenotype from one characteristic of epithelial cells to one characteristic of endothelial cells. Thus CTB, which target and transform the maternal arteriole network, adopt the adhesion molecule phenotype of the endothelial cells they replace. The net functional effect of this adhesion molecule switching is to tip the balance in favor of a more migratory and invasive phenotype for this subset of CTB.

III. MORPHOLOGY OF CYTOTROPHOBLAST–VESSEL WALL INTERACTIONS IN SITU: COMPLEX ADHESIVE INTERACTIONS

In normal pregnancy, CTB participating in endovascular invasion have two types of interactions with maternal arterioles. To demonstrate these interactions, we double-stained tissue sections with (1) an antibody that recognizes cytokeratin and thereby identifies CTB and (2) an antibody that reacts with von Willebrand factor and thereby identifies endothelial cells (Zhou et al., 1997b). In the first type of interaction, large aggregates of these fetal cells are found primarily inside the vessel lumen. These aggregates can either lie adjacent to the apical surface of the resident endothelium or replace it, such that they appear directly attached to the vessel wall. In the second type of interaction, CTB are found within the vessel wall rather than in the lumen. In this position, they colonize the smooth-muscle layer of the vessel and lie subjacent to the endothelium. These different types of interactions may be progressive stages in a single process, or indicative of different strategies by which CTB accomplish endovascular invasion. In either case, the stage in which fetal CTB cohabit with maternal endothelium in the spiral arterioles is transient. By late second trimester these vessels are lined exclusively by CTB, and endothelial cells are no longer visible in either the endometrial or the superficial portions of their myometrial segments.

IV. MORPHOLOGY OF THE FETAL–MATERNAL INTERFACE IN PREECLAMPSIA DEFECTIVE INVASION AND REMODELING OF THE MATERNAL VASCULATURE

Although the cause of preeclampsia is unknown, the accumulated evidence strongly implicates the placenta (Redman, 1991). Anatomic examination shows that the specific area of the placenta most affected by this syndrome is the fetal–maternal interface. CTB invasion of the uterine interstitium is shallow, and endovascular invasion does not proceed beyond the terminal portions of the spiral arterioles. Serial sections through placental bed biopsies of all the patients we studied showed that few of the spiral arterioles contained CTB. Instead, most CTB were in the interstitium at some distance from these vessels (see Figure 4). In locations where endovascular CTB were detected, invasion was limited to the portion of the vessel that spans the superficial decidua. Thus, the interactions of CTB with uterine veins and arterioles differed little in preeclampsia. Even if the CTB gained access to the lumen, they usually failed to form tight aggregates among themselves, or to spread out on the vessel wall, as was observed for CTB in control samples matched for gestational age. Instead, in preeclampsia CTB that reached the uterine arterioles tended to remain as indi-

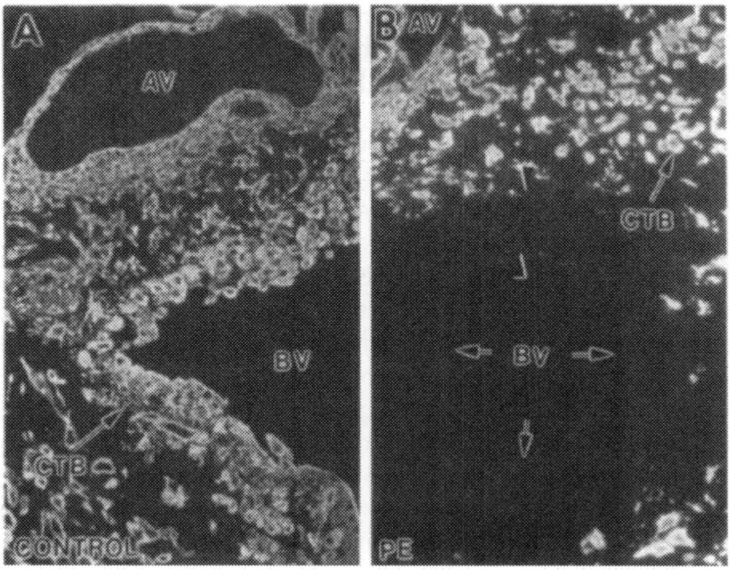

Figure 4. Section of second trimester placenta at the fetal–maternal interface stained for cytokeratin to label CTB. (**A**) In normal pregnancy, CTB leave the anchoring villus (AV) enter the uterine interstitium and colonize the walls of maternal spiral arterioles (long arrow) as far as the first third of the myometrium. (**B**) In preeclamptic pregnancies, CTB leave the AV and invade the superficial portions of the uterine interstitium (long arrow) and vasculature, but fail to colonize the spiral arteriole walls (BV; short arrows) or to remodel their muscular tunica media (bracketed by arrowheads). (Reproduced from The Journal of Clinical Investigation (1997) 99 (9): Cover figure, by copyright permission of The American Society for Clinical Investigation.)

vidual rounded cells, which suggested that they were poorly anchored to the vessel wall. Thus, CTB in preeclampsia not only have a limited capacity for endovascular invasion but also display an altered morphology in their interactions with maternal arterioles (Zhou et al., 1997b).

Because of these alterations in endovascular invasion, the maternal vessels of preeclampsia patients do not undergo the complete spectrum of physiologic changes that normally occur (e.g., loss of endothelial lining and musculoelastic tissue); the mean external diameter of the myometrial vessels is less than half that of similar vessels from uncomplicated pregnancies (Brosens et al., 1972; Gerretsen et al., 1981; Moodley and Ramsaroop, 1989). In addition, not as many vessels show evidence of CTB invasion (Khong et al., 1986; Zhou et al., 1993). Thus, the architecture of these vessels precludes an adequate response to gestation-related fetal demands for increased blood flow and likely accounts for the relatively hypoxic state of the fetal-maternal interface in preeclampsia.

V. NORMAL PREGNANCY: INVASIVE CYTOTROPHOBLASTS MIMICKING: VASCULAR ENDOTHELIAL CELLS

Our initial analysis of adhesion molecule expression focused on the β1 family of integrins. We and others found an intricate pattern of integrin switching as human CTB differentiate along the invasive pathway, (Damsky et al., 1992; Vicovac et al., 1995) that is analogous to the integrin switching that takes place in the peri-implantation period in the mouse, described earlier (Sutherland, et al., 1993). The chorionic villus stem cell CTB express primarily the laminin receptor, α6β4. This receptor is downregulated as differentiation proceeds, and the α5β1 fibronectin receptor and α1β1 laminin/collagen receptor complexes are upregulated. Furthermore, we found that anti-α1 antibodies significantly inhibit the ability of isolated human CTB to invade the basement membrane-like ECM, Matrigel (Damsky et al., 1994). In contrast, blocking α5β1 function stimulates CTB invasion. This suggests that CTB upregulate invasion-restraining as well as invasion-promoting adhesive mechanisms and balance them to help achieve the appropriate depth of invasion. These data suggest further that integrin switching has functional significance for trophoblast invasion.

In more recent studies, we have expanded this survey and tested the hypothesis that since invasive CTB can penetrate and line maternal blood vessel walls, they must be able to mimic broadly the adhesion phenotype of the endothelial cells they replace. We hypothesized further that these changes in adhesion phenotype also have the net effect of enhancing CTB motility and invasiveness. To test these hypotheses, we first stained tissue sections of the fetal–maternal interface for specific integrins, cadherins, and immunoglobulin families of adhesion receptors that are characteristic of endothelial cells and leukocytes. We then tested the functional consequences for CTB adhesion and invasion of expressing the particular adhesion receptors that were upregulated during CTB differentiation (Zhou et al., 1997a).

We were particularly interested in the distribution patterns of αV integrin family members because of their regulated expression on endothelial cells during angiogenesis (Brooks et al., 1994) and their upregulation on some types of metastatic tumor cells (Albelda et al., 1990). αV family members displayed unique and highly specific spatial staining patterns on CTB in anchoring villi and the placental bed. Staining for the αVβ5-complex was restricted to the CTB stem cell monolayer in chorionic villi and was not found in the syncytiotrophoblast layer or in CTB in cell columns and the placental bed. In contrast, anti-αVβ6 stained only those chorionic villus CTB that were at sites of column formation. The CTB layer still in contact with basement membrane at these column initiation sites stained brightly, while the first layer of the cell column showed reduced staining. The rest of the CTB in chorionic villi, those in more distal regions of cell columns, and those within placental bed and vasculature did not stain for αVβ6, documenting a specific association of this integrin with initiation of column formation. In yet a different pattern, staining for anti-αVβ3 was weak or not detected on villus CTB or on those in the initial layers of cell columns. However, strong staining was detected on CTB within the uterine wall and vasculature. Thus, individual members of the αV family, like those of the β1 family (Damsky et al., 1992), are spatially regulated during CTB differentiation. Of particular relevance to this study is the observation that αVβ3 integrin, expression of which is stimulated on endothelial cells by angiogenic factors (Brooks et al., 1994), is enhanced on CTB that have invaded the uterine wall and maternal vasculature.

Blocking αVβ3 function suppresses endothelial migration during angiogenesis (Brooks et al., 1994). To determine whether perturbing the function of αVβ3 also affects CTB invasion in vitro, freshly isolated first trimester CTB were plated for 48 hours on Matrigel-coated Transwell filters in the presence of control mouse IgG or the complex-specific anti-αVβ3 IgG, LM609. CTB invasion was evaluated by counting cells and cellular processes that had invaded the Matrigel barrier and extended through the holes in the Transwell filters (Librach et al., 1991). LM609 reduced CTB invasion by more than 75% in this assay, indicating that αVβ3, like the αlβ1 integrin (Damsky et al., 1994), contributes significantly to the invasive phenotype of CTB (Zhou et al., 1997a).

Next, we looked for evidence of cadherin switching during CTB differentiation in vivo. The CTB epithelial monolayer stained strongly for the ubiquitous epithelial cadherin, E-cadherin, in a polarized pattern. Staining was strong on the surfaces of CTB in contact with one another and with the overlying syncytiotrophoblast layer, and was absent at the basal surface of CTB in contact with basement membrane. In cell columns, E-cadherin staining intensity was reduced on CTB near the uterine wall and on CTB within the decidua. This reduction in staining was particularly pronounced in second trimester tissue (see Figure 5A). At this stage, E-cadherin staining was also very weak or undetectable on CTB that had colonized maternal blood vessels and on those in the surrounding myometrium (Zhou et al., 1997a). All locations of reduced E-cadherin staining were areas in which invasion is active during the first half of gestation. Interestingly, the staining intensity of E-cadherin was strong on CTB in all locations in term placentas, at which time CTB invasive activity is poor. Taken together, these data are consistent with the idea that CTB transiently reduce E-cadherin function at times and places of their greatest invasive activity.

Cadherin switching occurs frequently during embryonic development when significant morphogenetic events take place (Takeichi, 1995). We therefore stained sections of first and second trimester placental tissue with antibodies to other classical cadherins. These tissues did not stain with antibodies to P-cadherin but did stain with antibodies that recognize the endothelial cadherin, VE-cadherin. In chorionic villi, antibody to VE-cadherin did not stain villus CTB, although it stained the endothelium of fetal blood vessels within the villus stroma. In contrast, anti-VE-cadherin stained CTB in cell columns and in the decidua (see Figure 5B)—just the

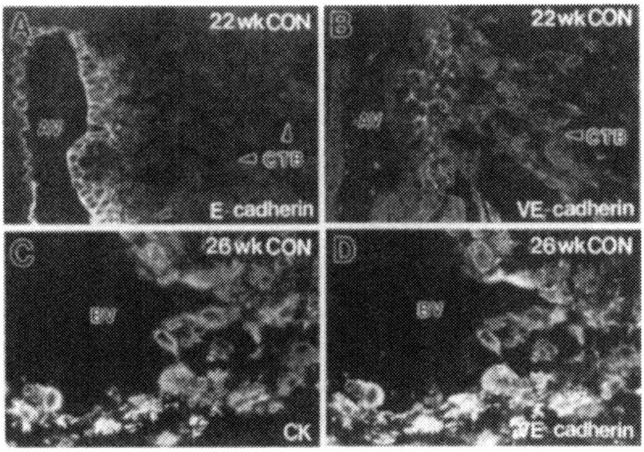

Figure 5. Staining patterns of E-cadherin. (**A**) and VE-cadherin (**B**,**D**) in sections of fetal–maternal interface from a normal second trimester placental bed biopsy. (**A**) E-cadherin stains CTB in the anchoring villus (AV) and initial layers of cell column strongly. Staining is decreased on CTB in decidua (arrow heads) and maternal vasculature (not shown). (**B**) Adjacent section stained with anti-VE-cadherin. Villus CTB are negative. VE-cadherin is first detected on column and decidual CTB (arrowhead). (**C**,**D**) Section of myometrium double stained for cytokeratin (CK: C) to identify CTB and VE-cadherin (**D**). Staining for VE-cadherin is strong on CTB in interstitium and on the wall of a large fully modified maternal blood vessel (BV). (Reproduced from J. Clin. Investig. 99 (9): **A**, **B**: Zhou et al., 1997a; **C**, **D**: Zhou et al., 1997b, by copyright permission of The American Society for Clinical Investigation.)

areas in which E—cadherin staining was reduced. In maternal vessels that had not yet been modified by CTB, VE-cadherin stained the endothelial layer strongly. Following endovascular invasion, CTB lining maternal blood vessels also stained strongly for VE-cadherin (see Figure 5C,D). Thus, CTB that invade the uterine wall and vasculature express a cadherin characteristic of endothelial cells.

To determine the functional consequences of VE-cadherin for CTB differentiation/invasion, it was important to show that CTB could modulate their cadherin repertoire in vitro. Interestingly, we found that freshly isolated CTB did not change their level of E-cadherin protein or mRNA as they differentiated in vitro, even though the staining intensity of E-cadherin declines dramatically on differentiating CTB in vivo. This suggests that the E-cadherin protein has a more diffuse cellular localization in the non-polarized differentiating/invading cells and is not organized in a polarized fashion at junctional complexes, where its high concentration would result in a strong staining signal. In contrast, the anti-VE-cadherin antibody did not detect its antigen in extracts of freshly isolated CTB. However, by 10 hours, this antibody recognized a single band at 140 kD, which increased in intensity by 36 hours. This band co-migrated with a 140 kD band recognized by anti-VE-cadherin in extracts of human umbilical vein endothelial cells. VE-cadherin mRNA was also dramatically upregulated.

Next, we used function-perturbing anti-cadherin antibodies, in conjunction with the Matrigel invasion assay, to assess the functional consequences of cadherin modulation for CTB invasiveness. By 48 hours, significant invasion by control CTB was evident. However, in cultures treated with anti-E-cadherin, CTB invasiveness increased more than threefold, suggesting that E-cadherin normally has a restraining effect on invasiveness. In contrast, antibody against VE-cadherin reduced the invasion of CTB to about 60% of control. This suggests that the presence of VE cadherin normally facilitates CTB invasion. Taken together, these functional data suggest that as they differentiate, the cells modulate their cadherin repertoire to one that contributes to their increased invasiveness (Zhou et al., 1997a). One hypothesis to explain this effect is that the upregulation of VE-cadherin in E-cadherin-containing CTB at sites of column formation interferes with the localization and function of E-cadherin, thereby reducing its ability to maintain the stable epithelial monolayer characteristic of villus CTB.

Our data thus far indicate that, as they differentiate, CTB downregulate expression or function of adhesion receptors highly characteristic of epithelial cells (integrins $\alpha6\beta4$, [Damsky et al., 1992], and $\alpha V\beta6$, and E-cadherin) and upregulate analogous receptors that are expressed on endothelial cells (integrins $\alpha1\beta1$ [Damsky et al., 1992] and $\alpha V\beta3$, and VE-cadherin). Further experiments showed that CTB also upregulate additional adhesion receptors characteristic of endothelial cells. Thus, villus CTB do not express VCAM-1, its counter receptor integrin $\alpha4\beta1$, or platelet-endothelial cell adhesion molecule (PECAM)-1. However, all three are detected on CTB in cell columns and the uterine interstitium and on endovascular CTB. Furthermore, the $\alpha4$-containing complexes on CTB were functional, as determined by the ability of isolated CTB to adhere to a substrate coated with VCAM-1-IgGFc in an in vitro adhesion assay. This interaction was blocked by both anti-$\alpha4$ and anti-VCAM-1. These data suggest that CTB could use their $\alpha4$ integrins to interact either with VCAM-1 on other CTB or with VCAM-1 on maternal endothelial cells (Zhou et al., 1997a). Taken together, these observations support our hypothesis that normal CTB undergo a comprehensive switch in phenotype so as to resemble the endothelial cells they replace during endovascular invasion.

Together, these results suggest that during normal pregnancy, invasive CTB mimic the adhesion receptor phenotype of endothelial cells. We hypothesize that this unusual phenomenon plays an important role in the process whereby these cells form vascular connections with the uterine vessels. Ultimately these connections are so extensive that the spiral arterioles become hybrid structures in which fetal CTB replace the maternal endothelium and much of the highly muscular tunica media. As a result, the diameter of the spiral arterioles increases dramatically, allowing blood flow to the placenta to keep pace with fetal growth. Circumstantial evidence suggests that several of the adhesion molecules whose expression we studied could play an important role in forming these novel vascular connections. For example, it is very interesting to find that CTB are the only cells, other than the endothelium, that express VE-cadherin. In addition, VE-cadherin and PECAM-1 are the first adhesion receptors expressed by differentiating endothelial cells during early development. $\alpha V\beta 3$ expression is upregulated on endothelial cells during angiogenesis by soluble factors that regulate this process. Thus, adhesion receptors that are upregulated as normal CTB differentiate/invade play vital roles in differentiation and expansion of the vasculature

VI. PREECLAMPSIA: INVASIVE CYTOTROPHOBLASTS FAIL TO MIMIC VASCULAR CELLS

In these studies we tested the hypothesis that preeclampsia negatively affects the ability of CTB to express adhesion molecules that are normally modulated during the unique epithelial-to-vascular transformation that occurs in normal pregnancy (Zhou et al., 1997b). Our initial studies (Zhou et al., 1993) showed that in preeclampsia, CTB failed to downregulate the major integrin of villus CTB $\alpha 6\beta 4$ and failed to upregulate the $\alpha 1\beta 1$ complex, which we have shown is required for normal CTB invasion in vitro (Damsky et al., 1994). Next, we found that CTB expression of all three αV-family members was altered in preeclamptic tissues. In particular, staining for $\beta 6$ was much brighter and extended beyond the column to include CTB within the superficial decidua. Of even greater interest, staining for $\beta 3$ was weak on CTB in all locations; those in the uterine wall of preeclampsia patients failed to show strong staining for $\beta 3$, as did normal CTB that penetrated the spiral arterioles. Thus, in preeclampsia, differentiating/invading CTB retain expression of $\alpha V\beta 6$, which is transiently expressed in many epithelial tissues during development and remodeling, and fail to upregulate $\alpha V\beta 3$, which is characteristic of angiogenic endothelium. Therefore, as was the case for integrin $\alpha 1$ (Zhou et al., 1993), our analyses of the expression of αV-family members suggest that in preeclampsia, CTB start to differentiate along the invasive pathway, as evidenced by their ability to turn on expression of $\alpha V\beta 6$ and $\alpha 5\beta 1$ (Damsky et al., 1992; Zhou et al., 1993), but cannot complete this process. The functional implications of the failure of CTB in preeclampsia to switch their integrin repertoire and, in particular, to upregulate the $\alpha 1$ and $\beta 3$ subunits were reinforced by our finding that CTB freshly isolated from preeclamptic tissues failed both to invade Matrigel and to upregulate expression of integrin $\alpha 1$ at mRNA or protein (Lim et al., 1997).

Preeclampsia also had a striking effect on CTB cadherin expression. In contrast to control samples, CTB in both the villi and the decidua of preeclamptic placentae showed strong reactivity with anti-E-cadherin (see Figure 6A) and staining remained strong even on CTB that had penetrated the superficial portions of uterine arterioles (Zhou et al., 1997b). Interestingly, in preeclampsia we consistently found that CTB within the uterine wall tended to exist as large aggregates, rather than as

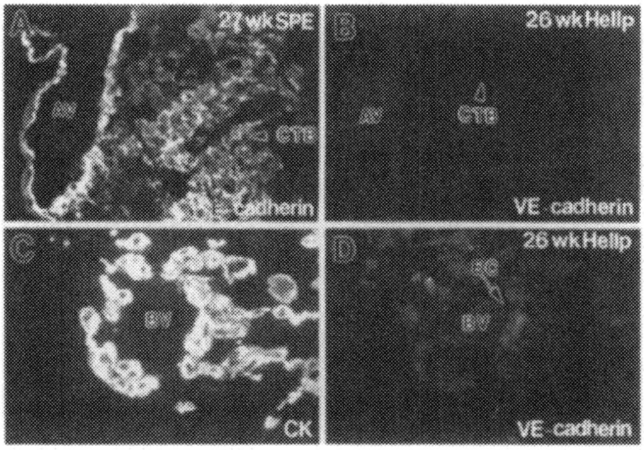

Figure 6. Staining patterns of E-cadherin. (**A**) and VE-cadherin (**B,D**) in sections of fetal–maternal interface from preeclamptic second trimester placental bed biopsies. (**A**) 27 wk SPE (severe preeclampsia). E-cadherin staining is strong on decidual CTB (arrowheads) as well as on villus CTB. (**B**) 26 wk HELLP (Hepatic involvement, Elevated Liver enzymes and Lowered Platelets): a very severe form of preeclampsia). VE-cadherin is not detected on CTB in anchoring villus (AV) or decidua (arrowhead). (**C,D**) Section of deeper decidual region of 26 wk HELLP placental bed biopsy double stained for CK and VE-cadherin. CK-positive CTB are present in the decidua and surrounding a small BV. However, these CTB do not stain for VE-cadherin. VE-cadherin staining is detected on the maternal endothelial cells (EC) of the unmodified BV. (Reproduced from J. Clin. Investig. 99 (9): Zhou et al., 19971, by copyright permission of The American Society for Clinical Investigation.)

smaller clusters and single cells, as is the case in normal pregnancy. This observation is in accord with the likelihood that E-cadherin mediates strong intercellular adhesion between CTB, as it does in all other normal epithelia examined.

Strikingly, no VE-cadherin staining was detected on CTB in any location in placental bed specimens obtained from preeclampsia patients (see Figure 6 B, D)—neither CTB in the cell columns, nor the few cells that were found in association with vessels in the superficial decidua expressed VE-cadherin. However, staining for this adhesion molecule was detected on maternal endothelium in the unmodified uterine vessels in placental bed biopsy specimens from preeclampsia patients (see Figure 6D). Thus, cadherin modulation by CTB in preeclampsia was defective, as shown by the persistence of strong E-cadherin staining and the absence of VE-cadherin staining on CTB in columns and in the superficial decidua. Similarly, when the pregnancy was complicated by preeclampsia, staining for VCAM-1 was not detected on CTB in the decidua. The few CTB that reached the termini of spiral arterioles showed only weak antibody reactivity. However, integrin $\alpha4$ expression on CTB in preeclampsia specimens was the same as in age-matched control tissue. Finally, preeclampsia

CTB also failed to stain for PECA M-1, whereas, maternal endothelial cells that lined the uterine vessels stained brightly.

VII. OXYGEN TENSION: AN IMPORTANT ENVIRONMENTAL FACTOR REGULATING CYTOTROPHOBLAST ADHESION RECEPTOR SWITCHING AND INVASION

Our results raise the interesting possibility that the failure of CTB in preeclampsia to express vascular-type adhesion molecules, as normal CTB do, impairs their ability to form connections with the uterine arterial vessels. This failure ultimately limits the supply of well-oxygenated blood to the fetal compartment and is likely to contribute to the relatively hypoxic environment at the fetal-maternal interface that persists throughout pregnancies complicated by preeclampsia (discussed in Genbacev et al., 1996, 1997) and is thought to be closely linked to the pathophysiology of this syndrome.

The possibility that a chronic reduction in oxygen tension at the fetal–maternal interface might limit the ability of CTB in preeclampsia to differentiate and/or invade led us to investigate whether exposing normal CTB to hypoxic conditions would alter their ability to differentiate, switch their adhesion phenotype, and invade Matrigel in vitro (Genbacev et al., 1996, 1997). Parallel cultures of freshly isolated normal first or second trimester CTB were cultured on Matrigel-coated coverslips or invasion chambers under ambient oxygen (20%) or hypoxic (2%) conditions for 24–72 hours. Most cellular functions were not altered. However, in contrast to CTB cultured in ambient oxyen, CTB in hypoxia maintained their ability to enter S-phase of the cell cycle. Furthermore, although CTB in hypoxia could carry out some aspects of the adhesion receptor switching ($\alpha5\beta1$ and $\alpha V\beta3$ were upregulated normally), they did not upregulate $\alpha1\beta1$ and did not invade Matrigel. Thus, normal CTB exposed to hypoxic conditions mimicked important features of the arrested differentiation phenotype expressed by CTB in preeclampsia. How oxygen tension regulates CTB differentiation, and whether CTB in normal and preeclamptic pregnancies respond differently to the large oxygen gradient that exists at the fetal-maternal interface during early pregnancy, are still unclear. However, our studies support the hypothesis that oxygen tension is one critical environmental factor regulating CTB differentiation. These studies also show that the expression of the $\alpha1$ integrin subunit, and perhaps of other adhesion receptors (Ginis et al., 1995; Klein et al., 1995; Burges et al., 1997), is oxygen sensitive.

We also hypothesize that the failure of CTB in preeclampsia to shift to a vascular cell adhesion phenotype might be part of a broader-spectrum defect in which the cells fail to function properly as endothelium. Such a failure would no doubt have important effects on the maintenance of vascular integrity at the maternal-fetal interface. Clearly, in preeclampsia undifferentiated CTB that fail to mimic the adhesion phenotype of endothelial cells are present in the termini of maternal spiral arterioles. Whether this phenomenon affects the phenotype of maternal endothelium in deeper segments of the same vessels and/or is linked to the systemic maternal endothelial pathology that is a hallmark of this disease is an interesting question that remains to be investigated.

VIII. ADHESION RECEPTORS AND IMPLANTATION: COMMON REGULATORY PATHWAYS IN THE DEVELOPMENTAL PROGRAMS FOR VASCULOGENESIS AND PLACENTATION

The expression on differentiating CTB of adhesion receptors that are very early markers of endothelial cell differentiation (e.g. PECAM-1 and VE-cadherin) suggests that there are at least some common regulatory pathways in these two developmental programs. In fact, CTB in cell columns and placental bed express Flt-1, a receptor for vascular endothelial growth factor (VEGF), as well as VEGF itself. VEGF is also expressed by fetal and maternal macrophages (Ahmed et al., 1995; Clark et al., 1996; Shiraishi et al., 1996), which suggests that CTB could respond in either an autocrine or a paracrine fashion to the VEGF signal. Thus, the wave of endovascular invasion by CTB, which peaks in the second trimester, resembles a vasculogenic/angiogenic response. The ability of CTB to adopt the adhesion phenotype of activated endothelial cells, along with their high expression of proteinases, may account for their virtually unique ability not only to invade the maternal vasculature, but to displace maternal endothelium and remodel the muscular walls of the uterine arterioles. Other cells, such as metastatic tumor cells, enter the bloodstream and then extravasate, leaving the vessel wall relatively intact.

Dissection of the pathways by which signaling through receptors for vasculogenic factors regulates expression of adhesion receptors is currently at a very early stage for both vasculogenesis/angiogenesis and CTB differentiation. However, one report suggests that Hox-D3 expression is required to upregulate $\alpha V\beta 3$ and urokinase in the angiogenic response to fibroblast growth factor in endothelial cells (Boudreau et al., 1997). Similarly, evidence for the identity of transcription factors involved in committment and differentiation of endothelial and trophoblast lineages, or in the regulation of adhesion receptor expression in these lineages, is also limited and largely circumstantial. For example, Hxt (Cross et al., 1995) and Mash-2 (Guillemot et al., 1994) have been identified as critical for regulating specific aspects of the trophoblast differentiation program in the mouse, and ets family transcription factors have been implicated in vasculogenesis (reviewed in Risau and Flamme, 1995), but there is little insight into their downstream targets. In another example, the PECAM-1 promoter has been shown to have GATA-2 binding activity that is required for its optimal activation in endothelial cells (Gumina et al.,

1997). Interestingly, GATA-2 and GATA-3 have been reported to regulate trophoblast-specific gene expression in vivo (Ma et al., 1997). Progress in unraveling the cascade that regulates adhesion molecule expression during either vasculogenesis or trophoblast differentiation should shed light on both processes, as well as on their interrelationship.

NOTE ADDED IN PROOF

Bader and colleagues (1998) describe a placental phenotype in αv integrin knockout mice: The labyrinthine layer was reduced and there was less intermingling of fetal blood vessels and maternal blood sinuses.

REFERENCES

Ahmed, A., Li, X. F., Dunk, C., Whittle, M. J., Rushton, D. I., & Rollason, T. (1995). Localisation of vascular endothelial growth factor and its Flt-1 receptor in human placenta. Growth Factors 12, 235-243.

Albelda, S., Mette, S. A., Elder, D. E., Stewart, R., Damjanovich, L., Herlyn, M., & Buck, C. (1990). Integrin distribution in malignant melanoma: association of the beta 3 subunit with tumor progression. Cancer Res. 50, 6757-6764.

Bader, B. L. Rayburn, H., Crowley, D., & Hynes, R. O. (1998). Extensive vasculogenesis, angiogenesis, and organogenesis precede lethality in mice lacking all αv integrins. Cell. 95, 507-520.

Blankenship, T. N., Enders, A. C., & King, B. F. (1993a). Trophoblastic invasion and the development of uteroplacental arteries in the macaque: immunohistochemical localization of cytokeratins, desmin, type IV collagen, laminin, and fibronectin. Cell Tissue Res. 272, 227-236.

Blankenship, T. N., Enders, A. C., & King B. F. (1993) Trophoblastic invasion and modification of uterine veins during development in macaques. Cell Tissue Res. 274, 135-144.

Boudreau, N. & Cheresh, D. A. (1997). Induction of the angiogenic phenotype by Hox D3. J. Cell Biol. in press.

Brooks, P. C., Clark, R. A. F., & Charesh, D. (1994). Requirement of aVp3 integrin for angiogenesis. Science 264, 569-571.

Brosens, I. A., Robertson, W. B., & Dixon, H. G. (1972). The role of the spiral arteries in the pathogenesis of preeclampsia. Obstet. Gynecol. Ann. 1, 177-191.

Burges A., Allmeling A., & Krombach F. (1997). Hyperoxia induces upregulation of CD11b and amplifies LPS-induced TNF-alpha release by alveolar macrophages. Euro. J. Med. Res. 2, 149-154.

Clark, D. E., Smith, S. K., Sharkey, A. M., & Charnock-Jones, D. S. (1996) Localization of vascular endothelial growth factor and expression of its receptors flt and KDR in human placenta throughout pregnancy. Hum. Reprod., 11, 1090-1098.

Cross, J. C., Werb, Z., & Fisher, S. J. (1994). Implantation and the placenta: key pieces of the development puzzle. Science 266, 1508-1518.

Cross J. C., Flannery, M. L., Blanar, M. A., Steingrimsson, E., Jenkins, N. A., Copeland, N. G., Rutter, W. J., & Werb, Z. (1995). Hxt encodes a basic helix-loop-helix transcription factor that regulates trophoblast cell development. Development. 121, 2513-2523.

Damsky, C. H., Fitzgerald, M. L., & Fisher, S. J. (1992). Distribution patterns of extracellular matrix components and adhesion receptors are intricately modulated during first trimester CTB differentiation along the invasive pathway in vivo. J. Clin. Invest. 89, 210-222.

Damsky, C. H., Sutherland, A. E., & Fisher, S. J., (1993) Adhesive interactions in early mammalian embryogenesis, implantation and placentation. FASEB J., 7, 1320-1328.

Damsky, C. H., Librach, C., Lim, K-H., Fitzgerald, M. L., McMaster, M. T., Janatpour, M., Zhou, Y., Logan, S. K., & Fisher, S. J. (1994). Integrin switching regulates normal trophoblast invasion. Development. 120, 3657-3666.

Genbacev, O., Joslin, R., Damsky, C. H., Polliotti, B. M., & Fisher S. J. (1996). Hypoxia alters early gestation human cytotrophoblast differentiation/invasion in vitro and models the placental defects that occur in preeclampsia. J. Clin. Invest. 97, 540-550.

Genbacev, O., Zhou, Y., Ludlow, J., & Fisher, S. (1997). Regulation of human placental development by oxygen tension. Science, 227, 1669-1672.

Ginis, I., Mentzer, S. J., Li, X., & Faller, D. V. (1995). Characterization of a hypoxia-responsive adhesion molecule for leukocytes on human endothelial cells. J. Immunol. 155, 802-810.

Gerretsen, G., Huisjes, H. J., & Elema, J. D. (1981). Morphological changes of the spiral arteries in the placental bed in relation to pre-eclampsia and fetal growth retardation. Br. J. Obstet. Gynaecol. 88, 876-881.

Guillemot, F., Nagy, A., Auerbach, A., Rossant, J., & Joyner, A. L. (1994). Essential role of Mash-2 in extraembryonic development. Nature 371, 333-336.

Gurtner G. C., Davis, V., Li, H., McCoy, M. J., Sharpe, A., & Cybulsky, M. I. (1995). Targeted disruption of the murine VCAM-1 gene: essential role of VCAM-1 in chorioallantoic fusion and placentation. Genes & Dev. 9, 1-14.

Jackson, D. & Linzer, D. I. (1997) Proliferin transport and binding in the mouse fetus. Endocrinology. 138, 149-55.

Khong, T. Y., De Wolf, F., Robertson, W. B, & Brosens, I. (1986). Inadequate maternal vascular response to placentation in pregnancies complicated by pre-eclampsia and by small-for-gestational age infants. Br. J. Obstet. Gynaecol. 93. 1049-1059.

Klein, C. L., Kohler, H., Bittinger, F., Otto, M., Hermanns, I., & Kirkpatrick, C. J. (1995). Comparative studies on vascular endothelium in vitro. 2. Hypoxia: influences on endothelial cell proliferation and expression of cell adhesion molecules. Pathobiology 63, 1-8.

Kwee, L., Baldwin, H. S., Shen, H. M., Stewart, C. L., Buck, C., Buck, C. A., & Labow, M. A. (1995). Defective development of the embryonic and extraembryonic circulatory systems in vascular cell adhesion molecule (VCAM-1) deficient mice. Development 121, 489-503.

Larue, L., Ohsugi, M., Hirchenhainj J., & Kemler, R. (1994) E-cadherin-null mutant embryos fail to form a trophectoderm epithelium. Proceedings of the National Academy of Sciences (USA) 91,8263-7.

Librach, C. L., Werb, Z., Fitzgerald, M. L., Chiu, K., Corwin, N. M., Esteves, R. A., Grobelny, D., Galardy, R., Damsky, C. H., & Fisher, S. J. (1991). 92-kD type IV collagenase mediates invasion of human cytotrophoblasts, J. Cell Biol., 113, 437-449.

Lim, K-H., Zhou, Y., Janatpour, M., McMaster, M., Bass, K., Chun, S-H., & Fisher, S. J. (1997) Human Cytotrophoblast differentiation/invasion is abnormal in preeclampsia. Am. J. Path. (in press).

Ma, M. G. T., Roth, M. E., Groskopf, J. C., Tsai, F. Y., Orkin, S. H., Grosveld, F., Engel, J. D., & Linzer, D. I. (1997). GATA-2 and GATA-3 regulate trophoblast-specific gene expression in vivo. Development 124, 907-914.

Moodley, J. & Ramsaroop, R. (1989). Placental bed morphology in black women with eclampsia. S. Afr. Med. J., 75, 376-378.

Redman, C. W. (1991). Current topic: pre-eclampsia and the placenta. Placenta 12, 301-308.

Risau, W. & Flamme, (1995) Vasculogenesis. Ann. Rev. Cell Dev. Biol., 11, 73-91.

Riethmacher, D., Brinkmann, V., & Birchmeier, C. (1995). A targeted mutation in the mouse E-cadherin gene results in defective preimplantation development. Proc. Natl. Acad. Sci. U.S.A. 92, 855-859.

Schultz, J. F. & Armant, D. R. (1995). Beta 1- and beta 3-class integrins mediate fibronectin binding activity at the surface of developing mouse pert-implantation blastocysts. Regulation by ligand-induced mobilization of stored receptor. J. Biol. Chem. 270, 11522- 11531.

Shiraishi, S. K., Nakagawa, K., Kinukawaf, N., Nakano, H., & Sueishi, K. (1996) Immunohistochemical localization of vascular endothelial growth factor in the human placenta. Placenta 17, 111-121.

Sutherland, A. E., Calarco, P. G., & Damsky, C. H. (1988). Expression and function of cell surface extracellular matrix receptors in mouse blastocyst attachment and outgrowth. J. Cell. Biol. 106, 1331-1348.

Sutherland, A. E., Calarco, P. G., & Damsky, C. H. (1993). Developmental regulation of integrin expression at the time of implantation in the mouse embryo. Development 119, 1175-1186.

Takeichi, M. (1995). Morphogenetic roles of classic cadherins. Curr. Opin. Cell Biol. 7, 619-627.

Vicovac, L., Jones, C. J., & Aplin, J. D. (1995). Trophoblast differentiation during formation of anchoring villi in a model of the early human placenta in vitro. Placenta 16, 41-56.

Yang, J. T, Rayburn, H., & Hynes, R. O. (1995). Cell adhesion events mediated by alpha 4 integrins are essential in placental and cardiac development. Development 121, 549-560.

Zhou, Y., Damsky, C. H., Chiu, K., Roberts, J. M., & Fisher, S. J. (1993). Preeclampsia is associated with abnormal expression of adhesion molecules by invasive cytotrophoblasts. J. Clin. Invest. 91, 950-960.

Zhou, Y., Fisher, S. J., Janatpour, M., Genbacev, O., Dejana, E., Wheelock, M., & Damsky, C. H. (1997a). Human cytotrophoblasts adopt a vascular phenotype as they differentiate. A strategy for successful endovascular invasion? [see comments]. J. Clin. Invest. 99, 2139-2151

Zhou, Y., Damsky, C. H., & Fisher, S. J. (1997b). Preeclampsia is associated with failure of human cytotrophoblasts to mimic a vascular adhesion phenotype. One cause of defective endovascular invasion in this syndrome? [see comments]. J. Clin. Invest. 99. 2152-2164.

INDEX

www.ingramcontent.com/pod-product-compliance
Ingram Content Group UK Ltd.
Pitfield, Milton Keynes, MK11 3LW, UK
UKHW020230290926
15470UKWH00002B/445

9 780762 3049